1500217578

AF598647

WITHDRAWN
University of Bristol
UNIVERSITY
OF BRISTOL
MATHEMATICS

Physics and Chemistry in Space
Volume 10

A. V. Gurevich

Nonlinear Phenomena in the Ionosphere

Translated by J. George Adashko

With 76 Figures

Springer-Verlag
New York Heidelberg Berlin

A. V. Gurevich

P. N. Lebedev Physics Institute, USSR Academy of Sciences/
Moscow, USSR

The illustration on the cover is adapted from Figure 63

ISBN 0-387-08605-6 Springer-Verlag New York Heidelberg Berlin
ISBN 3-540-08605-6 Springer-Verlag Berlin Heidelberg New York

Printed in the United States of America

9 8 7 6 5 4 3 2 1

Library of Congress Cataloging in Publication Data. Gurevich, Aleksandr Viktorovich. Nonlinear phenomena in the ionosphere. (Physics and chemistry in space ; 10) Translation of Nelineĭnye i͡avlenii͡a v ionosfere. Added t.p.: Nelineĭnye i͡avlenii͡a v ionosfere. Includes bibliographical references. 1. Ionospheric radio wave propagation. 2. Non-linear theories. 3. Plasma (Ionized gases) I. Title. II. Title: Nelineĭnye i͡avlenii͡a v ionosfere. III. Series. QC801.P46 vol. 10 [QC973.4.I6] 523.01'8 [551.5'27] 78-7280

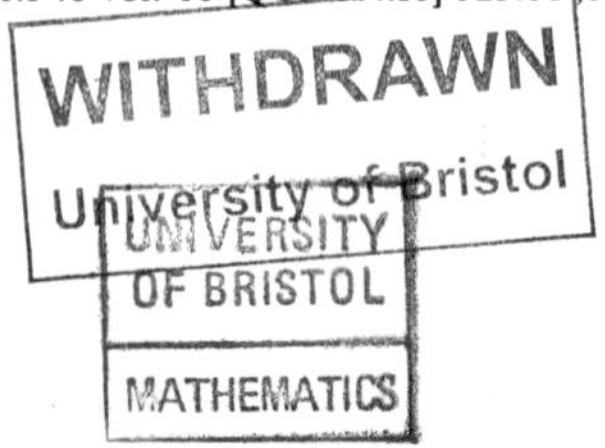

Preface

Nonlinear effects in the ionosphere (cross modulation of radio waves) have been known since the 1930s. Only recently, however, has the rapid increase in the power and directivity of the radio transmitters made it possible to alter the properties of the ionosphere strongly and to modify it artificially by applying radio waves. This has revealed a variety of new physical phenomena. Their study is not only of scientific interest but also undisputedly of practical interest, and is presently progressing very rapidly. This monograph is devoted to an exposition of the present status of theoretical research on this problem. Particular attention is paid, naturally, to problems in the development of which the author himself took part.

It is my pleasant duty to thank V. L. Ginzburg, L. P. Pitaevskii, V. V. Vas'kov, E. E. Tsedilina, A. B. Shvartsburg, and Ya. S. Dimant for useful discussions and for valuable remarks during various stages of the work on the problem considered in this book.

Contents

1. Introduction

1.1. Data on the Structure of the Ionosphere

The ionosphere is a part of the earth's upper atmosphere, extending in height from 60 to about 1000 km. In this region, the atmosphere is a partly ionized gas or plasma. The processes that occur in the ionospheric plasma are closely connected with the wave and corpuscular radiation of the sun, with events in the magnetosphere and variation of the earth's magnetic field, with motion of the upper atmosphere, and so on. This is why the ionosphere varies so greatly with time (with the time of day, with the season of the year, with the 11-year cycle of solar activity) and with geographic latitude.

The ionosphere is a transition layer between the nonionized upper atmosphere and the fully ionized hydrogen plasma of the magnetosphere. The structure and properties of the ionosphere, therefore, vary rapidly with height.

The lower region of the ionosphere, at heights $z = 50$ to 80 km, is usually called the D layer. This layer is ionized in day time. The region at heights from 80 to 130 km is called the E layer, and that above 150 km the F layer. A distinction is sometimes made between the F_1 layer, up to approximately 250 km, and the F_2 layer, above 250 km.

A model of the ionosphere at medium latitudes and at average solar activity is represented in Tables 1 and 2 (Harris et al., 1962; Al'pert et al., 1967). This model will be used here in estimates and numerical calculations.

The density of the upper atmosphere, as seen from Table 1, decreases rapidly with height. The molecular composition varies little up to 100–110 km. In the 100–120 km region, dissociation of the oxygen molecule, $O_2 \rightarrow O + O$, takes place. The nitrogen dissociates at ~ 300 km. At 500–600 km, the relative helium concentration increases rapidly, and the same occurs for hydrogen at $z \sim 1000$ km. At $z \sim 1500$ km, the hydrogen atoms H are in the majority.

The electron and ion densities increase up to 300–400 km (Table 2) and then decrease quite slowly at $z > 400$ km. The degree of plasma ionization, which is very low at small heights ($N/N_m = 10^{-8} - 10^{-4}$ at

Table 1. Molecular composition of the ionosphere

z, km	Day-time (12:00 noon), concentration N_m, cm^{-3}						T, K
	N_2	O_2	He	O	H	Total N_m	
60	$5.5 \cdot 10^{15}$	$1.5 \cdot 10^{15}$				$7.0 \cdot 10^{15}$	270
70	$1.6 \cdot 10^{15}$	$4.2 \cdot 10^{14}$				$2.0 \cdot 10^{15}$	200
80	$2.3 \cdot 10^{14}$	$6.2 \cdot 10^{13}$				$2.9 \cdot 10^{14}$	180
90	$3.1 \cdot 10^{13}$	$8.2 \cdot 10^{12}$				$3.9 \cdot 10^{13}$	190
100	$7.7 \cdot 10^{12}$	$1.9 \cdot 10^{12}$	$5.7 \cdot 10^{7}$	$2.0 \cdot 10^{11}$	$6.2 \cdot 10^{4}$	$9.6 \cdot 10^{12}$	210
110	$1.4 \cdot 10^{12}$	$3.5 \cdot 10^{11}$	$3.8 \cdot 10^{7}$	$1.4 \cdot 10^{11}$	$5.3 \cdot 10^{4}$	$1.9 \cdot 10^{12}$	270
120	$5.8 \cdot 10^{11}$	$1.2 \cdot 10^{11}$	$2.5 \cdot 10^{7}$	$7.6 \cdot 10^{10}$	$4.3 \cdot 10^{4}$	$7.8 \cdot 10^{11}$	360
130	$2.0 \cdot 10^{11}$	$3.8 \cdot 10^{10}$	$1.7 \cdot 10^{7}$	$3.7 \cdot 10^{10}$	$3.2 \cdot 10^{4}$	$2.8 \cdot 10^{11}$	460
150	$4.8 \cdot 10^{10}$	$7.2 \cdot 10^{9}$	$1.0 \cdot 10^{7}$	$1.35 \cdot 10^{10}$	$2.1 \cdot 10^{4}$	$6.9 \cdot 10^{10}$	670
200	$4.8 \cdot 10^{9}$	$5.9 \cdot 10^{8}$	$4.9 \cdot 10^{6}$	$3.0 \cdot 10^{9}$	$1.3 \cdot 10^{4}$	$8.4 \cdot 10^{9}$	1070
250	$1.1 \cdot 10^{9}$	$1.2 \cdot 10^{8}$	$3.4 \cdot 10^{6}$	$1.3 \cdot 10^{9}$	$1.0 \cdot 10^{4}$	$2.5 \cdot 10^{9}$	1250
300	$3.2 \cdot 10^{8}$	$2.7 \cdot 10^{7}$	$2.7 \cdot 10^{6}$	$5.9 \cdot 10^{8}$	$9.3 \cdot 10^{3}$	$9.3 \cdot 10^{8}$	1330
400	$3.5 \cdot 10^{7}$	$2.2 \cdot 10^{6}$	$1.9 \cdot 10^{6}$	$1.6 \cdot 10^{8}$	$8.3 \cdot 10^{3}$	$2.0 \cdot 10^{8}$	1390
500	$4.4 \cdot 10^{6}$	$2.1 \cdot 10^{5}$	$1.4 \cdot 10^{6}$	$5.0 \cdot 10^{7}$	$7.6 \cdot 10^{3}$	$5.6 \cdot 10^{7}$	1400
600	$5.9 \cdot 10^{5}$	$2.1 \cdot 10^{4}$	$1.0 \cdot 10^{6}$	$1.6 \cdot 10^{7}$	$7.1 \cdot 10^{3}$	$1.7 \cdot 10^{7}$	1400
700	$8.5 \cdot 10^{4}$	$2.3 \cdot 10^{3}$	$8.0 \cdot 10^{5}$	$5.2 \cdot 10^{6}$	$6.6 \cdot 10^{3}$	$6.0 \cdot 10^{6}$	1400
800	$1.3 \cdot 10^{4}$	$2.7 \cdot 10^{2}$	$6.1 \cdot 10^{5}$	$1.8 \cdot 10^{6}$	$6.2 \cdot 10^{3}$	$2.4 \cdot 10^{6}$	1400
900	$2.1 \cdot 10^{3}$	$3.3 \cdot 10^{1}$	$4.7 \cdot 10^{5}$	$6.2 \cdot 10^{5}$	$5.8 \cdot 10^{3}$	$1.1 \cdot 10^{6}$	1400
1000	$3.5 \cdot 10^{2}$	4.3	$3.7 \cdot 10^{5}$	$2.2 \cdot 10^{5}$	$5.4 \cdot 10^{3}$	$6.0 \cdot 10^{5}$	1400

z, km	Night-time (midnight), concentration N_m, cm^{-3}						T, K
	N_2	O_2	He	O	H	Total N_m	
60	$5.5 \cdot 10^{15}$	$1.5 \cdot 10^{15}$				$7.0 \cdot 10^{15}$	270
70	$1.6 \cdot 10^{15}$	$4.2 \cdot 10^{14}$				$2.0 \cdot 10^{15}$	200
80	$2.3 \cdot 10^{14}$	$6.2 \cdot 10^{13}$				$2.9 \cdot 10^{14}$	180
90	$3.1 \cdot 10^{13}$	$8.2 \cdot 10^{12}$				$3.9 \cdot 10^{13}$	190
100	$7.7 \cdot 10^{12}$	$1.9 \cdot 10^{12}$	$5.7 \cdot 10^{7}$	$2.2 \cdot 10^{11}$	$6.2 \cdot 10^{4}$	$9.6 \cdot 10^{12}$	210
110	$1.4 \cdot 10^{12}$	$3.5 \cdot 10^{11}$	$3.8 \cdot 10^{7}$	$1.4 \cdot 10^{11}$	$5.3 \cdot 10^{4}$	$1.9 \cdot 10^{12}$	270
120	$5.8 \cdot 10^{11}$	$1.2 \cdot 10^{11}$	$2.5 \cdot 10^{7}$	$7.6 \cdot 10^{10}$	$4.3 \cdot 10^{4}$	$7.8 \cdot 10^{11}$	360
130	$2.0 \cdot 10^{11}$	$3.7 \cdot 10^{10}$	$1.7 \cdot 10^{7}$	$3.7 \cdot 10^{10}$	$3.2 \cdot 10^{4}$	$2.7 \cdot 10^{11}$	470
150	$4.8 \cdot 10^{10}$	$7.0 \cdot 10^{9}$	$1.0 \cdot 10^{7}$	$1.4 \cdot 10^{10}$	$2.2 \cdot 10^{4}$	$6.9 \cdot 10^{10}$	650
200	$4.7 \cdot 10^{9}$	$5.6 \cdot 10^{8}$	$5.9 \cdot 10^{6}$	$3.2 \cdot 10^{9}$	$1.6 \cdot 10^{4}$	$8.4 \cdot 10^{9}$	850
250	$7.7 \cdot 10^{8}$	$7.1 \cdot 10^{7}$	$4.3 \cdot 10^{6}$	$1.2 \cdot 10^{9}$	$1.4 \cdot 10^{4}$	$2.0 \cdot 10^{9}$	910
300	$1.4 \cdot 10^{8}$	$1.0 \cdot 10^{7}$	$3.3 \cdot 10^{6}$	$4.4 \cdot 10^{8}$	$1.3 \cdot 10^{4}$	$5.9 \cdot 10^{8}$	930
400	$6.0 \cdot 10^{6}$	$2.8 \cdot 10^{5}$	$2.1 \cdot 10^{6}$	$7.0 \cdot 10^{7}$	$1.1 \cdot 10^{4}$	$7.8 \cdot 10^{7}$	940
500	$2.8 \cdot 10^{5}$	$8.6 \cdot 10^{3}$	$1.3 \cdot 10^{6}$	$1.2 \cdot 10^{7}$	$1.0 \cdot 10^{4}$	$1.3 \cdot 10^{7}$	950
600	$1.5 \cdot 10^{4}$	$2.9 \cdot 10^{2}$	$8.7 \cdot 10^{5}$	$2.3 \cdot 10^{6}$	$9.0 \cdot 10^{3}$	$3.2 \cdot 10^{6}$	950
700	$8.4 \cdot 10^{2}$	$1.1 \cdot 10^{1}$	$5.8 \cdot 10^{5}$	$4.4 \cdot 10^{5}$	$8.1 \cdot 10^{3}$	$1.0 \cdot 10^{6}$	950
800	$5.1 \cdot 10$	$4.6 \cdot 10^{-1}$	$3.9 \cdot 10^{5}$	$8.9 \cdot 10^{4}$	$7.3 \cdot 10^{3}$	$4.8 \cdot 10^{5}$	950
900	3.4	$2.1 \cdot 10^{-2}$	$2.6 \cdot 10^{5}$	$1.9 \cdot 10^{4}$	$6.7 \cdot 10^{3}$	$2.8 \cdot 10^{5}$	950
1000	$2.4 \cdot 10^{-1}$	$1.0 \cdot 10^{-3}$	$1.8 \cdot 10^{5}$	$4.2 \cdot 10^{3}$	$6.1 \cdot 10^{3}$	$1.9 \cdot 10^{5}$	950

Table 2. Structure of the ionosphere (electrons and ions)

z, km	Day-time					Night-time					Relative ion concentration $n = N_i/N$						
	N, cm^{-3}	T_e, K	T_i, K	ω_0, s^{-1}	Ω_0, s^{-1}	N, cm^{-3}	T_e, K	T_i, K	ω_0, s^{-1}	Ω_0, s^{-1}	NO^+	O_2^+	O^+	He^+	H^+	N^+	N_2^+
60	80	270	270	$5 \cdot 10^5$	$2.1 \cdot 10^3$	—	—	—	—	—	0.8	0.2	—	—	—	—	—
70	$2 \cdot 10^2$	200	200	$8 \cdot 10^5$	$3.3 \cdot 10^3$	—	—	—	—	—	0.8	0.2	—	—	—	—	—
80	10^3	180	180	$1.8 \cdot 10^6$	$7.5 \cdot 10^3$	10	180	180	$1.8 \cdot 10^5$	$7.5 \cdot 10^2$	0.7	0.3	—	—	—	—	—
90	$8 \cdot 10^3$	200	190	$5 \cdot 10^6$	$2.1 \cdot 10^4$	60	190	190	$4.4 \cdot 10^5$	$1.8 \cdot 10^3$	0.65	0.35	—	—	—	—	—
100	$8 \cdot 10^4$	240	210	$1.6 \cdot 10^7$	$6.7 \cdot 10^4$	$1.2 \cdot 10^3$	210	210	$1.9 \cdot 10^6$	$8 \cdot 10^3$	0.62	0.38	—	—	—	—	—
110	$1.2 \cdot 10^5$	320	270	$1.9 \cdot 10^7$	$8 \cdot 10^4$	$1.8 \cdot 10^3$	270	270	$2.4 \cdot 10^6$	10^4	0.55	0.45	—	—	—	—	—
120	$1.3 \cdot 10^5$	400	360	$2 \cdot 10^7$	$8.4 \cdot 10^4$	$2.1 \cdot 10^3$	360	360	$2.6 \cdot 10^6$	$1.1 \cdot 10^4$	0.39	0.60	0.01	—	—	—	—
130	$1.5 \cdot 10^5$	500	460	$2.2 \cdot 10^7$	$9.2 \cdot 10^4$	$2.2 \cdot 10^3$	480	470	$2.7 \cdot 10^6$	$1.1 \cdot 10^4$	0.42	0.55	0.02	—	—	—	0.01
150	$3 \cdot 10^5$	800	670	$3.1 \cdot 10^7$	$1.35 \cdot 10^5$	$2.4 \cdot 10^3$	670	650	$2.8 \cdot 10^6$	$1.2 \cdot 10^4$	0.45	0.41	0.13	—	—	—	0.01
200	$5 \cdot 10^5$	1300	1100	$4 \cdot 10^7$	$2.2 \cdot 10^5$	$3 \cdot 10^3$	900	850	$3.1 \cdot 10^6$	$1.7 \cdot 10^4$	0.045	0.045	0.90	—	—	$5 \cdot 10^{-3}$	$5 \cdot 10^{-3}$
250	$1.0 \cdot 10^6$	1700	1300	$5.6 \cdot 10^7$	$3.3 \cdot 10^5$	10^4	1000	910	$5.6 \cdot 10^6$	$3.3 \cdot 10^4$	$6 \cdot 10^{-3}$	$5 \cdot 10^{-3}$	0.98	—	—	$6 \cdot 10^{-3}$	$3 \cdot 10^{-3}$
300	$1.6 \cdot 10^6$	2000	1400	$7.1 \cdot 10^7$	$4.2 \cdot 10^5$	10^5	1200	930	$1.8 \cdot 10^7$	$1.1 \cdot 10^5$	—	—	0.99	—	—	0.01	—
400	$1.5 \cdot 10^6$	2400	1450	$6.9 \cdot 10^7$	$4.1 \cdot 10^5$	$3 \cdot 10^5$	1400	950	$3.1 \cdot 10^7$	$1.8 \cdot 10^5$	—	—	0.97	$5 \cdot 10^{-3}$	0.01	0.02	—
500	$9 \cdot 10^5$	2600	1600	$5.4 \cdot 10^7$	$3.2 \cdot 10^5$	$2 \cdot 10^5$	1500	1000	$2.5 \cdot 10^7$	$1.5 \cdot 10^5$	—	—	0.90	0.015	0.03	0.06	—
600	$4 \cdot 10^5$	2700	2100	$3.6 \cdot 10^7$	$2.2 \cdot 10^5$	$1.3 \cdot 10^5$	1600	1020	$2 \cdot 10^7$	$1.2 \cdot 10^5$	—	—	0.84	0.02	0.06	0.08	—
700	$2 \cdot 10^5$	2800	2200	$2.5 \cdot 10^7$	$1.6 \cdot 10^5$	$8 \cdot 01^4$	1700	1100	$1.6 \cdot 10^7$	$1.0 \cdot 10^5$	—	—	0.75	0.04	0.11	0.10	—
800	10^5	2870	2300	$1.8 \cdot 10^7$	$1.25 \cdot 10^5$	$5 \cdot 10^4$	1800	1200	$1.3 \cdot 10^7$	$9 \cdot 10^4$	—	—	0.61	0.06	0.21	0.12	—
900	$7 \cdot 10^4$	2940	2400	$1.5 \cdot 10^7$	$1.2 \cdot 10^5$	$3 \cdot 10^4$	1900	1300	$9.8 \cdot 10^6$	$7.7 \cdot 10^4$	—	—	0.41	0.09	0.40	0.10	—
1000	$5 \cdot 10^4$	3000	2500	$1.3 \cdot 10^7$	$1.2 \cdot 10^5$	$2 \cdot 10^4$	2000	1400	$8 \cdot 10^6$	$7.3 \cdot 10^4$	—	—	0.28	0.14	0.51	0.07	—

Symbols: N—concentration of ions and electrons in quasineutral plasma, N_i—ion concentration; T_e—electron temperature, T_i—ion temperature
$\omega_0 = (4\pi e^2 N/m)^{1/2}$, $\Omega_0 = (4\pi e^2 N/M)^{1/2}$—Langmuir frequencies of electrons and ions (m and M—masses of electron and ion)

$z = 100$–300 km), increases rapidly with increasing z and reaches 10% at $z \gtrsim 1000$ km. At greater heights the plasma is almost fully ionized. The temperatures of the electrons and ions increase quite smoothly with height, but vary strongly with the time of day. At altitudes 300–700 km, the electron temperature can exceed the ion temperature by 1.5–2 times.

The ionic structure of the ionosphere (Table 2) does not vary greatly with latitude θ up to $\theta = 55$–$60°$. At higher latitudes, the total ion density in the ionosphere decreases more rapidly with height. The relative concentration of the oxygen ions O^+ increases rapidly at larger heights. Owing to the variability of the properties of the ionosphere, the main parameters that characterize the propagation of the electromagnetic waves also vary strongly with height, time of day, etc.

Detailed information on the structure and properties of the ionosphere and on the methods used for its investigation can be found in the specialized monographs (Whitten and Poppoff, 1965; Bauer, 1973; Al'pert, 1974) and in the extensive periodical literature.

1.2. Features of Nonlinear Phenomena in the Ionosphere

1.2.1. Nonlinearity Mechanisms

One of the main features of a plasma is that nonlinear effects manifest themselves in it even in relatively weak and easily obtainable electric fields.

Depending on the condition in the plasma, a distinction can be made between two characteristic types of nonlinearity. The first is connected with the collisional heating of electrons in the electric field of the wave. This heating is easily produced because the electron mean free path is large and the electron can acquire from the field an appreciable energy in one run between collisions. In addition, the energy transfer from the electrons to the molecules, atoms, and ions by collision is hindered by the small ratio of the electron mass to the masses of these heavy particles. As a result, the plasma electrons are rapidly heated even in a comparatively weak electric field. The dielectric constant and the conductivity of the plasma then become dependent on the field intensity. In other words, the electric current is no longer proportional to the field $\boldsymbol{E}$. Consequently the electrodynamic processes in a plasma, particularly the propagation of electric waves, become nonlinear (the superposition principle is violated, etc.). This type of nonlinearity will be called *thermal*. It is connected with electron collisions and plays the principal role in those cases when the characteristic dimensions of the plasma region perturbed by the electric field are much larger than the electron mean free path.

The second type of nonlinearity is not connected with collisions. It plays the main role in collisionless plasma, when the dimensions of the field-perturbed plasma region are, to the contrary, much smaller than the electron mean free path. Several nonlinearity mechanisms operating in a collisionless plasma can be indicated. The most important of them is caused by the fact that the inhomogeneous alternating electric field of the wave exerts a pressure on the electrons (Gaponov and Miller, 1958; Pitaevskii, 1960) thereby compressing the plasma. The electron density, and with it the dielectric constant of the plasma, then becomes dependent on the electric field amplitude, and it is this which causes the nonlinearity of the electrodynamic process. This is frequently called the *striction* nonlinearity mechanism.

Let us estimate the relative role of the various nonlinearity mechanisms in the ionosphere. Thermal nonlinearity leads to a nonlinear perturbation of the complex dielectric constant (see Sect. 2.1.):

$$\Delta\varepsilon_T \sim (E_0/E_p)^2, \tag{1.1}$$

where E_0 is the amplitude of the electric field of the wave and E_p is a characteristic plasma field given by

$$E_p = \left[3T\frac{m}{e^2}\delta_0(\omega^2 + \nu_{e0}^2)\right]^{1/2} = 4.2 \cdot 10^{-5}[\delta_0 T(\omega^2 + \nu_{e0}^2)]^{1/2}\ \mathrm{mV/m}. \tag{1.2}$$

Here T is the plasma temperature (in degrees), ω is the cyclic frequency of the field, ν_{e0} is the effective frequency of the electron collisions with the molecules and ions, and δ_0 is the average fraction of the energy lost by the electron in one collision. The electric field in the numerical expressions is given in mV/m. (T is in K and ω is in s^{-1}.)

The values of $\Delta\varepsilon_T$ for radio waves of various frequencies and powers in the ionosphere are listed in Table 3. Here W_0 is the effective radiation power of the equivalent dipole.[1] We see that at realistic station powers the ratio $(E_0/E_p)^2 \sim \Delta\varepsilon_T$ can be of the order of unity or even much larger. The action of such strong fields alters the properties of the plasma appreciably. The nonlinearity then determines the entire character of the wave propagation. It is important also that the thermal nonlinearity increases strongly in resonance regions where the electron heating has an anomalous character.

---

[1] In estimates of the radiation power it is assumed henceforth that the radiating station is land-based. The electric field at a height z is calculated from the formula $E_0 = 300\sqrt{W_0}/z$, where $W_0 = PG$ is the power (in kW) radiated by the equivalent dipole (P is the actual power of the radiating station and G is the antenna gain) and z is the distance in kilometers; the field E_0 is expressed in mV/m.

Table 3. Comparative values of the nonlinear perturbations of the ionosphere: thermal ($\Delta\varepsilon_T$) and striction ($\Delta\varepsilon_s$)

		$z = 100$ km			$z = 300$ km		
ω, s^{-1}		10^6	10^7	10^8	$3 \cdot 10^7$	10^8	10^9
$W_0 = 10^2$ kW	$\Delta\varepsilon_T$	1.1	$1.2 \cdot 10^{-2}$	$1.2 \cdot 10^{-4}$	$4.1 \cdot 10^{-4}$	$3.8 \cdot 10^{-5}$	$3.8 \cdot 10^{-7}$
	$\Delta\varepsilon_s$	$8 \cdot 10^{-4}$	$8 \cdot 10^{-6}$	$8 \cdot 10^{-8}$	$1.5 \cdot 10^{-8}$	$1.4 \cdot 10^{-9}$	$1.4 \cdot 10^{-11}$
$W_0 = 10^4$ kW	$\Delta\varepsilon_T$	110	1.2	$1.2 \cdot 10^{-2}$	$4.1 \cdot 10^{-2}$	$3.8 \cdot 10^{-3}$	$3.8 \cdot 10^{-5}$
	$\Delta\varepsilon_s$	$8 \cdot 10^{-2}$	$8 \cdot 10^{-4}$	$8 \cdot 10^{-6}$	$1.5 \cdot 10^{-6}$	$1.4 \cdot 10^{-7}$	$1.4 \cdot 10^{-9}$
$W_0 = 10^6$ kW	$\Delta\varepsilon_T$	$1.1 \cdot 10^4$	120	1.2	4.1	0.38	$3.8 \cdot 10^{-3}$
	$\Delta\varepsilon_s$	8	$8 \cdot 10^{-2}$	$8 \cdot 10^{-4}$	$1.5 \cdot 10^{-4}$	$1.4 \cdot 10^{-5}$	$1.4 \cdot 10^{-7}$

Striction nonlinearity leads to a perturbation of the dielectric constant (Pitaevskii, 1960):

$$\Delta\varepsilon_s \sim e^2 E_0^2/8mT\omega^2. \tag{1.3}$$

The values of this quantity for radio waves in the ionosphere are also listed in Table 3.

It is seen from the table that the principal role is played in the ionosphere by the thermal nonlinearity due to heating of the electrons in the field of a strong electromagnetic wave. Principal attention will therefore be paid to it in this book. It must be noted at the same time that the striction nonlinearity is also significant under certain conditions, principally in nonstationary processes, e.g., for short radio pulses of duration less than the electron free path time.

1.2.2. Qualitative Character of Nonlinear Phenomena

Let us examine qualitatively the main phenomena that can occur when a high-power radio wave propagates in the ionosphere.

Change in the Absorption and Modulation of the Wave. Heating of the electrons in the field of a high-power electromagnetic wave leads, first, to a change of the collisions of the electrons with the ions and with the neutral molecules and atoms. This changes the radio-wave absorption. In the lower layers of the ionosphere the collision frequency increases with increasing electron temperature. Therefore, at $\omega^2 \gg \nu_e^2$ the absorption increases sharply with increasing wave power. The field of a strong wave, $E_0 \gg E_p$, cannot therefore penetrate into the interior of the plasma beyond a definite limit. The field ceases to depend on the power of the wave incident on the plasma boundary; a sort of *saturation* of the field is produced in the interior of the plasma, and the field of the wave reflected from the ionosphere even decreases with increasing radiation power. To the contrary, in the case $\omega^2 < \nu_e^2$ the absorption decreases with increasing wave power. The plasma becomes so to speak *transparent* to the high-power radio waves.

Thus, owing to the nonlinear change of absorption, the high-power wave field amplitude exhibits in the interior of a plasma an essentially nonlinear dependence on the field amplitude of the wave incident on the plasma boundary. Therefore, if the wave at the plasma boundary is amplitude-modulated, then its modulation in the interior of the plasma and the modulation of the reflected wave are distorted. For strong radio waves ($E_0 \gtrsim E_p$) this change of modulation can be appreciable (*autodemodulation* and *automodulation* of the radio waves). Analogous nonlinear distortions occur also in the waveform of the envelope of a strong pulse reflected from the ionosphere.

Change of Wave Refraction. The heating of the electrons changes their pressure in the plasma region through which the radio wave passes. In the course of time, the pressure becomes equalized by the outflow of the electrons from the heated region. The result is a decrease in the plasma density. The plasma density can also be changed because the overall ionization balance is disturbed by the heating of the plasma. All this leads to a nonlinear change of the refractive index of the wave.

The change of the refractive index of the wave in the plasma gives rise to new nonlinear effects. The beam trajectory is distorted and the point from which the wave is reflected is shifted. If the imbalance of the ionization is not significant, then the point of reflection of the radio waves shifts upwards in the ionosphere. Sufficiently powerful and narrow wave beams can in this case penetrate through the ionospheric plasma. At the same time, the structure of the ionosphere should become distorted in the region where the wave passes, namely, the electron density decreases, and a "hole" is punched, as it were, in the ionosphere layer. In the opposite case, when the change of the ionization balance plays the principal role, the plasma density in the perturbed region increases.

The nonlinearity connected with the change of the refractive index exerts a particular influence on the propagation of radio beams. Even a weak nonlinearity causes the beam trajectories to bend noticeably. This leads to a stratification of the beams, to the onset of an oscillatory structure in the distribution of the wave field intensity, and to perturbation of the plasma density (*self-focusing* or *modulation* instability). The stratification develops particularly rapidly in the region of wave reflection.

Wave Interaction. One of the most important manifestations of nonlinearity is the violation of the principle of superposition of the waves. If radio wave 2 of frequency ω_2 passes through the same region of the plasma as a high-power wave 1 of frequency ω_1, then the absorption and refraction of wave 2 will vary with the power of wave 1. The radio waves thus interact in the plasma. In particular, if the high-power wave 1 is amplitude-modulated, then the change of absorption can cause this modulation to be transferred to wave 2 passing through the same region of the ionosphere. This phenomenon, called *cross modulation*, has been thoroughly investigated in experiments and is of practical importance for radio broadcasting at medium wave lengths. Interaction between short pulses, an effect widely used recently to study the properties of the ionosphere, is similar in character.

The absorption change induced by strong radio waves $E_0 \gtrsim E_p$ is very large, and can enable a strong radio wave to suppress other radio waves propagating in the perturbed region.

Nonlinear interaction leads also to generation of new radio waves with combination frequencies $\omega_2 \pm \omega_1, \omega_2 \pm 2\omega_1$, etc. In particular, when a powerful high-frequency wave modulated in amplitude at a low frequency Ω propagates in the ionosphere, the nonlinearity can cause generation of a wave of frequency Ω (*detection effect of the ionosphere*).

Owing to nonlinearity, various normal components of the radio wave polarization interact with one another in a magnetoactive plasma. This can lead to a nonlinear rotation of the polarization ellipse, and to the onset of amplitude automodulation of the wave. The nonlinear interaction can be the reason why radio waves in the ionosphere generate other types of electromagnetic waves such as plasma waves, whistlers, ion-sound waves, magnetohydrodynamic waves, or electro-acoustic waves (*nonlinear wave transformation*).

Ionization. In the field of very powerful radio waves, the electrons become so strongly heated that electric breakdown of the gas takes place. The degree of ionization is then greatly increased. The nonlinearity associated with this process leads to a very fast increase of the wave absorption by the plasma, and to a rapid saturation of the wave field (p. 7). An equally strong increase takes place in the absorption of other waves passing through the ionization region; this attenuates the field of these waves, which are suppressed by the strong wave. Ionization also makes it possible to excite in the field of a high-power radio wave certain special thermo-ionization plasma oscillations, which lead, in particular, to automodulation of the wave. Rapid growth of the ionization can also be produced by the heating of the neutral gas in the lower layers of the ionosphere and in the upper atmosphere. The artificial-ionization regions produced in the atmosphere at heights 20–60 km can be effectively used to reflect UHF radio waves.

Excitation of Instability. High-power radio waves increase the temperature and change the electron and ion densities in the perturbed region of the ionosphere. The artificial inhomogeneity produced in the plasma in this manner can easily become unstable with respect to definite types of waves, such as flute, drift, ion-cyclotron waves, etc. Thus, the heating of the ionosphere in the field of high-power radio waves should excite new plasma oscillations and enhance those already present there, and lead to formation of an oscillating "turbulized" region. This influences strongly the conditions of radio wave propagation in the perturbed region of the plasma, causing them to become more strongly absorbed and scattered.

Of particular importance are the regions of *resonances*, where the wave frequency is close to some natural frequency of the plasma. In these regions, the electromagnetic wave excites energetically natural plasma oscillations in the small-scale plasma inhomogeneities. This process of linear wave transformation by the plasma inhomogeneities leads to the effective absorption of radio waves by the plasma. Moreover, the inhomogeneities can themselves be increased due to the absorption of the wave field—this is *resonant* instability. Such an instability is produced in the ionosphere in the region of reflection of the ordinary wave. It destroys the smooth ionospheric layer and produces small-scale inhomogeneities that are strongly elongated along the earth's magnetic field. These inhomogeneities effectively scatter the UHF and VHF radio waves in a wide frequency band.

Another important phenomenon that occurs in the plasma-resonance region is *parametric* instability, which manifests itself in the generation of plasma and ion–sound waves in the field of a strong radio wave. Nondissipative parametric instability is due to the striction effect; it develops within a short time—less than the electron free path time—and leads to effective nonlinear generation of noise and to absorption of the wave energy by the plasma. Parametric instability is accompanied by pulsations of the wave reflected from the ionosphere and by generation of fast electrons accelerated as a result of Landau absorption of energy from the plasma oscillations.

The electrons accelerated in the F region, and also the high-power low-frequency radiation coming from the earth, can influence the magnetospheric plasma strongly. On the other hand, the wave energy absorbed by the electrons is transferred via the collisions to the neutral molecules and atoms and causes a noticeable heating of the neutral atmosphere. All these perturbations, as well as interaction of the radio waves with ionospheric currents, fast electrons, or whistlers of the magnetosphere lead to a number of new nonlinear phenomena.

The processes indicated here, of course, do not account for all the possible nonlinear phenomena in the upper atmosphere. We see, however, that they constitute a rather extensive and varied class. This is due both to the manifold variety of nonlinear effects in the plasma and to the great differences in the physical conditions in the ionosphere, where the properties of the plasma vary rapidly with height. The principal role in the lower ionosphere is played by effects connected with the nonlinear change of the wave absorption, while in the upper ionosphere it is played by distortion of the wave refraction and by plasma instability.

1.2.3. Brief Historical Review

The nonlinear interaction of radio waves in the ionosphere (cross modulation) was discovered in 1930 by Telegen. The theory of this phenomenon, which subsequently served as the basis for the theory of nonlinear effects in the ionosphere, was developed by Bailey and Martin (1934). A more consistent approach to the theory of nonlinear effects in the lower ionosphere, based on the Boltzmann kinetic theory, was proposed by Ginzburg (1949). Cross modulation and the development of the theory in 1940–1950 were the subject of important works by Bailey, Huxley, Ratcliffe, Bell, Shaw and others. Cross-modulation resonance at gyrofrequencies was observed by Cutolo (1950), Bailey et al. (1952).

In 1955, Fejer observed the interaction of short pulses and proposed to use it to investigate the lower ionosphere. Fejer's method was developed in detail by Smith (1966), by Ferraro and Lee (1966), and others.

The possibility of self-action of radio waves is indicated in the papers of Vilenskii (1953) and Hibberd (1956). This phenomenon was observed experimentally in the ionosphere by King (1959) and by Vilenskii et al. (1962b).

The possibility of artificially ionizing the lower ionosphere by acceleration of electrons in the field of high-power radio waves was indicated by Bailey (1937). Detailed calculations by E. Ginzburg (1962), Lombardini (1965) and others, have shown that high radiation powers are needed to realize this effect. Another possibility of changing the balance of ionization by heating of electrons and producing high-ionization clouds in the upper atmosphere by overall radio-wave heating of the neutral gas was indicated by Gurevich (1972) and Utlaut (1975).

The possibility of nonlinear generation of waves with combination frequencies $\omega_2 \pm 2\omega_1$ in the ionosphere was indicated by Vilenskii (1953), while the feasibility of the nonlinear detection effect (frequencies $\omega_2 \pm \omega_1$) was indicated by Ginzburg (1958). The detection effect of the ionosphere was observed experimentally by Getmantsev et al. (1974), and the detailed theory of this effect was developed by Trakhtengerts and Kotik (1975). Kapustin et al. (1977) observed a strong enchancement of the detection effect in the polar regions of the ionosphere and its connection with the ionospheric current.

A new stage in the study of nonlinear phenomena in the ionosphere, initiated in 1960–1970, arose in connection with the advent of strong radio waves, i.e., radio waves that perturb strongly the lower ionosphere and modify effectively the upper ionosphere. In 1956, Gurevich introduced the concept of the plasma field E_p and developed the theory of self action and interaction of strong radio waves ($E_0 \gtrsim E_p$) in the lower ionosphere.

The nonlinear effects that are produced when strong radio waves propagate in the lower ionosphere ("saturation," "suppression," "overmodulation," etc.), were observed experimentally by Shlyuger (1961–1968).[2]

The feasibility of modifying the upper ionosphere, i.e., of significantly changing the temperature and concentration of the plasma in the F layer by means of radio waves, was indicated by Ginzburg and Gurevich (1960), Farley (1963), and Gurevich (1965, 1967). Detailed calculations of these effects were performed by Meltz, Le-Levier, Tomljanovich, and others (1970–1974).

Fundamental experiments on artificial modification of the upper ionosphere was realized by Utlaut in 1970a. In addition to the indicated changes in the temperature and concentration of the plasma, a number of important new phenomena were observed, connected with excitation of the instability of the ionosphere under the influence of the radio waves. First among them is the large-scale stratification of the ionospheric plasma, which leads to the onset of artificial sporadic F layer (Utlaut, Violette, 1970b). It was investigated by Wright (1973), Rufenach (1973), Belikovich et al. (1974), Thome (1974), and Bowhill (1974). The theory relates this phenomenon with self-focusing instability in the radio-wave reflection region (Vas'kov and Gurevich, 1974; Perkins and Valeo, 1974). The possibility of thermal self-focusing in the ionosphere was indicated by Litvak (1968) and by Georges (1970).

A small-scale stratification of the ionosphere, accompanied by the formation of inhomogeneities strongly elongated along the earth's magnetic field, also takes place and was observed by Fialer in 1971. These inhomogeneities effectively scatter the UHF and VHF waves (Fialer, 1974; Minkoff et al., 1974; Carpenter, 1974), a fact that can be used for radio communication (Barry, 1974). The theory relates this phenomenon with the resonant instability (Vas'kov, Gurevich, 1975) that leads simultaneously to scattering and to broad-band absorption of ordinary radio waves, as observed in experiment by Cohen and Whitehead (1970) and by Getmantsev et al. (1973). Other processes that could be essential are the dissipative parametric instability (Perkins, 1974; Grach and Trakhtengerts, 1975; Dimant, 1977) and drift instability (Borisov et al., 1976).

The possibility of nondissipative parametric instability which leads to intense excitation of plasma and ion sound waves in the ionosphere, was indicated by Perkins and Kaw (1971). The theory was subsequently developed by Fejer (1972), Valeo and Kruer (1972), Goldman (1972), Weinstock (1972), and others. In experiment, excitation of plasma waves in the ionosphere was observed by Wong and Taylor (1971), Gordon,

[2] These researches were published in 1974–1975.

Carlson, Showen (1972), and Kantor (1974), while the luminescence produced by the accelerated electrons was observed by Biondi et al. (1970) and by Megill and Haslett (1974). Self-modulation of a high-power pulse reflected from the F layer of the ionosphere, which was observed by Shlyuger (1967), is apparently connected with intense excitation of plasma waves. Getmantsev, Belikovich, Benediktov, et al. (1975) have observed in the F layer of the ionosphere spatially periodic perturbations due to striction nonlinearity. Wright (1975) has pointed out the possibility of effective precipitation of fast electrons from the plasmosphere when the ionosphere is modified by radio waves.

2. Plasma Kinetics in an Alternating Electric Field

2.1. Homogeneous Alternating Field in a Plasma (Elementary Theory)

Consider a plasma situated in a homogeneous electric field. Under the influence of this field, the electrons and ions no longer have equilibrium distribution functions—they are accelerated in the direction of the field. This accelerated motion is slowed down by collisions of the electrons with the ions, and also with the molecules and atoms. As a result of these two processes—acceleration by the field and deceleration by the collisions—a certain stationary nonequilibrium distribution of the electron velocities is established in the steady state. The task of the theory is to determine this distribution.

To solve this problem in the general case it is necessary to use the kinetic equations for the electron and ion distribution functions. The investigation of these distribution functions will be treated in Sections 2.2 to 2.4. To illustrate the physical picture, however, and frequently also to obtain sufficiently exact quantitative formulas, it is convenient and useful to use a much simpler albeit approximate theory which we shall call "elementary," following Chapman and Cowling, 1952, and Ginzburg, 1960. In the elementary theory the velocity distributions of the electrons and ions are neglected and a certain "average" electron (or ion) is considered. Accordingly the state of the plasma is characterized in the elementary theory by the directional velocities of the "average" electron and ion ($\boldsymbol{v}_e$ and $\boldsymbol{v}_i$) and by their average energies or temperatures T_e and T_i. In accordance with the definitions of the velocities $\boldsymbol{v}_e$ and $\boldsymbol{v}_i$, the total electric current density is connected with them by the relation

$$\boldsymbol{j} = -e\boldsymbol{v}_e N_e + \sum_k eZ_k N_{ik} \boldsymbol{v}_{ik}. \tag{2.1}$$

Here N_{ik} is the ion density ($\sum_k Z_k N_{ik} = N_e$); the summation is over the ions of different sorts k with charge eZ_k.

The electron and ion temperatures are defined in the elementary theory by the relations

$$\bar{\varepsilon}_e = \tfrac{3}{2} T_e, \qquad \bar{\varepsilon}_i = \tfrac{3}{2} T_i. \tag{2.2}$$

where $\bar{\varepsilon}_e$ and $\bar{\varepsilon}_i$ are the energies of the random motion of the average electron and ion. Since the electron and ion velocity distributions are far from always Maxwellian, it is more correct to call T_e and T_i the effective temperatures.

We shall derive below equations for $\boldsymbol{v}$ and T in the elementary theory. The question of the accuracy of the elementary theory and the character of the approximations with which it is connected can be consistently assessed only on the basis of a kinetic analysis (Sect. 2.3). In the present section the electron and ion densities in the plasma are assumed given. In the succeeding sections we shall consider ionization and recombination in the ionosphere and take into account the effect of the alternating electric field on the processes.

2.1.1. Electron Current—Electronic Conductivity and Dielectric Constant

The equations for the directional velocity of the "average" electron in the plasma can be easily obtained by starting from the following considerations. The electron is acted upon by the Lorentz force

$$\boldsymbol{F} = -e\boldsymbol{E} - \frac{e}{c}[\boldsymbol{v}_e \times \boldsymbol{H}], \tag{2.3}$$

where $\boldsymbol{H}$ is a constant magnetic field (we neglect the action of the magnetic field of the radio wave). Consequently, in the absence of friction, the equation for the velocity $\boldsymbol{v}_e$ takes the form

$$m\frac{d\boldsymbol{v}_e}{dt} = -e\boldsymbol{E} - \frac{e}{c}[\boldsymbol{v}_e \times \boldsymbol{H}]. \tag{2.4}$$

The friction is due to collisions between the electrons and other particles—molecules or ions (collisions between electrons do not change the total momentum of the electrons, i.e., they do not change the average velocity $\boldsymbol{v}_e$).

Under the influence of the collisions the velocity $\boldsymbol{v}_e$ should obviously decrease. The time during which the average momentum decreases by $m\boldsymbol{v}_e$ will be denoted $\tau_v = \nu_e^{-1}$, where ν_e is the collision frequency. Then the collision-induced friction force between the electrons and the particles of sort k moving with average velocity $\boldsymbol{v}_k$ is $-m\nu_{ek}(\boldsymbol{v}_e - \boldsymbol{v}_k)$, and Equation (2.4) takes the form:

$$m\frac{d\boldsymbol{v}_e}{dt} = -e\boldsymbol{E} - \frac{e}{c}[\boldsymbol{v}_e \times \boldsymbol{H}] - \sum_k m\nu_{ek}(\boldsymbol{v}_e - \boldsymbol{v}_k); \tag{2.5}$$

here ν_{ek} is the frequency of the collisions of the electrons with the particles of sort k (molecules, ions).

Of course, the change of momentum is different for different collisions of the electron with the heavy particles, since the impact parameters are different and the electrons are distributed in velocity. Therefore the time τ_v and correspondingly the collision frequencies ν_{ek} are certain average or effective values that describe the rate of change of the average electron momentum. The calculation of ν_{ek} is the task of the kinetic theory and calls for knowledge of the effective scattering cross sections and of the electron velocity distribution; this will be done later (Sect. 2.3). Here, on the other hand, it is important to emphasize that it can be assumed, in a very wide class of conditions, that the frequencies ν_{ek} do not depend on the velocity $\boldsymbol{v}_e$, but only on the electron temperature T_e. The reason is that in both a strong and a weak electric field the average directional velocity of the electrons is much less than their thermal (random) velocity. Therefore the collision frequency also depends only on the random velocity, i.e., on the effective electron temperature. For example, in collisions with molecules, assuming that the collision cross section q_{em} does not depend on the velocity, we have

$$\nu_{em} = \overline{q_{em} N_m v} = \nu_{em0}\sqrt{T_e/T}. \tag{2.6}$$

Here ν_{em0} is the number of electron collisions at $T_e = T$ (i.e., in the absence of an electric field). Analogously, for collisions with ions, assuming that q is inversely proportional to v^4 (Rutherford scattering), we obtain

$$\nu_{ei} = \nu_{ei0}(T_e/T_i)^{-3/2}. \tag{2.7}$$

The average velocities of the heavy particles (ions, molecules) can usually be neglected in comparison with the average electron velocity. Then Equation (2.5) assumes the simple form

$$m\frac{d\boldsymbol{v}_e}{dt} = -e\boldsymbol{E} - \frac{e}{c}[\boldsymbol{v}_e \times \boldsymbol{H}] - m\nu_e\boldsymbol{v}_e, \tag{2.8}$$

where $\nu_e = \sum_k \nu_{ek}$ is the effective frequency of the electron collisions.

Consider first the case of an isotropic plasma ($H = 0$). In the absence of an external electric field we obtain from Equation (2.8)

$$\boldsymbol{v}_e = \boldsymbol{v}_{e0} \exp(-\nu_e t). \tag{2.9}$$

We assume from now on that ν_e depends only on the electron temperature

T_e and that T_e is either constant or varies slowly with time, so that $|dT_e/dt| \ll \nu_e T_e$. It will be shown in the next section that this condition is usually well satisfied. It is seen from Equation (2.9) that the average electron velocity relaxes with time to zero, the relaxation time being $\tau_v = \nu_e^{-1}$. The same time determines the establishment of the steady-state velocity in an electric field.

In an alternating uniform electric field $\boldsymbol{E} = \boldsymbol{E} \exp(-i\omega t)$ we obtain from Equation (2.8) for the steady-state solution

$$\boldsymbol{v}_e = -\frac{e\boldsymbol{E}}{m} \frac{\nu_e + i\omega}{\nu_e^2 + \omega^2}. \tag{2.10}$$

The total electron current in the plasma [Eq. (2.1)] is now

$$\boldsymbol{j} = \frac{e^2 N \boldsymbol{E}}{m} \frac{\nu_e + i\omega}{\omega^2 + \nu_e^2}. \tag{2.11}$$

In linear macroscopic electrodynamics it is customary to introduce the dielectric constant ε and the conductivity σ

$$\boldsymbol{j} = \left(-i\omega \frac{\varepsilon - 1}{4\pi} + \sigma\right) \boldsymbol{E} = -\frac{i\omega}{4\pi}(\varepsilon' - 1)\boldsymbol{E}; \qquad \varepsilon' = \varepsilon + i\frac{4\pi\sigma}{\omega}. \tag{2.12}$$

Contributions to σ and ε are made in general by both the electrons and the ions. Confining ourselves to the electrons, we can speak of the electronic conductivity σ_e and of the electronic dielectric constant ε_e. Comparing Equations (2.11) and (2.12), we get

$$\varepsilon_e = 1 - \frac{4\pi e^2 N}{m(\omega^2 + \nu_e^2)}; \qquad \sigma = \frac{e^2 N \nu_e}{m(\omega^2 + \nu_e^2)}. \tag{2.13}$$

In the presence of magnetic field $\boldsymbol{H}$ the plasma becomes anisotropic. Then the conductivity and dielectric constant are tensors. The total current in an anisotropic medium is defined as

$$\begin{gathered} j_n = \sum_k \left[-\frac{i\omega}{4\pi}(\varepsilon_{enk} - \delta_{nk}) + \sigma_{enk}\right] E_k = -\frac{i\omega}{4\pi} \sum_k (\varepsilon'_{nk} - \delta_{nk}) E_k, \\ \varepsilon'_{nk} = \varepsilon_{enk} + i\frac{4\pi}{\omega}\sigma_{enk}, \end{gathered} \tag{2.14}$$

where ε_{enk} and σ_{enk} are the components of the tensors $\hat{\sigma}_e$ and $\hat{\varepsilon}_e$, and δ_{nk} is the Kronecker symbol ($\delta_{nk} = 1$ if $n = k$ and $\delta_{nk} = 0$ if $n \neq k$).

The velocity $\boldsymbol{v}_e$ in a field $\boldsymbol{E} = \boldsymbol{E}_0 \exp(-i\omega t)$, determined from Equation (2.8), takes in this case the form

$$\boldsymbol{v}_e = -\frac{e}{m[\omega_H^2 + (-i\omega + \nu_e)^2]}\left\{\boldsymbol{E}(-i\omega + \nu_e) + \frac{\omega_H^2 \boldsymbol{H}(\boldsymbol{E}\boldsymbol{H})}{H^2(-i\omega + \nu_e)} - \omega_H \frac{[\boldsymbol{E} \times \boldsymbol{H}]}{H}\right\}. \tag{2.15}$$

Here $\omega_H = eH/mc$ is the gyromagnetic frequency.

Substituting this equation for $\boldsymbol{v}_e$ in Equation (2.14), we obtain the components of the tensors ε_{enk} and σ_{enk}. Choosing the z axis in the direction of $\boldsymbol{H}$, we obtain

$$\begin{aligned}
\varepsilon_{xx} &= \varepsilon_{yy} = 1 - \frac{\omega_0^2}{2\omega}\left[\frac{\omega - \omega_H}{(\omega - \omega_H)^2 + \nu_e^2} + \frac{\omega + \omega_H}{(\omega + \omega_H)^2 + \nu_e^2}\right], \\
\varepsilon_{xy} &= -\varepsilon_{yx} = i\frac{\omega_0^2}{2\omega}\left[\frac{\omega - \omega_H}{(\omega - \omega_H)^2 + \nu_e^2} - \frac{\omega + \omega_H}{(\omega + \omega_H)^2 + \nu_e^2}\right], \\
\varepsilon_{zz} &= 1 - \frac{\omega_0^2}{\omega^2 + \nu_e^2}, \quad \varepsilon_{xz} = \varepsilon_{zx} = \varepsilon_{yz} = \varepsilon_{zy} = 0, \\
\sigma_{xx} &= \sigma_{yy} = \frac{\omega_0^2 \nu_e}{8\pi}\left[\frac{1}{(\omega - \omega_H)^2 + \nu_e^2} + \frac{1}{(\omega + \omega_H)^2 + \nu_e^2}\right], \\
\sigma_{xy} &= -\sigma_{yx} = -i\frac{\omega_0^2 \nu_e}{8\pi}\left[\frac{1}{(\omega - \omega_H)^2 + \nu_e^2} - \frac{1}{(\omega + \omega_H)^2 + \nu_e^2}\right], \\
\sigma_{zz} &= \frac{\omega_0^2 \nu_e}{4\pi(\omega^2 + \nu_e^2)}, \quad \sigma_{xz} = \sigma_{zx} = \sigma_{yz} = \sigma_{zy} = 0.
\end{aligned} \tag{2.16}$$

It is seen from Equation (2.16) that when the frequency ω of a high-frequency electric field ($\omega^2 \gg \nu_e^2$) is close to the gyrofrequency there is a resonant increase of the electronic conductivity, usually called gyromagnetic (or cyclotron) resonance.

Equations (2.13) and (2.16) describe, within the framework of the elementary theory, the electronic conductivity and the dielectric constant of the plasma. The effective electron collision frequency is determined in this case by Equations (2.6) and (2.7). A more accurate determination of ν_e is obtained in the kinetic theory. A consistent kinetic analysis leads also to corrections to Equations (2.13) and (2.16); these corrections are expressed with the aid of special coefficients K_σ and K_ε (Sect. 2.3).

2.1.2. Electron Temperature

The equation for the effective electron temperature [Eq. (2.2)] can be obtained by starting with the energy conservation law. The electric field performs in a unit time work equal to $\boldsymbol{j} \cdot \boldsymbol{E} = -eN\boldsymbol{v}_e \cdot \boldsymbol{E}$ on the plasma electrons or $-e\boldsymbol{v}_e \cdot \boldsymbol{E}$ on one electron. On the other hand, the electrons lose energy in collisions with the heavy particles. This energy, per unit time, can be expressed in the form

$$\frac{3}{2}\sum_k \delta_{ek}\nu_{ek}(T_e - T_k).$$

Here ν_{ek} is the frequency of the electron collisions with the heavy particles of sort k (ions, atoms, molecules), and δ_{ek} is the average fraction of the energy transferred in one collision. The electrons give up their energy to the heavy particles only if $T_e > T_k$; the reverse process occurs at $T_e < T_k$.

We can now write down the energy balance for the electrons in the plasma, in the form

$$\frac{d}{dt}\left(\frac{3}{2}NT_e\right) = \boldsymbol{j}\boldsymbol{E} - \frac{3}{2}\sum_k \delta_{ek}\nu_{ek}N(T_e - T_k). \tag{2.17}$$

If $dN/dt = 0$ and all the heavy particles have the same temperature, then Equation (2.17) simplifies to

$$\frac{dT_e}{dt} = -\frac{2}{3}e\boldsymbol{v}_e\boldsymbol{E} - \delta\nu_e(T_e - T), \tag{2.18}$$

where

$$\delta = \frac{1}{\nu_e}\sum_k \delta_{ek}\nu_{ek}, \qquad \nu_e = \sum_k \nu_{ek}. \tag{2.19}$$

Here $\boldsymbol{v}_e$ is the average electron velocity defined by Equation (2.8). Equations (2.8) and (2.18) constitute a closed system that determines the average directional velocity of the electrons $\boldsymbol{v}_e$ and their temperature T_e. It is very important that under stationary conditions in the plasma δ is always less than unity.[3] As a result, even in a strong electric field, the stationary thermal

[3] In a weakly ionized plasma in inert gases, at low electron temperatures ($T_e \lesssim 1$ eV) we usually have $\delta = 2m/M \sim 10^{-4}$ to 10^{-5}. Under the same conditions, but in molecular gases, $\delta \sim 10^{-3}$. With increasing electron temperature, δ also increases (owing to the increased role of the inelastic collisions). At the same time, however, the degree of ionization is increased. Collisions with the ions then assume an ever-increasing role, and δ is therefore again decreased. δ is discussed in greater detail in Sections 2.3 and 3.1.

(random) electron velocity $v_{Te} = \sqrt{2T_e/m}$ is always much larger than its directional velocity $\boldsymbol{v}_e$.

Let us consider certain particular solutions to Equations (2.18) and (2.8).

In the absence of an electric field, if the product $\delta\nu_e$ does not depend on T_e, we have

$$T_e = T + (T_{e0} - T)\exp(-\delta\nu_e t). \tag{2.20}$$

Here T_{e0} is the temperature of the electrons at the initial instant of time $t = 0$. It is seen from Equation (2.20) that the electron-temperature relaxation time is $\tau_T = 1/\delta\nu_e$. We note that according to Equation (2.9) the relaxation time of the average directional velocity of the electrons is $\tau_v = 1/\nu_e$. Since $\delta \ll 1$, the electron directional–velocity relaxation time is always much shorter than the temperature relaxation time.

We consider now an isotropic plasma ($H = 0$) in a constant electric field $\boldsymbol{E}$. In this case the average electron velocity, according to Equation (2.10), is

$$\boldsymbol{v}_e = -e\boldsymbol{E}/m\nu_e,$$

so that Equation (2.18) for the electron temperature T_e takes the form

$$\frac{dT_e}{dt} = \frac{2}{3}\frac{e^2E^2}{m\nu_e} - \delta\nu_e(T_e - T).$$

Its solution, if δ and ν_e are independent of T_e, is

$$T_e = T + \frac{2e^2E^2}{3m\,\delta\nu_e^2} + \left(T_{e0} - T - \frac{2e^2E^2}{3m\,\delta\nu_e^2}\right)\exp(-\delta\nu_e t).$$

We see therefore that at $t \gg \tau_T$ the plasma electron temperature assumes a stationary value that does not depend on the initial temperature T_{e0}, namely

$$T_e = T + \frac{2e^2E^2}{3m\,\delta\nu_e^2}. \tag{2.21}$$

The steady-state electron temperature increases in proportion to the square of the electric field intensity. It increases also with decreasing fraction δ of the energy lost by the electrons.

In an alternating electric field

$$\boldsymbol{E} = \boldsymbol{E}_0 \cos\omega t \tag{2.22}$$

(it is more convenient to change over to real quantities) we obtain from Equations (2.18) and (2.10) for the average velocity $\boldsymbol{v}_e$, at ν_e independent of T_e,

$$\boldsymbol{v}_e = -\frac{e\boldsymbol{E}_0}{m(\omega^2 + \nu_e^2)}(\nu_e \cos \omega t + \omega \sin \omega t). \tag{2.23}$$

Substituting Equations (2.23) and (2.22) in Equation (2.18), we obtain for the electron temperature

$$\frac{dT_e}{dt} = \frac{e^2 E_0^2}{3m(\omega^2 + \nu_e^2)}(\nu_e + \nu_e \cos 2\omega t + \omega \sin 2\omega t) - \delta\nu_e(T_e - T). \tag{2.24}$$

The steady-state solution of this equation, for δ and ν_e independent of T_e, is

$$T_e = T + \frac{e^2 E_0^2}{3m\,\delta(\omega^2 + \nu_e^2)} + \frac{e^2 E_0^2}{3m(\omega^2 + \nu_e^2)}\left[\frac{\delta\nu_e^2 - 2\omega^2}{4\omega^2 + \delta^2\nu_e^2}\cos 2\omega t + \frac{2\omega\nu_e}{4\omega^2 + \delta^2\nu_e^2}\sin 2\omega t\right]. \tag{2.25}$$

In the case of very low frequencies

$$2\omega \ll \delta\nu_e, \qquad \omega\tau_T \ll 1 \tag{2.26}$$

this yields, accurate to small terms of order $\omega/\delta\nu_e$,

$$T_e = T + \frac{2e^2 E_0^2 \cos^2 \omega t}{3m\,\delta\nu_e^2}. \tag{2.27}$$

Comparing this expression with Equation (2.21) for the case of a constant electric field, we see that under Equation (2.26) the temperature of the electrons is quasistationary. This is as it should be, since Equation (2.26) means that the electron temperature relaxation time $1/\delta\nu_e$ is much shorter than the time $1/\omega$ that characterizes the rate of change of the field.

In the other limiting case

$$\omega \gg \delta\nu_e \tag{2.28}$$

we have accurate to small terms of the order of $\delta\nu_e/\omega$ and δ:

$$T_e = T + \frac{e^2 E_0^2}{3m\,\delta(\omega^2 + \nu_e^2)}. \tag{2.29}$$

Thus, in first-order approximation, the electron temperature is constant in time in the case of Equation (2.28). The alternating component of T_e (of frequency 2ω) has a small amplitude, smaller by a factor $\delta\nu_e/\omega$ or δ than the dc component. This fact is quite understandable. Indeed, the relaxation time for the electron temperature in the plasma is $\tau_T = 1/\delta\nu_e$, and under Equation (2.28) the electron temperature cannot change significantly within the time $1/\omega$ during which the electron field changes. Therefore, the electron temperature settles on a certain average level independent of the time, and the deviations from this level are small.

It is now easily seen that in the steady state the directional velocity of the electron in the plasma, at an arbitrary value of the electric field, is much smaller than its random velocity. Indeed, according to Equations (2.21) and (2.29) we have

$$v_{Te} = \sqrt{\frac{2T_e}{m}} \gtrsim \frac{eE_0}{m\sqrt{\delta(\omega^2 + \nu_e^2)}}, \tag{2.30}$$

whereas the directional velocity is

$$|\boldsymbol{v}_e| = \frac{eE_0}{m\sqrt{\omega^2 + \nu_e^2}} \lesssim \sqrt{\delta} v_{Te}. \tag{2.31}$$

At $\delta \ll 1$, consequently, we always have $|\boldsymbol{v}_e| \ll v_{Te}$, regardless of the value of the field.[4]

It was assumed above that δ and ν_e do not depend on the electron temperature T_e. When account is taken of the dependencies of δ and ν_e on T_e, it is easy to obtain the solution of Equations (2.8) and (2.18) by expanding in terms of the small parameters $\omega/\delta\nu_e$ and δ [under Eq. (2.26)] or $\delta\nu_e/\omega$ and δ [under Eq. (2.28)]. In either case, the electron temperature is determined in first-order approximation by Equations (2.21) and (2.29), which must be regarded now as equations for T_e. For example, under Equation (2.28) the electron temperature is constant, as before, and is given by

$$T_e = T + \frac{e^2E_0^2}{3m\,\delta(T_e)[\omega^2 + \nu_e^2(T_e)]}. \tag{2.32}$$

[4] This is true, of course, only for steady–state distributions. In strong fields this relation may not be satisfied during the transient process. In certain cases there may be no steady–state solution at all (see below).

We rewrite this equation in the form

$$\frac{T_e}{T} = 1 + \left(\frac{E_0}{E_p}\right)^2 \frac{\delta_0}{\delta(T_e)} \frac{\omega^2 + \nu_{e0}^2}{\omega^2 + \nu_e^2(T_e)}. \tag{2.33}$$

Here $\nu_{e0} = \nu_e(T_{e0})$ and $\delta_0 = \delta(T_{e0})$ are respectively the number of collisions and the fraction of energy at $T_e = T_{e0} = T$; E_p is the characteristic plasma field [Eq. (1.2)]. From Equation (2.33), we see that if the amplitude of the electric field intensity is much lower than that of the plasma field, $E_0 \ll E_p$, then the electron temperature is only insignificantly changed by the action of the field:

$$\Delta T_e = T_e - T = e^2 E_0^2 / 3m\, \delta_0(\omega^2 + \nu_{e0}^2). \tag{2.32a}$$

The changes of the electron collision frequency, and consequently also the changes of the conductivity and of the dielectric constant of the plasma, are also insignificant in this case.

Thus, an electric field $E_0 \ll E_p$ perturbs the plasma weakly, and we shall call such a field *weak*. On the other hand, if $E_0 \gtrsim E_p$, then the electron temperature and the other plasma parameters (ν_e, σ_e, ε_e) are significantly altered by the electric field. Such fields, as already indicated, will be called *strong*.

Let us assume that δ is independent of T_e. Then solving the algebraic Equation (2.33) in the case of collisions with the molecules, with a collision cross section [Eq. (2.6)] independent of the velocity, we obtain

$$T_e = T\left[1 + \frac{\omega^2 + \nu_{e0}^2}{2\nu_{e0}^2}\left(\sqrt{1 + \frac{4\nu_{e0}^2}{\omega^2 + \nu_{e0}^2}\left(\frac{E_0}{E_p}\right)^2} - 1\right)\right]. \tag{2.34}$$

The dependence of T_e on E_0/E_p is shown in Figure 1. We see that T_e increases in this case monotonically with increasing E_0. At high frequencies $\omega^2 \gg \nu_e^2$, as is clear from Equation (2.32), we have

$$T_e = T\left(1 + \frac{e^2 E_0^2}{3Tm\,\delta\omega^2}\right). \tag{2.35}$$

This expression for T_e is independent of ν_e. It is consequently valid regardless of the character of the collisions.

At low frequencies $\omega^2 \ll \nu_e^2$, an interesting singularity appears in the behavior of the electron temperature, namely, the connection of T_e with

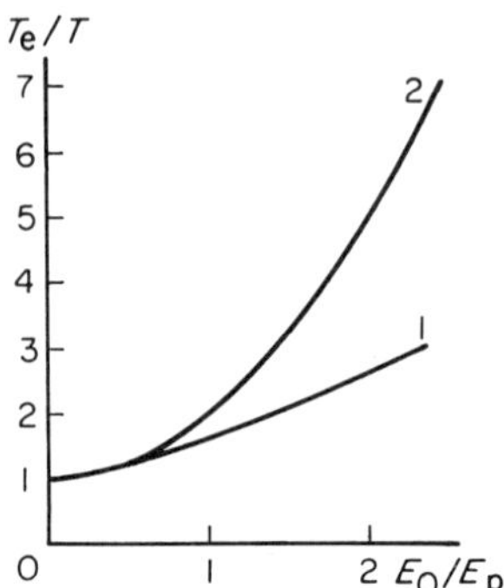

Fig. 1. Temperature of electrons in collisions with molecules: 1. $\omega^2 \ll \nu_{e0}^2$; 2. $\omega^2 \gg \nu_{e0}^2$

$(E_0/E_p)^2$ may no longer be single-valued. More accurately, in a definite range of field amplitudes $E_{c2} < E_0 < E_{c1}$, one value of E_0 may correspond not to one stationary value of the electron temperature, as is customary, but to three different values (Gurevich, 1958c). The plot of T_e against E_0/E_p then becomes *S*-shaped, as shown in Figure 2.

Let us find a criterion for the appearance of an *S*-shaped plot of T_e/T against E_0/E_p. As seen from Figure 2, in this case the derivative $d(T_e/T)/d(E_0/E_p)$ should become infinite, or

$$\frac{d(E_0/E_p)}{d(T_e/T)} = 0. \tag{2.36}$$

Expressing E_0/E_p in terms of T_e/T with the aid of Equation (2.33) and differentiating with respect to T_e, we rewrite the criterion [Eq. (2.36)]

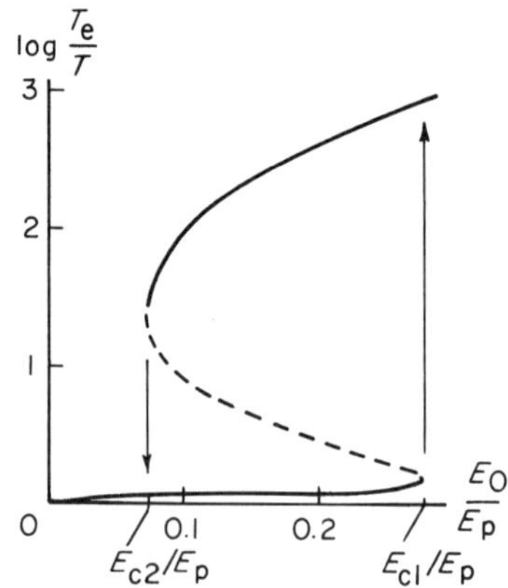

Fig. 2. Temperature of electrons colliding with ions; $\omega = 0.01\nu_{e0}$

in the form

$$\delta[\omega^2 + \nu_e^2] + \left(\frac{T_e}{T} - 1\right)\frac{d}{dT_e}[\delta(\omega^2 + \nu_e^2)] = 0. \tag{2.37}$$

It follows, therefore, that at constant δ the S-shaped plot can occur only at low frequencies $\omega \ll \nu_e$, and the collision frequency $\nu_e(T_e)$ should decrease with increasing T_e at a rate faster than $(T_e - T)^{-1/2}$. Equation (2.37) can be satisfied in regions where there is a decrease in the lost energy $\delta(T_e)$ in molecular gases (Al'tshuler, 1963), or where the electron-atom collision cross section decreases (the region of the Ramsauer effect), and also in a strongly ionized plasma, when the decisive role is played by collisions with ions under Equation (2.7).

Let us examine the last case in greater detail. Equation (2.33) takes in this case the form

$$\frac{E_0^2}{E_p^2} - \left(\frac{T_e}{T_i} - 1\right)\left[\frac{\omega^2}{\nu_{e0}^2} + \left(\frac{T_e}{T_i}\right)^{-3}\right]\left(\frac{\omega^2}{\nu_{e0}^2} + 1\right)^{-1} \tag{2.38}$$

The plot of T_e/T_i against E_0^2/E_p^2 as defined by Equation (2.38) is S-shaped at $\omega < 0.2\nu_{e0}$ (Fig. 2). In the range $E_{c2} \leq E_0 \leq E_{c1}$, one value of the field corresponds to three stationary values of the electron temperature. However, only two of them, corresponding to the upper and lower curves in Figure 2, are stable. The state corresponding to the middle curve is unstable. It is shown dashed in the figure. The transition from the low-temperature stationary state to the high-temperature one is shown in the figure by the arrow. The critical field E_{c1} at which this transition takes place is determined from Equation (2.37). At $\omega \ll \nu_{e0}$ it does not depend on the frequency ω. Indeed, from Equation (2.37) at $\omega^2 \ll \omega_{e0}^2$ and $\nu_e = \nu_{e0}(T_e/T_i)^{-3/2}$ we obtain $(T_e/T_i)_{c1} = 3/2$, and from Equation (2.38) we get

$$E_{c1} = \frac{2}{\sqrt{27}} E_p. \tag{2.39}$$

The absence of a "low-temperature" stationary state at $E_0 \geq E_{c1}$ is due to the fact that the energy imparted to the electron by the low-frequency electric field increases very rapidly with increasing electron temperature, $\boldsymbol{E}\boldsymbol{j} \sim \nu_e^{-1} \sim (T_e/T_i)^{3/2}$, whereas the energy transferred to the ions by the electrons decreases, $\delta\nu_e(T_e - T_i) \sim T_e^{-1/2}$. In a sufficiently strong electric field $E_0 \geq E_{c1}$ the electrons can, therefore, no longer transfer to the ions all the energy they absorb from the field, and the electron temperature begins to increase. This phenomenon is usually called *overheat instability* or *electron temperature runaway*.

With increasing temperature, the collision frequency decreases. After it becomes lower than the field frequency, the low-frequency condition $\omega < \nu_e$ is violated. It is this which makes possible the second "high-temperature" stable state [Eq. (2.35)] for a strongly heated electron gas, when $\nu_e^2(T_e) \ll \omega^2$. The reverse transition from the high-temperature to the low-temperature state takes place in a field

$$E_{c2} = 3^{1/2} 2^{-1/3} (\omega/\nu_{e0})^{2/3} E_p.$$

The field E_{c2} is much weaker than E_{c1} at $\omega/\nu_{e0} \ll 1$. This leads to hysteresis in the dependence of the stationary electron temperature on the amplitude of the alternating electric field.[5]

Overheat instability takes place in the case of collision with ions in a constant electric field. The corresponding critical field value is

$$E_c = E_{c1}/\sqrt{2}. \tag{2.40}$$

In contrast to the case of an alternating field, there is no high-temperature state here [since Equation (2.35) cannot be realized at $\omega = 0$]. Therefore at $E > E_{c1}$, in the case of collisions with the ions, the electron temperature increases continuously. In a constant field, collisions with neutral particles (molecules, atoms) can play a stabilizing role, and because of them a high-temperature state for T_e could be realized in a constant field.

We note, in addition, that in the considered case of collisions with ions in a very strong constant electric field stronger than the Dreicer critical field E_D (Dreicer, 1959),

$$E > E_D \approx \frac{E_p}{\sqrt{\delta}} \sim \sqrt{Tm}\,\frac{\nu_{e0}}{e} \tag{2.41}$$

the average electron directional velocity also experiences an instability that increases continuously. One can then no longer assume that the average directional velocity is much lower than the random velocity (as is always

[5] We note that the critical fields E_{c1} and E_{c2} change somewhat when account is taken of the kinetic effects. This is seen from the curve of Figure 2, which was plotted with the kinetic corrections taken into account.

the case under stationary conditions). As a result, the number of collisions of the electrons with the ions begins to depend essentially on the directional velocity $\boldsymbol{v}_e$, and ν_e decreases with increasing $|\boldsymbol{v}_e|$ in proportion to $|\boldsymbol{v}_e|^{-3}$. In a very strong field $E > E_D$ the average directional electron velocity increases so strongly that the role of the collisions becomes negligibly small and the electrons begin to be uniformly accelerated by the field (*runaway* electrons).

We have considered above only a monotonic dependence of ν_e on T_e. In a weakly ionized plasma, the electron–molecule collision cross sections can have also a more complicated character. A nonmonotonic dependence of ν_e on T_e leads, accordingly, to a complicated dependence of the electron current j_e on the electric field intensity E. In addition to the already considered S-shaped plot, we can have, for example, an N-shaped plot of j_e against E. The middle part of the N-shaped curve is unstable (just as for the S-curve; see review by Volkov and Kogan, 1969).

Analogous expressions for T_e can be obtained also if a magnetic field is present in the plasma. In particular, at $\omega \gg \nu_e$, the temperature of the electrons is constant in the first-order approximation. It is given by

$$\frac{T_e}{T} = 1 + \frac{E_0^2}{E_p^2}\varphi_p. \tag{2.42}$$

$$\varphi_p = \alpha_{\parallel}^2 + (\omega^2 + \nu_e^2)\left[\frac{\alpha_{\perp -}^2}{(\omega - \omega_H)^2 + \nu_e^2} + \frac{\alpha_{\perp +}^2}{(\omega + \omega_H)^2 + \nu_e^2}\right].$$

Here φ_p is a polarization factor, $\alpha_{\parallel}$, $\alpha_{\perp +}$, and $\alpha_{\perp -}$ are the coefficients of the resolution of the alternating electric field in components parallel to $\boldsymbol{H}$ ($\alpha_{\parallel}$) and components that rotate with and against the motion of the electron in a plane perpendicular to $\boldsymbol{H}$ ($\alpha_{\perp -}$ and $\alpha_{\perp +}$, respectively). The coefficients α depend on the polarization of the wave and satisfy the relation $\alpha_{\parallel}^2 + \alpha_{\perp -}^2 + \alpha_{\perp +}^2 = 1$. Their values in the general case of high-frequency field are given for example by Vas'kov and Gurevich (1976). In particular, for a plane polarized field we have $\alpha_{\parallel}^2 = \cos^2\beta$, $\alpha_{\perp -}^2 = \sin^2\beta/2$, and $\alpha_{\perp +}^2 = \sin^2\beta/2$, where β is the angle between $\boldsymbol{E}$ and $\boldsymbol{H}$. It is seen from Equation (2.42) that a resonant increase of the electron temperature takes place in the case of a high-frequency field $\omega^2 \gg \nu_e$, if the frequency ω is close to the gyrofrequency ω_H (Fig. 3). This increase is a consequence of the already noted resonant increase of the conductivity.

If the electric field is circularly polarized in a plane perpendicular to $\boldsymbol{H}$, and the electric field vector $\boldsymbol{E}$ rotates in the same direction as the plasma electrons (extraordinary wave), then $\alpha_{\parallel} = 0$, $\alpha_{\perp +} = 0$, $\alpha_{\perp -} = 1$ and

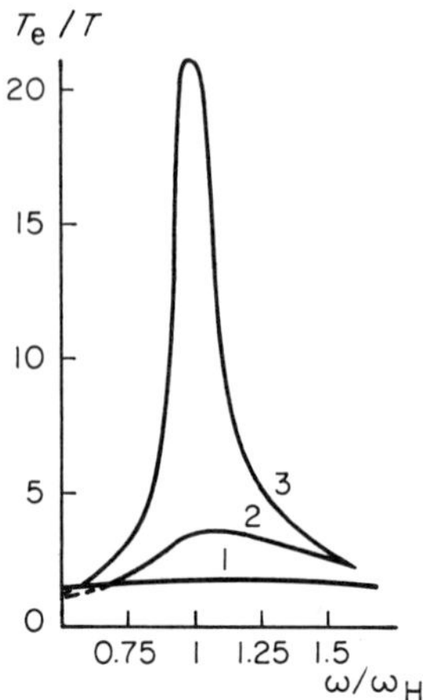

Fig. 3. Frequency dependence of electron temperature; $E_0^2/E_p^2 = 0.4$. 1. $\nu_e/\omega_H = 1$; 2. $\nu_e/\omega_H = 0.316$; 3. $\nu_e/\omega_H = 0.1$

Equation (2.42) for the electron temperature takes the form

$$\frac{T_e}{T} = 1 + \frac{e^2 E_0^2}{3m\,\delta(T_e)[(\omega - \omega_H)^2 + \nu_e^2]}.$$

This equation is identical with Equation (2.32) except that the field frequency ω is replaced by the difference ω–ω_H. It follows, therefore, that in a strongly ionized plasma the behavior of T_e, as a function of the field amplitude E_0 has near the gyromagnetic resonance ($\omega \approx \omega_H$) the same singularities as in the already considered case of a low–frequency electric field at $\boldsymbol{H} = 0$. Namely, the same ambiguity and the same hysteresis in the dependence of T_e on E_0^2 are obtained near the gyrofrequency. The case of a constant electric field is now equivalent to the case of strict cyclotron resonance $\omega = \omega_H$. No second stationary state is produced then in a fully ionized plasma, so that the electron temperature increases constantly at $E_0 > E_{c1}$. Moreover, in a sufficiently strong electric field $E_0 > E_D$ [Eq. (2.41)] there is likewise no stable transverse electron directional velocity at gyromagnetic resonance, and this velocity increases continuously. Runaway electrons are produced here just as in a constant electric field.

If Equation (2.28) is satisfied, the electron temperature T_e in an alternating electric field is constant in first-order approximation and is given by Equation (2.32). However, alternating corrections ΔT_e of frequency 2ω also arise; this is seen, for example, from Equation (2.25).

When Equation (2.28) is satisfied, the amplitude of the alternating corrections is always small in comparison with the value of the stationary temperature $\overline{T}_e$. Using this fact, we can easily obtain a general expression for ΔT_e. For example, in the absence of a magnetic field we obtain from Equations (2.8) and (2.18)

$$T_e = \overline{T}_e + \Delta T_e, \qquad \boldsymbol{v}_e = \overline{\boldsymbol{v}}_e + \Delta \boldsymbol{v}_e;$$

$$\Delta T_e = \delta \frac{E_0^2}{E_p^2} \overline{T}_e \left[\frac{\delta \nu_e^2 - 2\omega^2}{4\omega^2 + \delta^2 \nu_e^2} \cos 2\omega t + \frac{2\omega \nu_e}{4\omega^2 + \delta^2 \nu_e^2} \sin 2\omega t \right]. \tag{2.43}$$

Accordingly, the correction to the current (at $\omega \gg \nu_e \sqrt{\delta}$) is

$$\Delta \boldsymbol{v}_e = -\frac{\delta}{4} \frac{e\boldsymbol{E}_0}{m\omega(\omega^2 + \nu_e^2)} \frac{E_0^2}{E_p^2} \overline{T}_e \left(\frac{\partial \nu_e}{\partial T_e} \right)_{\overline{T}_e} \cdot \left\{ \frac{\omega(3\omega^2 - 5\nu_e^2)}{9\omega^2 + \nu_e^2} \cos 3\omega t \right.$$
$$\left. + \frac{\nu_e(\nu_e^2 - 7\omega^2)}{9\omega^2 + \nu_e^2} \sin 3\omega t + \nu_e \sin \omega t - \omega \cos \omega t \right\}. \tag{2.44}$$

Here, $\overline{T}_e$ and $\overline{\boldsymbol{v}}_e$ are defined by Equations (2.32) and (2.23). We see that nonlinearity necessitates the introduction of corrections to the electron current at triple the frequency. In addition, nonlinear corrections of frequency ω arise in the current. The correction to the active component ($\sim \cos \omega t$) describes the additional nonlinear change of the conductivity, and the correction to the inductive component ($\sim \sin \omega t$) describes the change of the dielectric constant of the plasma. Comparing these corrections with the main perturbations of ε_e and σ_e due to the change of $\overline{T}_e$ [Eqs. (2.32) and (2.23)], we see that they are small, of the order of $\delta \nu_e / \omega$.

2.1.3. Ion Current—Heating of Electrons and Ions

Let a plasma situated in an alternating electric field consist of electrons, of several sorts of ions with charges eZ_k and masses M_k, and of neutral molecules. Within the framework of the elementary theory, the equation for the average ion velocity can be easily obtained by reasoning similar to that advanced in Section 2.1.1 in the derivation of the equation for $\boldsymbol{v}_e$. It takes the form

$$M_k \frac{d\boldsymbol{v}_{ik}}{dt} = eZ_k \boldsymbol{E} + \frac{eZ_k}{c} [\boldsymbol{v}_{ik} \times \boldsymbol{H}] - m\nu_{ek}(\boldsymbol{v}_{ik} - \boldsymbol{v}_e) - M_k \sum_{p} \nu_{kp}(\boldsymbol{v}_{ik} - \boldsymbol{v}_p) \tag{2.45}$$

The term $m\nu_{ek}(\boldsymbol{v}_e - \boldsymbol{v}_{ik})$ describes here the interaction of the ions with the electrons, while the last term describes the interaction with the heavy plasma particles such as the other ions, molecules, and atoms; ν_{kp} is a collision frequency, and $\boldsymbol{v}_p$ is the average velocity of these particles.

In the case of a three-component plasma, i.e., a plasma consisting of electrons, singly charged ions of one sort, and molecules, Equations (2.8) and (2.45) for $\boldsymbol{v}_e$ and $\boldsymbol{v}_i$ assume the simple form

$$\begin{aligned} m\frac{d\boldsymbol{v}_e}{dt} &= -e\boldsymbol{E} - \frac{e}{c}[\boldsymbol{v}_e \times \boldsymbol{H}] - m\nu_{ei}(\boldsymbol{v}_e - \boldsymbol{v}_i) - m\nu_{em}\boldsymbol{v}_e; \\ M\frac{d\boldsymbol{v}_i}{dt} &= e\boldsymbol{E} + \frac{e}{c}[\boldsymbol{v}_i \times \boldsymbol{H}] - m\nu_{ei}(\boldsymbol{v}_i - \boldsymbol{v}_e) - M\nu_{im}\boldsymbol{v}_i. \end{aligned} \tag{2.46}$$

It is assumed here that the molecule gas is at rest—its velocity is zero. As before, we assume that the collision frequencies do not depend on the directional velocity. Then, in an isotropic plasma situated in an alternating electric field $\boldsymbol{E} = \boldsymbol{E}_0 \exp(-i\omega t)$ we have

$$\begin{aligned} \boldsymbol{v}_e &= -\frac{e\boldsymbol{E}}{m}\,\frac{M(-i\omega + \nu_{im})}{m\nu_{ei}(-i\omega + \nu_{em}) + M(-i\omega + \nu_{im})(-i\omega + \nu_{ei} + \nu_{em})}; \\ \boldsymbol{v}_i &= e\boldsymbol{E}\,\frac{-i\omega + \nu_{em}}{m\nu_{ei}(-i\omega + \nu_{em}) + M(-i\omega + \nu_{im})(-i\omega + \nu_{ei} + \nu_{em})}. \end{aligned} \tag{2.47}$$

Separating now the imaginary and real parts, we can obtain the electronic and ionic conductivities and the dielectric constants of the plasma. It is important to take it into account here that usually $M\nu_{im} \gg m\nu_{em}$. As a result, the first term in the denominators of Equation (2.47) can be neglected. The expressions for σ_e and ε_e then coincide with Equation (2.13) obtained with the ion motion neglected. For σ_i and $\Delta\varepsilon_i$ we have

$$\begin{aligned} \sigma_i &= \frac{\Omega_0^2}{4\pi}\,\frac{\nu_{em}\nu_{im}(\nu_{ei} + \nu_{em}) + \omega^2(\nu_{ei} + \nu_{im})}{[-\nu_{im}(\nu_{ei} + \nu_{em}) + \omega^2]^2 + \omega^2(\nu_{ei} + \nu_{em} + \nu_{im})^2}, \\ \Delta\varepsilon_i &= -\frac{\Omega_0^2[\omega^2 - \nu_{im}\nu_{ei} + \nu_{em}(\nu_{ei} + \nu_{em})]}{[-\nu_{im}(\nu_{ei} + \nu_{em}) + \omega^2]^2 + \omega^2(\nu_{ei} + \nu_{em} + \nu_{im})^2}. \end{aligned} \tag{2.48}$$

Here $\Omega_0^2 = 4\pi e^2 N/M$. Comparing Equations (2.48) with (2.13), we see that in the absence of a magnetic field the ion conductivity is always much lower than the electron conductivity.

This statement is valid also in the presence of a constant magnetic field in the plasma, provided only that the frequency of the alternating field is not too low, [see Eq. (2.50)]. On the other hand, if $\omega \lesssim (\Omega_H \omega_H)^{1/2}$, where $\Omega_H = eH/Mc$, then the ion motion becomes important. In the case of a constant field ($\omega = 0$), the electronic and ionic plasma conductivity tensor components determined from Equation (2.46) take the form

$$
\begin{aligned}
\sigma_{ezz} &= \frac{\omega_0^2}{4\pi(\nu_{em} + \nu_{ei})},\\
\sigma_{exx} = \sigma_{eyy} &= \frac{\omega_0^2}{4\pi A}\left[\nu_{ei} + \nu_{em}\left(1 + \frac{\Omega_H^2}{\nu_{im}^2}\right)\right],\\
\sigma_{exy} = -\sigma_{eyx} &= -\frac{\omega_0^2\omega_H}{4\pi A}\left[1 + \frac{m\nu_{ei}}{M\nu_{im}} + \frac{\Omega_H^2}{\nu_{im}^2}\right];\\
\sigma_{izz} &= \frac{\Omega_0^2\nu_{em}}{4\pi\nu_{im}(\nu_{em} + \nu_{ei})},\\
\sigma_{ixx} = \sigma_{iyy} &= \frac{\Omega_0^2}{4\pi A\nu_{im}}(\nu_{em}^2 + \nu_{em}\nu_{ei} + \omega_H^2),\\
\sigma_{ixy} = -\sigma_{iyx} &= \frac{\Omega_0^2\Omega_H}{4\pi A\nu_{im}^2}\left(\nu_{em}^2 + \frac{M}{m}\nu_{ei}\nu_{im} + \omega_H^2\right);\\
A &= (\nu_{em} + \nu_{ei})^2 + \omega_H^2\left(1 + 2\frac{m\nu_{ei}}{M\nu_{im}} + \frac{\Omega_H^2}{\nu_{im}^2}\right).
\end{aligned}
\tag{2.49}
$$

We have written out here the components of the total conductivity tensor $\sigma'_{nk} = \sigma_{nk} - i\omega/4\pi\ (\varepsilon_{nk} - \delta_{nk})$. We see that the ion current component along the magnetic field is much smaller than the electronic component. The transverse components of the ion current, on the other hand, are small only if $\nu_{im}(\nu_{ei} + \nu_{em}) \gg \omega_H \Omega_H$. When the collision frequency is sufficiently low in an adequately strong magnetic field (in the ionosphere, at $z \gtrsim 100$ km) this condition is not satisfied and the transverse current of the ions can be larger here than the transverse electron current. In an alternating electric field at $\omega > \Omega_H$ the transverse components of the ion conductivity decrease; at

$$
\omega^2 \gg \omega_H^2 \frac{m}{M}\frac{\nu_{em}}{\nu_{im}}
\tag{2.50}
$$

they are always smaller than the corresponding components of electron conductivity. In other words, in an alternating electric field, if Equation (2.50) is satisfied, the ion current is small in comparison with the electron current.

Equation (2.2) for the effective ion temperature can be easily obtained by using the energy conservation laws, as was done above for the electron temperature. It takes the form

$$\frac{dT_{ik}}{dt} = \frac{2}{3} eZ_k \boldsymbol{v}_{ik}\boldsymbol{E} - \delta_{ei}\nu_{ek}(T_{ik} - T_e) - \sum_{p} \delta_p \nu_{pk}(T_{ik} - T_p). \tag{2.51}$$

The term $\delta_{ei}\nu_{ek}(T_{ik} - T_e)$ determines here the energy transferred by the collisions from the ions to the electrons, and the last term represents the energy transferred to the plasma particles; ν_{pk} is the collision frequency; δ_p is the average fraction of the energy lost by the ion in one impact.

Consider, by way of example, the heating of a three-component isotropic plasma ($H = 0$). The ion current in this case is always much smaller than the electron current. Therefore the work performed by the ion current can be neglected. Equations (2.18) for the electron and ion temperatures then become

$$\frac{dT_e}{dt} = -\frac{2}{3} e\boldsymbol{v}_e\boldsymbol{E} - \delta_{ei}\nu_{ei}(T_e - T_i) - \delta_{em}\nu_{em}(T_e - T), \tag{2.52}$$

$$\frac{dT_i}{dt} = \delta_{ei}\nu_{ei}(T_e - T_i) - \nu_{im}(T_i - T). \tag{2.53}$$

Here $\delta_{im} = \delta_p$ is set equal to unity for simplicity. T is the temperature of molecules, which we assume to be constant and independent of the time.

In a high-frequency electric field ($\omega \gg \delta_{em}\nu_{em} + \delta_{ei}\nu_{ei}$) the electron and ion temperatures are stationary in first-order approximation in δ and $\delta\nu/\omega$ [cf. Eq. (2.32)] and are determined by the transcendental equations

$$\frac{e^2 E_0^2}{3m} = [\delta_{ei}\nu_{ei}(T_e - T_i) + \delta_{em}\nu_{em}(T_e - T)]\frac{\omega^2 + (\nu_{ei} + \nu_{em})^2}{\nu_{ei} + \nu_{em}}, \tag{2.54}$$

$$\delta_{ei}\nu_{ei}(T_e - T_i) = \nu_{im}(T_i - T). \tag{2.55}$$

The last equation can be rewritten in the form

$$T_i = \frac{\delta_{ei}\nu_{ei}T_e + \nu_{im}T}{\delta_{ei}\nu_{ei} + \nu_{im}}. \tag{2.56}$$

It follows therefore that the ion temperature T_i falls between the electron temperature T_e and the neutral–molecule temperature T. In particular, if $\nu_{im} \gg \delta_{ei}\nu_{ei}$, then T_i is close to T:

$$T_i = T + \Delta T_i; \qquad \Delta T_i = \frac{\delta_{ei}\nu_{ei}}{\nu_{im}}(T_e - T). \tag{2.56a}$$

The dependence of the electron or ion temperature on the field amplitude E_0, as given by Equations (2.54) and (2.55), has in the main a monotonically rising plot (of the type shown in Fig. 1). The curves, however, can also be S-shaped. The latter case is realized at a low frequency ω and at a sufficiently high degree of plasma ionization. Indeed, assume for simplicity that δ_{em}, ν_{em}, and ν_{im} do not depend on the electron and ion temperatures. Then the ion temperature is determined by Equation (2.56). Substituting this expression in Equation (2.54), we obtain

$$\frac{e^2 E_0^2}{3m} = (T_e - T)\left[\frac{\delta_{ei}\nu_{ei}\nu_{im}}{\delta_{ei}\nu_{ei} + \nu_{im}} + \delta_{em}\nu_{em}\right]\frac{\omega^2 + (\nu_{ei} + \nu_{em})^2}{\nu_{ei} + \nu_{em}}. \tag{2.57}$$

Recognizing that $\nu_{ei}(T_e) = \nu_{ei}(T)(T_e/T)^{-3/2}$ [see Eq. (2.7)], we find therefore that at a sufficiently high degree of plasma ionization (when $\delta_{ei}\nu_{ei}(T) \gg \nu_{im}$ and $\nu_{ei}(T) \gg \nu_{em}$) the plot of the electron temperature against the field amplitude is indeed S-shaped. In the low-temperature state at $\omega \ll \nu_{ei}$, $\nu_{ei} \gg \nu_{em}$, and $\delta_{ei}\nu_{ei} \gg \nu_{im}$ the temperature T_e is then given by the equation

$$\left(\frac{T_e}{T} - 1\right)\left(\frac{T}{T_e}\right)^{3/2} = \frac{e^2 E_0^2}{3mT(\nu_{im} + \delta_{em}\nu_{em})\nu_{ei}(T)}. \tag{2.58}$$

Equating, as before, the derivative dE_0^2/dT_e to zero, we obtain the critical value of the electron temperature and the electric-field amplitude at which the low-temperature state is unstable:

$$T_e = 3T; \qquad E_{c1} = \frac{\sqrt{2}}{3^{1/4}}\sqrt{\frac{mT}{e^2}(\nu_{im} + \delta_{em}\nu_{em})\nu_{ei}(T)}. \tag{2.59}$$

The temperature of the ions in the low-temperature state is close to the electron temperature. In the high-temperature state it is much lower than T_e. With increasing E_0, the ion temperature decreases and tends at large E_0 to the temperature of the neutral particles. At the same time, the electron temperature increases in proportion to E_0^2.

In conclusion we note that in this section it was assumed that for the ions, as for the electrons, the collision frequency depends only on the temperature (i.e., on the random velocity) and does not depend in first-order approximation on the average directional velocity $\boldsymbol{v}_i$. In the case of electrons this was due to the smallness of the fraction δ of the average energy lost by the electron in one collision [see Eq. (2.31)]. For the ions, the fraction of the lost energy is not small. However, owing to the large mass of the ion, its directional velocity in an electric field is small. At the same time, the random velocity of the ions increases with increasing field amplitude E_0 owing to the heating of the ions in collisions with the electrons. Consequently, the indicated assumption $|\boldsymbol{v}_\mathrm{i}| \ll |v_{T\mathrm{i}}|$ turns out to be valid in an alternating field if Equation (2.50) is satisfied, and valid in a constant field if $E < E_\mathrm{D}$ [Eq. (2.41)], i.e., under the same conditions as for the electrons. If the condition $E < E_\mathrm{D}$ is not satisfied, then the dependence of the collision frequency on the directional velocity of the ions in a constant field becomes appreciable, and this leads to the appearance of "runaway" ions. A similar effect arises in a magnetic field at the gyromagnetic resonance for the ions ($\omega = \Omega_H$).

2.2. The Kinetic Equation

In the kinetic theory, the behavior of an electron or ion gas (as well as of gases of neutral atoms and molecules) in electric and magnetic fields is described by the distribution functions

$$f(\boldsymbol{r}, \boldsymbol{v}, t), \qquad f_\mathrm{i}(\boldsymbol{r}, \boldsymbol{v}, t), \qquad F_\mathrm{m}(\boldsymbol{r}, \boldsymbol{v}, t).$$

Here $\boldsymbol{v}$ and $\boldsymbol{r}$ are the velocity and coordinates of the particles. The distribution function constitutes the particle density in the $(\boldsymbol{r}, \boldsymbol{v})$ configuration space. In other words, the average number of particles in the configuration space $d\boldsymbol{r} \cdot d\boldsymbol{v} = dx\, dy\, dz\, dv_x\, dv_y\, dv_z$ is equal to $f\, d\boldsymbol{r} \cdot d\boldsymbol{v}$. It follows therefore that the density N, the flux $\boldsymbol{j}$, and the average particle energy $\bar{\varepsilon}$ can be expressed in terms of the function f in the following manner:

$$N = \int f(\boldsymbol{r}, \boldsymbol{v}, t)\, d\boldsymbol{v}, \qquad \boldsymbol{j} = \int \boldsymbol{v} f(\boldsymbol{r}, \boldsymbol{v}, t)\, d\boldsymbol{v},$$
$$\bar{\varepsilon} = \frac{1}{N} \int \frac{mv^2}{2} f(\boldsymbol{r}, \boldsymbol{v}, t)\, d\boldsymbol{v} \tag{2.60}$$

Here m is the mass of the particles in question.

The Boltzmann kinetic equation for the electron distribution function takes the form

$$\frac{\partial f}{\partial t} + \boldsymbol{v}\frac{\partial f}{\partial \boldsymbol{r}} - \frac{e}{m}\left(\boldsymbol{E} + \frac{1}{c}[\boldsymbol{v} \times \boldsymbol{H}]\right)\frac{\partial f}{\partial \boldsymbol{v}} + S = 0. \tag{2.61}$$

Here S is the Boltzmann collision integral, which describes the change of the electron distribution function f following collisions of the electrons with one another as well as with all plasma particles:

$$S = \int\int d\boldsymbol{v}_1 \, d\Omega \, q(u, \theta) u \{ f(\boldsymbol{v}) F(\boldsymbol{v}_1) - f(\boldsymbol{v}') F(\boldsymbol{v}'_1) \}. \tag{2.62}$$

Here $\boldsymbol{v}_1$ is the velocity of the particle that collides with the electron (we call it particle 1); $u = |\boldsymbol{v} - \boldsymbol{v}_1|$, $q(u, \theta)$ is the differential effective scattering cross section; $\boldsymbol{v}'$ and $\boldsymbol{v}'_1$ are the velocity of the electron and of particle 1 prior to the collision (their velocities after the impact are respectively $\boldsymbol{v}$ and $\boldsymbol{v}_1$); and F is the distribution function of the particle 1. The integration in Equation (2.62) is over the velocities of the particle 1 ($d\boldsymbol{v}_1$) and over the scattering angles $d\Omega = \sin\theta \, d\theta \, d\varphi$, where θ is the angle between $\boldsymbol{v} - \boldsymbol{v}_1$ and $\boldsymbol{v}' - \boldsymbol{v}'_1$.

For the plasma electrons, an important role is played by elastic and inelastic collisions with the molecules and with the ions, by electron–electron collisions, and by collisions with various types of waves excited in the plasma. When collisions between electrons are considered we have $F = f$; Equation (2.61) is made nonlinear by the electron–electron collisions.

The kinetic equation for the distribution function of neutral molecules (if they have no appreciable electric or magnetic moment) takes the form of Equation (2.61), but without the third term.

The interaction of the particles is governed by pair collisions only if the plasma has not deviated too strongly from the equilibrium state. This is the case that will be considered here. If, however, the deviation from equilibrium is large, then an important role is assumed by the interaction of the particles with radiation—waves of various types excited in the plasma. Readers interested in these questions can refer to special monographs devoted to plasma kinetics (Klimontovich, 1964; Shkarofsky et al., 1966; Tsitovich, 1970; Silin, 1972).

2.2.1. Simplification of the Kinetic Equation for Electrons

The kinetic Equation (2.61) can be simplified by using the main features of the behavior of electrons in a plasma. It was shown in the elementary treatment (Sect. 2.1) that the thermal (random) electron velocity is usually much larger than its average directional velocity. Accordingly one can expect the distribution function of the electrons to depend, under the same conditions, mainly on the absolute value of the velocity and not on its direction. It is, therefore, convenient to separate in the distribution function $f(\boldsymbol{v}, \boldsymbol{r}, t)$ its principal (symmetrical) part $f_0(|\boldsymbol{v}|, \boldsymbol{r}, t)$, which depends only on the absolute value of the velocity, and the directional part $\boldsymbol{f}_1$ (Allis, 1956). In other words, it is convenient to expand the angular part of the distribution function in velocity space in a series in spherical functions.

Consider first, for simplicity, the case of an isotropic plasma ($\boldsymbol{H} = 0$), and assume that the spatial gradient of the distribution function is directed along the z axis, which is parallel to the electric field $\boldsymbol{E}$. Then there is only one preferred direction of $\boldsymbol{E}$ (i.e., the z axis). Consequently, the distribution function is $f(\boldsymbol{v}, \boldsymbol{r}, t) = f(v, \cos\theta_1, z, t)$, where $v = |\boldsymbol{v}|$ and θ_1 is the angle between the velocity $\boldsymbol{v}$ and the z axis. It can, therefore, be expanded in spherical functions of zero order, i.e., in Legendre polynomials $P_k(\cos\theta_1)$:

$$f(\boldsymbol{v}, \boldsymbol{r}, t) = \sum_{k=0}^{\infty} P_k(\cos\theta_1) f_k(v, z, t). \tag{2.63}$$

We now substitute the expansion (2.63) in Equation (2.64):

$$\sum_{k=0}^{\infty} P_k(\cos\theta_1)\frac{\partial f_k}{\partial t} + v\cos\theta_1 \sum_{k=0}^{\infty} P_k(\cos\theta_1)\frac{\partial f_k}{\partial z}$$
$$-\frac{eE}{m}\cos\theta_1 \sum_{k=0}^{\infty} P_k(\cos\theta_1)\frac{\partial f_k}{\partial v} - \frac{eE}{mv}\sin^2\theta_1 \sum_{k=0}^{\infty} f_k \frac{\partial P_k}{\partial(\cos\theta_1)} + S = 0. \tag{2.64}$$

We have taken into account here the fact that

$$\boldsymbol{E}\frac{\partial f}{\partial \boldsymbol{v}} = E\cos\theta_1 \frac{\partial f}{\partial v} + \frac{E\sin^2\theta_1}{v}\frac{\partial f}{\partial(\cos\theta_1)}$$

Multiplying now Equation (2.64) by $P_k(\cos\theta_1)$ and integrating with respect to $d\Omega_1 = \sin\theta_1\, d\theta_1\, d\varphi$, we obtain the following equation for the function f_k:

$$\frac{\partial f_k}{\partial t} + v\left[\frac{k}{2k-1}\frac{\partial f_{k-1}}{\partial z} + \frac{k+1}{2k+3}\frac{\partial f_{k+1}}{\partial z}\right]$$
$$-\frac{eE}{m}\left\{\frac{k+1}{(2k+3)v^{k+2}}\frac{\partial}{\partial v}(v^{k+2}f_{k+1}) + \frac{kv^{k-1}}{2k-1}\frac{\partial}{\partial v}(v^{1-k}f_{k-1})\right\} + S_k = 0 \tag{2.65}$$

$$S_k = \frac{2k+1}{4\pi}\int S P_k(\cos\theta_1)\, d\Omega_1$$

We have taken into account here the orthogonality of the polynomials

$$\int_{-1}^{1} P_k(x)P_l(x)\,dx = \frac{2}{2k+1}\,\delta_{kl}$$

and the recurrence relations

$$xP_k(x) = \frac{k+1}{2k+1}\,P_{k+1} + \frac{k}{2k+1}\,P_{k-1},$$

$$(1-x^2)\frac{dP_x}{dx} = \frac{k(k+1)}{2k+1}\,[P_{k-1}(x) - P_{k+1}(x)].$$

When collisions of electrons with other particles are considered, the collision integral S is linear in the electron distribution function f. In this case, if the distribution function of particles 1 is symmetrical, i.e., $F = F(v_1)$, then the integral S_k depends only on the function f_k. In fact, substituting the expansion [Eq. (2.63)] in the collision integral [Eq. (2.62)], multiplying it by $P_k(\cos\theta_1)$, and integrating with respect to $d\Omega_1$, we obtain

$$S_k = \frac{2k+1}{4\pi}\int d\boldsymbol{v}_1\,d\Omega \int q(u,\theta)uP_k(\cos\theta_1)$$
$$\cdot\left\{F\sum_k P_k(\cos\theta_1)f_k - F'\sum_k P_k(\cos\theta_1')f'_k\right\}d\Omega_1. \qquad (2.66)$$

Recognizing that $\cos\theta_1' = \cos\theta\,\cos\theta_1 + \sin\theta\,\sin\theta_1\,\cos\varphi_1$($\theta_1'$ is the angle between $\boldsymbol{v}'$ and $\boldsymbol{E}$, and the difference between $\boldsymbol{v}_1'$ and $\boldsymbol{v}_1$ is neglected), and taking into account the addition theorem for spherical functions:

$$P_k(\cos\theta_1') = P_k(\cos\theta)P_k(\cos\theta_1)$$
$$+ 2\sum_{m=1}^{k}\frac{(k-m)!}{(k+m)!}\,P_k^m(\cos\theta)P_k^m(\cos\theta_1)\cos m\varphi_1,$$

we integrate in Equation (2.66) with respect to $d\Omega_1$. We then obtain

$$S_k = \int d\boldsymbol{v}_1\,d\Omega\,q(u,\theta)u\{f_kF - F'f'_kP_k(\cos\theta)\}. \qquad (2.67)$$

Putting now in Equation (2.65) in succession $k = 0, 1, 2, \ldots$, we obtain instead of Equation (2.65) the following chain of equations

$$\frac{\partial f_0}{\partial t} + \frac{v}{3}\frac{\partial f_1}{\partial z} - \frac{eE}{3mv^2}\frac{\partial}{\partial v}(v^2 f_1) + S_0 = 0,$$

$$\frac{\partial f_1}{\partial t} + v\left(\frac{\partial f_0}{\partial z} + \frac{2}{5}\frac{\partial f_2}{\partial z}\right) - \frac{eE}{m}\left[\frac{\partial f_0}{\partial v} + \frac{2}{5v^3}\frac{\partial}{\partial v}(v^3 f_2)\right] + S_1 = 0,$$

$$\frac{\partial f_2}{\partial t} + v\left(\frac{2}{3}\frac{\partial f_1}{\partial z} + \frac{3}{7}\frac{\partial f_3}{\partial z}\right) - \frac{eE}{m}\left[\frac{2}{3}v\frac{\partial}{\partial v}\left(\frac{1}{v}f_1\right) + \frac{3}{7v^4}\frac{\partial}{\partial v}(v^4 f_3)\right] + S_2 = 0 \tag{2.68}$$

etc. The obtained chain of equations can be terminated after the first two equations if the function f_2 can be neglected in comparison with f_0, or more accurately if

$$\frac{\partial f_0}{\partial v} \gg \frac{2}{5v^3}\frac{\partial}{\partial v}(v^3 f_2), \qquad \frac{\partial f_0}{\partial z} \gg \frac{2}{5}\frac{\partial f_2}{\partial z}.$$

We consider these conditions first for the stationary spatially homogeneous problem. Neglecting f_3 and recognizing that in this case

$$\frac{\partial f_1}{\partial t} = -i\omega f_1, \qquad \frac{\partial f_2}{\partial t} = -i\omega f_2, \qquad S_1 = \nu f_1, \qquad S_2 = \nu_2 f_2$$

where $\nu_2 \sim \nu$, we obtain from Equation (2.68):

$$f_1 = \frac{eE}{m(-i\omega + \nu)}\frac{\partial f_0}{\partial v}, \qquad f_2 = \frac{2eEv}{3m(-i\omega + \nu_2)}\frac{\partial}{\partial v}\left[\frac{eE}{m(-i\omega + \nu)v}\frac{\partial f_0}{\partial v}\right].$$

The condition for terminating the chain [Eq. (2.65)] is therefore

$$\left|\frac{\partial f_0}{\partial v}\right| \gg \left|\frac{4}{15v^3}\frac{\partial}{\partial v}\left\{\frac{eEv^4}{m(-i\omega + \nu_2)}\frac{\partial}{\partial v}\left[\frac{eE}{m(-i\omega + \nu)v}\frac{\partial f_0}{\partial v}\right]\right\}\right|. \tag{2.69}$$

If we confine ourselves to the use of this condition for the average velocity $\bar{v} \sim \sqrt{T_e/m}$ and put $\partial/\partial v \sim 1/v$, then Equation (2.69) reduces to the requirement

$$\frac{e^2 E_0^2}{mT_e[\omega^2 + \nu_e^2(T_e)]} \ll 1. \tag{2.70}$$

Using Equation (2.32) for the electron temperature, we rewrite this condition in the form

$$\delta(T_e)\frac{T_e - T}{T_e} \ll 1, \tag{2.71}$$

where $\delta(T_e)$ is the average fraction of energy lost by the electron in one impact. Equation (2.71) is thus actually identical with the initial requirement $\delta \ll 1$, which, as already indicated in Section 2.1, is always satisfied. Of course, this conclusion pertains only to the average electron velocity. Equation (2.69) may, for example, be violated if the gradient of the distribution function $f_0(v)$ with respect to the velocity v is very large. At low frequencies, and especially in a constant electric field, Equation (2.69) may not be satisfied at velocities $v \gg v_{Te} = \sqrt{2T_e/m}$ if the collision frequency ν decreases with the increasing electron velocity v. In particular, in a fully ionized plasma we have $\nu \sim v^{-3}$ and Equation (2.69) is not satisfied in a constant electric field if

$$v > (E_D/E)^{1/4} v_{Te}, \tag{2.72}$$

where E_D is the Dreicer critical field, defined by Equation (2.41). At large velocities $v \gg (E_D/E)^{1/4} v_{Te}$, the electronic distribution function in a constant electric field acquires a sharply directional character (Gurevich, 1960; Lebedev, 1965).

It was assumed above that the plasma is stationary and homogeneous. If it is nonstationary and inhomogeneous, Equation (2.69) must be supplemented by the requirement

$$\left|\frac{\partial f_0}{\partial t}\right| \ll \nu f_0, \qquad \frac{v}{\sqrt{\omega^2 + \nu^2}}\left|\frac{\partial^2 f_1}{\partial z^2}\right| \ll \left|\frac{\partial f_0}{\partial z}\right|, \tag{2.73}$$

where $\nu = \nu(v)$ is the collision frequency of an electron with velocity v.

A similar expansion of the electron distribution function is carried out also in the presence of a constant magnetic field $\boldsymbol{H}$ in the plasma, as well as for an arbitrary direction of the spatial gradient of the distribution function. Separating in this case, as before, the symmetrical part $f_0(v, \boldsymbol{r}, t)$ of the distribution function (which depends only on the absolute value of the velocity) from its directional part $(\boldsymbol{v}/v)\boldsymbol{f}_1(v, \boldsymbol{r}, t)$ and neglecting the remaining terms (i.e., putting $f = f_0 + (\boldsymbol{v}/v)\boldsymbol{f}_1$), we can reduce Equation (2.61) to the following system of equations for the function f_0 and $\boldsymbol{f}_1$ (Davydov, 1937):

$$\begin{gathered}
\frac{\partial f_0}{\partial t} + \frac{v}{3}\operatorname{div}_r \boldsymbol{f}_1 - \frac{e}{3mv^2}\frac{\partial}{\partial v}(v^2 \boldsymbol{E}\boldsymbol{f}_1) + S_0 = 0, \\
\frac{\partial \boldsymbol{f}_1}{\partial t} + v \operatorname{grad}_r f_0 - \frac{e\boldsymbol{E}}{m}\frac{\partial f_0}{\partial v} - \frac{e}{mc}[\boldsymbol{H} \times \boldsymbol{f}_1] + \boldsymbol{S}_1 = 0.
\end{gathered} \tag{2.74}$$

In the absence of a magnetic field and at $\operatorname{grad}_r f$ parallel to $\boldsymbol{E}$, Equations (2.74) are identical with the first two equations of the system [Eqs. (2.68)].

2.2.2. Transformation of the Electron Collision Integral

Proceeding to the analysis of electron collisions, we note their most important distinguishing feature: in most cases the principal role is played by collisions in which the change of the electron energy, and sometimes also the change of its momentum, is very slight. In such situations the collision integral can be simplified by using the Fokker–Planck method and can be represented in differential form. We now consider in succession the various types of electron collisions.

Elastic Collisions with Neutral Particles. In an elastic impact of a light particle (electron) against a heavy one (atom or molecule), the velocity of the light particle can readily change direction, but the absolute value of the velocity or the energy of the electron can change only insignificantly. Using this circumstance, we can assume in Equation (2.67), in first-order approximation, that $v' = v$ and $\boldsymbol{v}_1' = \boldsymbol{v}_1$. Recognizing also that the electron velocity is much larger than that of the molecule (it is assumed that $T \ll (M/m)T_e$) and consequently $u = |\boldsymbol{v} - \boldsymbol{v}_1| \approx v$, we obtain from Equation (2.67):

$$\begin{aligned} S_{mk} &= \int d\boldsymbol{v}_1\, d\Omega\, q(u, \theta)u\{f_k(v)F_m(\boldsymbol{v}_1) - P_k(\cos\theta)f_k(v')F_m(\boldsymbol{v}_1')\} \\ &= f_k(v)\int d\boldsymbol{v}_1\, F_m(\boldsymbol{v}_1)vq(v, \theta)[1 - P_k(\cos\theta)]\, d\Omega = \nu_{mk}(v)f_k(v). \quad (2.75) \end{aligned}$$

Here $q(v, \theta)$ is the differential effective cross section for elastic scattering of the electron,

$$\nu_{mk}(v) = N_m v \int q(v, \theta)[1 - P_k(\cos\theta)]\, d\Omega, \qquad N_m = \int F_m(v_1)\, d\boldsymbol{v}_1 \quad (2.76)$$

N_m is the concentration of the molecules. In particular,

$$S_{m1} = \nu_m(v)f_1, \quad \nu_m(v) = N_m v\sigma_t, \quad \sigma_t = \int q(v, \theta)(1 - \cos\theta)\, d\Omega. \quad (2.77)$$

$\nu_m(v)$ is the electron collision frequency, and σ_t is the effective transport cross section. In a collision with a hard sphere of radius a (elastic collisions of electrons with neutral particles can be tentatively simulated in this manner if q is constant), as is well known, we have $q = a^2/4$. Then

$$\nu_m = \pi a^2 N_m v, \qquad \sigma_t = \pi a^2. \quad (2.78)$$

Accordingly, $\nu_{mk}(v) = \nu_m(v)$ in this case, since $\int_{-1}^{1} P_k(\cos\theta)\, d\Omega = 0$, $k \neq 0$, and the differential scattering cross section q does not depend on the angle θ.

It follows from Equation (2.75) that $S_{m0} = 0$, inasmuch as $P_0(\cos\theta) = 1$. This should be the case if energy exchange is completely neglected. To calculate S_{m0} it is necessary to take into account the change of the electron and molecule energy. It is well known from the elastic-impact laws that the electron velocity prior to the collision is

$$\boldsymbol{v}' = \frac{m\boldsymbol{v} + M\boldsymbol{v}_1}{m + M} + \boldsymbol{n}\frac{M}{M + m}|\boldsymbol{v} - \boldsymbol{v}_1|. \tag{2.79}$$

Here $\boldsymbol{v}$ and $\boldsymbol{v}_1$ are the velocities of the electron and of the molecule after the impact and $\boldsymbol{n}$ is a unit vector parallel to $\boldsymbol{v}' - \boldsymbol{v}_1'$. Recognizing that $v \gg v_1 \gg (m/M)v$, we obtain from Equation (2.79):

$$v'^2 \simeq u^2 + 2u\boldsymbol{n}\boldsymbol{v}_1 + v_1^2 - 2\frac{m}{M}u(u - \boldsymbol{n}\boldsymbol{u}), \qquad \boldsymbol{u} = \boldsymbol{v} - \boldsymbol{v}_1.$$

It follows therefore that

$$\Delta v^2 = v'^2 - v^2 \simeq 2vv_1[-(1 - \cos\theta)\cos\psi + \sin\theta\sin\psi\cos(\varphi - \varphi_1)] + 2(1 - \cos\theta)\left[v_1^2 - \frac{m}{M}v^2\right]. \tag{2.80}$$

Here ψ is the angle between $\boldsymbol{v}$ and $\boldsymbol{v}_1$, while θ and φ are the scattering angles (i.e., the angles between $\boldsymbol{n}$ and $\boldsymbol{u} = \boldsymbol{v} - \boldsymbol{v}_1$), while φ_1 is the azimuthal angle of the velocity $\boldsymbol{v}_1$ in a coordinate system with its axis along $\boldsymbol{v}$. In the derivation of Equation (2.80) it was taken into account that if the vector $\boldsymbol{v}_1$ has angular coordinates $\tilde{\psi}$ and $\tilde{\varphi}_1$ in a coordinate system with axis along $\boldsymbol{v} - \boldsymbol{v}_1$, and angular coordinates ψ and φ_1 respectively in a coordinate system with axis along $\boldsymbol{v}$, then $\cos\tilde{\psi} \simeq \cos\psi - (v_1/v)\sin^2\psi$, $\sin\tilde{\psi} \simeq \sin\psi + (v_1/v)\sin\psi\cos\psi$, and $\tilde{\varphi}_1 \simeq \varphi_1$. Using the law of energy conservation in the collision

$$\frac{mv'^2}{2} + \frac{Mv_1'^2}{2} = \frac{mv^2}{2} + \frac{Mv_1^2}{2},$$

we obtain

$$\Delta v_1^2 = v_1'^2 - v_1^2 = -\frac{m}{M}\Delta v^2. \tag{2.81}$$

The collision integral takes the form

$$S_{m0} = \int d\Omega\, d\boldsymbol{v}_1\, q(u, \theta)u\{f_0(v)F_m(\boldsymbol{v}_1) - f_0(v')F_m(\boldsymbol{v}_1')\}. \tag{2.82}$$

Assuming, for simplicity, the molecule distribution function to be symmetrical, i.e., dependent only on the modulus of the velocity $\boldsymbol{v}_1$, and recognizing that the energy is changed little by the collision, $v^2 \approx v'^2$ and $v_1^2 \approx v_1'^2$, we can expand the functions

$$f_0(x') = f_0(x + \Delta x), \qquad F_{\rm m}(x_1') = F_{\rm m}(x_1 + \Delta x_1)$$
$$x = v^2, \quad x' = v'^2; \qquad x_1 = v_1^2, \quad x_1' = v_1'^2$$

in a Taylor series. We then obtain in place of Equation (2.82)

$$S_{\rm m0} = -\int d\Omega\, d\boldsymbol{v}_1 \left\{ vq(v, \theta) - \frac{\partial}{\partial v}\left[vq(v, \theta)v_1 \cos\psi\right]\right]$$
$$\cdot \left\{ f_0(v)\left[\Delta x_1 \frac{dF_{\rm m}}{dx_1} + \frac{(\Delta x_1)^2}{2}\frac{d^2F_{\rm m}}{dx_1^2}\right]\right.$$
$$\left. + \frac{\partial f_0}{\partial x}\left[\Delta x F_{\rm m} + (\Delta x\, \Delta x_1)\frac{dF_{\rm m}}{dx_1}\right] + \frac{\partial^2 f_0}{\partial x^2}\frac{(\Delta x)^2}{2}F_{\rm m}\right\}. \tag{2.83}$$

Substituting here Equations (2.80) and (2.81) for Δx and Δx_1, we integrate with respect to $d\Omega\, d\boldsymbol{v}_1 = \sin\theta\, d\theta\, d\varphi \times v_1^2 \sin\psi\, dv_1\, d\psi\, d\varphi_1$. We then obtain (Davydov, 1937; Allis, 1956).

$$S_{\rm m0} = -\frac{1}{2v^2}\frac{\partial}{\partial v}\left\{v^2\, \delta_{\rm el}\nu_{\rm m}(v)\left[\frac{T_{\rm ef}}{m}\frac{\partial f_0}{\partial v} + vf_0\right]\right\}, \tag{2.84}$$

$$T_{\rm ef} = M\int_0^\infty v_1^4 F_{\rm m}(v_1)\, dv_1\left[3\int_0^\infty v_1^2 F_{\rm m}(v_1)\, dv_1\right]^{-1} \tag{2.85}$$

Here $\nu_{\rm m}(v)$ is the electron-molecule collision frequency, determined by Equation (2.77); $\delta_{\rm el} = 2m/M$ is the average fraction of the energy lost by the electron in one elastic impact; $T_{\rm ef}$ is the effective temperature of the molecules ($T_{\rm ef} = \frac{2}{3}\bar{\varepsilon}_{\rm m}$, where $\bar{\varepsilon}_{\rm m}$ is the average kinetic energy); in the case of a Maxwellian distribution of the molecules we have $T_{\rm ef} = T$.

Equation (2.84) obtained for the collision integral $S_{\rm m0}$ has a differential character and can be represented in the form

$$S_{\rm m0} = \frac{1}{v^2}\frac{\partial}{\partial v}(v^2 j_{\rm m0}). \tag{2.86}$$

Here $j_{\rm m0}(v)$ is the flux in the space of the modulus of the velocity (or energy) and is due to the elastic collisions of the electron with the molecules

$$j_{\rm m0} = -\frac{1}{2}\delta_{\rm el}\nu_{\rm m}(v)\left[\frac{T_{\rm ef}}{m}\frac{\partial f_0}{\partial v} + vf_0\right]. \tag{2.87}$$

The kinetic equation for the function f_0 in a spatially homogenous plasma in the absence of perturbing fields is given by

$$\frac{\partial f_0}{\partial t} + \operatorname{div} \boldsymbol{j}_{m0} = 0 \tag{2.88}$$

and constitutes the usual continuity equation in velocity space. Equation (2.87) for the flux in velocity space $\boldsymbol{j}_{m0}(v)$, due to electron-molecule collisions, has a lucid physical meaning, namely, the flux $\boldsymbol{j}_{m0}$ consists, first, of the "transport" or "deceleration" flux

$$\frac{1}{2}\,\delta v v f_0 = \overline{\left\langle \frac{dv}{dt} \right\rangle} f_0,$$

which describes the average electron energy loss in the collisions, and the "diffusion" flux

$$\frac{1}{2}\,\delta v\,\frac{T_{ef}}{m}\frac{\partial f_0}{\partial v} = \overline{\left\langle \frac{dv^2}{dt} \right\rangle}\frac{\partial f_0}{\partial v},$$

which appears in the presence of a gradient in the electron velocity distribution and is due to the fact that the particles with which the electron collides are themselves in motion.

Comparing Equations (2.77) and (2.84), we can verify that the corrections to the collision integrals (due to the changes in the electron energy) are small, of the order of $\delta_{el} = 2m/M$. Therefore, the corresponding corrections to the integrals [Eqs. (2.75) and (2.76)] can always be neglected. They are important only in the expression for S_{m0}, and exclusively by virtue of the fact that the integral S_{m0} vanishes in first-order approximation, in which these corrections are neglected.

Elastic Collisions with Ions. The general expressions derived above for the integral of elastic collisions with molecules are fully applicable for the description of elastic collisions of electrons with ions, since they have been derived for an arbitrary form of the cross section $q(u, \theta)$ (all that was used in the derivation was that $m \ll M$, and this is always true also for the ions). It is thus necessary only to calculate the number $\nu_i(v)$ of electron-ion collisions. To this end it is necessary to substitute in Equation (2.77) the Rutherford formula for the effective cross section of the scattering of an electron by an ion

$$q_i(v, \theta) = \left(\frac{Ze^2}{2mv^2}\right)^2 \sin^{-4}\frac{\theta}{2}. \tag{2.89}$$

We then have

$$\nu_{\mathrm{i}}(v) = vN_{\mathrm{i}} \int q_{\mathrm{i}}(v, \theta)(1 - \cos\theta)\, d\Omega = \frac{2\pi N_{\mathrm{i}} Z^2 e^4}{m^2 v^3} \ln\left[1 + \mathrm{ctg}^2 \frac{\theta_{\min}}{2}\right] \tag{2.90}$$

where N_{i} is the density of the ions and eZ is their charge. If we consider the scattering of an electron by a free ion, then the integration in Equation (2.90) must be carried out from 0 to π. The collision frequency then diverges logarithmically at small θ. The ions in a plasma, however, are not quite free: owing to the interaction between them and the electrons, the field of each ion is a Coulomb field only up to distances on the order of the Debye radius $D = (T_{\mathrm{e}}/4\pi e^2 N)^{1/2}$. At distances larger than D, screening causes the Coulomb field of the ion to decrease rapidly exponentially; see, for example, Sect 4, Ginzburg, 1960. Consequently, we can assume that D is the maximum distance at which an appreciable interaction between the electron and ion is still present, i.e., the maximal impact parameter. In terms of this parameter we can express the minimum scattering angle:

$$\theta_{\min} = 2 \,\mathrm{arctg}\left(\frac{e^2 Z}{mv^2 D}\right). \tag{2.91}$$

Consequently

$$\nu_{\mathrm{i}}(v) = \frac{2\pi N_{\mathrm{i}} e^4 Z^2}{m^2 v^3} \ln\left(1 + \frac{D^2 m^2 v^4}{Z^2 e^4}\right). \tag{2.92}$$

For the collision frequencies $\nu_{\mathrm{i}k}(v)$ [Eq. (2.76)] we obtain in the case of collisions with ions

$$\nu_{\mathrm{i}k}(v) = N_{\mathrm{i}} v \int q_{\mathrm{i}}(v, \theta)[1 - P_k(\cos\theta)]\, d\Omega = \frac{k(k+1)}{2} \nu_{\mathrm{i}}(v). \tag{2.93}$$

It is important that in a gas plasma the parameter

$$\Lambda = \frac{2}{\theta_{\min}} \approx \frac{DT_{\mathrm{e}}}{e^2} \approx \left(\frac{T_{\mathrm{e}}}{e^2 N^{1/3}}\right)^{3/2} \gg 1 \tag{2.94}$$

is always a large quantity (the condition $T_{\mathrm{e}}/e^2 N^{1/3} \gg 1$ defines a gas plasma, i.e., a plasma in which the average kinetic energy of the particles T_{e} is much larger than the average potential energy of the interaction between them $e^2 N^{1/3}$). This means that the second term in the logarithm is always the principal one. The main contribution to the number of collisions of the electron with the ions is therefore made by weak scattering—small-angle scattering. One such impact changes insignificantly not only the energy but also the momentum of the electron. In fact, the energy lost in the elastic scattering of the electron through an angle θ is $\Delta\varepsilon = (2m/M)(1 - \cos\theta)\varepsilon$. Recognizing that the principal role is played by collisions that lead to scattering through a small angle on the order of $\theta_{\min}$, we obtain

$$\frac{\Delta\varepsilon}{\varepsilon} \approx \frac{2m}{M}\frac{\theta_{\min}^2}{2} \approx \frac{m}{M}\left(\frac{e^2 N^{1/3}}{T_e}\right)^3 \ll 1.$$

Analogously, the change Δp of the momentum is given by

$$\frac{\Delta p}{p} = \frac{|\boldsymbol{p} - \boldsymbol{p}'|}{|\boldsymbol{p}|} \approx (1 - \cos\theta_{\min}) \approx \left(\frac{e^2 N^{1/3}}{T_e}\right)^3 \ll 1.$$

It must be emphasized that although the change of momentum in one impact is small, the change of the energy is much smaller

$$\frac{p\,\Delta\varepsilon}{\varepsilon\,\Delta p} \approx \frac{m}{M}.$$

The Landau Collision Integral. It was shown above that in electron-ion collisions the main contribution is made by collisions that alter insignificantly the energy as well as the momentum of the electron. The reason is that the Coulomb forces decrease slowly with increasing distance. Consequently the main contribution to the total scattering cross section is made by the interaction of the particles at large distances, which leads only to weak scattering.

Understandably, the momentum changes little not only in collisions between electrons and ions, but also in collisions between electrons or between ions, i.e., in all types of Coulomb collisions. This circumstance makes it possible, by using the Fokker-Planck method, to simplify the form of the collision integral [Eq. (2.62)] in the case of Coulomb interaction of the particles, and representing this integral in the Landau differential form (Landau, 1936)

$$S = \operatorname{div} \boldsymbol{j},$$

$$j_i = 2\pi \ln \Lambda \frac{e^4 Z^2 Z_1^2}{m} \int \frac{d\boldsymbol{v}_1}{u}\left(\delta_{ik} - \frac{u_i u_k}{u^2}\right)\left[\frac{f(\boldsymbol{v})}{m_1}\frac{\partial F(\boldsymbol{v}_1)}{\partial v_{1k}} - \frac{F(\boldsymbol{v}_1)}{m}\frac{\partial f}{\partial v_k}\right] \tag{2.95}$$

$$u = |\boldsymbol{v} - \boldsymbol{v}_1|, \qquad \Lambda = \frac{2}{\theta_{\min}} = \frac{mDv^2}{e^2 Z Z_1}.$$

Here eZ and eZ_1 are the charges of the colliding particles, m and m_1 are their masses, and $\boldsymbol{j}$ is the particle flux, in velocity space, due to the collisions. It is convenient to express it in vector form

$$\boldsymbol{j} = \frac{1}{2}\int d\boldsymbol{v}_1\ \nu(u)\left\{u^2\left[\frac{m}{m_1}f\frac{\partial F}{\partial \boldsymbol{v}_1} - F\frac{\partial f}{\partial \boldsymbol{v}}\right] - \boldsymbol{u}\left[\frac{m}{m_1}f\left(\boldsymbol{u}\frac{\partial F}{\partial \boldsymbol{v}_1}\right) - F\left(\boldsymbol{u}\frac{\partial f}{\partial \boldsymbol{v}}\right)\right]\right\},$$

$$\nu(u) = \frac{4\pi Z^2 Z_1^2 e^4 \ln \Lambda}{m^2 u^3}. \tag{2.96}$$

Assume that there is one preferred direction $\boldsymbol{e}$ in the plasma. Accordingly $f(\boldsymbol{v}) = f(v, \theta)$, where v is the absolute value of the velocity and θ is the angle between $\boldsymbol{v}$ and $\boldsymbol{e}$. In this case we have

$$\operatorname{div} \boldsymbol{j} = \frac{1}{v^2}\frac{\partial}{\partial v}(v^2 j_v) + \frac{1}{v \sin\theta}\frac{\partial}{\partial \theta}(\sin\theta\ j_\theta), \tag{2.97}$$

and the flux components j_v and j_θ are determined, according to Equation (2.96), by the following expressions:

$$j_v = \frac{1}{2}\left[-vfA_1 - \frac{\partial f}{\partial v}A_2 - fA_3 - \frac{1}{v}\frac{\partial f}{\partial \theta}A_4\right],$$

$$j_\theta = \frac{1}{2}\left[vf(B_1 + B_3) - \frac{\partial f}{\partial v}B_2 - \frac{1}{v}\frac{\partial f}{\partial \theta}B_4\right];$$

$$A_1 = -\frac{m}{m_1}\int v_1\frac{\partial F}{\partial v_1}(1 - \cos^2\theta')\nu(u)\, d\boldsymbol{v}_1,$$

$$A_2 = \int F v_1^2(1 - \cos^2\theta')\nu(u)\, d\boldsymbol{v}_1,$$

$$A_3 = \frac{m}{m_1}\int \frac{\partial F}{\partial \theta_1}\nu(u)(v\cos\theta' - v_1)\cos\psi_1\, d\boldsymbol{v}_1, \tag{2.98}$$

$$A_4 = \int F v_1(v - v_1\cos\theta')\nu(u)\cos\xi_1\, d\boldsymbol{v}_1;$$

$$B_1 = \frac{m}{m_1}\int \frac{\partial F}{\partial v_1}(v - v_1\cos\theta')\nu(u)\cos\xi_1\, d\boldsymbol{v}_1,$$

$$B_2 = \int F v_1(v - v_1\cos\theta')\nu(u)\cos\xi_1\, d\boldsymbol{v}_1,$$

$$B_3 = \frac{m}{m_1}\int \frac{\partial F}{\partial \theta_1}\left(\frac{u^2}{vv_1}\cos\gamma_1 + \cos\xi_1\cos\psi_1\right)\nu(u)\, d\boldsymbol{v}_1,$$

$$B_4 = \int F(u^2 - v_1^2\cos^2\xi_1)\nu(u)\, d\boldsymbol{v}_1$$

Here θ_1 is the angle between the velocity $\boldsymbol{v}_1$ and the preferred direction $\boldsymbol{e}$: φ_1 is a second (azimuthal) angle characterizing the vector $\boldsymbol{v}_1$. The remaining angles are connected with them by the relations[6]

$$\begin{aligned}
\cos\theta' &= \cos\theta\cos\theta_1 + \sin\theta\sin\theta_1\cos\varphi_1,\\
\cos\xi_1 &= -\sin\theta\cos\theta_1 + \cos\theta\sin\theta_1\cos\varphi_1,\\
\cos\psi_1 &= -\cos\theta\sin\theta_1 + \sin\theta\cos\theta_1\cos\varphi_1,\\
\cos\gamma_1 &= \sin\theta\sin\theta_1 + \cos\theta\cos\theta_1\cos\varphi_1.
\end{aligned} \tag{2.99}$$

Consider, for example, collisions between electrons and ions. The ion distribution function f_i is assumed to be Maxwellian

$$f_\mathrm{i} = N_\mathrm{i}\left(\frac{M}{2\pi T_\mathrm{i}}\right)^{3/2}\exp\left(-\frac{Mv_1^2}{2T_\mathrm{i}}\right) \tag{2.100}$$

Then, recognizing that $v \ll v_1$, i.e., $u \approx v$, we obtain from Equations (2.98) and (2.96)

$$\nu_\mathrm{i}(v) = \frac{4\pi e^4 Z^2 N_\mathrm{i}\ln\Lambda}{m^2 v^3},$$

$$A_1 = \frac{2m}{M}\nu_\mathrm{i}(v), \qquad A_2 = \frac{2T_\mathrm{i}}{M}\nu_\mathrm{i}(v), \qquad B_4 = v^2\nu_\mathrm{i}(v), \tag{2.101}$$

$$A_3 = A_4 = B_1 = B_2 = B_3 = 0.$$

Substituting these expressions in Equations (2.97) and (2.98), and neglecting small terms of order m/M, we obtain

$$S_\mathrm{ei} = -\frac{1}{2v^2}\frac{\partial}{\partial v}\left[\frac{2m}{M}v^2\nu_\mathrm{i}(v)\left(\frac{T_\mathrm{i}}{m}\frac{\partial f}{\partial v} + vf\right)\right] - \frac{\nu_\mathrm{i}(v)}{2\sin\theta}\frac{\partial}{\partial\theta}\left[\sin\theta\frac{\partial f}{\partial\theta}\right] \tag{2.102}$$

Expanding $f(v, \theta)$ in the Legendre polynomials [Eq. (2.63)], we obtain from Equation (2.102) the Equations (2.84), (2.76), and Equation (2.93) for $S_{\mathrm{ei}k}(f_k)$.

[6] We have taken into account here the fact that the vector $\partial f/\partial\boldsymbol{v}$ has two components, one directed along $\boldsymbol{v}$ and the other along a vector $\boldsymbol{\tau}$ orthogonal to $\boldsymbol{v}$ in the $\boldsymbol{ev}$ plane; the vector $\partial F/\partial\boldsymbol{v}_1$ has components along $\boldsymbol{v}_1$ and the vector $\boldsymbol{\tau}_1$ orthogonal to $\boldsymbol{v}_1$ in the $\boldsymbol{ev}_1$ plane. Accordingly, θ' is the angle between $\boldsymbol{v}$ and $\boldsymbol{v}_1$, ψ_1 is the angle between $\boldsymbol{v}$ and $\boldsymbol{\tau}_1$, ξ_1 is the angle between $\boldsymbol{\tau}$ and $\boldsymbol{v}_1$, and γ_1 is the angle between $\boldsymbol{\tau}$ and $\boldsymbol{\tau}_1$. Equations (2.99) express these angles in terms of θ, θ_1, and φ_1.

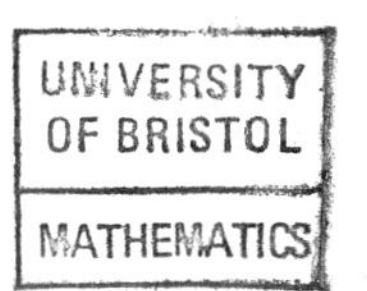

In the derivation of the Landau collision integral, only small–angle scattering is taken into account. Large-angle scattering adds to the number of collisions a term of the order of unity, which is small in comparison with the main logarithmic term ($\ln \Lambda \gg 1$). A similar correction is arrived at in an equilibrium plasma by solving exactly the problem of scattering in a Debye field, and also by considering interactions, not accounted for [Eq. (2.95)], at distances larger than the Debye radius (the so-called collective effects). In the case of large deviations from equilibrium, particularly in a turbulent plasma, the role of the collective effects becomes decisive (Shkarofsky et al., 1966; Tsitovich, 1967; Silin, 1972).

It should also be noted that Equation (2.95) is obtained under the assumption that the classical theory is valid. Quantum effects become noticeable at $T_e \gtrsim mZ^2e^4/\hbar^2$. Allowance for these effects leads to corrections under the logarithm sign. Corrections of the same type arise in a high-frequency electric field $\omega > \omega_0$ and in a strong magnetic field $\omega_H > \omega_0$. These corrections become appreciable at $\omega \gg \omega_0$ and $\omega_H \gg \omega_0$ (Silin, 1972).

Electron-Electron Collisions. In the case of collisions between electrons, the collision integral is described by Equations (2.95) and (2.98). It is necessary only to take into account that particles 1 are also electrons, i.e., the distribution function $F(v_1)$ is equal to $f(v_1)$.

If the distribution function of the electrons is symmetrical and depends only on the modulus of the velocity, $f = f_0(v)$, then Equations (2.97) and (2.98) become simpler: the current $j_\theta = 0$, the coefficients $A_3 = A_4 = 0$, and it is possible to integrate with respect to the angles in the expressions for the coefficients A_1 and A_2. Recognizing that

$$d\boldsymbol{v}_1 = v_1^2 \sin\theta_1 \, d\theta_1 \, d\varphi_1 = v_1^2 \sin\theta' \, d\theta' \, d\varphi'$$

(θ' is the angle between $\boldsymbol{v}$ and $\boldsymbol{v}_1$), we obtain

$$\begin{aligned} A_1 &= -\int_0^\infty v_1 \frac{\partial f_0}{\partial v_1} 2\pi v_1^2 \, dv_1 \int_{-\pi}^{\pi} (1 - \cos^2\theta')\nu(u) \sin\theta' \, d\theta' \\ &= \frac{8\pi\nu_e(v)}{N} \int_0^v v_1^2 f_0(v_1) \, dv_1, \end{aligned} \tag{2.103}$$

$$\nu_e(v) = \frac{4\pi N e^4 \ln \Lambda}{m^2 v^3}.$$

Here $\nu_e(v)$ is the frequency of the electron–electron collisions. In the derivation of Equation (2.103) we took into account the fact that

$$u = |\boldsymbol{v} - \boldsymbol{v}_1| = (v^2 + v_1^2 - 2vv_1 \cos\theta')^{1/2},$$

$$\int_{-\pi}^{\pi} \frac{(1 - \cos^2\theta')\sin\theta'\, d\theta'}{(v^2 + v_1^2 - 2vv_1\cos\theta')^{3/2}} = \frac{4}{3}\begin{cases} v_1^{-3} & \text{if} \quad v \le v_1, \\ v^{-3} & \text{if} \quad v \ge v_1. \end{cases} \tag{2.104}$$

We obtain analogously also the second coefficient

$$A_2 = \frac{8\pi \nu_e(v)}{3N}\left(\int_0^v v_1^4 f_0(v_1)\, dv_1 + v^3 \int_v^\infty v_1 f_0(v_1)\, dv_1\right). \tag{2.105}$$

For the function $f_0(v)$, the electron–electron collision integral thus takes the form

$$S_{0e}(f_0) = -\frac{1}{2v^2}\frac{\partial}{\partial v}\left\{v^2\left[A_1(f_0)vf_0 + A_2(f_0)\frac{\partial f_0}{\partial v}\right]\right\}, \tag{2.106}$$

where the coefficients $A_1(f_0)$ and $A_2(f_0)$ are given by Equations (2.103) and (2.105).

For fast ($v \gg v_{Te}$) and slow ($v \ll v_{Te}$) electrons, Equation (2.106) becomes much simpler. In the former case it is possible to replace v by ∞ in the integration limit we have:

$$A_1 = 2\nu_e(v) = \frac{8\pi N e^4 \ln \Lambda}{m^2 v^3}, \quad A_2 = \frac{4}{3}\frac{\bar{\varepsilon}}{m}\nu_e(v), \quad \bar{\varepsilon} = \frac{m}{2N}\int v^2 f_0\, d\boldsymbol{v} \tag{2.107}$$

$\bar{\varepsilon}$ is the average energy of the scattered electrons. In the case of a Maxwellian velocity distribution we have $\bar{\varepsilon} = \frac{3}{2}T_e$.

In the latter case we can, on the contrary, assume $v \to 0$ in the integration limit. Then

$$A_1 = \frac{32 \cdot \pi^2 e^4 \ln \Lambda}{3m^2} f_0(0), \qquad A_2 = \frac{32 \cdot \pi^2 e^4 \ln \Lambda}{3m^2}\bar{u}; \qquad \bar{u} - \int_0^\infty v_1 f_0\, dv_1 \tag{2.108}$$

For a Maxwellian velocity distribution we have

$$f_0(0) = \left(\frac{m}{2\pi T_e}\right)^{3/2} N, \qquad \bar{u} = N m^{1/2}(2\pi)^{-3/2} T_e^{-1/2}.$$

We note that if we multiply Equation (2.106) for S_{0e} by v^2 or v^4 and integrate it over all the velocities v (from 0 to ∞), then the corresponding integral vanishes identically regardless of the form of the function $f_0(v)$:

$$\int_0^\infty S_{0e} v^2 \, dv = 0, \qquad \int_0^\infty S_{0e} v^4 \, dv = 0. \tag{2.109}$$

These relations reflect the conservation of the total number of particles and the conservation of the energy in collisions between electrons. In the case of a Maxwellian function $f_0(v)$, the collision integral [Eq. (2.106)] vanishes identically.

We have considered above the electron–electron collision integral only for the symmetrical part of the distribution function $f_0(v)$. When the function $f(\boldsymbol{v})$ is expanded in the Legendre polynomials [Eq. (2.63)], owing to nonlinearity, the integrals S_{ke} [Eq. (2.66)] depend, generally speaking, not only on the function f_k. However, if $f_0 \gg f_1$, then the integral S_{0e} depends only on f_0, and the integral for the function f_1 (just as for the succeeding functions f_k) can be linearized. Substituting directly in Equation (2.62) the distribution function in the form

$$f = f_0(v) + f_1(v) \cos \theta \tag{2.110}$$

we arrive at the following expression for the collision integral S_{1e}:

$$\begin{aligned} S_{1e} = \frac{3}{4\pi} \int q_e(u, \theta) u \{ & f_1(v) f_0(v_1) \cos \theta + f_1(v_1) f_0(v) \cos \theta_1 \\ & - f_1(v') f_0(v'_1) \cos \theta' - f_1(v'_1) f_0(v') \cos \theta'_1 \} \, dv_1 \, d\Omega \, d\Omega_1, \end{aligned} \tag{2.111}$$

where $q_e(v, \theta)$ is determined by Equation (2.89) at $Z = 1$. It is also possible to use Equations (2.97)–(2.99) and integrate over the angles. The collision integral S_{1e} is then written in integro-differential form (Shkarofsky et al., 1966).

2.2.3. Inelastic Collisions

Inelastic-Collision Integral. Inelastic collisions of electrons with neutral molecules, atoms, and ions are accompanied by excitation of rotational, vibrational, and optical levels, by ionization, by radiation in free–free transitions, and by various types of recombination processes, such as radiative recombination, dissociative recombination, recombination in ternary collisions, attachment of electrons to molecules, etc. An important role is played also by the so-called impacts of second kind, when the energy of the excited state of the molecule is transferred to the incident electron. Exact allowance for all these inelastic processes is a very complicated task, and in only a few cases are their cross sections known accurately enough.

The inelastic-collision integral for the symmetrical part of the distribution function f_0 is represented in the form (Holstein, 1946)

$$S_{0\,\mathrm{in}} = -\frac{2}{mv}\sum_{k,j} N_k[(\varepsilon + \varepsilon_{kj})f_0(\varepsilon + \varepsilon_{kj})\sigma_{kj}(\varepsilon + \varepsilon_{kj}) - \varepsilon f_0(\varepsilon)\sigma_{kj}(\varepsilon)$$

$$+ (\varepsilon - \varepsilon_{k,-j})f_0(\varepsilon - \varepsilon_{k,-j})\sigma_{k,-j}(\varepsilon - \varepsilon_{k,-j}) - \varepsilon f_0(\varepsilon)\sigma_{k,-j}(\varepsilon)], \quad (2.112)$$

To abbreviate the notation, we have changed over from the modulus of the velocity v to the electron energy $\varepsilon = mv^2/2$; of course, in this case $f_0(\varepsilon) = f_0(v)$. Further, ε_{kj} is the energy of the transition of the molecule (atom, ion) from the state k to the state $k + j$, $\sigma_{kj} = \int q_{kj}\, d\Omega$ is the total cross section for the excitation of the $k \to k + j$ transition, and N_k is the concentration of the molecules in the state k. The first two terms of Equation (2.112) describe the excitation of the molecules: the departure and arrival of the electrons as a result of excitation of the quantum ε_{kj}. The last two terms describe impacts of the second kind: de-excitation of the molecules with transfer of an energy $\varepsilon_{k,-j}$ to the electron and transition of the molecules from the state k to the state $k - j$; $\sigma_{k,-j}$ is the total cross section of this process. Equation (2.112) follows directly from Equation (2.82) if the molecule motion is neglected and it is recognized that the electron energy changes in one impact by ε_{kj} or $\varepsilon_{k,-j}$.

The inelastic-collision integral for the function f_1 takes the form

$$S_{1\,\mathrm{in}} = \nu_{\mathrm{in}} f_1(\varepsilon) - \sum_{k,j} \nu_{kj} f_1(\varepsilon + \varepsilon_{kj})$$

$$\nu_{\mathrm{in}} = N_{\mathrm{m}} v \sum_{k,j} \sigma_{kj}; \qquad \nu_{kj} = \frac{2(\varepsilon + \varepsilon_{kj})}{mv} N_{\mathrm{m}} \int q_{kj}(\varepsilon + \varepsilon_{kj}, \theta) \cos\theta \, d\Omega, \quad (2.113)$$

where q_{kj} is the differential inelastic-scattering cross section.

It is possible to separate two important limiting cases when the energy of the electrons is much larger than the transition energy and, on the contrary, when the electron energy is only insignificantly higher than the excitation energy. In the former case

$$\varepsilon_{kj}, \varepsilon_{k,-j} \ll \varepsilon \quad (2.114)$$

Expanding Equation (2.112) in a Taylor series we convert the collision integral to a differential form:

$$S_{0\,\mathrm{in}} = -\frac{2}{mv}\sum_{k,j} N_k \frac{\partial}{\partial \varepsilon}\Big\{[\varepsilon(\varepsilon_{kj}\sigma_{kj} - \varepsilon_{k,-j}\sigma_{k,-j}) + \frac{1}{2}(\varepsilon_{kj}^2\sigma_{kj} + \varepsilon_{k,-j}^2\sigma_{k,-j})]f + \frac{\varepsilon}{2}\frac{\partial}{\partial \varepsilon}[(\varepsilon_{kj}^2\sigma_{kj} + \varepsilon_{k,-j}^2\sigma_{k,-j})f]\Big\}. \tag{2.115}$$

$$S_{1\,\mathrm{in}} = \nu_{\mathrm{int}} f_1(\varepsilon), \qquad \nu_{\mathrm{int}} = N_{\mathrm{m}} v \sum_{k,j} \sigma_{tkj}, \qquad \sigma_{tkj} = \int q_{kj}(1 - \cos\theta)\, d\Omega$$

In the case of high excitation energy ($\varepsilon - \varepsilon_{kj} \ll \varepsilon$), if the collisions of the second kind can be neglected ($\varepsilon_{kj} \gg T$), then the collision integral assumes the simple form:

$$S_{0\,\mathrm{in}} = \nu_{\mathrm{in}} f_0, \qquad S_{1\,\mathrm{in}} = \nu_{\mathrm{in}} f_1, \qquad \nu_{\mathrm{in}} = N_{\mathrm{m}} v \sum_{k,j} \sigma_{kj}. \qquad \varepsilon > \varepsilon_{kj}, \qquad \varepsilon - \varepsilon_{kj} \ll \varepsilon \tag{2.116}$$

We now consider concretely some inelastic processes that exist in the ionospheric plasma.

Excitation of Rotational Levels. The principal molecules in the lower ionosphere, N_2 and O_2, are diatomic, consisting of identical atoms. Such molecules have no dipole moment, and therefore their interaction with slow electrons is of the quadrupole type. The allowed transitions are rotational with change of angular momentum $\Delta J = \pm 2$, i.e.,

$$\sigma_{kj} = \sigma_J \delta(J, J+2), \qquad \sigma_{k,-j} = \sigma_{-J}\delta(J, J-2). \tag{2.117}$$

The total cross sections are in this case (Gerjouy and Stein, 1955)

$$\sigma_J = \sigma_0 \frac{(J+2)(J+1)}{(2J+3)(2J+1)}\left[1 - \frac{2(2J+3)B_0}{\varepsilon}\right]^{1/2},$$
$$\sigma_{-J} = \sigma_0 \frac{J(J-1)}{(2J-1)(2J+1)}\left[1 + \frac{2(2J-1)B_0}{\varepsilon}\right]^{1/2}, \tag{2.118}$$
$$\sigma_0 = 8\pi Q^2 a_0^2/15.$$

Here $a_0 = \hbar^2/me^2$ is the Bohr radius, B_0 is the rotational constant, and Q is a quadrupole constant of the molecule ($Q \sim 1$). The energy of the rotational quanta is

$$\varepsilon_J = B_0 J(J+1). \tag{2.119}$$

It follows from Equations (2.117) and (2.119) that

$$\varepsilon_{kj} = \varepsilon_{J+2} - \varepsilon_J = (4J+6)B_0; \qquad \varepsilon_{k,-j} = \varepsilon_J - \varepsilon_{J-2} = (4J-2)B_0. \tag{2.120}$$

Under the conditions of Equation (2.114), using Equations (2.117)–(2.120) in Equation (2.115), we have

$$S^{\mathrm{r}}_{0\ \mathrm{in}} = -\frac{2}{mv}\sum_J N_J \frac{\partial}{\partial \varepsilon}\left\{\varepsilon\left[4B_0\sigma_0 f + 4B_0^2\sigma_0(J^2+J+3)\frac{\partial f}{\partial \varepsilon}\right]\right\}. \tag{2.121}$$

In the summation over the levels we take into account the distribution of the excited molecules (Frost and Phelps, 1962):

$$N_J = \frac{N_{\mathrm{m}}}{P}\exp(-\varepsilon_J/T)(2J+1)x_J, \qquad P = \sum_J x_J(2J+1)\exp(-\varepsilon_J/T) \tag{2.122}$$

where N_{m} and T are the total concentration and temperature of the molecules, P is a normalization factor, and x_J is a constant that is different for even and odd values of J. Then, as can be easily seen,

$$\sum_J N_J = N_{\mathrm{m}}, \qquad \sum_J N_J(J^2+J+3) \simeq \frac{T}{B_0}N_{\mathrm{m}}.$$

In the last expression it is assumed that $T \gg B_0$. Furthermore, taking account of the fact that $\varepsilon = mv^2/2$, we reduce Equation (2.121) to the form

$$S^{\mathrm{r}}_{0\ \mathrm{in}} = \frac{1}{2v^2}\frac{\partial}{\partial v}\left\{v^2 R_{\mathrm{r}}(v)\left[\frac{T}{m}\frac{\partial f_0}{\partial v} + vf_0\right]\right\},$$
$$R_{\mathrm{r}}(v) = \frac{8B_0\sigma_0 N_{\mathrm{m}}}{mv}. \tag{2.123}$$

We note that Equation (2.123) is valid also in the general case, regardless of the form of the collisions, if the energy lost by the electron in one impact is small: $\varepsilon_{kj} \ll \varepsilon$, $\varepsilon_{kj} \ll T$ [all that changes is the loss function $R(v)$].

In our case the condition for the applicability of Equation (2.123) takes the form $\varepsilon \gg B_0$, $T \gg B_0$. This condition is well satisfied in the ionosphere, since in molecular nitrogen we have $B_0^{N_2} = 2.48 \cdot 10^{-4}$ eV $\approx 2.9^\circ$, and in oxygen $B_0^{O_2} = 1.79 \cdot 10^{-4}$ eV $\approx 2.1^\circ$ (Herzberg, 1950). The quadrupole constants are $Q^{N_2} \approx 1.04$ and $Q^{O_2} \approx 1.8$. The expression for $R_r(v)$ contains the product $B_0 Q^2$, and for air (or the lower ionosphere) we have $B_0 Q^2 \approx 3.3 \cdot 10^{-4}$ eV.

Vibrational Levels. The excitation of the vibrational levels is described by the general Equation (2.112). This formula becomes simpler in a low-temperature plasma, where the molecule temperature T is much lower than the energy of the vibrational quanta

$$T \ll \hbar\omega_1. \tag{2.124}$$

In this case the number of excited molecules is small, the collisions with them are negligible in first-order approximation, and the integral [Eq. (2.112)] takes the form:

$$S_{0\text{ in}}^{v} = \frac{2N_{\text{m}}}{mv}\left[\varepsilon f_0(\varepsilon)\sum_{v^*}\sigma_{v^*}(\varepsilon) - \sum_{v^*}(\varepsilon + \varepsilon_{v^*})f_0(\varepsilon + \varepsilon_{v^*})\sigma_{v^*}(\varepsilon + \varepsilon_{v^*})\right]. \tag{2.125}$$

Here ε_{v^*} is the energy of the vibrational quanta

$$\varepsilon_{v^*} = \hbar\omega_1(v^* + \tfrac{1}{2}), \tag{2.126}$$

and σ_{v^*} is the total excitation cross section. If the vibrational losses are small in comparison with the electron energy [Eq. (2.114)], then the integral [Eq. (2.125)] takes the simple form [Eq. (2.115)]:

$$\begin{gathered} S_{0\text{ in}}^{v} = -\frac{1}{2v^2}\frac{\partial}{\partial v}\{v^3 R_v f_0\}, \\ R_v = \frac{2N_{\text{m}}}{mv}\sum_{v^*}\varepsilon_{v^*}\sigma_{v^*} = \frac{2N_{\text{m}}\hbar\omega_1}{mv}\sum_{v^*}\left(v^* + \frac{1}{2}\right)\sigma_{v^*}. \end{gathered} \tag{2.127}$$

In molecular nitrogen, vibrational levels are effectively excited by electrons with energies 2 eV $\lesssim \varepsilon \lesssim$ 3 eV. The process has a clearly pronounced resonant character; the excitation is connected with the temporary attachment of an electron. The maximum cross sections are very large, $\sigma_{v^*} \sim 5 \cdot 10^{-16}$ cm^2 (Schulz, 1962a, 1964). At low energies, $\varepsilon \lesssim 1$ eV, the cross sections, to the contrary, are small (Engelhardt et al., 1964). On the whole, the picture in molecular oxygen is similar. The region of effective excitation is here $0.4\ \mathrm{eV} \lesssim \varepsilon \lesssim 2\ \mathrm{eV}$ and the maximum cross sections are $\sigma_{v^*} \sim 10^{-17}\ \mathrm{cm}^2$ (Hake and Phelps, 1967).

The vibrational quantum values are $\hbar\omega_e = 0.195$ eV in oxygen and $\hbar\omega_e = 0.292$ eV in nitrogen. Therefore the condition [Eq. (2.124)] is well satisfied in the atmosphere and in the lower ionosphere. On the other hand, the condition [Eq. (2.114)] is not satisfied, therefore, the integral $S^v_{0\ \mathrm{in}}$ is described by the general Equation (2.125). However, upon excitation of the first levels $v^* = 1$ to 2, in the principal region of losses in nitrogen, the ratio [Eq. (2.114)] is small: $\varepsilon_{v^*}/\varepsilon \sim 0.2$ to 0.3. The contribution of these levels at an electron energy $\lesssim 2$ eV turns out to be the principal one. As a rough approximation we can therefore use in this energy region the differential expression [Eq. (2.127)]. The form of the loss function R_v for the lower ionosphere (air) is shown in Figure 4.

Another approximate representation for σ_{v^*}, which takes into account the resonant character of the excitation of the vibrational levels, is

$$\sigma_{v^*} = \sum_k \sigma^k_{v^*}\delta(\varepsilon - \varepsilon^k_{v^*}). \tag{2.128}$$

The values of the parameters $\sigma^k_{v^*}$, $\varepsilon^k_{v^*}$, and ε_{v^*} are listed in Table 4.

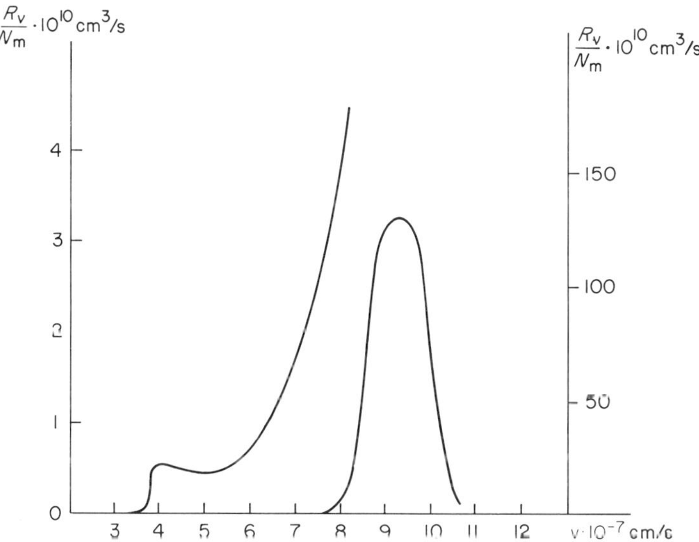

Fig. 4. Electron energy loss function in air (lower ionosphere) for excitation of vibrational levels. The *left-* and *right-hand scales* of R_v pertain to $v < 8 \cdot 10^7$ and $v > 8 \cdot 10^7$ cm/s, respectively

Table 4. Excitation cross section constants of vibrational levels

		N_2			O_2		
v^*	k	$\sigma^k_{v^*}$, cm^2 eV · 10^{18}	$\varepsilon^k_{v^*}$, eV	ε_{v^*}, eV	σ_{v^*}, cm^2 eV · 10^{18}	$\varepsilon^k_{v^*}$, eV	ε_{v^*}, eV
1	0	65	2.03		1.4	0.39	
	1	50	2.32	0.44			0.29
	2	39	2.62				
2	0	44	2.05		0.91	0.59	
	1	35	2.42	0.73			0.49
	2	19	2.74				
3	0	45	2.20		0.57	0.77	
	1	18	2.60	1.02			0.68
4	0	38	2.27	1.31	1.15	0.97	0.88
5	0	30	2.42	1.60	1.62	1.18	1.07
6	0	26	2.55	1.90	1.15	1.36	1.27
7	0	14	2.70	2.20	0.75	1.55	1.46
8	0	10	2.85	2.48	0.92	1.75	1.65

Optical Levels. In the excitation of optical levels, the integral $S^{op}_{0\,\text{in}}$ is described by an expression of the type of Equation (2.125), since impacts of the second kind are immaterial if $T \ll \varepsilon_{kj}$. The summation in Equation (2.125) now extends over all the allowed optical transitions from the ground state. The corresponding cross sections, and also the ionization and attachment cross sections, are given by Frost and Phelps (1962), Schulz (1964), Engelhardt et al. (1964), Hake and Phelps (1967), Nighan (1970), and Kiefer (1969, 1971). Figure 5 shows the summary cross sections for the excitation of optical and vibrational levels in air.

Ionization and Recombination. In the ionospheric plasma, the principal role is played by ionization by external sources (photoionization, corpuscular ionization) and dissociative recombination (see Sect. 2.5.1.). Their contribution to the inelastic collision integral is described by the expression

$$S^{\text{ir}}_{0\,\text{in}} = -q_{\text{ion}}(v) + \nu_{\text{r}}(v) f_0, \tag{2.129}$$

Here $q_{\text{ion}}(v)$ is the number of electrons with velocity v generated in one

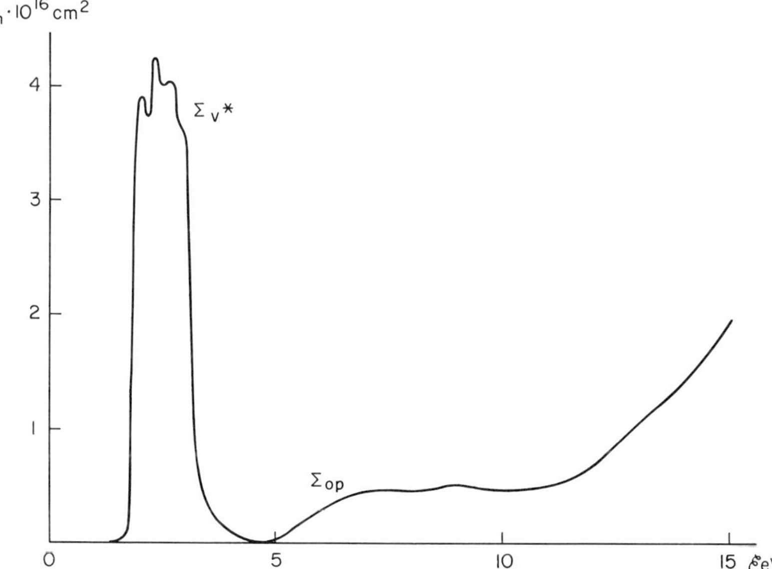

Fig. 5. Summary cross section for the excitation of vibrational and optical levels in air (lower ionosphere)

$$\sum_{\text{in}} = \sum_{v^*} + \sum_{\text{op}}; \qquad \sum_{v^*} = \sum_{v^*} \sigma_{v^*}(\varepsilon); \qquad \sum_{\text{op}} = \sum_{k,j} \sigma_{kj}^{op}(\varepsilon)$$

second by the ionizing source in a unit configuration-space 1 cm³ (cm/s)³, and $v_r(v)$ is the frequency of the recombination of electrons having a velocity v.

In the case of ionization by fast electrons we get

$$S_{0\ \text{in}}^i = -\int_{\sqrt{2\varepsilon_i/m}}^{\infty} v_{\text{ion}}(v', v) f_0(v') v'^2 \, dv' + \Delta S_{0\ \text{in}}^i, \tag{2.130}$$

$$v_{\text{ion}}(v, v') = N_m v' \int q_{\text{ion}}(v, v', \theta) \, d\Omega.$$

Here $v_{\text{ion}}(v, v')$ is the ionization frequency, i.e., the number of ionizations produced in one second by an electron having a velocity v' and leading to the appearance of a new electron having a velocity v; ε_i is the ionization energy. For simplicity we consider only ionization from the ground level. The term $\Delta S_{0\ \text{in}}^i$ describes the inelastic-energy loss by the ionizing electron. It has the same structure as the usual inelastic-collision integral [Eq. (2.125)]. It must be recognized that only the spectrum of the states is now continuous, i.e., it is necessary to replace summation over v^* in Equation (2.125) by integration with respect to v'.

2.3. Electron Distribution Function

The distribution function of the electrons in a plasma, as shown above, can be represented in the form

$$f(\boldsymbol{v}, \boldsymbol{r}, t) = f_0(v, \boldsymbol{r}, t) + \frac{\boldsymbol{v}}{v} \boldsymbol{f}_1(v, \boldsymbol{r}, t).$$

The system of equations for the functions f_0 and $\boldsymbol{f}_1$, with allowance for the expressions obtained for the collision integral in the preceding section, takes the form

$$\begin{aligned} &\frac{\partial f_0}{\partial t} + \frac{v}{3} \operatorname{div}_{\boldsymbol{r}} \boldsymbol{f}_1 - \frac{e}{3mv^2} \frac{\partial}{\partial v} (v^2 \boldsymbol{E} \boldsymbol{f}_1) \\ &= \frac{1}{2v^2} \frac{\partial}{\partial v} \left\{ v^2 \left[\delta_{\mathrm{el}} \nu_{\mathrm{m}}^{el} \left(\frac{T}{m} \frac{\partial f_0}{\partial v} + v f_0 \right) + \delta_{\mathrm{el}}^i \nu_{\mathrm{i}} \left(\frac{T_{\mathrm{i}}}{m} \frac{\partial f_0}{\partial v} + v f_0 \right) \right] \right\} \\ &\quad - S_{0\,\mathrm{in}}(f_0) + \frac{1}{2v^2} \frac{\partial}{\partial v} \left\{ v^2 \left[A_1(f_0) v f_0 + A_2(f_0) \frac{\partial f_0}{\partial v} \right] \right\}. \end{aligned} \tag{2.131}$$

$$\frac{\partial \boldsymbol{f}_1}{\partial t} + v \operatorname{grad}_{\boldsymbol{r}} f_0 - \frac{e\boldsymbol{E}}{m} \frac{\partial f_0}{\partial v} - \frac{e}{mc} [\boldsymbol{H} \times \boldsymbol{f}_1] = -\nu(v) \boldsymbol{f}_1 - S_{\mathrm{ie}}(\boldsymbol{f}_1). \tag{2.132}$$

Here, $\delta_{\mathrm{el}} = 2m/M$ and $\nu_{\mathrm{m}}^{el}(v)$ is the number of elastic collisions of the electron with the molecules [Eq. (2.77)]; $\nu_{\mathrm{i}}(v)$ is the number of collisions of the electrons with ions [Eq. (2.92); $\nu(v) = \nu_{\mathrm{m}}^{el}(v) + \nu_{\mathrm{i}}(v) + \nu_{\mathrm{in}}(v)$ is the total number of collisions, where $\nu_{\mathrm{in}}(v)$ is the number of inelastic collisions of the electron with the molecules, determined by Equations (2.115) and (2.116); $S_{0\,\mathrm{in}}(f_0)$ is the inelastic-collision integral for the function f_0; $A_1(f_0)$, $A_2(f_0)$ and $S_{1\mathrm{e}}(\boldsymbol{f}_1)$ are integral expressions describing the changes of the functions f_0 and $\boldsymbol{f}_1$ due to collisions between the electrons [Eqs. (2.103), (2.105), and (2.111)].

Proceeding to the solution of Equations (2.131) and (2.132), we shall dwell first on one feature that will be extensively used from now on. As noted in Section 2.1, the electron energy, relaxation time $\tau_T = 1/\delta\nu_{\mathrm{e}}$ is always much larger than the time of relaxation for their momentum $\tau_v = 1/\nu_e$. Therefore, the relaxation time is much longer for the function f_0 than for the function f_1. As a result, the function f_0 always changes more slowly with time than the function f_1; consequently, when integrating Equation (2.132) for the function f_1, we can assume, in first-order approximation, the function f_0 to be constant and independent of the time. This greatly facilitates the integration of Equation (2.132). The resultant simple expression for f_1 is valid accurate to terms smaller than or of the order of δ, i.e., to the same degree of accuracy to which Equations (2.131) and (2.132) themselves are valid. The problem thus reduces the integration of only one equation for the function f_0.

In the right-hand side of the equation for the function f_0, the last term, which is due to the collisions between the electrons, is of the order of $\nu_{ee} f_0$, where ν_{ee} is the frequency of the electron–electron collisions. The remaining terms, which describe the collisions of the electrons with heavy particles, are of the order of $\delta\nu_e f_0$, where $\delta\nu_e = \delta_{em}\nu_{em} + \delta_{ei}\nu_{ei}$. It is clear that, depending on the ratio of ν_{ee} and $\delta\nu_e$, the form of the function f_0 is determined either by the electron–electron collisions or by the collisions of the electrons with the heavy particles. We shall, therefore, consider first these two essential limiting cases separately, namely, the case of a *strongly ionized* plasma, when $\nu_{ee} \gg \delta\nu_e$, and the case of *weakly ionized* plasma, when $\nu_{ee} \ll \delta\nu_e$. The solution of the problem at any electron density, i.e., at any ratio of ν_{ee} and $\delta\nu_e$, is the subject of Section 2.3.3, where a criterion is established for the validity of the formulas obtained in each of the limiting cases indicated above. We note that in a fully ionized plasma we always have $\nu_{ee} \gg \delta\nu_e$, since $\delta \ll 1$ and $\nu_{ee} \sim \nu_{ei}$; on the contrary, at a very low degree of ionization, when the electron concentration is low enough, we have $\nu_{ee} \ll \delta\nu_e = \delta_{em}\nu_{em}$. In the ionosphere $\delta\nu_e \sim \nu_{ee}$ at a height 90–100 km in day time and 110–150 km at night. The ionosphere plasma should thus be regarded as weakly ionized at heights z lower than 100 km and as strongly ionized at $z > 150$ km. The terms "strongly ionized" and "weakly ionized" are, of course, relative.

2.3.1. Strongly Ionized Plasma

Distribution Function (Maxwellian Distribution). In a strongly ionized plasma, when $\nu_{ee} \gg \delta\nu_e$, the form of the function f_0 is determined by the electron–electron collisions. The solution of Equation (2.131) must be sought in this case by the method of successive approximations: $f_0 = f_{00} + f_{01} + \cdots$, taking into account in the zeroth approximation, naturally, only collisions between electrons. For a uniform plasma we then obtain from Equation (2.131) the following chain of equations:

$$S_{0e}(f_{00}) = -\frac{1}{2v^2}\frac{\partial}{\partial v}\left\{v^2\left[A_1(f_{00})vf_{00} + A_2(f_{00})\frac{\partial f_{00}}{\partial v}\right]\right\} = 0; \quad (2.133a)$$

$$\frac{1}{2v^2}\frac{\partial}{\partial v}\left\{v^2\left[A_1(f_{01})vf_{00} + A_1(f_{00})vf_{01} + A_2(f_{01})\frac{\partial f_{00}}{\partial v} + A_2(f_{00})\frac{\partial f_{01}}{\partial v}\right]\right\}$$

$$= \frac{\partial f_{00}}{\partial t} - \frac{e}{3mv^2}\frac{\partial}{\partial v}(v^2 \boldsymbol{E} f_{10})$$

$$- \frac{1}{2v^2}\frac{\partial}{\partial v}\left\{v^2\,\delta_{el}(\nu_m^{el} + \nu_i)\left[\frac{T}{m}\frac{\partial f_{00}}{\partial v} + vf_{00}\right]\right\} + S_{0\,in}(f_{00}). \quad (2.133b)$$

We have assumed here that $\delta_{el}^i = \delta_{el}$ and $T_i = T$. It is immediately seen from Equation (2.133a) that the zeroth approximation f_{00} is Maxwellian:

$$f_{00} = N\left(\frac{m}{2\pi T_e}\right)^{3/2}\exp\left(-\frac{mv^2}{2T_e}\right), \quad (2.134)$$

since it is precisely for a Maxwellian distribution that the electron–electron collision integral [Eq. (2.106)] vanishes. Physically this result is perfectly understandable: owing to the electron–electron collisions, a Maxwellian distribution should set in within a time $1/\nu_{ee}$; if $\nu_{ee} \gg \delta\nu_e$, this process is much faster than the energy transfer to the heavy particles, meaning that the function f_0 should be close to Maxwellian.

The electron density N and the electron temperature T_e in Equation (2.134) need not necessarily be constant, but are certain functions of time. They are determined from the condition that Equation (2.133b) should have a solution for the next (first) approximation. In fact, as indicated in Section 2.2.2 [see Eq. (2.109)], if the left-hand side of Equation (2.133b) is multiplied by v^2 or v^4 and integrated with respect to velocity, then the corresponding integral vanishes identically (regardless of the form of the function f_{01}). Consequently, the right-hand side of Equation (2.133b) must likewise vanish in this case. It is this fact which leads to the equations for the density and temperature of the electrons.

Multiplying Equation (2.133b) by $4\pi v^2$ and integrating with respect to dv, we obtain

$$4\pi\frac{d}{dt}\left(\int_0^\infty v^2 f_{00}\,dv\right) - 4\pi\int_0^\infty v^2 S_{0\,in}(f_{00})\,dv = 0.$$

Taking Equation (2.129) into account, we obtain from this the ionization balance equation

$$\frac{dN}{dt} + q_i(N) + q_r(N) = 0, \tag{2.135}$$

where $q_i(N)$ and $q_r(N)$ are the number of ionization and recombination events in a unit volume per second.

Perfectly analogously, multiplying Equation (2.133b) by $(mv^2/2)\cdot 4\pi v^2$ and integrating it with respect to dv, we obtain

$$\frac{d}{dt}\left(2\pi m \int_0^\infty v^4 f_{00}\, dv\right) + \frac{4\pi e}{3} \boldsymbol{E} \int_0^\infty v^3 \boldsymbol{f}_{10}\, dv$$
$$+ \delta_{el} 2\pi m \int_0^\infty v^3 (\nu_m^{el} + \nu_i)\left(\frac{T}{m}\frac{\partial f_{00}}{\partial v} + v f_{00}\right)$$
$$+ 2\pi m \int_0^\infty v^4 S_{0\,in}(f_{00})\, dv = 0.$$

Taking Equations (2.134) and (2.135) into account, we rewrite this equation in the form

$$\frac{dT_e}{dt} + \delta(T_e)\nu_e(T_e)(T_e - T) = -\frac{2e}{3N} \boldsymbol{Ej}(T_e) + \frac{2}{3}\frac{Q}{N}. \tag{2.136}$$

Here $\nu_e(T_e)$ denotes a parameter defined by the relation

$$\nu_e(T_e) = \frac{4\pi m}{3NT_e} \int_0^\infty v^4 \nu(v) f_{00}\, dv$$
$$= \frac{\sqrt{2}}{3\sqrt{\pi}}\left(\frac{m}{T_e}\right)^{5/2} \int_0^\infty v^4 \nu(v) \exp\left(-\frac{mv^2}{2T_e}\right) dv. \tag{2.137}$$

where $\nu(v) = \nu_m^{el}(v) + \nu_i(v) + \nu_{in}(v)$ is the number of the collisions of the electrons with the heavy particles; it is natural to call $\nu_e(T_e)$ the effective number or the effective frequency of electron collisions. Further, $\delta(T_e)$ is another characteristic parameter with the meaning of the average relative energy fraction lost by the electron during the time $1/\nu_e$ (cf. Sect. 2.1):

$$\delta(T_e) = \delta_{el} \frac{\nu_e^{el}(T_e)}{\nu_e} + \frac{1}{\nu_e(T_e - T)}\left[\frac{4\pi m}{3N} \int_0^\infty v^4 S_{0\,in}(f_{00})\, dv\right]. \tag{2.138}$$

Here $\delta_{el} = 2m/M$; ν_e^{el} is the effective number of elastic collisions calculated from Equation (2.137), where $\nu(v)$ takes into account only elastic collisions with the molecules $\nu_{\mathrm{m}}^{el}(v)$ and with the ions $\nu_{\mathrm{i}}(v)$; $S_{0\ \mathrm{in}}(f_{00})$ is that part of the collision integral which describes the inelastic collisions of the electrons with the molecules. Finally, $\boldsymbol{j} = -e\boldsymbol{j}_{\mathrm{e}}$ is the total electron-current density as defined in Equation (2.60), and Q/N is the average energy received by the electron from other heat sources [see, e.g. Eq. (2.129)].

An estimate of the first-approximation function shows that $f_{01} \sim (\delta\nu_{\mathrm{e}}/\nu_{\mathrm{ee}})f_0$. Consequently, in a strongly ionized plasma, the zeroth (Maxwellian) approximation suffices for the symmetrical part of the distribution function f_0, accurate to terms of order $\delta\nu_{\mathrm{e}}/\nu_{\mathrm{ee}}$.

Effective Collision Frequency. If the effective electron collision cross section does not depend on the electron velocity (hard-sphere collisions), then $\nu_{\mathrm{m}}(v)$ is given by Equation (2.78). Substituting this formula in Equation (2.137), we have:

$$\nu_{\mathrm{em}} = \frac{8}{3}\sqrt{\frac{2T_{\mathrm{e}}}{\pi m}}\,\pi a^2 N_{\mathrm{m}} = \nu_{\mathrm{em}0}\sqrt{\frac{T_{\mathrm{e}}}{T}}, \tag{2.139}$$

where $\nu_{\mathrm{em}0}$ is the effective number of collisions of the electrons with the molecules in a weak field when $T_{\mathrm{e}} = T$ [cf. Eq. (2.6)].

Actually, the effective cross sections for the collisions of electrons with molecules usually depend on the electron velocity. They are shown in Figure 6 for the principal gases making up the lower ionosphere, N_2 (Engelhardt et al., 1964) and O_2 [curves 1 and 2—from the data of Crompton and Sutton (1952) and Hake and Phelps (1967), respectively]. The dashed curves in this figure are the result of interpolation by the formula $q = 3.3 \cdot 10^{-7}v(1 + 10^{-8}v)^{-1} \cdot 10^{-16}\ \mathrm{cm}^2$ for nitrogen and the formula $q = (1.8 + 10^{-7}v) \cdot (1 + 8.3 \cdot 10^{-9}v)^{-1} \cdot 10^{-16}\ \mathrm{cm}^2$ for oxygen.

The values of ν_{em} calculated from Equation (2.137) for N_2 and O_2, using the approximation represented by the dashed curves in Figure 6, are given in Section 2.5.2.

Figure 7 shows the effective cross section for the collisions of an electron with air molecules. The dashed curve in this figure corresponds to the value $4\pi q = Cv^{2/3}$. It approximates sufficiently well the actual cross section at $v \lesssim 10^8$ cm/s, i.e., at electron energies $\varepsilon \lesssim 2$ eV. Consequently, for air (and in the lower ionosphere) at $v \lesssim 10^8$ cm/s, we can assume that

$$\begin{gathered} \nu_{\mathrm{m}}(v) = CN_{\mathrm{m}}v^{5/3}, \qquad C \approx 3.5 \cdot 10^{-16}\ \mathrm{cm}^2\ (10^7\ \mathrm{cm/s})^{-2/3} \\ \nu_{\mathrm{em}}(T_{\mathrm{e}}) = \frac{4 \cdot 2^{5/6}}{3\sqrt{\pi}}\,CN_{\mathrm{m}}\left(\frac{T_{\mathrm{e}}}{m}\right)^{5/6}\Gamma\left(\frac{10}{3}\right), \end{gathered} \tag{2.140}$$

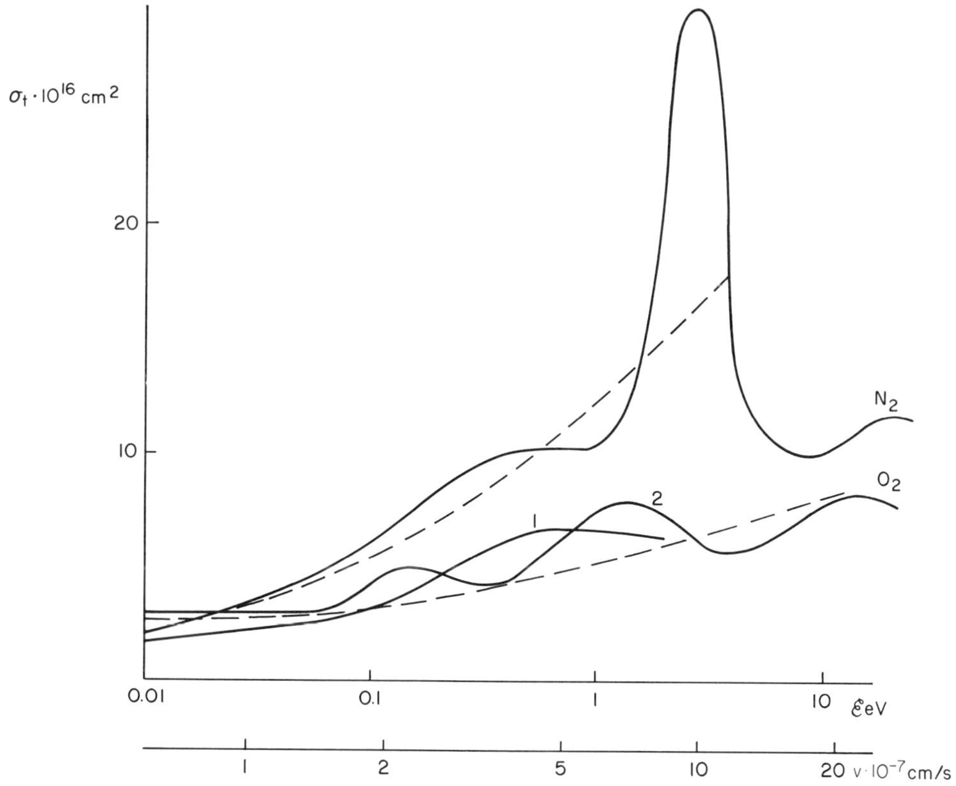

$\sigma_t \cdot 10^{16}$ cm^2
20
10
N$_2$
O$_2$
1
2
0.01
0.1
1
10
$\mathscr{E}$eV
1
2
5
10
20
$v \cdot 10^{-7}$ cm/s

Fig. 6. Effective transport cross section for electron collisions in molecular nitrogen and oxygen

$\sigma_t \cdot 10^{16}$ cm^2
20
10
1
2
5
10
20
$v \cdot 10^{-7}$ cm/s

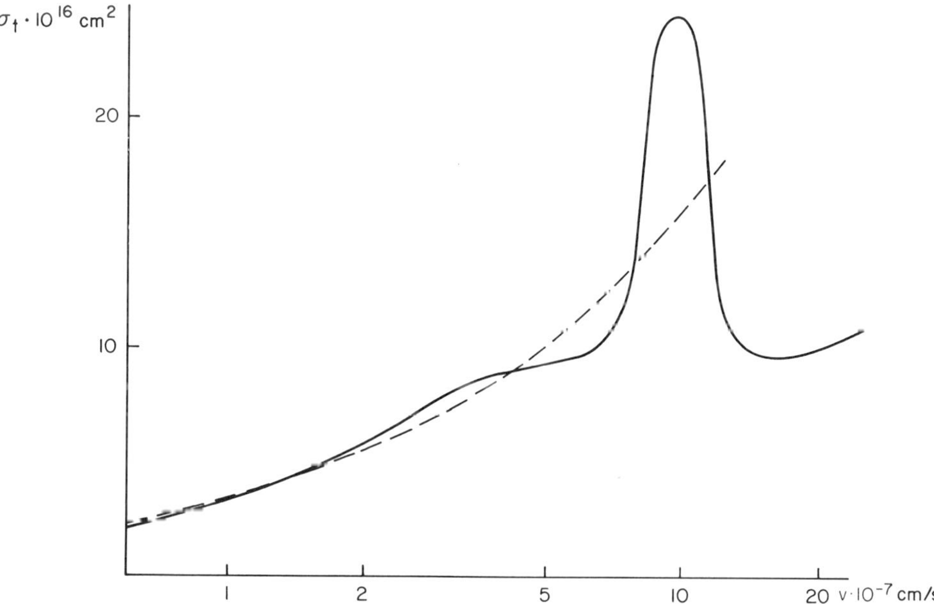

Fig. 7. Effective transport cross section for collisions of electrons with air molecules (in the lower ionosphere)

where $\Gamma(x)$ is the gamma function. In the case of collisions with ions, substituting in Equation (2.137) Equation (2.92) for $\nu_i(v)$, we obtain

$$\nu_{ei} = \frac{4\sqrt{2\pi}}{3} \frac{e^4 N_i}{m^{1/2} T_e^{3/2}} \ln\left(\frac{T_e D}{e^2}\right). \tag{2.141}$$

The quantity $\Lambda = T_e D/e^2$ under the logarithm sign is always much larger than unity [Eq. (2.94)]. As a result, the logarithm changes only insignificantly even at relatively large changes of the electron temperature; it is therefore usually possible to assume that

$$\nu_{ei} = \nu_{ei0}(T_i/T_e)^{3/2}, \tag{2.142}$$

where ν_{ei0} is the effective number of collisions of the electrons with the ions in a weak field where $T_e = T_i$ [cf. Eq. (2.7)].

Fraction of Lost Energy. It is clear from Equation (2.138) that $\delta = \delta_{el} = 2m/M$ in the case of only elastic collisions of electrons with heavy particles (molecules and ions). We now consider also inelastic collisions. We assume that the main contribution is made by collisions in which the fraction of the electron energy lost in one impact is small. To calculate δ it is then necessary to use Equation (2.115) for $S_{0\,in}$. Substituting Equation (2.115) in Equation (2.138) and integrating by parts, we obtain

$$\delta = \delta_{el} \frac{\nu_e^{el}}{\nu_e} + \frac{8\pi}{3N\nu_e(T_e - T)} \sum_{k,j} N_k \int_0^\infty v^3 \alpha_1 f_{00}\, dv$$

$$\alpha_1 = \varepsilon_{kj}\sigma_{kj} - \varepsilon_{k,-j}\sigma_{k,-j}.$$

Here σ_{kj} and $\sigma_{k,-j}$ are the cross sections for the excitation of the quanta ε_{kj} and $\varepsilon_{k,-j}$ [Eq. (2.112)]. If $S_{0\,in}$ is described by the canonical expression [Eq. (2.123)], then it follows from Equation (2.138) that

$$\delta = \delta_{el} \frac{\nu_e^{el}}{\nu_e} + \frac{4\pi m}{3T_e\nu_e N} \int_0^\infty v^4 R(v) f_{00}\, dv, \tag{2.143a}$$

where $R(v)$ is the loss function [see Eqs. (2.123) or (2.127)]. In particular, for the fraction of the energy lost to the excitation of only rotational levels [Eq. (2.123)] we have

$$\delta_r = \frac{4}{3\sqrt{\pi}} \frac{R_r(v_{Te})}{\nu_e} = \frac{16\sqrt{2}}{3\sqrt{\pi}} \frac{B_0\sigma_0 N_m}{\nu_e\sqrt{T_e m}}. \tag{2.143b}$$

Here B_0 is the rotational constant, σ_0 is defined in accordance with Equation (2.118), and $\nu_e \simeq \nu_e^{el}$. A characteristic feature of δ_r is its rather sharp decrease with increasing electron temperature. For excitation of vibrational levels, using for σ_{v*} the approximate Equation (2.128), we obtain from Equation (2.138):

$$\delta_v = \frac{16\pi N_m}{3Nm^2\nu_e T_e} \sum_{v*,\, k} \varepsilon_{v*}\varepsilon_{v*}^k \sigma_{v*}^k f_{00}(\varepsilon_{v*}^k) \tag{2.144}$$

Here ε_{v*}, ε_{v*}^k, σ_{v*}^k, and $f_{00}(\varepsilon_{v*}^k)$ are defined in accordance with Equations (2.126), (2.128), (2.134) and Table 4. The summary value is $\delta = \delta_{el} + \delta_r + \delta_v + \delta_{op} + \delta_{ion}$, where δ_{op} and δ_{ion} are the fractions of the energy lost to excitation of optical levels and ionization, respectively.

The values of the parameter δ in the ionosphere will be discussed in Section 2.5.2.

Electron Current—Dielectric Constant and Conductivity of Plasma. Substituting the distribution function $f_0(v) + (\boldsymbol{v}/v)\boldsymbol{f}_1(v)$ in Equation (2.60), we obtain the electron current:

$$\boldsymbol{j} = -\frac{4\pi e}{3} \int_0^\infty v^3 \boldsymbol{f}_1 \, dv. \tag{2.145}$$

To calculate the current it is necessary, therefore, to determine the function $\boldsymbol{f}_1$, i.e., to solve Equation (2.132).

Substituting the function f_{00} in this equation, we take into account that the dependence of f_{00} on the time t can be neglected (see the start of this section). If, furthermore, $\nu_m(v) + \nu_i(v) \gg \nu_{ee}(v)$, then the electron–electron collisions are insignificant and Equation (2.132) becomes much simpler. Its solution in this case, as can be easily verified by direct substitution, is

$$\boldsymbol{f}_{10} = -\boldsymbol{u} \frac{\partial f_{00}}{\partial v}, \tag{2.146}$$

where $\boldsymbol{u}$ is the velocity of the directional motion of the electron, determined by the equation

$$\frac{d\boldsymbol{u}}{dt} + \nu(v)\boldsymbol{u} = -\frac{e\boldsymbol{E}}{m} - \frac{e}{mc}[\boldsymbol{u} \times \boldsymbol{H}]. \tag{2.147}$$

We note that the equation for $\boldsymbol{u}$ is perfectly analogous to Equation (2.8) for the directional velocity in the elementary theory, the only difference being that ν in Equation (2.147) depends on the velocity v of the random motion. Therefore, the expressions for the function $\boldsymbol{u}(v, t)$ are perfectly analogous to the expressions obtained above in the elementary theory, except that ν_e must be replaced by $\nu(v)$. For example, in an isotropic plasma situated in an alternating electric field of frequency ω, it follows from Equation (2.10) that

$$\boldsymbol{u}(v, t) = -\frac{e\boldsymbol{E}}{m[-i\omega + \nu(v)]} = -\frac{e\boldsymbol{E}}{m}\frac{\nu(v) + i\omega}{\omega^2 + \nu^2(v)} \tag{2.148}$$

Substituting the obtained function $\boldsymbol{f}_{10}$ in Equation (2.145) and integrating with respect to the velocity v, we obtain an expression for the current $\boldsymbol{j}$, and consequently also for the conductivity and the dielectric constant of the plasma, since $\boldsymbol{j} = [\sigma - i\omega(\varepsilon - 1)/4\pi]\boldsymbol{E}$. Changing over to integration with respect to the dimensionless velocity $z = v\sqrt{m/2T_e}$, we obtain

$$\varepsilon = 1 - \frac{8\omega_0^2}{3\sqrt{\pi}}\int_0^\infty \frac{z^4 e^{-z^2}\,dz}{\omega^2 + \nu^2(v)}, \qquad \sigma = \frac{2\omega_0^2}{3\pi\sqrt{\pi}}\int_0^\infty \frac{\nu(v) z^4 e^{-z^2}\,dz}{\omega^2 + \nu^2(v)}. \tag{2.149}$$

Here $\nu(v) = \nu(z\sqrt{2T_e/m})$ is the collision frequency. The expressions for ε and σ are conveniently written in the form

$$\varepsilon = 1 - \frac{4\pi e^2 N}{m(\omega^2 + \nu_e^2)} K_\varepsilon\left(\frac{\omega}{\nu_e}\right), \qquad \sigma = \frac{e^2 N \nu_e}{m(\omega^2 + \nu_e^2)} K_\sigma\left(\frac{\omega}{\nu_e}\right); \tag{2.150}$$

$$K_\sigma(x) = \frac{8}{3\sqrt{\pi}}(x^2 + 1)\int_0^\infty \frac{\nu_e(v)}{\nu_e}\frac{z^4 e^{-z^2}\,dz}{x^2 + \nu^2(v)/\nu_e^2},$$
$$K_\varepsilon(x) = \frac{8}{3\sqrt{\pi}}(x^2 + 1)\int_0^\infty \frac{z^4 e^{-z^2}\,dz}{x^2 + \nu^2(v)/\nu_e^2}, \qquad x = \frac{\omega}{\nu_e}. \tag{2.151}$$

Here ν_e is the effective collision frequency, determined in accordance with Equation (2.137), and $K_\varepsilon(x)$ and $K_\sigma(x)$ are correction coefficients which show the extent to which the values of ε and σ calculated in the kinetic theory, differ from the corresponding values obtained with the aid of the elementary formulas of [Eq. (2.13)]. They are listed in Table 5 for collisions with molecules at $q = \text{const}$ [Eq. (2.78)] and for collisions in air [Eq. (2.140). We see that the coefficients K_ε and K_σ do not differ very strongly from unity in this case. In collisions with ions, the coefficients K_ε and K_σ are larger, especially in the region of low frequencies $\omega \lesssim \nu_e$.

Table 5. Correction coefficients for σ and ε

$\frac{\omega}{\nu_e}$	Collisions with molecules, q = const		Collisions with air molecules, $q \sim v^{2/3}$		Collisions with ions: without allowance for *ee* collisions		Collisions with ions: with allowance for *ee* collisions	
	K_ε	K_σ	K_ε	K_σ	K_ε	K_σ	K_ε	K_σ
0	1.51	1.13	3.71	1.42	19.8	3.39	4.59	1.95
0.01	1.51	1.13	3.58	1.41	19.5	3.38	4.59	1.95
0.05	1.50	1.13	3.17	1.38	15.8	2.76	4.51	1.92
0.1	1.48	1.12	2.77	1.33	11.1	2.12	4.34	1.86
0.2	1.40	1.09	2.22	1.21	5.47	1.40	3.79	1.65
0.5	1.19	1.02	1.46	0.984	2.44	0.90	2.30	1.07
1.0	1.07	0.94	1.10	0.877	1.52	0.68	1.41	0.72
2.0	0.98	0.95	0.993	0.895	1.15	0.59	1.05	0.62
4.0	1.0	0.98	0.989	0.954	1.01	0.67	0.97	0.73
6.0	1.0	0.99	0.994	0.976	0.97	0.72	0.98	0.82
10.0	1.0	1.0	0.997	0.991	0.98	0.78	0.99	0.92
35.0	1.0	1.0	1.0	0.999	0.99	0.91	1.0	0.99
∞	1	1	1	1	1	1	1	1

We note that the function K_ε and K_σ depend on one variable $x = \omega/\nu_e$ only in the case of a power-law dependence of ν on v (i.e., at $\nu \sim v^\alpha$). If the dependence of ν on v is more complicated, the coefficients K_ε and K_σ become functions of two variables, ω/ν_e and T_e.

In the definition of the function f_1 above, no account was taken of the electron–electron collisions. This is correct if the collision frequencies of the electrons with the ions and molecules [Eqs. (2.77) and (2.92)] are much higher than the electron–electron collision frequency [Eq. (2.103)]. This condition is not satisfied at higher degrees of plasma ionization. It is then necessary to take the electron–electron collisions [Eq. (2.111)] into account in Equation (2.132), which now becomes an integro-differential equation. The solution of Equation (2.132) is obtained by expanding the function f_1 in Laguerre polynomials (Chapman and Cowling, 1952; Shkarofsky et al., 1966; Hochstim and Massel, 1969).

A calculation of ε and σ with allowance for the electron–electron collisions shows that these quantities can be represented as before in the form of Equation (2.150) (Gurevich, 1958c). All that changes in this case are the functions K_ε and K_σ; they are listed in Table 5 for a fully singly ionized plasma. It is seen that allowance for the collisions between the electrons lowers the values of the functions K_ε and K_σ in a fully ionized plasma, but they still remain appreciably different from unity. We note also that at high frequencies $\omega^2 \gg \nu_e^2$ the functions K_ε and K_σ are close to unity also when account is taken of the electron–electron collisions, i.e., the influence of the electron–electron collisions in the case of high frequencies is insignificant. In a partially ionized plasma the functions K_ε and K_σ depend already not only on ω/ν_e but also on the degree of ionization of the plasma, or more accurately on the relation between ν_{ei} and ν_{em} (Shkarofsky et al., 1966).[7]

With the aid of the same functions K_ε and K_σ it is possible to express also the components of the tensors ε_{ij} and σ_{ij} in an anisotropic plasma, i.e., in the presence of a constant magnetic field $\boldsymbol{H}$. Equation (2.150) remains valid as before for the components ε_{zz} and σ_{zz} of the tensors ε_{ij} and σ_{ij} in the direction parallel to the magnetic field; in the plane perpendicular to $\boldsymbol{H}$ (the x, y plane) we have

$$\begin{aligned}
\varepsilon_{xx} = \varepsilon_{yy} &= 1 - \frac{\omega_0^2}{2\omega}\left\{\frac{\omega - \omega_H}{(\omega - \omega_H)^2 + \nu_e^2} K_\varepsilon\left(\frac{|\omega - \omega_H|}{\nu_e}\right)\right.\\
&\quad \left. + \frac{\omega + \omega_H}{(\omega + \omega_H)^2 + \nu_e^2} K_\varepsilon\left(\frac{\omega + \omega_H}{\nu_e}\right)\right\},\\
\varepsilon_{xy} = -\varepsilon_{yx} &= \frac{i\omega_0^2}{2\omega}\left\{\frac{\omega - \omega_H}{(\omega - \omega_H)^2 + \nu_e^2} K_\varepsilon\left(\frac{|\omega - \omega_H|}{\nu_e}\right)\right.\\
&\quad \left. - \frac{\omega + \omega_H}{(\omega + \omega_H)^2 + \nu_e^2} K_\varepsilon\left(\frac{\omega + \omega_H}{\nu_e}\right)\right\}.
\end{aligned} \tag{2.152}$$

$$\begin{aligned}
\sigma_{xx} = \sigma_{yy} &= \frac{\omega_0^2 \nu_e}{8\pi}\left\{\frac{1}{(\omega - \omega_H)^2 + \nu_e^2} K_\sigma\left(\frac{|\omega - \omega_H|}{\nu_e}\right)\right.\\
&\quad \left. + \frac{1}{(\omega + \omega_H)^2 + \nu_e^2} K_\sigma\left(\frac{\omega + \omega_H}{\nu_e}\right)\right\},\\
\sigma_{xy} = -\sigma_{yx} &= -i\,\frac{\omega_0^2 \nu_e}{8\pi}\left\{\frac{1}{(\omega - \omega_H)^2 + \nu_e^2} K_\sigma\left(\frac{|\omega - \omega_H|}{\nu_e}\right)\right.\\
&\quad \left. - \frac{1}{(\omega + \omega_H)^2 + \nu_e^2} K_\sigma\left(\frac{\omega + \omega_H}{\nu_e}\right)\right\}.
\end{aligned} \tag{2.153}$$

[7] Shkarofsky et al. (1966) tabulated the functions $g(\omega/\nu_e)$ and $h(\omega/\nu_e)$. The functions K_σ and K_ε are connected with them by the relations

$$K_\sigma = \frac{g[1 + (\omega/\nu_e)^2]}{g^2 + (\omega/\nu_e)^2 h^2}, \qquad K_\varepsilon = \frac{h[1 + (\omega/\nu_e)^2]}{g^2 + (\omega/\nu_e)^2 h^2}.$$

These expressions for ε_{ij} and σ_{ij} differ only by the factors K_ε and K_σ from the corresponding expressions [Eq. (2.16)] obtained in the elementary theory. Therefore, in particular, a resonant increase of the conductivity σ_{ij} takes place as before near the gyrofrequency (at $\omega \sim \omega_H$). The value of K_σ then influences the height and the width of the resonance curve.

Electron Temperature. Substituting in Equation (2.136) the obtained expressions for the effective collision frequency ν_e, for the relative fraction δ of the transferred energy, and for the current $\boldsymbol{j}$ we obtain a closed equation for the electron temperature. It is important that this equation for T_e is close to Equation (2.18) of elementary theory. It differs from Equations (2.18) and (2.10) only in the coefficient K_σ in the expression for the electron current, and also, in that the number ν_e of collisions and in the fraction of the lost energy, which remained undetermined in the elementary theory, can now be exactly obtained from Equations (2.137) and (2.138).

2.3.2. Weakly Ionized Plasma

In a weakly ionized plasma the electron–electron collisions have little effect on function f_0 (since $\nu_{ee} \ll \delta\nu_e$) and they can be disregarded in first-order approximation. They are even less important in the equation for $\boldsymbol{f}_1$, since $\nu_{ee} \ll \delta\nu_e \ll \nu_e$. The function $\boldsymbol{f}_1$ for a homogeneous weakly ionized plasma is therefore always given, accurate to terms of order δ, by Equation (2.146): $\boldsymbol{f}_1 = -\boldsymbol{u}(\partial f_0/\partial v)$, where the velocity $\boldsymbol{u} = \boldsymbol{u}(v, t)$ is defined by Equation (2.147). Substituting this value of $\boldsymbol{f}_1$ in Equation (2.131), we arrive ultimately at the following expression for the function f_0:

$$\frac{\partial f_0}{\partial t} - \frac{1}{2v^2}\frac{\partial}{\partial v}\left\{v^2\left[\left(\delta_{el}\nu_m^{el}\frac{T}{m} + \delta_{el}^i\nu_i\frac{T_i}{m} - \frac{2e\boldsymbol{E}\boldsymbol{u}}{3m}\right)\frac{\partial f_0}{\partial v} + (\delta_{el}\nu_m^{el} + \delta_{el}^i\nu_i)vf_0\right]\right\}$$
$$+ S_{0\,in}(f_0) = 0. \quad (2.154)$$

Since $\nu_i \sim \nu_{ee}$, the collisions with the ions can usually be neglected here. Depending on the ratio of the time $1/\omega$ required for a significant change of the electric field to the relaxation time $1/\delta\nu_e$ of the function f_0, we can distinguish between cases of slowly ($\omega \ll \delta\nu_e$) and rapidly ($\omega \gg \delta\nu_e$) alternating fields (just as in Sect. 2.1). In the former case, which is quasi-stationary, the time dependence of f_0 in Equation (2.154) can be disregarded, particularly of course in the case of a constant electric field.

On the other hand, in a rapidly alternating electric field $\omega \gg \delta\nu_e$ the function f_0 cannot follow the fast changes of the field. It settles therefore at a certain average time-independent level, the alternating deviations from which are small—with an amplitude on the order of $\delta\nu_e/\omega$ (just as the oscillations of the electron temperature in the elementary theory; see Sect. 2.1.2). As a result, we can neglect in first-order approximation the term $\partial f_0/\partial t$ in Equation (2.154) in either case, and by the same token, in fact, get rid of the time variable. This enables us to find a simple analytic solution of Equation (2.154) in a number of cases.

The plasma of the lower ionosphere is weakly ionized at heights $z \lesssim 100$ to 120 km. Its molecular composition is close to that of air (Table 1). We shall, therefore, deal henceforth with just such a molecular plasma.

The Distribution Function. Consider first the region of low electron energies ($\varepsilon \lesssim \varepsilon_r \approx 0.4$ eV), when an important role is played by inelastic collisions accompanied by excitation of rotational levels, and the simple expression [Eq. (2.123)] can be used for the inelastic–collision integral. Equation (2.154) is then written under stationary conditions in the form

$$\frac{1}{v^2}\frac{\partial}{\partial v}(v^2 j_v) = \frac{1}{v^2}\frac{\partial}{\partial v}\left\{v^2\left[\left((v\delta_{el} + R_r)\frac{T}{m} - \frac{2e\boldsymbol{E}\boldsymbol{u}}{3m}\right)\frac{\partial f_0}{\partial v} + v(\delta_{el}\nu + R_r)f_0\right]\right\} = 0. \tag{2.155}$$

Here $\nu = \nu_m(v)$ [see Eq. (2.140)]. According to Equation (2.147), $\boldsymbol{u} = -e\boldsymbol{E}/m\nu$ in a constant electric field $\boldsymbol{E}$ (at $H = 0$). Multiplying Equation (2.155) by v^2 and integrating from zero to v, we verify that $j_v = 0$ (since $v^2 j_v|_{v=0} = 0$ in the absence of an electron source). Integrating now the equation $j_v = 0$, we obtain

$$f_0 = C\exp\left\{-\int_0^v \frac{mv\,dv}{T + 2e^2E^2/3m\nu(\delta_{el}\nu + R_r)}\right\}. \tag{2.156}$$

In a weak field this yields a Maxwellian distribution, but in a strong field the distribution function f_0 can differ significantly from Maxwellian, therefore the collision frequency and the loss function R_r depend on v. For example, in a strong electric field we obtain from Equation (2.156)

$$f_0 = C\exp\left\{-\frac{3m^2}{2e^2E^2}\int_0^v v\nu(\delta_{el}\nu + R_r)\,dv\right\}. \tag{2.157}$$

The distribution function [Eq. (2.157)] is of the Druyvestein type: it decreases more rapidly than the Maxwellian function $\ln f_0 \sim -v^{1+\alpha}$, where $\alpha \approx \frac{5}{3}$ [see Eq. (2.140)] and $R_r \gg \delta_{el}\nu$ [Eq. (2.123)].

In a quasi-stationary electric field $\omega \ll \delta v_e$ the distribution function f_0 is determined as before by Equations (2.156) and (2.157). It is only necessary to recognize that E varies with the time t. It has a perfectly analogous form also in a rapidly alternating electric field $\omega \gg \delta v_e$. Indeed, by seeking in this case the function f_0 in the form of a series

$$f_0 = f_0(v) + f_{02}^{\pm} \exp(\pm 2i\omega t) + \cdots$$

we arrive at the stationary Equation (2.155) for $f_0(v)$, where the product $\boldsymbol{Eu}$ is replaced by its time-averaged value $\langle \boldsymbol{Eu} \rangle$ (Margenau, 1946) [cf. Eq. (2.24)]:

$$e\langle \boldsymbol{Eu} \rangle = -\frac{e^2 E_0^2 \nu}{2m(\omega^2 + \nu^2)}. \tag{2.158}$$

The alternating corrections of the function $f_0(v)$ are small: $f_{02}/f_0 \sim \delta v_e/\omega$. This result is perfectly understandable: just as the electron temperature (see Sect. 2.1.2), the symmetrical part of the distribution function f_0 does not manage to follow the field frequency in a rapidly alternating electric field $\omega \gg \delta v_e$. It settles therefore at a certain average time-independent level, deviations from which are small.

The function $f_0(v)$ in a rapidly alternating field, as follows from Equations (2.155) and (2.158), is of the form

$$f_0 = C \exp\left\{ -\int_0^v \frac{mv\,dv}{T + e^2 E_0^2/3m(\omega^2 + \nu^2)(\delta_{el} + R_r/\nu)} \right\}. \tag{2.159}$$

The constant C is determined by the normalization conditions of Equation (2.60). A characteristic feature of the distribution function [Eq. (2.159)] in a high-frequency field $\omega \gg \nu$ is its slowed-down decrease with increasing v. In weak fields E_0 this leads to formation of an elongated "tail" of the function f_0 (Engelhardt et al., 1964). The reason is easily understood. The electron energy loss to excitation of rotational levels decreases with increasing electron velocity $R_r \sim 1/v$ [Eq. (2.123)]. At the same time, the energy acquired by the electron under the influence of the high-frequency field increases in proportion to the collision frequency $\nu(v)$, i.e., in our case it is proportional to $v^{5/3}$ [Eq. (2.140)]. Therefore the fast electrons acquire energy intensively, randomly, and this leads to formation of a "tail" in the distribution function. This phenomenon is similar to the formation of the "tail" for runaway electrons (Connor and Hastie, 1975).

Neglecting $\delta_{el}\nu$ in Equation (2.159) in comparison with R_r, and assuming $\nu(v)$ to be given by Equation (2.140), we have at $\omega \gg \nu$

$$f_0(v) = C \exp\left\{\frac{3}{2\sqrt{2}a^{3/4}}\left[\frac{1}{2}\ln\frac{1+\sqrt{2}z+z^2}{1-\sqrt{2}z+z^2} - \operatorname{arctg}(\sqrt{2}z-1) - \operatorname{arctg}(\sqrt{2}z+1)\right]\right\}. \quad (2.160)$$

$$z = a^{1/4}\left(\frac{mv^2}{2T}\right)^{1/3}, \quad a = \left(\frac{E_0}{E_p}\right)^2 = \frac{e^2E_0^2\nu(v_T)}{3mT\omega^2R_r(v_T)}, \quad \delta_0 = \frac{R_r(v_T)}{\nu(v_T)}, \quad v_T = \sqrt{\frac{2T}{m}} \quad (2.161)$$

It is seen from Equation (2.160) that at $z \gg 1$ we have $f_0(v) \to \text{const} = C\exp(-3\pi/2\sqrt{2}a^{3/4})$. In fact, this asymptotic value is never reached, for the distribution function is cut off at $\varepsilon \gtrsim \varepsilon_r$ because of vibrational loss.

In the vibrational-loss region $\varepsilon > 0.4$ eV, using the approximate differential expression [Eq. (2.127)] for $S^v_{0\,\text{in}}$, it is easy to obtain formulas analogous to Equations (2.157) and (2.159) for the distribution function. Thus, in a rapidly alternating electric field $\omega \gg \delta\nu_e$ we have in place of Equation (2.159)

$$f_0 = C\exp\left\{-\int_0^v \frac{mv\,dv[1+R_v/(R_r+\delta_{el}\nu)]}{T+e^2E_0^2/3m(\omega^2+\nu^2)(\delta_{el}+R_r/\nu)}\right\}, \quad (2.162)$$

where the function R_v is defined in accordance with Equation (2.127) (see Fig. 4). We note that a comparison with the exact numerical calculations shows that the approximate Equation (2.162) is valid only in a sufficiently strong field $E_0 \gtrsim E_p$ and at low electron energies $\varepsilon \lesssim 2$ eV.

At high electron energies, $\varepsilon > 2$–3 eV, the cross sections for the excitation of both the vibrational and the optical levels are large (see Fig. 5), as are the quanta of the lost energy ε_{kj}. In a not very strong field E_0, the following condition is therefore usually satisfied:

$$\frac{6(\omega^2+\nu^2)}{e^2E_0^2}\frac{S}{\sigma_t} \gg 1, \quad (2.163)$$

$$S = \int_{\varepsilon-\bar{\varepsilon}_{kj}}^{\varepsilon} \Sigma_{\text{in}}\,d\varepsilon, \quad (2.164)$$

where Σ_{in} is the summary cross section of the inelastic processes, and σ_t is the transport cross section [Eq. (2.77)]. Under Equation (2.163), the second term of expression [Eq. (2.125)] for the inelastic-collision integral $S_{0\,in}$ is negligible and the integral takes the simple form [Eq. (2.116)]:

$$S_{0\,in} = \nu_{in} f_0, \qquad \nu_{in} = N_m v \sum_{k,j} \sigma_{kj} = N_m v\, \Sigma_{in}. \tag{2.165}$$

The stationary Equation (2.154) with the collision integral $S_{0\,in}$ [Eq. (2.165)] is similar to the stationary Schrödinger equation, which describes the penetration of particles through a potential barrier. The role of the barrier is played here by the total cross section of the inelastic collisions (Fig. 5). Under the conditions of Equation (2.163), the quasi-classical approximation is valid. Seeking therefore the solution of Equation (2.154) in the quasi–classical form $f_0 = C(v) \exp(-\int K\, dv)$, we find that

$$f_0 = \frac{C_0}{v(Kb)^{1/2}} \exp\left\{-\int_{v_0}^{v} K\, dv\right\}, \tag{2.166}$$

where

$$b = e^2 E_0^2 \nu / 6m^2(\omega^2 + \nu^2), \tag{2.167}$$

$$K = \sqrt{\nu_{in}/b} = [6(\omega^2 + \nu^2)\nu_{in} m^2 / e^2 E_0^2 \nu]^{1/2}, \tag{2.168}$$

$\nu = \nu_m^{el}(v) + \nu_{in}$ is the electron collision frequency [Eqs. (2.77) and (2.113)]. For simplicity we have assumed here the field to be rapidly alternating, $\omega \gg \delta\nu_e$, and have neglected the elastic and rotational losses ($E_0 > E_p$). Equation (2.166) describes the distribution of the electrons in the high-energy "tail" of the distribution function. The constant C_0 is determined by matching Equation (2.166) at the boundary v_0 to the distribution function in the principal velocity region. It follows directly from Equations (2.166)–(2.168) that under the conditions of Equation (2.163) we have $f_0(\varepsilon + \bar{\varepsilon}_{kj}) \ll f_0(\varepsilon)$; consequently, the second term in the integral [Eq. (2.125)] is much smaller than the first, a fact already used in the derivation of Equation (2.165).

The electron distribution function in a high–frequency electric field $\omega^2 \gg \nu_e^2$ in the molecular plasma of the lower ionosphere (air) is shown in Figure 8. At $E_0/E_p \sim 1$ the distribution function decreases gradually in the rotational region $v \lesssim 4.10^7$ cm/s (i.e., $\varepsilon \lesssim 0.4$ eV) and rapidly in the region of vibrational losses. In strong fields $E_0/E_p \gtrsim 5$ the distribution function is almost constant up to an energy $\varepsilon \sim 2$ eV.

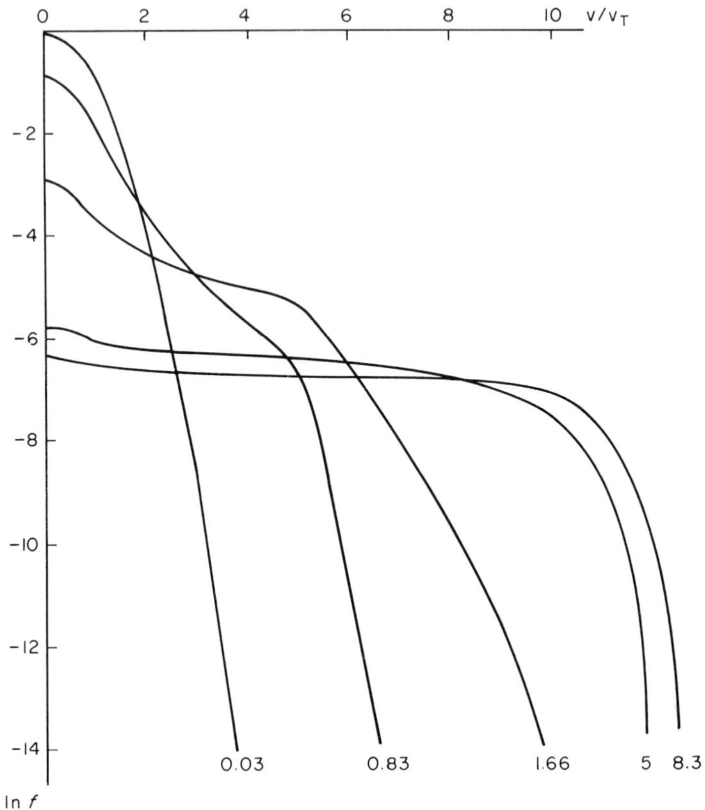

Fig. 8. Electron distribution function in air (lower ionosphere); $\omega \gg \nu_e$, $T = 185^\circ$, $v_T = 0.76 \cdot 10^7$ cm/s. The values of E_0/E_p are indicated next to the curves ($\delta_0 = 5 \cdot 10^{-3}$)

We note that under ionospheric conditions the "tail" of the distribution function can be strongly influenced by the fast electrons produced by ionization [Eq. (2.129)]. When Equation (2.129) is taken into account, the distribution function in the tail is given by

$$f(v) \approx \frac{q_{\text{ion}}(v)}{\nu_{\text{in}}(v)} + f_0(v)$$

where $f_0(v)$ is the distribution function [Eq. (2.166)].

We did not take into account above the constant magnetic field $\boldsymbol{H}$. It is easy to see, however, that allowance for $\boldsymbol{H}$ leads only to a change in the velocity $\boldsymbol{u}$ in Equation (2.154). As a result, the expressions obtained above for $f_0(v)$ also change little (Allis, 1956; Ginzburg and Gurevich,

1960). In particular, Equations (2.159) and (2.162) are valid in an alternating field $\omega \gg \delta\nu_e$, except that it is necessary to substitute

$$E_0^2 \to E_0^2\varphi_p(v), \; \varphi_p(v) = \alpha_{\parallel}^2 + (\omega^2 + \nu^2(v))\left[\frac{\alpha_{\perp -}^2}{(\omega - \omega_H)^2 + \nu^2(v)} + \frac{\alpha_{\perp +}^2}{(\omega + \omega_H)^2 + \nu^2(v)}\right]. \tag{2.169}$$

Here $\varphi_p(v)$ is a polarization factor analogous to Equation (2.42) [of course, in place of the effective collision frequency ν_e in Equation (2.42), Equations (2.169) and (2.159) contain $\nu(v)$]. It is seen from Equation (2.169) that with changing frequency of the extraordinary wave near the electron gyrofrequency $\omega \approx \omega_H$ the distribution function in the rotational region experiences an appreciable distortion at $E_0/E_p \sim 1$: from a function with a "tail" of the type of Equations (2.159) and (2.160) at $|\omega - \omega_H| \gg \nu$ it changes at $\omega \simeq \omega_H$ into a function of the type of Equation (2.157), which decreases more rapidly than a Maxwellian function with increasing v.

Average Electron Energy, Electron Current. Using the expressions obtained above for the distribution function, we can easily determine the average electron energy [Eq. (2.60)]:

$$\bar{\varepsilon} = \frac{2\pi m}{N}\int_0^\infty v^4 f_0 \, dv. \tag{2.170}$$

It is shown in Figure 9 for the case of the high-frequency field ($\omega^2 \gg \nu^2$) at $T = 185° \simeq 0.016$ eV. We see that at $E_0 \lesssim 0.5E_p$ the average electron energy is close to the molecule energy $\varepsilon_0 = \frac{3}{2}T$. At $E_0 \sim (1\text{–}4)E_p$, the energy of the electrons increases rapidly to values $\bar{\varepsilon}/\bar{\varepsilon}_0 \sim 30\text{–}40$, and then its growth is slowed down by the increased losses. The presence of the region of rapid growth of $\bar{\varepsilon}(E_0)$ is connected with the decrease of the rotational losses $R_r(v)$, and with the appearance of the "tail" of the electron distribution function in the high-frequency electric field.

The components of the dielectric tensor and the plasma conductivity, as follows from Equations (2.145)–(2.147), are determined by the expressions

$$\varepsilon_{ij} = -\frac{4\pi}{3N}\int_0^\infty v^3 \frac{\partial f_0}{\partial v}\varepsilon_{ij}(v)\, dv, \qquad \sigma_{ij} = -\frac{4\pi}{3N}\int_0^\infty v^3 \frac{\partial f_0}{\partial v}\sigma_{ij}(v)\, dv. \tag{2.171}$$

Here $\varepsilon_{ij}(v)$ and $\sigma_{ij}(v)$ are determined by the elementary-theory Equations

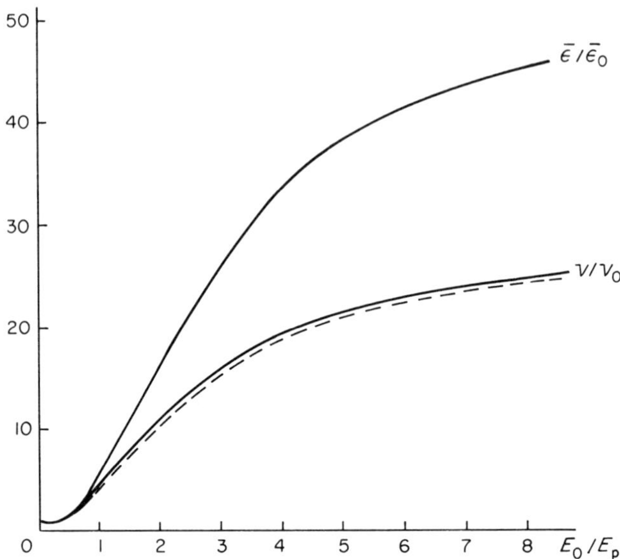

Fig. 9. Average energy ε and effective electron collision frequency ν_e in air (lower ionosphere); $\omega \gg \nu_e$, $T = 185°$. *Dashed curve* is a plot of ν_e in accordance with Equation (2.140)

(2.13) and (2.16), where ν_e is replaced by $\nu(v)$. Equations (2.171) are discussed, for example, by Shkarofsky et al. (1966). In particular, in a high-frequency electric field with $\omega^2 \gg \nu^2$ and $(\omega - \omega_H)^2 \gg \nu^2$ the elementary-theory formulas are valid for ε and σ, with

$$\nu_e = -\frac{4\pi}{3N}\int_0^\infty v^3\nu(v)\frac{\partial f_0}{\partial v}\,dv = \frac{4\pi(3+\alpha)}{3N}\int_0^\infty v^2\nu(v)f_0\,dv. \qquad (2.172)$$

In the last expression it is assumed that $\nu \sim v^\alpha$. The dependence of the collision frequency ν_e at $\alpha = \frac{5}{3}$ on the field amplitude E_0/E_p is shown in Figure 9. We see that in the region $E_0/E_p \sim 1$–4 the collision frequency also increases rapidly. The dashed curve in the figure shows a plot of $\nu_e \sim (T_e/T)^{5/6} = (\bar{\varepsilon}/\bar{\varepsilon}_0)^{5/6}$, which is valid for the Maxwellian distribution function [Eq. (2.140)]. We see that the result of the exact kinetic calculation of ν_e is sufficiently close to the result of the calculation by the simple Equation (2.140). It can be noted that even in the general case allowance for the exact form of the distribution function $f_0(v)$, i.e., allowance for its deviation from the Maxwellian function [Eq. (2.134)], has usually little effect on the average quantities $\bar{\varepsilon}$, ν_e, σ, and ε. On the contrary, quantities such as the ionization frequency or the optical-level excitation coefficients, which are determined by the "tail" of the distribution function, depend very strongly on its exact form.

Ionization Frequency. The ionization frequency, i.e., the total number of ionizations per second per plasma electron, is given by

$$\nu_{\text{ion}} = \frac{N_{\text{m}}}{N} \int v q_{\text{ion}} f \, d\mathbf{v} = \frac{4\pi N_{\text{m}}}{N} \int_0^\infty v^3 f_0 q_{\text{ion}}(v) \, dv. \tag{2.173}$$

Here $q_{\text{ion}}(v)$ is the total cross section for molecule ionization by collision with an electron of velocity v (i.e., of energy $\varepsilon = mv^2/2$). The cross section is $q_{\text{ion}} = 0$ at $\varepsilon < \varepsilon_{\text{i}}$, where ε_{i} is the ionization energy (see Table 12). At $\varepsilon \geq \varepsilon_{\text{i}}$, the cross section q_{ion} first increases abruptly and then (at $\varepsilon \gtrsim (4\text{–}5)\varepsilon_{\text{i}}$) decreases slowly with increasing ε. Up to $\varepsilon \sim 1.5\varepsilon_{\text{i}}$, the cross section q_{ion} usually increases linearly

$$q_{\text{ion}} = \begin{cases} 0, & \text{if} \quad v < v_{\text{i}} = \sqrt{2\varepsilon_{\text{i}}/m} \\ Q(v - v_{\text{i}}), & \text{if} \quad v > v_{\text{i}} \end{cases} \tag{2.174}$$

For oxygen O_2 and nitrogen N_2 we have according to Engelhardt et al. (1964) and Hake and Phelps (1967)

$$\begin{aligned} Q_{\text{O}_2} &= 0.5 \cdot 10^{-24} \text{ cm} \cdot \text{sec} & Q_{\text{N}_2} &= 0.4 \cdot 10^{-24} \text{ cm} \cdot \text{sec} \\ v_{\text{iO}_2} &= 2.06 \cdot 10^8 \text{ cm/s}, & v_{\text{iN}_2} &= 2.34 \cdot 10^8 \text{ cm/s}. \end{aligned} \tag{2.175}$$

We substitute in Equation (2.173) the electron distribution function [Eq. (2.166)] and integrate with respect to dv, recognizing that the main contribution to the integral [Eq. (2.173)] is made by the region $v \approx v_{\text{i}}$, where $f_0(v) \simeq f_0(v_{\text{i}}) \exp[-K(v_{\text{i}})(v - v_{\text{i}})]$. We then obtain

$$\nu_{\text{ion}} = 4\pi N_{\text{m}} v_{\text{i}}^3 Q K^{-2}(v_{\text{i}}) f_0(v_{\text{i}})/N. \tag{2.176}$$

Here $f_0(v_{\text{i}})$ and $K(v_{\text{i}})$ are the values of the distribution function [Eq. (2.166)] and of the parameter $K(v)$ [Eq. (2.168)] at $v = v_{\text{i}}$. Equation (2.176) can be rewritten with sufficient accuracy in the form

$$\begin{aligned} \nu_{\text{ion}} &= \nu_{\text{ion}}^0 x_{\text{i}}^{-2} \exp(-x_{\text{i}}), & x_{\text{i}} &= \frac{\sqrt{6m}}{eE_0} (A_{\text{i}}^2 \omega^2 + B_{\text{i}}^2)^{1/2} K_{\text{i}}, \\ A_{\text{i}} &= \int_{v_0}^{v_{\text{i}}} (\nu_{\text{in}}/v)^{1/2} \, dv, & B_{\text{i}} &= \int_{v_0}^{v_{\text{i}}} (\nu_{\text{in}} v)^{1/2} \, dv, \end{aligned} \tag{2.177}$$

$$\nu_{\text{ion}}^0 = 4\pi N_{\text{m}} v_{\text{i}}^3 Q A_{\text{i}}^2 f_0(v_0)/N$$

Here $\nu = \nu(v)$ is the electron collision frequency, $\nu_{in}(v)$ is the frequency of the inelastic collision [Eq. (2.165)], v_0 is the point where the distribution function in the tail [Eq. (2.166)] is matched to the distribution function in the main region, and $f(v_0)$ is the value of $f(v)$ at this point. The coefficient

$$K_i = \frac{eE_0}{\sqrt{6}m(A_i^2\omega^2 + B_i^2)^{1/2}} \int_{v_0}^{v_i} K\, dv$$

is always close to unity.

The exponential term in Equation (2.177) is determined by the collision frequencies $\nu_{in}(v)$ [Eq. (2.165)] and $\nu(v)$ [Eqs. (2.168), (2.77), and (2.113)]. For air, the total inelastic-collision cross section Σ_{in} is shown in Figure 5, and the total transport cross section σ_t in Figure 7. It is natural to assume the boundary of v_0 at $\varepsilon = \varepsilon_0 \sim 2$ eV, where the cross section of the inelastic processes increases abruptly. In the calculation of the pre-exponential term we recognize that the losses are small in the region $\varepsilon < \varepsilon_0$. Therefore, in a sufficiently strong electric field $E_0 \gtrsim 5E_p$ we can approximately assume that the distribution function changes little in the region $\varepsilon < \varepsilon_0$ (see Fig. 8), i.e., $f(v_0) \approx 3N/4\pi v_0^3$. In addition, it must be recognized that the principal part is played by the ionization of the oxygen. For the constants A, B, and ν_{ion}^0 in Equation (2.177) we then obtain

$$\begin{aligned} A_i &= 3.0 \cdot 10^7 \text{ cm/s} \\ B_i &= 5.1 N_m \text{ cm}^4 \text{ s}^2 \\ \nu_{ion}^0 &= 2.6 \cdot 10^{-7} N_{O_2} \text{ cm}^3/\text{s} \end{aligned} \tag{2.178}$$

Here N_m and N_{O_2} are the concentrations of the air and oxygen molecules, respectively. For the ionization of nitrogen, the corresponding constants are $A_i = 4.1 \cdot 10^7$ cm/s, $B_i = 7.6N_m$ cm^4/s, and $\nu_{ion}^0 = 2.0 \cdot 10^{-7} N_{N_2}$ cm^3 s. Equations (2.177) and (2.178) describe approximately the ionization frequency in air and in the lower ionosphere. It is possible to obtain similarly the coefficients for the excitation of the optical levels.

We note that the presence of nitrogen oxide NO can influence the initial state of ionization of the air. To be sure, at low gas temperatures $T \lesssim 300$—500 K the number of NO molecules is small and their role is insignificant. At $T \gtrsim 1000$ K, however, their number increases greatly and, since the ionization energy is low ($\varepsilon_i = 9.3$ eV), ionization of these molecules can turn out to be significant. The values of the constants A_i, B_i, and ν_{ion}^0 for NO are $A_i = 2.4 \cdot 10^7$ s, $B_i = 4.0N_m$ cm^4/s^2, and $\nu_{ion}^0 = 2.5 \cdot 10^{-7} N_{NO}$ cm^3/s.

We note that at large values of the field amplitude, close to critical breakdown fields E_c [see Eq. (2.253)], the growth of the limiting velocity v_0 becomes essential. This decreases on the quantities ΔA, ΔB the integral coefficients A_i and B_i. At $E_0 \simeq E_c$, in particular, we have $\Delta B \simeq -0.14B_i$ and $\Delta A \simeq -0.16A_i$. We note also that the expressions obtained for ν_{ion} are valid not only in a rapidly alternating ($\omega \gg \delta\nu$) but also in a constant electric field E, if the amplitude E_0 is replaced by $E\sqrt{2}$.

Coefficient of Dissociative Attachment. A prominent role is played in the ionization balance of a plasma in air by dissociative attachment of electrons to the oxygen molecules, $O_2 + e \rightarrow O^- + O$. The cross section q_a of this process has a sharp peak at an electron energy $\varepsilon \approx 6.7$ eV (Schulz, 1962b); it can be represented approximately in the form

$$q_a(v) \simeq q_m \begin{cases} \dfrac{v - v_a}{v_m - v_a}, & v_a \leq v \leq v_m \\ \exp\left[-b_a(v - v_m)\right], & v > v_m \end{cases} \tag{2.174a}$$

where the constants v_a, v_m, b_a, and q_m are equal to:

$$v_a = 1.34 \cdot 10^8 \text{ cm/s}; \qquad v_m = 1.54 \cdot 10^8 \text{ cm/s};$$

$$b_a = 7.6 \cdot 10^{-8} \text{ s/cm}; \qquad q_m = 1.3 \cdot 10^{-18} \text{ cm}^2$$

The frequency of the dissociative attachment is determined by an expression analogous to Equation (2.173)

$$\nu_a = \frac{4\pi N_m}{N} \int_{v_a}^{\infty} v^3 f_0(v) q_a(v)\, dv$$

Using Equation (2.166) for the electron distribution function and Equation (2.174a) for the attachment cross section, and integrating, we obtain

$$\nu_a = 4\pi N_{O_2} \frac{f_0(v_0)}{N} q_m v_m^3 \left[\frac{1}{b_a + K_m} + \frac{1 - \exp\left[-K_m(v_m - v_a)/2\right]}{K_m}\right] \exp(-x_a)$$

$$x_a = \frac{\sqrt{6m}}{eE_0}(A_a^2\omega^2 + B_a^2)^{1/2} K_a, \qquad K_m = K(v_m)$$

Here A_a and B_a are the integral functions [Eq. (2.177)] with the upper integration limit v_i replaced by v_a. In the calculation of the integrals it is also assumed that $K(v)$ increases linearly in the region $v_a \leq v \leq v_m$ (see

Fig. 5). The coefficient K_a differs from unity by at most 5%; at $\omega \gg B_a/A_a$ and at $\omega \ll B_a/A_a$ we have the coefficient $K_a = 1$.

Assuming now, just as in the preceding section, that $f(v_0) \approx 3N/4\pi v_0^3$, and substituting the numerical values of v_0, v_m, b_a, and q_m, we obtain the coefficient of dissociative attachment k_a

$$k_a = \nu_a/N_{O_2} \tag{2.177a}$$

where the constants in the exponential are

$$A_a = 1.5 \cdot 10^7 \text{ cm/s}, \qquad B_a = 2.5 \cdot N_m \text{ cm}^4/\text{s}^2.$$

We note that the expression obtained for ν_a is valid only in not too strong fields, so long as $x_a \gtrsim 6$. At $x_a < 6$, the corrections ΔA and ΔB indicated in the preceding section become important. Consequently, the increase of k_a with the amplitude E_a is strongly retarded. The maximum value of k_a is (Hake and Phelps, 1967)

$$k_{am} \approx 2.4 \cdot 10^{-11} \text{ cm}^3 \text{ s}^{-1}. \tag{2.178a}$$

It is reached in fields close to the critical breakdown fields, $E_0 \sim (0.5\text{–}1)E_c$.

Nonstationary Perturbation of the Distribution Function. Consider now the nonstationary perturbation of the electron distribution function by an intense high-frequency field ($\omega^2 \gg \nu^2$). We assume that the time elapsed from the instant when the field is turned on is large enough, $\Delta t \gg 1/\nu$; then the main part of the distribution function of the electrons is symmetrical $f = f_0(v, t)$ and is described by Equation (2.154). In a strong electric field

$$E_0^2/E_p^2 \gg 1 \tag{2.179}$$

the relaxation terms in Equation (2.154) are negligible during the initial period of the perturbation, and this equation takes the form

$$\frac{\partial f_0}{\partial t} - \frac{e^2 E_0^2}{6m^2\omega^2 v^2}\frac{\partial}{\partial v}\left(v^2 \nu(v)\frac{\partial f_0}{\partial v}\right) = 0. \tag{2.180}$$

We recognize that when the distribution function is perturbed by a strong electric field [Eq. (2.179)], the initial thermal scatter of the electron velocities is insignificant. In this case $f_0(v, 0) \sim \delta(v)$ and at $\nu(v) = \nu_0 v^\alpha$ there are no characteristic parameters with the dimension of the velocity v either in the initial conditions of the problem or in Equation (2.180) itself. This means that the sought nonstationary solution of Equation (2.180) should be self-similar: it can depend only on the ratio v^β/t. We seek, therefore, the distribution function in the form

$$f_0(v, t) = t^\gamma F(\zeta), \qquad \zeta = v^\beta/t. \tag{2.181}$$

Substituting Equation (2.181) in Equation (2.180) we find at $\nu(v) = \nu_0 v^\alpha$ that $\gamma = -3/(2-\alpha)$, $\beta = 2-\alpha$, and

$$F(\zeta) = \exp(-\zeta/D), \qquad D = \frac{(2-\alpha)^2 e^2 E_0^2 \nu_0}{6m^2\omega^2}. \tag{2.182}$$

Finally, taking into account the normalization conditions of Equation (2.60), we have

$$f_0(v, t) = \frac{(2-\alpha)N}{4\pi(Dt)^{3/(2-\alpha)}\Gamma\left(\dfrac{3}{2-\alpha}\right)} \exp\{-|v|^{2-\alpha}/Dt\}. \tag{2.183}$$

An interesting feature of the distribution function [Eq. (2.183)] is its slow decrease with increasing electron energy ε. Thus, under the conditions of the lower ionosphere (air) at $\varepsilon \lesssim 2$ eV, according to Equation (2.140), we have $\alpha = \frac{5}{3}$, and consequently $f_0 \sim \exp(-\varepsilon^{1/6}/t)$; in the region $\varepsilon \gtrsim 2$ eV we have $\alpha \approx 1$ and $f_0 \sim \exp(-\varepsilon^{1/2}/t)$. The reason is that the fast electrons collide more frequently and therefore acquire energy from the alternating electric field more rapidly [see Eq. (2.180)]. It is this which leads to their effective acceleration. Of course, Equation (2.180) and accordingly Equation (2.183) for the electron distribution function is valid only if relaxation processes are neglected, i.e., so long as

$$f_0(v, t) \ll f_{0s}(v), \tag{2.184}$$

where $f_{0s}(v)$ is the stationary distribution function defined above. As $f_0(v, t)$ approaches $f_{0s}(v)$, the relaxation processes become appreciable and the function $f_0(v, t)$ gradually assumes a stationary value. It is clear from Equation (2.183) that the stationary solution sets in most rapidly at high velocities.

Relaxation of Distribution Function. The process of relaxation of the distribution function after the perturbing electric field is turned off is determined completely by the collisions. In the high-energy region, where the excitation of the optical and vibrational levels is appreciable, we have

$$f_0(v, t) = f_0(v, 0) \exp(-\nu_{in}(v) \cdot t). \tag{2.185}$$

Here $f_0(v, 0)$ is the distribution function at the instant of the start of relaxation $t = 0$, and $\nu_{in}(v)$ is the combined frequency of the inelastic collisions, determined by Equation (2.165). The relaxation proceeds here rapidly, except for the energy region $\varepsilon \approx 4.5$–5 eV, where $\nu_{in}(v)$ is greatly decreased (see Fig. 5).

In the low-energy region, Equation (2.154) takes the form

$$\frac{\partial f_0}{\partial t} - \frac{1}{2v^2}\frac{\partial}{\partial v}(v^3 R_r f_0) = 0. \tag{2.186}$$

We have neglected here $\delta_{el}\nu$ in comparison with R_r and neglected the term $\sim \partial f_0/\partial v$, since we are considering electrons with $\varepsilon \gg T$. The solution of Equation (2.186) is

$$f_0(v, t) = \left(1 + \frac{A}{2v} t\right)^2 f_0\left(v + \frac{A}{2} t, 0\right), \qquad A = \frac{8B_0\sigma_0 N_m}{m}. \tag{2.187}$$

We have taken into account here the fact that $R_r(v)$ is described by Equation (2.123). It is seen from Equation (2.187) that the characteristic relaxation time of the distribution function in the rotational region is

$$\Delta t_r = \frac{v_r m}{4B_0\sigma_0 N_m}, \tag{2.188}$$

where $v_r = (2\varepsilon_r/m)^{1/2}$ is the limit of the rotational region. In air we have $v_r \approx 4 \cdot 10^7$ cm/s. The time Δt_r is much longer here than the relaxation time in the high-energy region [Eq. (2.185)].

2.3.3. Arbitrary Degree of Ionization—Concerning the Elementary Theory

We have considered above the limiting case of a weakly ionized plasma, when the electron–electron collisions are negligible, and the case of a strongly ionized plasma, when, to the contrary, the form of the function $f_0(v)$ is determined precisely by the interelectron collisions. We now consider an intermediate case, when the form of the function f_0 is greatly influenced both by the interelectron collisions and by the collisions between the electrons and the heavy particles. In the equation for the function f_1 it is possible in this case to neglect the interelectron collisions, since $\nu_{ee} \sim \delta\nu_{em} \ll \nu_{em}$. The problem thus reduces to an analysis of equation for the function f_0:

$$\frac{\partial f_0}{\partial t} - \frac{1}{2v^2}\frac{\partial}{\partial v}\left\{v^2\left[\delta_{el}\nu_m^{el}\frac{T}{m} - \frac{2e(\boldsymbol{Eu})}{3m} + A_2(f_0)\right]\frac{\partial f_0}{\partial v} + v^3[\delta_{el}\nu_m^{el} + A_1(f_0)]f_0\right\} + S_{0\,in}(f_0) = 0 \quad (2.189)$$

The coefficients A_1 and A_2 are integrals that depend on the function f_0 [see Eqs. (2.103) and (2.105)].

Under stationary conditions (constant or rapidly alternating electric field) the first term in Equation (2.189) can be neglected, by averaging the product $\boldsymbol{E} \cdot \boldsymbol{u}$ over the time $t \gg 1/\omega$. The solution of the remaining nonlinear integro-differential equation can be obtained by an iteration method. This method gives good convergence, for as the function f_0 changes on going from a weakly ionized to a strongly ionized plasma the integral coefficients $A_1(f_0)$ and $A_2(f_0)$ change relatively little.

We choose as the zeroth approximation $f_0(v)$ a Maxwellian electron distribution function with the temperature given by Equation (2.136). Substituting it in Equation (2.103) we obtain, in the same approximation, the integral coefficient A_1:

$$A_{10} = 8\pi\nu_{ee}\int_0^v v_1^2 \exp\left\{-\frac{mv_1^2}{2T_e}\right\}\left(\frac{m}{2\pi T_e}\right)^{3/2} dv$$
$$= 2\nu_{ee}\left[\Phi(z) - \frac{2z}{\sqrt{\pi}}\exp(-z^2)\right] \quad (2.190)$$

$$\Phi(z) = \frac{2}{\sqrt{\pi}}\int_0^z \exp(-t^2)\,dt, \qquad \nu_{ee}(v) = \frac{4\pi e^4 N \ln \Lambda}{m^2 v^3}.$$

Analogously from Equation (2.105) we have

$$A_{20} = \frac{2T_e}{m}\nu_{ee}\left[\Phi(z) - \frac{2z}{\sqrt{\pi}}\exp(-z^2)\right] = \frac{T_e}{m}A_{10}. \quad (2.191)$$

Substituting the coefficients A_{10} and A_{20} in Equation (2.189) and integrating this expression, we find that the electron distribution function in the next (first-order) approximation is given by

$$f_0^{(1)} = C \exp\left\{-\int_0^v \frac{v[R(v) + A_{10}]\,dv}{\dfrac{T}{m}(\delta_{el}\nu + R_r) + A_{20} - \dfrac{2e\boldsymbol{E}\boldsymbol{u}}{3m}}\right\}. \tag{2.192}$$

$$R(v) = \delta_{el}\nu + R_r + R_v, \qquad \nu = \nu_m^{el}.$$

It is assumed here for simplicity that the integral of the inelastic collisions takes the form of Equations (2.123) and (2.127). In a strongly ionized plasma, the principal role in Equation (2.192) is played by the coefficients A_{10} and A_{20}, so that the function $f_0^{(1)}$ is in this case Maxwellian with an electron temperature T_e determined by Equation (2.136). In a weakly ionized plasma, to the contrary, the coefficients A_{10} and A_{20} can be neglected and the function $f_0^{(1)}$ takes the form of Equations (2.159), (2.164), etc. With increasing degree of plasma ionization, the coefficients A_{10} and A_{20} become more and more appreciable and the distribution function gradually goes over into a Maxwellian one. This transition depends on the parameter (Gurevich, 1959)

$$\begin{aligned} g &= \frac{\nu_{ee}(v_{Te})}{R(v_{Te})} = \frac{N}{N_m}\,\frac{\pi e^4 \ln \Lambda}{\sigma_t\,\delta T_e^2} \\ &\simeq 5 \cdot 10^7 \frac{N}{N_m}\left(\frac{1000^\circ}{T_e}\right)^2 \left(\frac{10^{-15}\ \mathrm{cm}^2}{\sigma_t}\right)\left(\frac{10^{-3}}{\delta}\right). \end{aligned} \tag{2.193}$$

$$\delta = R(v_{Te})/\nu(v_{Te}), \qquad \nu = \nu_m^{el}.$$

Here $v_{Te} = \sqrt{2T_e/m}$ and T_e is the electron temperature. At $g \gg 5$ the distribution is Maxwellian and the plasma can be regarded as fully ionized, i.e., the formulas obtained in Section 2.3.1 can be used; at $g \ll 5$, the plasma is weakly ionized.

It should be noted that the transition region is quite broad, especially at high velocities, i.e., in the tail of the distribution function. For example, the deviations from Maxwellian distribution are appreciable in the tail even at $g \sim 10^2$. In general, the distribution in the tail (i.e., at large v) can be regarded as Maxwellian only when g is larger than $mv^2/2T_e$.

Calculating the coefficients $A_1^{(1)}$ and $A_2^{(1)}$ with the aid of the distribution function $f_0^{(1)}(v)$ by means of Equations (2.103) and (2.105), and then again integrating Equation (2.189), we can find the second iteration $f_0^{(2)}(v)$, etc. The calculation shows that the second and succeeding iterations differ little from the function $f_0^{(1)}$, so that in fact the first iteration [Eq. (2.192)] is sufficient.

Concerning the Elementary Theory. We consider the simplest case, when the electron collision frequency ν and the energy fraction do not depend on the electron velocity: $\nu \neq \nu(v)$, $\delta \neq \delta(v)$. The solution of the kinetic Equations (2.131) and (2.132) takes in this case the form

$$f_0 = N\left(\frac{m}{2\pi T_e}\right)^{3/2} \exp\left(-\frac{mv^2}{2T_e}\right), \qquad f_1 = -\boldsymbol{u}\frac{\partial f_0}{\partial v}, \tag{2.194}$$

where the temperature T_e and the average electron directional velocity $\boldsymbol{u}$ are defined by the equations

$$\frac{dT_e}{dt} + \delta\nu(T_e - T) = -\frac{2}{3}e\boldsymbol{E}\boldsymbol{u}, \qquad \frac{d\boldsymbol{u}}{dt} + \nu\boldsymbol{u} = -\frac{e}{m}\left(\boldsymbol{E} + \frac{1}{c}[\boldsymbol{u} \times \boldsymbol{H}]\right). \tag{2.195}$$

These equations for $\boldsymbol{u}$ and T_e are identical with the Equations (2.8) and (2.18) of the elementary theory at constant ν_e and δ. In other words, the elementary theory conforms rigorously to the assumption that ν and δ do not depend on v.

Actually, however, ν and $\delta = R/\nu$ depend on the electron velocity v. In the elementary theory this is taken into account by introducing $\nu_e(T_e)$ and $\delta(T_e)$. Since $\nu_e(T_e)$ and $\delta(T_e)$ are not defined at all in the elementary theory, it is natural to introduce them to satisfy the condition for best agreement with the more rigorous kinetic analysis. In particular, $\nu_e(T_e)$ and $\delta(T_e)$ for a strongly ionized plasma, where the distribution function is Maxwellian [Eq. (2.134)], must be defined in accordance with Equations (2.137) and (2.138). The electron temperature is then given by the same equation as in the elementary theory. The only difference lies in the expressions obtained for the average electron directional velocity $\boldsymbol{u}$, i.e., for the conductivity and the dielectric constant of the plasma. This difference is determined by the coefficients K_σ and K_ε, which are given in Table 5. (In the elementary theory we have $K_\sigma = 1$ and $K_\varepsilon = 1$). We emphasize that in the important case of high frequency electric field $\omega \gg \nu_e$ the coefficients K_σ and K_ε are close to unity regardless of the form of the function $\nu(v)$. The elementary analysis then turns out to be exact in this case. The same pertains also to the case of a plasma in a magnetic field far from resonance, when $(\omega - \omega_H)^2 \gg \nu_e^2$ and $\omega^2 \gg \nu_e^2$.

In a weakly ionized plasma, the electron distribution function is not Maxwellian. The effective collision frequency and the fraction of the lost energy can be determined here, as before, by Equations (2.137) and (2.138), except that the Maxwellian function $f_{00}(v)$ must be replaced by the distribution function $f_0(v)$ of the weakly ionized plasma. The plot of δ against

T_e constructed in this manner for air is shown in Figure 10 [Eq. (2.138)] determines the dependence of δ on the field amplitude E_0, and the dependence of $T_e = \frac{2}{3}\bar{\varepsilon}$ on E_0 is determined from Equation (2.170) (Fig. 9). The effective collision frequency for the same case is shown in Figure 9. With this choice of $\delta(T_e)$ and $\nu_e(T_e)$, the elementary theory leads to sufficiently exact quantitative results even for a weakly ionized plasma.

We emphasize, however, that the values of the parameters $\delta(T_e)$ and $\nu_e(T_e)$ in a weakly ionized plasma depend essentially on the form of the electron distribution function $f_0(v)$. Therefore $\delta(T_e)$ and $\nu_e(T_e)$ are different, for example, in a high-frequency ($\omega^2 \gg \nu_e^2$) and in a low-frequency ($\omega^2 \ll \nu_e^2$) electric field, for stationary and non-stationary processes, etc. Note, that $\nu_e(T_e)$ is usually described with sufficient accuracy by Equation (2.140) (dashed curve in Fig. 9). On the other hand, the values of $\delta(T_e)$ as functions of the form of $f_0(v)$ can differ strongly (see Fig. 10).

We emphasize also that effects such as an S-shaped dependence of T_e on E_0, which have been considered in Section 2.1.2, arise only in a strongly

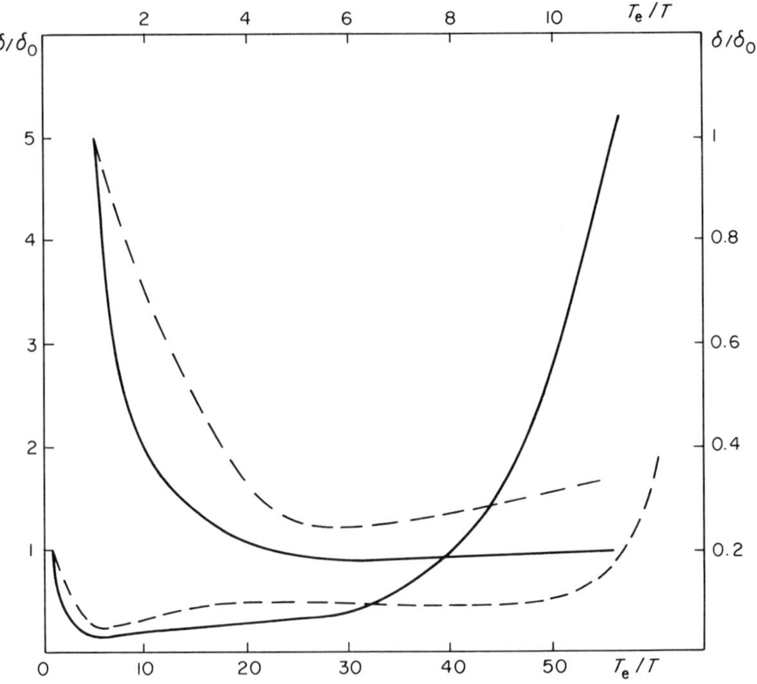

Fig. 10. Fraction of energy lost by electron in a weakly-ionized plasma in air; $T = 185°$, $\delta_0 = 5 \cdot 10^{-3}$. *Solid curve*—$\omega \gg \nu$; *dashed*—$\omega \ll \nu$. The initial section of the curves (at $T_e/T < 10$) is shown separately in enlarged scale (this scale is shown *on the top* and *on the right side*)

ionized plasma. Indeed, a multiply-valued dependence of the electron temperature on the field is possible only in a nonlinear system. The nonlinearity in the kinetic Equation (2.61) is determined by the electron–electron collisions. In a weakly ionized plasma, the dependence of f_0 on E_0^2 is linear, meaning that there is always a one–to–one correspondence between the average electron energy $\bar{\varepsilon}$ and E_0^2. The region in which the plot is S-shaped gives way in a weakly ionized plasma to a region in which $\bar{\varepsilon}$ increases very rapidly with E_0^2 (Engelhardt et al., 1964; Gurevich et al., 1975).

2.4. Ion Distribution Function

2.4.1. Simplification of the Kinetic Equation

The Boltzmann kinetic equation for the ion distribution function also takes the form of Equation (2.61), where the distribution function, charge, and mass of the electrons (f, $-e$, and m) must be replaced by the corresponding quantities (f_i, eZ, and M_i) for the ions. The simplification of the kinetic equation for the distribution function of the electrons, carried out in Section 2.2.2, was based on the fact that the velocity of the random (thermal) motion of the electrons is always much larger than their average directional velocity. A similar relation usually holds also for ions (see Sect. 2.1.3). Consequently, in the ion distribution function $f_i(\boldsymbol{v}, \boldsymbol{r}, t)$ it is also convenient to separate the symmetrical part f_{0i} from the directional part $\boldsymbol{f}_{1i}$, i.e., to represent the distribution function in the form

$$f_i(\boldsymbol{v}, \boldsymbol{r}, t) = f_{0i}(v, \boldsymbol{r}, t) + \frac{\boldsymbol{v}}{v}\boldsymbol{f}_{1i}(v, \boldsymbol{r}, t). \tag{2.196}$$

The equations for the functions f_{0i} and $\boldsymbol{f}_{1i}$ take a form which is quite analogous to Equations (2.74):

$$\frac{\partial f_{0i}}{\partial t} + \frac{v}{3}\operatorname{div}_{\boldsymbol{r}} \boldsymbol{f}_{1i} + \frac{eZ}{3M_i v^2}\frac{\partial}{\partial v}(v^2 \boldsymbol{E}\boldsymbol{f}_{1i}) + S_{0i} = 0, \tag{2.197}$$

$$\frac{\partial \boldsymbol{f}_{1i}}{\partial t} + v\,\operatorname{grad}_{\boldsymbol{r}} f_{0i} + \frac{eZ\boldsymbol{E}}{M_i}\frac{\partial f_{0i}}{\partial v} + \frac{eZ}{M_i c}[\boldsymbol{H} \times \boldsymbol{f}_{1i}] + \boldsymbol{S}_{1i} = 0. \tag{2.198}$$

The collision integrals S_{0I} and $\boldsymbol{S}_{1I}$ for the symmetrical and directional parts of the ion distribution function are determined by the general

Equation (2.67). In elastic collisions between ions and molecules and in the case of charge exchange, the collision integrals are in fact not simplified and retain the form of Equation (2.67). In ion-electron and ion–ion collisions it is possible to use the Landau collision integral [Eqs. (2.95) and (2.96)]. In this case the expressions for S_{0i} and $\boldsymbol{S}_{1i}$ take the same form as for the electron-ion and electron–electron collisions [see Eqs. (2.97), (2.98), and (2.106)], except that the masses and the charges of the interacting particles must be changed.

2.4.2. Distribution Function

The principal role in Equation (2.197) is played by the ion collision integral. It is therefore natural to seek the solution of this equation in the form of a series

$$f_{0i} = f_{00} + f_{01} + \cdots, \qquad \boldsymbol{f}_{1i} = \boldsymbol{f}_{10} + \boldsymbol{f}_{11} + \cdots. \tag{2.199}$$

Substituting this expansion in Equation (2.197) we obtain the following chain of equations

$$S_{0ii}(f_{00}) + S_{0ie}(f_{00}) + S_{0im}(f_{00}) = 0, \tag{2.200}$$

$$\frac{\partial \boldsymbol{f}_{10}}{\partial t} + v\,\mathrm{grad}_{\boldsymbol{r}}\, f_{00} + \frac{eZ\boldsymbol{E}}{M_i}\frac{\partial f_{00}}{\partial v} + \frac{eZ}{M_i c}[\boldsymbol{H} \times \boldsymbol{f}_{10}] + \boldsymbol{S}_{1i}(f_{00}, \boldsymbol{f}_{10}) = 0, \tag{2.200a}$$

$$\frac{\partial f_{00}}{\partial t} + \frac{v}{3}\,\mathrm{div}_{\boldsymbol{r}}\boldsymbol{f}_{10} + \frac{eZ}{3M_i v^2}\frac{\partial}{\partial v}(v^2 \boldsymbol{E}\boldsymbol{f}_{10}) + S_{0i}(f_{01}) = 0, \tag{2.200b}$$

Equation (2.200) takes into account the fact that the collision integral for the symmetrical part of the ion distribution function is determined by contributions due to the ion–ion collisions (S_{0ii}), electron-ion collisions (S_{0ie}), and ion-molecule collisions (S_{0im}). Each of the integrals is characterized by its own relaxation time:

$$S_{0ie} \sim \delta^{i}_{el}\nu_{ei} f_{0i}, \quad S_{0im} \sim \nu_{im} f_{0i}; \qquad S_{0ii} \sim \nu_{ii} f_{0i} \tag{2.201}$$

Here, as before, $\delta^{i}_{el} = 2m/M_i$, ν_{ei} is the electron-ion collision frequency [Eq. (2.141)], ν_{ii} is the ion–ion collision frequency, which is also given by Equation (2.141) with the electron mass, temperature, and charge replaced by the corresponding values for the ions, and ν_{im} is the collision frequency between the ions and the neutral molecules [Eq. (2.212)]. Depending on the ratio of the frequencies ν_{ii}, $\delta^i_{\text{el}}\nu_{\text{ei}}$, or ν_{im}, the decisive role is played by one of these types of collision. It is important that

$$\frac{\delta^i_{\text{el}}\nu_{\text{ei}}}{\nu_{\text{ii}}} = 2\sqrt{\frac{mT_{\text{i}}^3}{M_{\text{i}}T_{\text{e}}^3}} \ll 1, \tag{2.202}$$

i.e., we have practically always $\nu_{\text{ii}} \gg \delta^i_{\text{el}}\nu_{\text{ei}}$. Consequently, collisions with the electrons exert no significant influence on the form of the symmetrical part of the ion distribution function.[8]

What is important is, therefore, only the ratio of the frequencies ν_{ii} and ν_{im}. If $\nu_{\text{ii}} \gg \nu_{\text{im}}$, then the ion distribution function f_{00} is Maxwellian in first approximation (cf. Sect. 2.3.1). On the other hand, if $\nu_{\text{im}} \gtrsim \nu_{\text{ii}}$, then the interaction of the ions with the molecules becomes decisive. It is natural to assume for the molecules an equilibrium and Maxwellian distribution function. Then if $\nu_{\text{im}} \gtrsim \nu_{\text{ii}}$ the ion distribution function is Maxwellian with temperature $T_{\text{i}} \simeq T_{\text{m}}$. Indeed, its perturbations are determined by the interaction of the ions with the electrons, but $\delta^i_{\text{el}}\nu_{\text{ei}} \ll \nu_{\text{ii}} \lesssim \nu_{\text{im}}$, i.e., the deviations of the ion distribution function from the molecule distribution function at $\nu_{\text{im}} \gtrsim \nu_{\text{ii}}$ are small, on the order of $\delta^i_{\text{el}}\nu_{\text{ei}}/\nu_{\text{im}}$. Thus, in first-order approximation the ion distribution function turns out to be always Maxwellian:

$$f_{00} = N_{\text{i}}\left(\frac{M_{\text{i}}}{2\pi T_{\text{i}}}\right)^{3/2} \exp\left(-\frac{M_{\text{i}}v^2}{2T_{\text{i}}}\right). \tag{2.203}$$

This statement calls for a special explanation. First, it is valid if the expansion [Eq. (2.199)] is valid; the corresponding conditions will be indicated in Equations (2.219) and (2.220) below. In addition, of course, it is necessary that the lifetime τ_N of the ions be much longer than their average free path time, $\tau_N \gg 1/(\nu_{\text{ii}} + \nu_{\text{im}})$. In a weakly ionized plasma $\nu_{\text{ii}} \ll \nu_{\text{im}}$ it is also important that the molecule distribution be in equilibrium (Maxwellian). The point is that a major role can be played by inelastic collisions, accompanied by radiation, between the ions and the

[8] The foregoing, of course, does not mean that the collisions of the ions with the electrons are not important at all. They determine, for example, the heating of the ions [see Eq. (2.56)]. However, these collisions have little effect on the form of the distribution function. It must be emphasized that this pertains only to ions from the main velocity region, whose energy does not differ very strongly from the average ion energy $\bar{\varepsilon} = (\frac{3}{2})T_{\text{i}}$. For fast ions with energy $\varepsilon > T_{\text{i}}(M_{\text{i}}/m)^{1/3}$, the interaction with the electrons turns out to be decisive.

molecules. In this case the interaction of the ions with the molecules and with the radiation quanta has a complicated character and the distribution function of the molecules is Maxwellian only if equilibrium is established for all the components (molecules at the ground state level, excited molecules, radiation quanta), the interaction with which is of importance for the ions. On the other hand, if the mean free path of the quanta of any level that plays an important role in the interaction with the ions (say a resonant level) is larger than the system dimensions, then the equilibrium is upset and the distribution function of the excited molecules, and consequently also of the ions, can greatly differ from Maxwellian.

Finally, the foregoing arguments are valid only for ions in the main velocity region $v \sim v_{Ti} = (2T_i/M_i)^{1/2}$. In the region of the "tail" (at $v \gg v_{Ti}$) the ion distribution can differ appreciably from Equation (2.203), even if all the conditions indicated above are satisfied. Indeed, consider by way of example a fully ionized plasma. Equation (2.200) then takes the form

$$-\frac{1}{2v^2}\frac{\partial}{\partial v}\left\{v^2\left[\left(\delta_{el}^{i}\nu_{ei}\frac{T_e}{M_i}+A_{2i}(f_{00})\right)\frac{\partial f_{00}}{\partial v}+v(\delta_{el}^{i}\nu_{ei}+A_{1i}(f_{00})\right)f_{00}\right]\right\}=0. \tag{2.204}$$

Use is made here of the concrete expressions for the collision integrals of the electrons with the ions and of the ions with the ions [Eq. (2.98)]. The integration in these expressions was carried out with respect to angle, just as in the derivation of Equations (2.103) and (2.105). The terms proportional $\delta_{el}^{i}\nu_{ei}$ describe the collisions of the ions with the electrons; here $\delta_{el}^{i} = 2m/M_i$ and ν_{ei} is the collision frequency defined by Equation (2.141). The terms $A_{1i}(f_{00})$ and $A_{2i}(f_{00})$ describe the collisions of the ions with the ions. They are given by Equations (2.103) and (2.105), in which it is only necessary to replace the electron–electron collision frequency $\nu_e(v)$ by the ion–ion collision frequency $\nu_i(v)$:

$$\begin{aligned} A_{1i} &= \frac{8\pi\nu_i(v)}{N_i}\int_0^v v_1^2 f_{00}(v_1)\,dv_1, \\ A_{2i} &= \frac{8\pi\nu_i(v)}{3N_i}\left\{\int_0^v v_1^2 f_{00}(v_1)\,dv_1 + v^3\int_v^\infty v_1 f_{00}(v_1)\,dv_1\right\}. \end{aligned} \tag{2.205}$$

$$\nu_i(v) = 4\pi e^4 N_i Z^4 \ln \Lambda / M_i^2 v^3.$$

The solution of Equation (2.204), just as in Section 2.3.3, is obtained by iteration. Assuming the Maxwellian ion distribution function [Eq. (2.203)] in the zeroth approximation, we obtain in the next approximation[9]:

$$f_{00}^{(1)} = C \exp\left\{-\int_0^v \frac{v[\delta_{el}^i \nu_{ei} + A_{10}(v)]\,dv}{\delta_{el}^i \nu_{ei} T_e/M_i + A_{20}(v)}\right\}, \qquad C \simeq \left(\frac{M_i}{2\pi T_i}\right)^{3/2} N_i \tag{2.206}$$

$$A_{20} = \frac{T_i}{M_i} A_{10} = \frac{T_i}{M_i} \nu_i(v)\left[\Phi(z) - \frac{2z}{\sqrt{\pi}} \exp(-z^2)\right], \qquad z = (M_i v^2/2T_i)^{1/2} \tag{2.207}$$

$\Phi(z)$ is the probability integral. In the main velocity region (at $z \sim 1$) we have $A_{10} \gg \delta_{ei}^i \nu_{ei}$, $A_{20} \gg \delta_{el}^i \nu_{ei} T_e/M_i$, and the distribution function [Eq. (2.207)] is Maxwellian with temperature T_i, i.e., it coincides with Equation (2.203). With increasing velocity v, however, the coefficients A_{10} and A_{20} decrease in proportion to v^{-3} and at $v > (M_i/m)^{1/6} v_{Ti}$ they became smaller than $\delta_{el}^i \nu_{ei}$. This changes the form of the function $f_{00}(v)$ at large velocities. In particular, if $T_e \gg T_i$, then the distribution function [Eq. (2.206)] decreases at large values of v much more slowly than Equation (2.203). At $v \gg (M_i/m)^{1/6} v_{Ti}$ we have

$$f_{00}^{(1)} \simeq C \exp(-M_i v^2/2T_e). \tag{2.208}$$

2.4.3. Ion Temperature, Ion Current

Equation (2.200) determines only the form of the distribution function f_{00}. The ion temperature T_i and concentration N_i are determined by the equations obtained from the conditions under which Equation (2.200b) has a solution for the next higher approximation f_{01} (cf. Sect. 2.3.1). Indeed, we multiply Equation (2.200b) in succession by v^2 and $M_i v^4/2$ and integrate it with respect to dv. Then, taking into account the conservation of the number of particles and of the energy in the collisions between the ions, we arrive at the following equations for the ion concentration and temperature in a spatially homogeneous plasma:

$$\frac{dN_i}{dt} = q_i(N_i) - q_r(N_i), \tag{2.209}$$

where q_i and q_r are the numbers of ionization and recombination events

[9] The succeeding iterations lead to functions that practically coincide with Equation (2.206).

per unit volume per second,

$$\frac{dT_i}{dt} = \frac{2eZ}{3N_i} \boldsymbol{E j}_i + \delta^i_{el} \nu_{ei}(T_e - T_i) - \delta_{im} \nu_{im}(T_i - T) - \frac{2}{3}(q_i - q_r)\frac{T_i}{N_i}. \tag{2.210}$$

The first term in the right-hand side of this equation describes the ohmic heating of the ions. Here $\boldsymbol{j}_i$ is the ion current

$$\boldsymbol{j}_i = \frac{4\pi eZ}{3} \int_0^\infty v^3 \boldsymbol{f}_{10}\, dv. \tag{2.211}$$

In the elementary-theory approximation, the current $\boldsymbol{j}_i$ is determined by Equations (2.48) and (2.49).

To calculate the ion current in the kinetic theory, as is clear from Equation (2.211), it is necessary to find the function $\boldsymbol{f}_{10}$, i.e., to solve Equation (2.200a). The integral $\boldsymbol{S}_{1\,im}(f_{00}, \boldsymbol{f}_{10})$ of the collisions with the molecules is in this equation a linear integral operator for a function $\boldsymbol{f}_{10}$ of the type Equation (2.111). Thus, Equation (2.200a) is a linear integro-differential equation.[10] The solution of this equation for a weakly ionized plasma is obtained by the Enskog-Chapman method (expansion in terms of Laguerre or Sonine polynomials) or by the Grad method (expansion in the moments of the distribution function). The corresponding solution is described in detail in a number of books (Chapman and Cowling, 1952; Shkarofsky et al., 1966; Hochstim and Massel, 1969; Silin, 1972). It is important that the expression for $\boldsymbol{j}_i$ can be represented, just as for the electrons, in Equation (2.150), by replacing m by M_i, e by eZ, and ν_{em} by ν_{im}, and by introducing the corresponding kinetic correction coefficients $K_{\sigma i}(\omega/\nu_{im})$ and $K_{\varepsilon i}(\omega/\nu_{im})$. In the hard-sphere collision model (the ions are spheres of radius a_1 and mass M_1, while the molecules are spheres of radius a_2 and mass M_2) we have in this case

$$\nu_{im} = \frac{8\sqrt{2\pi}}{3}(a_1 + a_2)^2 \frac{N_m M_2}{M_1 + M_2}\left(\frac{T_1}{M_1} + \frac{T_2}{M_2}\right)^{1/2}. \tag{2.212}$$

If $T_1 = T_2 = T$, $a_1 = a_2 = a$, and $M_1 = M_2 - M$, then

$$\nu_{im} = \frac{32\sqrt{\pi}}{3} a^2 N_m (T/M)^{1/2}. \tag{2.213}$$

[10] At high degrees of ionization $\nu_{ei} \gg \nu_{em}$ the mutual influence of the electrons and ions is already significant, so that it is necessary to solve simultaneously the equations for $\boldsymbol{f}_{1i}$ and $\boldsymbol{f}_{1e}$ [the equation for $\boldsymbol{f}_{1i}$ contains the integral $\boldsymbol{S}_1(f^i_{00}, \boldsymbol{f}^e_{10})$, while the equation for $\boldsymbol{f}_{1e}$ contains $\boldsymbol{S}_1(f^e_{00}, \boldsymbol{f}^i_{10})$].

The effective collision frequencies for the other function $\nu_{im}(v)$ are given by Chapman and Cowling (1952).

In the case of collisions of ions (mass M_1, charge eZ_1) with other ions (mass M_2, charge eZ_2) we have

$$\nu_{ii} = \frac{4\sqrt{2\pi}}{3} \frac{e^4 Z_1^2 Z_2^2 N_{i2}}{T^{3/2}} \left[\frac{M_2}{M_1(M_1 + M_2)}\right]^{1/2} \ln \Lambda, \tag{2.214}$$

where Λ is the Coulomb logarithm [Eq. (2.94)].

The kinetic correction coefficients $K_{\sigma i}$ and $K_{\varepsilon i}$ are much closer to unity in the case of the ion current than the corresponding coefficients for the electron current. Thus, for example, in the case of collisions between ions (mass M_1) and molecules (mass M_2) the maximum value of $K_{\sigma i} = K_{\sigma i}(0)$ is

$$K_{\sigma i}(0) = 1 + \frac{M_2^2}{13M_2^2 + 30M_1^2 + 16M_1M_2}, \tag{2.215}$$

i.e., at $M_1 = M_2$ we have $K_{\sigma i} = 1.02$. For collisions of ions (mass M_1) with ions (mass M_2) we have

$$K_{\sigma i}(0) = 1 + \frac{9M_2^2}{13M_2^2 + 30M_1^2 + 16M_1M_2}, \tag{2.216}$$

i.e., at $M_1 = M_2$ we have $K_{\sigma i} = 1.15$. Thus, the elementary-theory formulas can be used with rather sufficient accuracy to calculate the ion current.

It is important to emphasize that in a high-frequency alternating electric field, as indicated in Section 2.1.3, the ion current is much smaller than the electron current. Therefore the influence of the ion current on the heating of the ions can usually be neglected. An exception is a constant (or quasi-stationary $\omega^2 \lesssim \Omega_H \omega_H \nu_{em}/\nu_{im}$) electric field E perpendicular to the magnetic field. Such a field produces an ion current that is comparable in magnitude with the electron current, and in this case the term $\boldsymbol{E} \cdot \boldsymbol{j}_i$ [Eq. (2.210)], which determines the heating of the ions in the plasma, cannot be neglected.

Next, the term $\delta'_{el} \nu_{ei}(T_e - T_i)$ in Equation (2.210) determines the heating of the ions as a result of collisions with the electrons. The term $\delta_{im} \nu_{im}(T_i - T)$ determines the energy transfer as the ions collide with the molecules. Here ν_{im} is the effective collision frequency [Eq. (2.212)], and δ_{im} is the fraction of the energy lost in one collision. In elastic collisions we have

$$\delta_{im} = 2M_i/(M_i + M). \tag{2.217}$$

At $M_i = M$ we have

$$\delta_{im} = 1. \tag{2.218}$$

Finally, the last term of Eq. (2.210) is the average energy lost by one ion per unit time as a result of ionization–recombination processes.

Substituting the Maxwellian function f_{00} [Eq. (2.203)] in Equation (2.200b), we can determine the function f_{01}. Estimates show that $f_{01} \ll f_{00}$ if the following conditions are satisfied:

$$\frac{\partial f_{00}}{\partial t} \ll \nu_i f_{00}, \qquad \left(\frac{v}{\nu_i}\right)^2 \frac{\partial^2 f_{00}}{\partial x^2} \ll f_{00}, \tag{2.219}$$

$$\frac{e^2 Z^2 E^2}{3M_i^2 v^2(\nu_i^2 + \omega'^2)} \ll 1, \qquad \omega' = |\omega - \Omega_H| \quad \text{or} \quad \omega' = \omega. \tag{2.220}$$

The conditions of Equation (2.219) indicate the degree of nonstationarity and inhomogeneity of the plasma, at which the expansion [Eq. (2.199)] is valid. The condition of Equation (2.220) limits the amplitude of the electric field, and is not satisfied in a constant or quasi-stationary electric field ($\omega \ll \nu_i$ or $|\omega - \Omega_H| \ll \nu_i$), the value of which is larger than critical E_D in Equation (2.41). In this case the directional velocity of the ions in a fully ionized plasma increases, i.e., ion runaway takes place. The electron runaway occurs under the same conditions (see Sect. 2.1.2).

2.5. Action of Radio Waves on the Ionosphere

In this section we discuss the perturbation of the temperature and of the electron concentration in the ionosphere by the action of a homogeneous alternating field. Transport processes are disregarded here. This approach is valid in the lower layers of the ionosphere, where the transport processes are of little significance. In the upper layers, to the contrary, they usually play an important role, and furthermore various instabilities readily evolve here. These processes will be investigated later on (Chaps. 5 and 6).

2.5.1. Ionization Balance in the Ionosphere

The processes of ionization and recombination in the ionosphere are quite varied, owing to the variety of the chemical compositions of the ionosphere and of the ionization agents. These questions have recently been the subject of the many studies, the results of which are summarized in a number of monographs (Whitten and Poppoff, 1965; Ivanov-Kholodnyi and Nikol'skii, 1969; Bauer, 1973). We confine ourselves here to a brief qualitative description of the main processes and present only the simplest ionization-balance equations.

Ionization and Recombination. The daytime ionosphere is formed mainly under the influence of the UV and X radiation of the sun ($\lambda \lesssim$ 130 nm). The emission spectrum of the sun is continuous in this region, but does have a number of strongly pronounced lines: the hydrogen Lyman series (particularly L_α and L_β), and lines of certain multiply-charged ions (He II, C III, O VI, Si III, Si II, O II, O III, Mg X, He I). The UV spectrum is quite stable. It determines the ionization in the E and F layers of the ionosphere. Figure 11 shows the intensity of the ionization for the sun at a zenith angle 30° (Banks, 1969). Here q_i is the number of the O^+, N_2^+ and O_2^+ ions produced in 1 cm^3 per second. The maximum rate of ion production occurs at a height of 100–200 km.

The X radiation varies strongly with the activity of the sun. It makes an appreciable contribution to the ionization of the lower part of the E layer and of the D layer ($\lambda \sim$ 1–10 nm in the E layer and $\lambda \sim$ 0.1–1 nm in the D layer). The main contribution to the ionization of the lower part of the D layer (below 60 km) is made by corpuscular streams. At polar latitudes, the contribution of the corpuscular streams to the ionization of the ionosphere is more appreciable. The night-time ionization in the E layer is maintained by the UV radiation scattered in the geocorona, by the influx of plasma from some higher layers of the ionosphere and also by the meteoritic and corpuscular ionization.

The UV and X radiation of the sun produces in the E and F layers mainly the principal ions O^+, O_2^+, and N_2^+. A number of ion-atomic reactions then transform the principal ions into the usually observed ions O^+, $NO^+ \cdot N^+$ etc. (see Table 2).

The main recombination process in the ionosphere is dissociative recombination (Bates and Dalgarno, 1962), wherein a molecular ion is

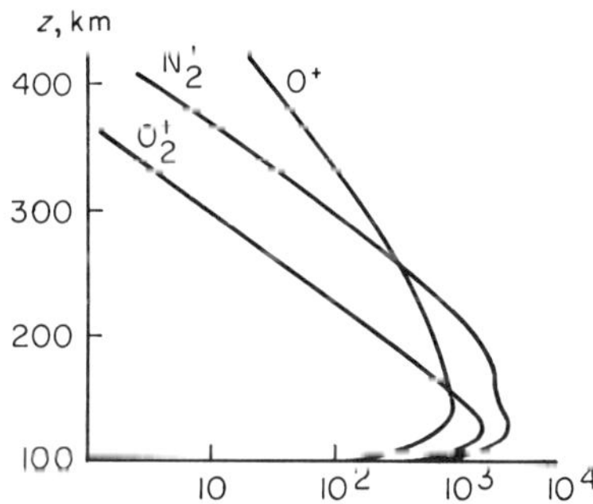

Fig. 11. Intensity of photoionization in the ionosphere

combined with an electron and then dissociates into atoms. The most significant dissociative reactions in the ionosphere are

$$O_2^+ + e \to O + O, \qquad NO^+ + e \to N + O. \tag{2.221}$$

The number of electrons recombining per cm^3 per second can be represented in the form αN^2. The ionization-balance equation for the electrons then takes the form:

$$\frac{\partial N}{\partial t} = q_i - \alpha N^2 = q_i - q_r. \tag{2.222}$$

Here $q_i = q_{iO} + q_{iO_2} + q_{iN_2}$ is the total intensity of the ionization (q_{iO}, q_{iO_2}, and q_{iN_2} are the ionization intensities of the atoms O, O_2 and N_2), and α is the effective coefficient of dissociative recombination. Since the recombination is determined by the processes in Equation (2.221), it follows that

$$\alpha = \alpha_1 n_{NO^+} + \alpha_2 n_{O_2^+}, \tag{2.223}$$

where α_1 and α_2 are the coefficients of dissociative recombination of the ions NO^+ and O_2^+; $n_{NO^+} = N_{NO^+}/N$ and $n_{O_2^+} = N_{O_2^+}/N$ are the relative concentrations of the ions NO^+ and O_2^+. The latter are determined by the ion-balance equations (Nicolet and Swidder, 1963). Taking into account the reactions between the principal ions and the molecules N_2 and O_2,

$$N_2 + O^+ \to NO^+ + N; \qquad O_2 + O^+ \to O_2^+ + O \tag{2.224}$$

we write down the ion-balance equation in the form

$$\frac{\partial N_{O_2^+}}{\partial t} = q'_{iO_2} + \beta_2 N_{O_2} N_{O^+} - \alpha_2 N_{O_2^+} N, \tag{2.225}$$

$$\frac{\partial N_{O^+}}{\partial t} = q'_{iO} - \beta_1 N_{O^+} N_{N_2} - \beta_2 N_{O^+} N_{O_2}, \tag{2.226}$$

$$\frac{\partial N_{NO^+}}{\partial t} = \beta_1 N_{O^+} N_{N_2} - \alpha_1 N_{NO^+} N + q'_{iNO}, \tag{2.227}$$

where

$$N = N_{O^+} + N_{O_2^+} + N_{NO^+},$$

$$q'_{iO} = q_{iO} + p_1 N_O, \qquad q'_{iO_2} = q_{iO_2} + p_2 N_{O_2}, \qquad q'_{iNO} = p_3 N_O, \tag{2.227a}$$

$$p_k = \gamma_k q_{iN_2}/[(\gamma_1 + \gamma_3) N_O + \gamma_2 N_{O_2}].$$

We have taken into account here the reactions of the ions N_2^+ with the molecules O_2 and O:

$$N_2^+ + O \rightarrow N_2 + O^+; \qquad N_2^+ + O_2 \rightarrow N_2 + O_2^+; \qquad N_2^+ + O \rightarrow N + NO^+, \tag{2.228}$$

where are characterized by the coefficients γ_1, γ_2, and γ_3:

$$\frac{\partial N_{N_2^+}}{\partial t} = q_{iN_2} - \gamma_1 N_O N_{N_2^+} - \gamma_2 N_{O_2} N_{N_2^+} - \gamma_3 N_O N_{N_2^+}.$$

These reactions are rapid, since the coefficients γ_1, γ_2, and γ_3 are large; the coefficient γ_2 thus exceeds β_1 and β_2 by two orders of magnitude. Consequently, the N_2^+ ion concentration is always close to stationary:

$$N_{N_2^+} = q_{iN_2}/[(\gamma_1 + \gamma_3)N_O + \gamma_2 N_{O_2}].$$

The N_2^+ concentration is low (because the coefficients γ are large), i.e., the reactions in Equation (2.228) actually lead to a vanishing of the N_2^+ ions in the ionosphere (see Table 2). On the other hand the same reactions in Equation (2.228) lead to an appearance of the additional ions O^+, O_2^+, and NO^+. It is this which is taken into account by the re-normalized ion-production coefficients q'_{iO}, q'_{iO_2} and q'_{iNO} in Equations (2.225)–(2.227). In particular, in the E layer the principal role is played by the second process in Equation (2.228), so that $q'_{iO_2} \simeq q_{iO_2} + q_{iN_2}$. Of course, in the general case

$$q'_{iO} + q'_{iO_2} + q'_{iNO} = q_{iO} + q_{iO_2} + q_{iN_2},$$

i.e., the total intensity of the ionization remains unchanged.

Next, β_1 and β_2 in Equations (2.225)–(2.227) are the coefficients of the ionic reactions [Eq. (2.224)]. For example, $\beta_1 N_{O^+} N_{N_2}$ determines the number of NO^+ ions produced in 1 cm^3 per second as the result of the first reaction of Equation (2.224). In the F layer, according to Yonezawa (1966), $\beta_1 = 3 \cdot 10^{-12}$ cm^3 s^{-1} and $\beta_2 = 2 \cdot 10^{-12}$ cm^3 s^{-1}. In the E layer the values of the coefficients β are much lower, of the order of 10^{-13} cm^3 s^{-1} (Whitten and Poppoff, 1965; Yonezawa, 1966).

According to Yonezawa (1966), the coefficients α are $\alpha_1 = 10^{-7}$ cm^3/s and $\alpha_2 = 2 \cdot 10^{-7}$ cm^3/s in the E layer and $\alpha_1 = (3\text{–}5) \cdot 10^{-8}$, $\alpha_2 = (2\text{–}5) \cdot 10^{-8}$ in the F layer.

The dependence of the coefficients α on the temperature was investigated both theoretically and experimentally. The decrease of α with increasing T_e at $T_e \sim 10^3$–10^4 K is determined by the expression $\alpha \sim T_e^{-\lambda}$, where λ

ranges approximately from $\frac{1}{2}$ to $\frac{3}{2}$. According to the DASA Rate Handbook (1970).

$$\alpha_1 \approx 5 \cdot 10^{-7}\,(300^\circ/T_e)^{1,2}; \qquad \alpha_2 \approx 2.2 \cdot 10^{-7}\,(300^\circ/T_e)^{0.7} \tag{2.229}$$

We note that different authors give noticeably different values of the coefficients α and β. Thus, according to Bauer (1973), at $T = 300$ K we have

$$\begin{gathered} \beta_1 = 1 \cdot 10^{-12}\ \mathrm{cm}^3/\mathrm{s}, \qquad \beta_2 = 2 \cdot 10^{-11}\ \mathrm{cm}^3/\mathrm{s}, \\ \text{and} \qquad \alpha_1 \sim T_e^{-1/3} \end{gathered} \tag{2.229a}$$

We have considered here processes with participation of the principal ions at heights $z \gtrsim 80$ km. Processes with production of other ions such as H^+ and He^+ were investigated by Nicolet and Swidder (1963). In the D layer ($z < 80$ km), an important role is played by recombination processes with production of heavy hydrated water cluster ions (Bauer, 1973).

Ionization Balance. Equations (2.222) and (2.225)–(2.227) determine the ionization balance in the E and F layers of the ionosphere. Under stationary conditions, we determine from Equations (2.225)–(2.227) the ion concentrations:

$$\begin{gathered} N_{O^+} = \frac{q'_{iO}}{\beta_1 N_{N_2} + \beta_2 N_{O_2}}, \qquad N_{NO^+} = \frac{q'_{iO}\beta_1 N_{N_2}}{\alpha_1 N(\beta_1 N_{N_2} + \beta_i N_{O_2})} + \frac{q'_{iNO}}{\alpha_1 N}, \\ N_{O_2^+} = \frac{q'_{iO}\beta_2 N_{O_2}}{\alpha_2 N(\beta_1 N_{N_2} + \beta_2 N_{O_2})} + \frac{q'_{iO_2}}{\alpha_2 N}, \end{gathered} \tag{2.230}$$

and from Equation (2.227a) we determine the electron concentration

$$N = \frac{1}{2}\left\{N_{O^+} + \sqrt{N_{O^+}^2 + 4\left[\frac{q'_{iNO} + \beta_1 N_{N_2} N_{O^+}}{\alpha_1} + \frac{q'_{iO_2} + \beta_2 N_{O_2} N_{O^+}}{\alpha_2}\right]}\right\} \tag{2.231}$$

where the concentration N_{O^+} is determined by Equation (2.230). As seen from Equation (2.230), the quantity in the square brackets under the square root is the product $N(N_{O_2^+} + N_{NO^+})$. In the F layer, this term is much smaller than $N_{O^+}^2$, so that in this case

$$N \approx N_{O^+} = \frac{q'_{iO}}{\beta_1 N_{N_2} + \beta_2 N_{O_2}}.$$

The ionization balance Equation (2.222) then takes the form

$$\frac{\partial N}{\partial t} = q'_{iO} - \beta N, \qquad \beta = \beta_1 N_{N_2} + \beta_2 N_{O_2}. \tag{2.232}$$

In the E layer, on the contrary, $N_{O^+} \ll N_{O_2^+} + N_{NO^+}$ (see Table 2), and here we have

$$N = \left[\frac{q'_{iO_2}}{\alpha_2} + \frac{q'_{iNO}}{\alpha_1} + \frac{q'_{iO}(\beta_1 N_{N_2}/\alpha_1 + \beta_2 N_{O_2}/\alpha_2)}{\beta_1 N_{N_2} + \beta_2 N_{O_2}}\right]^{1/2}. \tag{2.233}$$

The dissociative-recombination coefficients α decrease with increasing electron temperature in accordance with Equation (2.229). It is therefore seen from Equation (2.233) that the electron concentration in the E layer increases with increasing electron temperature.

The derived expressions describe, naturally, only the local ionization balance. They are valid if the ionization transport via diffusion, thermal diffusion, or drift can be neglected. In the F layer the transport processes turn out to be significant. It is they which determine the region of the maximum of the concentration in this layer: transport plays the decisive role above the maximum, and the ionization balance, below.

2.5.2. Effective Frequency of Electron and Ion Collisions—Fraction of Lost Energy

Effective Frequency of Electron Collisions. The effective frequency ν_{em} of the collisions of the electrons with the neutral molecules is determined by Equations (2.137) and (2.77). Using in Equation (2.137) the cross sections $\sigma_t(v)$ shown in Figures 6 and 7, we arrive at the following approximate expressions for ν_{em}:

$$\nu_{eN_2} = 2.5 \cdot 10^{-11} N_{N_2} T_e (1 + 0.9 \cdot 10^{-2} T_e^{1/2})^{-1}, \tag{2.234}$$

$$\nu_{eO_2} = 1.6 \cdot 10^{-10} N_{O_2} T_e^{1/2} (1 + 4 \cdot 10^{-2} T_e^{1/2}), \tag{2.235}$$

$$\nu_{em} = \begin{cases} 5.8 \cdot 10^{-11} N_m T_e^{5/6}, & \text{if} \quad T_e \lesssim 10^4, \\ 1.0 \cdot 10^{-9} N_m T_e^{1/2}, & \text{if} \quad T_e \gtrsim 10^4. \end{cases}$$

$$\nu_{eO} = 2.8 \cdot 10^{-10} N_O T_e^{1/2}, \qquad \nu_{eH} = 4.5 \cdot 10^{-10} N_H T_e^{1/2}, \tag{2.235a}$$

$$\nu_{eHe} = 4.6 \cdot 10^{-10} N_{He} T_e^{1/2}.$$

Table 6. Effective electron collision frequencies in the ionosphere

z, km	Day-time; effective collision frequencies ν, s^{-1}							l_{ei}, cm
	ν_{eN_2}	ν_{eO_2}	ν_{eHe}	ν_{eO}	ν_{eH}	ν_{em}	ν_{ei}	
60	$3.1 \cdot 10^7$	$6.1 \cdot 10^6$	—	—	—	$3.7 \cdot 10^7$	0.95	$1.1 \cdot 10^7$
70	$7.0 \cdot 10^6$	$1.3 \cdot 10^6$	—	—	—	$8.3 \cdot 10^6$	3.4	$2.6 \cdot 10^6$
80	$9.2 \cdot 10^5$	$2.0 \cdot 10^5$	—	—	—	$1.1 \cdot 10^6$	19	$4.4 \cdot 10^5$
90	$1.3 \cdot 10^5$	$2.7 \cdot 10^4$	—	—	—	$1.6 \cdot 10^5$	$0.12 \cdot 10^3$	$7.3 \cdot 10^4$
100	$4.0 \cdot 10^4$	$7.2 \cdot 10^3$	0.41	$1.0 \cdot 10^3$	$4.4 \cdot 10^{-4}$	$4.8 \cdot 10^4$	$0.84 \cdot 10^3$	$1.1 \cdot 10^4$
110	$9.6 \cdot 10^3$	$1.6 \cdot 10^3$	0.31	$7.0 \cdot 10^2$	$4.3 \cdot 10^{-4}$	$1.2 \cdot 10^4$	$0.85 \cdot 10^3$	$1.3 \cdot 10^4$
120	$4.9 \cdot 10^3$	$6.5 \cdot 10^2$	0.25	$4.2 \cdot 10^2$	$4.1 \cdot 10^{-4}$	$6.2 \cdot 10^3$	$0.58 \cdot 10^3$	$2.4 \cdot 10^4$
130	$2.2 \cdot 10^3$	$2.4 \cdot 10^2$	0.19	$2.5 \cdot 10^2$	$3.6 \cdot 10^{-4}$	$2.7 \cdot 10^3$	$0.44 \cdot 10^3$	$3.5 \cdot 10^4$
150	$7.1 \cdot 10^2$	84	0.14	$1.1 \cdot 10^2$	$2.9 \cdot 10^{-4}$	$9.1 \cdot 10^2$	$0.48 \cdot 10^3$	$3.9 \cdot 10^4$
200	$1.1 \cdot 10^2$	9.6	$8.5 \cdot 10^{-2}$	31	$2.2 \cdot 10^{-4}$	$1.5 \cdot 10^2$	$0.44 \cdot 10^3$	$5.3 \cdot 10^4$
250	30	2.3	$6.5 \cdot 10^{-2}$	15	$1.9 \cdot 10^{-4}$	47	$0.65 \cdot 10^3$	$4.0 \cdot 10^4$
300	10	0.59	$5.6 \cdot 10^{-2}$	7.3	$1.9 \cdot 10^{-4}$	18	$0.81 \cdot 10^3$	$3.4 \cdot 10^4$
400	1.3	$5.5 \cdot 10^{-2}$	$4.3 \cdot 10^{-2}$	2.2	$1.8 \cdot 10^{-4}$	3.5	$0.59 \cdot 10^3$	$5.2 \cdot 10^4$
500	0.17	$5.5 \cdot 10^{-3}$	$3.2 \cdot 10^{-2}$	0.69	$1.7 \cdot 10^{-4}$	0.90	$0.30 \cdot 10^3$	$1.8 \cdot 10^5$
600	$0.24 \cdot 10^{-1}$	$5.6 \cdot 10^{-4}$	$2.4 \cdot 10^{-2}$	0.23	$1.6 \cdot 10^{-4}$	0.27	$0.14 \cdot 10^3$	$2.3 \cdot 10^5$
700	$0.35 \cdot 10^{-2}$	$6.4 \cdot 10^{-5}$	$1.9 \cdot 10^{-2}$	$7.5 \cdot 10^{-2}$	$1.5 \cdot 10^{-4}$	0.098	69	$4.8 \cdot 10^5$
800	$0.55 \cdot 10^{-3}$	$7.8 \cdot 10^{-6}$	$1.5 \cdot 10^{-2}$	$2.7 \cdot 10^{-2}$	$1.5 \cdot 10^{-4}$	$4.2 \cdot 10^{-2}$	34	$9.8 \cdot 10^5$
900	$0.9 \cdot 10^{-4}$	$9.8 \cdot 10^{-7}$	$1.2 \cdot 10^{-2}$	$1.2 \cdot 10^{-2}$	$1.4 \cdot 10^{-4}$	$2.4 \cdot 10^{-2}$	24	$1.4 \cdot 10^6$
1000	$0.15 \cdot 10^{-4}$	$1.3 \cdot 10^{-7}$	$9.4 \cdot 10^{-3}$	$3.3 \cdot 10^{-3}$	$1.3 \cdot 10^{-4}$	$1.3 \cdot 10^{-2}$	17	$2.0 \cdot 10^6$

z, km	Night-time; effective collision frequencies ν, s^{-1}							l_{ei}, cm
	ν_{eN_2}	ν_{eO_2}	ν_{eHe}	ν_{eO}	ν_{eH}	ν_{em}	ν_{ei}	
60	—	—	—	—	—	—	—	—
70	—	—	—	—	—	—	—	—
80	$9.2 \cdot 10^5$	$2.0 \cdot 10^5$	—	—	—	$1.1 \cdot 10^6$	0.23	$3.6 \cdot 10^7$
90	$1.3 \cdot 10^5$	$2.7 \cdot 10^4$	—	—	—	$1.6 \cdot 10^5$	1.1	$7.4 \cdot 10^6$
100	$3.5 \cdot 10^4$	$6.5 \cdot 10^3$	0.39	$8.3 \cdot 10^2$	$4.2 \cdot 10^{-4}$	$4.2 \cdot 10^4$	18	$5.2 \cdot 10^5$
110	$8.0 \cdot 10^3$	$1.5 \cdot 10^3$	0.30	$6.8 \cdot 10^2$	$4.2 \cdot 10^{-4}$	$1.0 \cdot 10^4$	16	$6.6 \cdot 10^5$
120	$4.4 \cdot 10^3$	$5.5 \cdot 10^2$	0.23	$4.2 \cdot 10^2$	$3.9 \cdot 10^{-4}$	$5.4 \cdot 10^3$	13	$9.9 \cdot 10^5$
130	$2.0 \cdot 10^3$	$2.3 \cdot 10^2$	0.18	$2.3 \cdot 10^2$	$3.3 \cdot 10^{-4}$	$2.5 \cdot 10^3$	9.8	$1.4 \cdot 10^6$
150	$6.5 \cdot 10^2$	56	0.12	$1.0 \cdot 10^2$	$2.6 \cdot 10^{-4}$	$8.0 \cdot 10^2$	6.7	$2.0 \cdot 10^6$
200	72	6.5	$8.2 \cdot 10^{-2}$	27	$2.2 \cdot 10^{-4}$	$1.0 \cdot 10^2$	5.8	$3.2 \cdot 10^6$
250	13	0.90	$6.3 \cdot 10^{-2}$	10	$2.0 \cdot 10^{-4}$	25	16	$1.2 \cdot 10^6$
300	2.9	0.15	$5.4 \cdot 10^{-2}$	4.3	$2.0 \cdot 10^{-4}$	7.4	115	$1.9 \cdot 10^5$
400	0.14	$4.6 \cdot 10^{-3}$	$3.4 \cdot 10^{-2}$	0.73	$1.9 \cdot 10^{-4}$	0.90	270	$8.8 \cdot 10^4$
500	$6.9 \cdot 10^{-3}$	$1.4 \cdot 10^{-4}$	$2.3 \cdot 10^{-2}$	0.13	$1.7 \cdot 10^{-4}$	0.16	160	$1.4 \cdot 10^5$
600	$3.9 \cdot 10^{-4}$	$5.0 \cdot 10^{-6}$	$1.6 \cdot 10^{-2}$	$2.5 \cdot 10^{-2}$	$1.6 \cdot 10^{-4}$	$4.1 \cdot 10^{-2}$	100	$2.5 \cdot 10^5$
700	$2.2 \cdot 10^{-5}$	$2.1 \cdot 10^{-7}$	$1.1 \cdot 10^{-2}$	$5.0 \cdot 10^{-3}$	$1.5 \cdot 10^{-4}$	$6.3 \cdot 10^{-3}$	56	$4.6 \cdot 10^5$
800	$1.4 \cdot 10^{-6}$	$9.2 \cdot 10^{-9}$	$7.6 \cdot 10^{-3}$	$1.0 \cdot 10^{-3}$	$1.4 \cdot 10^{-4}$	$1.9 \cdot 10^{-3}$	34	$7.8 \cdot 10^5$
900	$1.0 \cdot 10^{-7}$	$4.4 \cdot 10^{-10}$	$5.3 \cdot 10^{-3}$	$2.3 \cdot 10^{-4}$	$1.3 \cdot 10^{-4}$	$8.9 \cdot 10^{-4}$	19	$1.4 \cdot 10^6$
1000	$7.4 \cdot 10^{-9}$	$2.2 \cdot 10^{-11}$	$3.7 \cdot 10^{-3}$	$5.2 \cdot 10^{-5}$	$1.2 \cdot 10^{-4}$	$5.5 \cdot 10^{-4}$	12	$2.3 \cdot 10^6$

Here T_e is in K, N_m in cm^{-3}, ν_e in s^{-1}, and ν_{em} is the frequency of the collisions with air molecules ($N_m = N_{N_2} + N_{O_2}$). The electron collision cross sections in air at $T_e \gtrsim 10^4$ K in Equation (2.235) and with the oxygen hydrogen, and helium atoms in Equation (2.235a) were assumed to be constant, equal respectively to $\sigma_{tm} = 1.1 \cdot 10^{-15}$ cm^2, $\sigma_{tO} = 3.3 \cdot 10^{-16}$ cm^2, $\sigma_{tH} = 5.4 \cdot 10^{-16}$ cm^2 and $\sigma_{tHe} = 5.5 \cdot 10^{-16}$ cm^2, while the collision frequency was determined in the same manner as for collisions of elastic spheres (Eq. (2.139) with $\pi a^2 = \sigma_t$.

The total frequency of electron collisions in a multicomponent plasma is equal to the sum of the effective frequencies of the collisions with the different components:

$$\nu_e = \sum_k \nu_{ek}$$

The contribution of the collisions of the electrons with different components of the ionosphere and the total effective frequency of the collisions for different heights are given in Table 6. The models of the night-time and day-time ionosphere represented in Tables 1 and 2 were used in the calculations. We see that at heights up to 150 km in the day-time, and up to 250 km at night, the principal role is played by collisions with molecules. To the contrary, collisions with ions always predominate at heights $z \gtrsim 300$ km. In Table 6 are given also the mean free paths l_{ei} of the electrons for e–i collisions.

The dependence of the effective collision frequency of the electrons on the temperature T_e for different heights in the ionosphere is given in Table 7. With increasing T_e the frequency of the collisions with the ions decreases, and the boundary separating the regions $\nu_{ei} < \nu_{em}$ from $\nu_{ei} > \nu_{em}$ shifts upwards.

Table 7. Dependence of the collision frequency ν (s^{-1}) on the electron temperature in the ionosphere (day-time)

		$z = 100$ km		$z = 150$ km		$z = 200$ km		$z = 300$ km	
T_e, K	T_e, eV	ν_{em}	ν_{ei}	ν_{em}	ν_{ei}	ν_{em}	ν_{ei}	ν_{em}	ν_{ei}
240	0.0207	$4.8 \cdot 10^4$	850						
350	0.03	$6.2 \cdot 10^4$	520						
580	0.05	$9.8 \cdot 10^4$	260						
1160	0.1	$1.8 \cdot 10^5$	97	$1.1 \cdot 10^3$	350				
2300	0.2	$3.2 \cdot 10^5$	38	$2.0 \cdot 10^3$	130	$2.2 \cdot 10^2$	220	20	670
3500	0.3	$4.5 \cdot 10^5$	21	$2.8 \cdot 10^3$	75	$3.1 \cdot 10^2$	120	27	350
4600	0.4	$5.8 \cdot 10^5$	14	$3.5 \cdot 10^3$	50	$3.9 \cdot 10^2$	82	36	250
5300	0.5	$6.9 \cdot 10^5$	11	$4.2 \cdot 10^3$	37	$4.7 \cdot 10^2$	61	38	190

Average Fraction of Energy Lost by the Electron in Collisions. In the lower ionosphere ($z \lesssim 100$ km), the electron energy loss is determined by the inelastic collisions with the N_2 and O_2 molecules. The dependence of the average fraction δ of lost energy on the effective electron temperature T_e in molecular nitrogen and oxygen is shown in Table 8. These values of δ agree with the laboratory data of Engelhardt et al. (1964) and Hake and Phelps (1967). The plasma of the lower ionosphere is weakly ionized. In this case, as indicated in Section 2.3.3., the values of $\delta(T_e)$ are different in a high-frequency ($\omega^2 \gg \nu_e^2$) and low-frequency ($\omega^2 \ll \nu_e^2$) electric field. In Table 8 they are given for a neutral-molecular temperature $T = 185$ K (at $T_e = T$ we have $\delta \approx 4.8 \times 10^{-3}$).

We note that the abrupt increase of $\delta(T_e)$ in a high-frequency field at $T_e \gtrsim (6\text{–}7) \cdot 10^3$ K is connected with the excitation of the vibrational levels of the nitrogen by the fast electrons. In a low-frequency field there are fewer fast electrons (see Sect. 2.3.2); therefore an analogous growth of the losses begins here at higher temperatures ($T_e \gtrsim 10^4$ K) (see Fig. 10).

At great heights, an important role is assumed by collisions with the ions. They have an elastic character, so that

$$\delta_{ei} = \delta_{el}^{i} = 2m/M_i. \tag{2.236}$$

Table 8. Average fraction of energy lost by electron in one collision

T_e^0	$\delta \cdot 10^3$		$\delta \cdot 10^3$		$\delta \cdot 10^4$
			lower ionosphere $z \lesssim 100$ km		F-layer, day-time $z = 300$ km
	N_2	O_2	$\omega^2 \gg \nu_e^2$	$\omega^2 \ll \nu_e^2$	
200	8.1	18	4.0	4.6	—
400	2.8	7.6	1.7	3.2	—
600	1.8	5.2	1.2	2.3	—
800	1.1	3.5	1.0	1.6	—
1000	0.8	2.9	0.9	1.2	
2000	0.5	6.1	1.0	1.6	0.85
3000	0.4	9.4	1.2	2.2	1.0
4000	0.4	9.8	1.4	2.3	1.2
5000	0.4	9.3	1.6	2.2	1.4
6000	0.5	8.6	2.3	2.1	1.8
7000	0.6	8.0	3.9	2.1	2.3
8000	0.7	7.4	6.1	2.1	2.9
9000	1.1	6.8	11	2.3	4.5
10000	1.8	6.2	20	2.8	7.6

In addition, the molecular composition of the ionosphere changes strongly with height. At $z > 200$ km the main constituent of the neutral component is monatomic oxygen. When slow electrons interact with this oxygen there is an appreciable excitation of optical fine-structure transitions. In this case we have (Dalgarno and Degges, 1968):

$$\delta_{eO} \approx 0.82 \cdot 10^2/TT_e^{1/2}. \tag{2.237}$$

The temperatures T of the neutral particles and T_e of the electrons are expressed here in K.

Table 8 lists the average fraction of the energy lost in the region of the maximum of the F layer ($z \sim 300$ km), determined from the relation

$$\delta = (\delta_{em}\nu_{em} + \delta_{ei}\nu_{ei})/(\nu_{em} + \nu_{ei}). \tag{2.238}$$

Here ν_{em} and δ_{em} are the collision frequency and the energy fraction lost in collisions with the neutral molecules (O, N_2, O_2). The increase of δ with increasing T_e is due in this case to the increased contribution of the collisions with the neutrals (see Table 7).

Ion Collision Frequency. When ions interact with neutral molecules at low temperatures $T \sim 300$ K, the principal role is played by the polarization of the molecules. In this case the collision frequency is independent of temperature in first-order approximation and takes the form (Banks, 1969):

$$\nu_{im} = \beta_{im}N_m,$$

where N_m is concentration of the molecules. The values of the coefficient β_{im} for different interactions are given in Table 9.

In the interaction of ions and molecules of the same element (for example O^+ and O), an important role is played by the charge-exchange

Table 9. Coefficients β_{im} for various interactions

	$\beta_{im} \cdot 10^{10}$, cm^3/s		$\beta_{im} \cdot 10^{10}$, cm^3/s		$\beta_{im} \cdot 10^{10}$, cm^3/s
O^+, N_2	5.1	NO^+, N_2	9.0	H^+, N_2	2.4
O^+, O_2	4.5	NO^+, O	7.6	H^+, He	4.3
O^+, He	2.2	NO^+, H	21	He^+, O	4.5
O^+, H	2.6	NO^+, He	6.4	He^+, H	7.7
NO^+, O_2	8.3	H^+, O_2	2.7	He^+, N_2	4.1

Table 10. Coefficients β^0_{im} for various charge exchanges

	O^+, O	O_2^+, O_2	H^+, H	He^+, He	N_2^+, N_2	N^+, N
$\beta^0_{im} \cdot 10^{11}$ cm^3/s · deg$^{1/2}$	1.7	1.1	10	3.0	2.1	1.7

process. The effective collision frequency for this process can be approximately represented in the form of Equation (2.212):

$$\nu_{\mathrm{im}} = \beta^0_{\mathrm{im}}(T_{\mathrm{i}} + T)^{1/2} N_{\mathrm{m}}.$$

The values of β^0_{im} for different types of charge-exchange are given in Table 10.

The effective collision frequencies for the ions NO^+ and O^+ in the ionosphere at the different heights are given in Table 11. In the E layer, the principal role is played by elastic collisions of ions and molecules, and in the F layer by charge-exchange of the O^+ ions. The table gives also the mean free paths l_{im} and l_{ii} of the ions for i–m and i–i collisions.

Table 11. Effective frequencies ν (s^{-1}) of ion collisions in the ionosphere

z km	Day-time					Night-time	
	ν_{NO^+m}	ν_{O^+m}	l_{im}, cm	ν_{ii}	l_{ii}, cm	ν_{ii}	l_{ii}, cm
60	$6.0 \cdot 10^6$	$3.5 \cdot 10^6$	10^{-2}	0.003	$2.1 \cdot 10^7$	—	—
70	$1.8 \cdot 10^6$	$1.0 \cdot 10^6$	$3 \cdot 10^{-2}$	0.01	$5.1 \cdot 10^6$	—	—
80	$2.6 \cdot 10^5$	$1.4 \cdot 10^5$	$2 \cdot 10^{-1}$	0.05	$8.9 \cdot 10^5$	0.001	$6.9 \cdot 10^7$
90	$3.5 \cdot 10^4$	$2.0 \cdot 10^4$	1.8	0.39	$1.3 \cdot 10^5$	0.004	$14.5 \cdot 10^6$
100	$7.3 \cdot 10^3$	$4.9 \cdot 10^3$	7.5	3.0	$1.8 \cdot 10^4$	0.055	$9.8 \cdot 10^5$
110	$1.7 \cdot 10^3$	$9.3 \cdot 10^2$	36	3.2	$1.9 \cdot 10^4$	0.058	$1 \cdot 10^6$
120	$6.8 \cdot 10^2$	$3.9 \cdot 10^2$	$1 \cdot 10^2$	2.3	$3.0 \cdot 10^4$	0.045	$1.6 \cdot 10^6$
130	$2.4 \cdot 10^2$	$1.4 \cdot 10^2$	$3.9 \cdot 10^2$	1.9	$4.2 \cdot 10^4$	0.035	$2.2 \cdot 10^6$
150	58	36	$1.7 \cdot 10^3$	2.0	$4.8 \cdot 10^4$	0.022	$4.4 \cdot 10^6$
200	7.1	5.2	$1.3 \cdot 10^4$	2.5	$6.8 \cdot 10^4$	0.020	$6 \cdot 10^6$
250	2.1	1.8	$4.3 \cdot 10^4$	3.9	$4.8 \cdot 10^4$	0.078	$2 \cdot 10^6$
300	0.76	0.75	$1.1 \cdot 10^5$	5.5	$3.5 \cdot 10^4$	0.67	$2.3 \cdot 10^5$
400	0.15	$1.8 \cdot 10^{-1}$	$1 \cdot 10^6$	5.0	$3.9 \cdot 10^4$	1.9	$8.3 \cdot 10^4$
500	$4.3 \cdot 10^{-2}$	$5.6 \cdot 10^{-2}$	$3.6 \cdot 10^6$	2.7	$7.7 \cdot 10^4$	1.2	$1.4 \cdot 10^5$
600	$1.3 \cdot 10^{-2}$	$1.8 \cdot 10^{-2}$	$1.2 \cdot 10^7$	0.82	$2.8 \cdot 10^5$	0.77	$2.1 \cdot 10^5$
700	$4.5 \cdot 10^{-3}$	$5.8 \cdot 10^{-3}$	$4 \cdot 10^7$	0.43	$5.9 \cdot 10^5$	0.40	$3.9 \cdot 10^5$
800	$1.8 \cdot 10^{-3}$	$2.1 \cdot 10^{-3}$	$1 \cdot 10^8$	0.22	$1.2 \cdot 10^6$	0.25	$7.3 \cdot 10^5$
900	$7.7 \cdot 10^{-4}$	$7.9 \cdot 10^{-4}$		0.20	$2 \cdot 10^6$	0.21	$1.4 \cdot 10^6$
1000	$4 \cdot 10^{-4}$	$3.3 \cdot 10^{-4}$		0.16	$2.9 \cdot 10^6$	0.13	$2.3 \cdot 10^6$

2.5.3. Electron and Ion Temperatures in the Ionosphere

Electron Heating by Photoionization. At the instant of ionization of the atom by the solar uv radiation, the knocked-out electron acquires a rather appreciable energy ε_f:

$$\varepsilon_f = \hbar\omega - \varepsilon_i. \tag{2.239}$$

Here ε_i is the ionization energy and $\hbar\omega$ is the energy quantum. The energy ε_f is lost by the electron in collisions with the neutral atoms, and also by transfer to the plasma electrons, and this determines the heating of the electron gas. As a result, the temperature of the electrons and ions in the day time ionosphere is maintained higher than the temperature of the neutral molecules. The ionization energy of the principal components of the ionosphere and the corresponding wave lengths are given in Table. 12.

The energy acquired by the electron via ionization from the ground level is determined by Equation (2.239). However, other ionization states are also possible, wherein part of the quantum energy goes to excitation of the ion. Taking into account the spectrum of the solar ultraviolet radiation, its absorption, and the probabilities of various ionization states, we can calculate, for a chosen model of the ionosphere, the number of the knocked-out photoelectrons and their energy spectrum (Tohmatsu et al., 1965). On the average (roughly speaking) at heights above 150 km the spectrum turns out to be Maxwellian with $T_e \approx 5$ eV. The deviations from the average curve, however, are appreciable. In particular, a characteristic peak appears in the photoelectron distribution at $\varepsilon_f \sim 20$–30 eV, connected with the 30.4 nm solar resonance line of He II. Below 150 km, the character of the photoelectron spectrum changes significantly, and high-energy electrons with $\varepsilon_f \approx 20$–30 eV assume an ever-increasing role

The energy acquired by the electrons by photoionization is consumed partially in heating of the electron gas and partially in excitation of the molecules.

Figure 12 shows the energy $\Delta\varepsilon_f$ transferred on the average by one photoelectron to the plasma electrons (Banks, 1969). Multiplying $\Delta\varepsilon_f$ by

Table 12. Ionization energies and corresponding wavelengths of principal ionosphere components

	NO	O_2	O	H	N	N_2	He
λ_i, Å	1340	1026	911	910	853	796	504
ε_i, eV	9.25	12.1	13.6	13.6	14.5	15.6	24.8

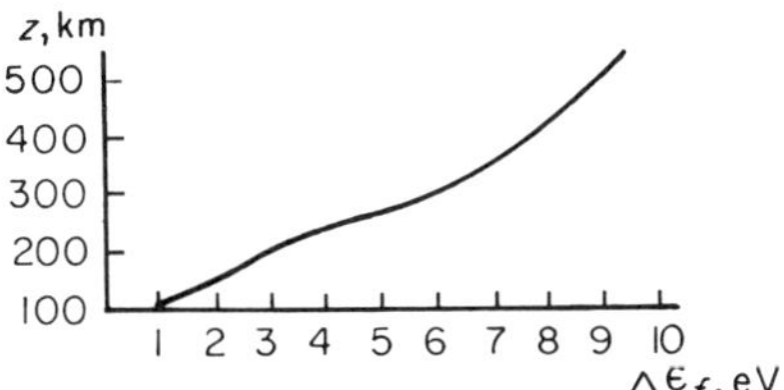

Fig. 12. Energy transferred on the average by one photoelectron to the electrons of the ionosphere

the total number q_i of ions produced in 1 cm^3 per second, we obtain the average energy transferred by the photoelectrons to the main plasma, i.e., used to heat the plasma electrons:

$$Q = q_i \, \Delta\varepsilon_f \tag{2.240}$$

The variation of Q as a function of the height z is shown by way of example in Figure 13. As seen from the figure, the intensity Q of the electron heating always has a maximum in the region of 150 km, due to the variation of the photoionization intensity; at larger heights the ionization becomes weaker, and at lower ones the fraction of the energy transferred by the photoelectron to the plasma electrons decreases rapidly. At heights above 300 km, the photoelectron mean free path is so large that it becomes necessary to take into account the inhomogeneity of the plasma.

Temperatures of the Electrons and of the Ions. When the atmosphere electrons and ions are heated only by the photoelectrons, their temperatures are given by Equations (2.52) and (2.53) or by Equations (2.136)

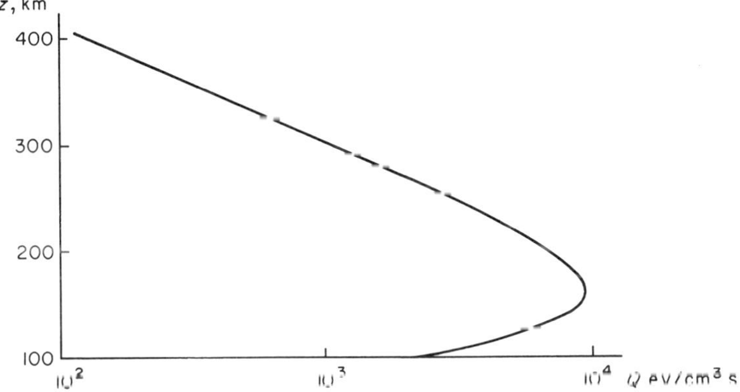

Fig. 13. Energy released by the photoelectrons in the ionosphere

and (2.210), except that the ohmic heating $\boldsymbol{E} \cdot \boldsymbol{j}$ must be replaced by the heating Q due to the photoelectrons. Under stationary conditions we have

$$\begin{aligned} \delta_{ei}\nu_{ei}(T_e - T_i) + \delta_{em}\nu_{em}(T_e - T) = 2Q/3N, \\ \delta_{ei}\nu_{ei}(T_e - T_i) - \nu_{im}(T_i - T) = 0. \end{aligned} \tag{2.241}$$

It is assumed here that $\delta_{im} = 1$ [Eq. (2.218)]. By solving the Equations (2.241) we can determine the temperatures of the electrons and of the ions. At small heights, the temperatures of the electrons (up to ≈ 150 km) and of the ions (up to ≈ 600 km) are close to the temperature of the neutral molecules. In this case they can be determined from the formulas

$$\begin{aligned} T_e = T + \Delta T_e, \qquad T_i = T + \Delta T_i; \\ \Delta T_i = \frac{\delta_{ei}\nu_{ei}}{\delta_{ei}\nu_{ei} + \nu_{im}} \Delta T_e, \\ \Delta T_e = \frac{2Q}{3N} \frac{\delta_{ei}\nu_{ei} + \nu_{im}}{\delta_{ei}\nu_{ei}\nu_{im} + \delta_{em}\nu_{em}(\delta_{ei}\nu_{ei} + \nu_{im})}. \end{aligned} \tag{2.242}$$

At heights above 300–400 km, the thermal conductivity begins to make an appreciable contribution.

We considered here only the heating of the electrons and ions of the ionosphere as a result of photoionization. There exist also other heating sources, namely the precipitation of the high-energy particles from the region of the radiation belts or from the tail of the magnetosphere, as well as heating induced by stationary or quasi-stationary electric field.

2.5.4. Heating of the Ionosphere in an Alternating Electric Field

Heating in the D *and* E *Layers.* In the lower layers of the ionosphere, the heating of the ions is of little importance, since here $\nu_{im} \gg \delta_{ei}\nu_{ei}$ and the ions transfer to the molecules, very energetically, the energy acquired from the electrons. In these regions $T_i = T + \Delta T_i$, where ΔT_i is defined by Equation (2.56a). Equation (2.136) for T_e takes the form

$$\frac{dT_e}{dt} = \frac{e^2 E_0^2 \nu_{em}\varphi_p}{3m(\omega^2 + \nu_{em}^2)} K_\sigma\left(\frac{\omega}{\nu_{em}}\right) - \delta_{em}\nu_{em}(T_e - T) + \frac{2Q}{3N}. \tag{2.243}$$

We have taken into account here the fact that the plasma is weakly ionized at the considered heights, so that the transfer of the electron energy to the ions can be neglected. Further, φ_p is the polarization factor [Eqs. (2.42), (2.169)]. Taking into account the kinetic coefficients K_σ in Equation (2.42), it is necessary to replace $\alpha_{\perp-}^2$ by $\alpha_{\perp-}^2 K_\sigma(|\omega - \omega_H|/\nu_{em})/K_\sigma(\omega/\nu_{em})$ and $\alpha_{\perp+}^2$ by $\alpha_{\perp+}^2 K_\sigma((\omega + \omega_H)/\nu_{em})/K_\sigma(\omega/\nu_{em})$.

The dependence of $T_e = 2\bar{\varepsilon}/3$ on E_0/E_p in the lower ionosphere at $\varphi_p = 1$, determined with the aid of Equation (2.243), is shown in Figure 9. It is seen from the figure that at $E_0/E_p \sim 1$ the effective temperature of the electrons increases energetically with increasing field amplitude. Subsequently the growth of T_e becomes weaker because of the increase of the energy losses connected with the excitation of the vibrational levels of the nitrogen.

Let us consider the propagation of a wave along the earth's magnetic field $\boldsymbol{H}$. In this case, the electric field $\boldsymbol{E}$ is circularly polarized in a plane perpendicular to $\boldsymbol{H}$, and in Equation (2.243) we have

$$\varphi_p = \frac{\omega^2 + \nu_{em}^2}{(\omega \pm \omega_H)^2 + \nu_{em}^2} \frac{K_\sigma(|\omega \pm \omega_H|/\nu_{em})}{K_\sigma(\omega/\nu_{em})}, \tag{2.244}$$

where the plus and minus signs stand for the ordinary and extraordinary waves. If $\omega \approx \omega_H$, then gyromagnetic resonance of the extraordinary wave takes place. The dependence of T_e on E_0/E_{p0} under the gyromagnetic resonance condition $\omega = \omega_H$ is shown in Figure 14. Here E_{p0} is the

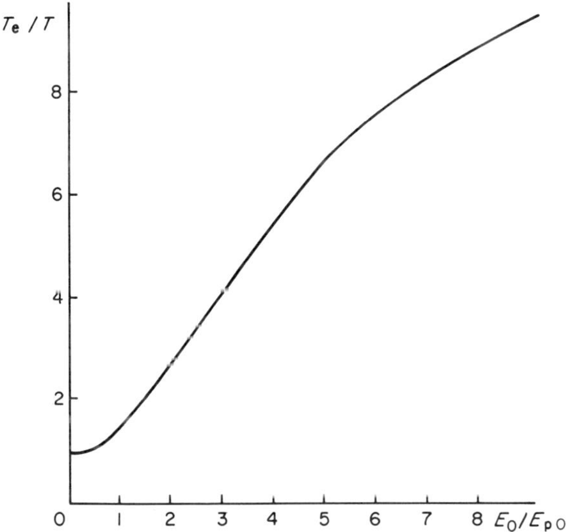

Fig. 14. Temperature of electrons in the lower ionosphere at gyroresonance $\omega = \omega_H$; $E_{p0} = E_p\ (\omega = 0)$

plasma field [Eq. (1.2)] at $\omega = 0$. We see that even very weak fields can lead to appreciable heating, since $E_{p0} \sim 1$ mV/m at $z \approx 100$ km. However, the production of such fields at the gyromagnetic frequency in the interior of the ionosphere is made difficult by the strong damping of the radio waves. The character of the dependence of T_e/T on the wave frequency ω in the vicinity of the gyroresonance is perfectly analogous to that shown in Figure 3.

Heating in the F *Layer.* In the upper layers of the ionosphere it is necessary in the general case to take into account also the heating of the ions. Equations (2.136), (2.210) or (2.52) and (2.53) then take the form

$$\frac{dT_e}{dt} = \frac{e^2 E_0^2 \nu_e K_\sigma \varphi_p}{3m(\omega^2 + \nu_e^2)} - \delta_{em}\nu_{em}(T_e - T) - \delta_{ei}\nu_{ei}(T_e - T_i) + \frac{2Q}{3N},$$
$$\frac{dT_i}{dt} = \delta_{ei}\nu_{ei}(T_e - T_i) - \nu_{im}(T_i - T). \tag{2.245}$$

The result of the calculation of the stationary temperatures of the electrons and ions as a function of the electric field intensity at $\omega \gg \omega_H$ for $z = 300$ km is shown in Figure 15. The quantity Q is chosen to be $1.7 \cdot 10^4$ eV/cm^3 s. Then $T_{e0} = 2000$ K and $T_{i0} = 1350$ K. It is seen from the figure that an appreciable heating of the electrons at $z = 300$ km is possible in fields $E_0 \sim E_p$. The temperature of the ions increases weakly up to $z \sim 600$ km; at heights $z > 600$ km it grows to values close to T_e.

In a weak electric field, the stationary perturbations of the electron temperature, as follows from Equation (2.245), are determined by the expression

$$\frac{\Delta T_e}{T_{e0}} = \left(\frac{E_0}{E_p}\right)^2 \varphi, \qquad \varphi = \varphi_p/\varphi_T \tag{2.246}$$

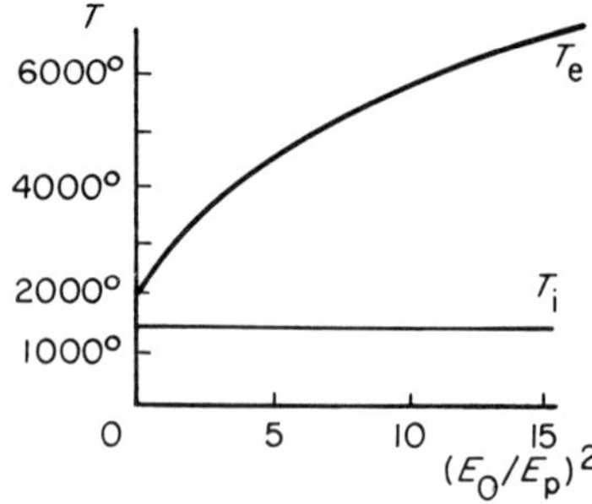

Fig. 15. Temperatures of electrons and ions in F layer

Table 13. Electron temperature and concentration relaxation times, τ_T and τ_N(s)

z, km	day-time		night-time	
	τ_T	τ_N	τ_T	τ_N
80	$2 \cdot 10^{-4}$		$2 \cdot 10^{-4}$	
100	$5.6 \cdot 10^{-3}$	52	$5.9 \cdot 10^{-3}$	$2 \cdot 10^{3}$
120	$9 \cdot 10^{-2}$	41	$9 \cdot 10^{-2}$	$1.5 \cdot 10^{3}$
150	1.1	35	1.1	$1.6 \cdot 10^{3}$
200	5.2	32	6.0	60
250	9.1	$1.4 \cdot 10^{2}$	25	$4 \cdot 10^{2}$
300	14.2	$4.9 \cdot 10^{2}$	71	$2.2 \cdot 10^{3}$
400	23	$4.5 \cdot 10^{3}$	50	$5 \cdot 10^{4}$
500	43	$3.7 \cdot 10^{4}$	77	$1.1 \cdot 10^{6}$

Here φ_p is the polarization factor [Eqs. (2.42) and (2.244)], and φ_T is the non-isothermy factor

$$\varphi_T = 1 + \frac{T_{e0} - T}{(\delta\nu_e)_{T_{e0}}} \left[\frac{d}{dT_e} (\delta\nu_e) \right]_{T_{e0}}, \qquad \delta\nu_e = \delta_{ei}\nu_{ei} + \delta_{em}\nu_{em}. \quad (2.246a)$$

In the last expression it is assumed that $\nu_{im} \gg \delta_{ei}\nu_{ei}$ (this condition is satisfied at $z \lesssim 600$ km).

We emphasize that in the case of heating in the F layer, an important role is usually played by transport processes (see Sect. 5.3). We note also that we have considered above stationary distributions of the electron temperature. The time of establishment of the temperature [Eq. (2.20)] is

$$\tau_T = 1/\delta\nu_e. \quad (2.247)$$

For the ionosphere it is represented in Table 13.

2.5.5. Perturbations of the Electron and Ion Concentrations

An alternating electric field of radio waves raises the electron temperature in the perturbed region of the ionosphere. With increasing T_e, a change takes place also in the effective recombination coefficient α. In particular α decreases with increasing T_e in the E layer of the ionosphere. Consequently, the concentration of the electrons and ions under the influence of the alternating electric field increases. This growth of N is limited by the influence of the transport processes, the role of which in the region of the maximum of the F layer turns out to be more significant than the change of the recombination coefficient (see Sect. 5.3).

We consider now the perturbation at heights 80 km $\lesssim z \lesssim$ 200 km, when the heating of the electrons and the change of the concentration have the same character as in an homogeneous plasma. The perturbation of the electron temperature was considered earlier. The perturbation of the electron densities when the heating is not very strong is connected with the change of the recombination coefficient $\alpha = \alpha(T_e)$.

Taking Equation (2.230) into account, we can rewrite Equation (2.231) in the form

$$\frac{N(T_e)}{N(T_{e0})} = \frac{1}{2}\left\{n_{O^+} + \sqrt{n_{O^+}^2 + 4n_{O_2^+}\frac{\alpha_2(T_{e0})}{\alpha_2(T_e)} + 4n_{N_2^+}\frac{\alpha_1(T_{e0})}{\alpha_1(T_e)}}\right\}. \quad (2.248)$$

Here T_{e0}, $N(T_{e0}) = N_0$, n_{O^+}, $n_{O_2^+}$, and $n_{N_2^+}$ are the temperature and concentration of the electrons and the relative concentrations of the ions in the unperturbed plasma.

Equation (2.248) determines the change of the concentration of the electrons when the electrons are heated in the field of strong radio waves, neglecting transport processes. In particular, in the case of a weak perturbation, the change of the electron temperature in the ionosphere is determined by Equation (2.246). For the perturbation of the concentration we obtain from Equation (2.248)

$$\frac{\Delta N}{N_0} = \gamma_1 \frac{\Delta T_e}{T_{e0}}, \qquad \gamma_1 = \frac{\lambda_1 n_{NO^+} + \lambda_2 n_{O_2^+}}{2 - n_{O^+}}, \quad (2.249)$$

where, in accordance with Equation (2.229), $\lambda_1 = 1.2$ and $\lambda_2 = 0.7$. The dependence of the coefficient γ_1 on the height z for the day time ionosphere is shown in Table 14. It is seen from the table that γ_1 remains close to 0.5 up to 160–170 km; it then decreases rapidly and at $z > 200$ km it

Table 14. Height dependence of γ_1, day-time ionosphere

z, km	γ_1	z, km	γ_1	z, km	γ_1
60	0.55	110	0.49	160	0.42
70	0.55	120	0.45	170	0.36
80	0.53	130	0.45	180	0.27
90	0.51	140	0.445	190	0.16
100	0.505	150	0.44	200	0.08

is practically equal to zero. It can be assumed approximately that $\gamma_1 = 0.5$ at $z < 180$ km and $\gamma_1 = 0$ at $z > 180$ km. Equation (2.249) can be used up to values $\Delta T_e/T_{e0} \sim 1$. We note that if Equation (2.229a) is correct, then γ_1 is approximately half as large, $\gamma_1 \approx 0.25$.

Thus, the change in the ionization-recombination balance as the result of the heating of the electrons is significant only at heights $z < 200$ km. The reason is that the electron heating can change only the number of the NO^+ and O_2^+ ions. At heights $z > 180$–200 km, the relative concentration of these ions is negligible. The principal role is played here by the O^+ ions, the concentration of which does not depend on the electron temperature [Eq. (2.230)]. Therefore the change of T_e has little effect in this region on the total number N of the ions or electrons.[11]

In the D region of the ionosphere, $z < 80$ km, the recombination is connected with formation of clusters of ions and negative ions. Heating in an electric field produces here appreciable perturbations of the electron concentrations. Particularly large changes in the ionization balance are produced by strong action of radio waves on the ionospheric plasma, when the decisive role is assumed by new processes, namely dissociative attachment to oxygen and ionization by electron impact (see Sect. 2.5.6).

The time of establishment of the stationary value of the concentration (the electron lifetime) is

$$\tau_N = 1\bigg/\left(\frac{\partial q_r}{\partial N}\right), \tag{2.250}$$

where q_r is the recombination intensity [Eq. (2.222)]. The value of τ_N in the ionosphere is given in Table 13.

2.5.6. Artificial Ionization of the Ionosphere—Heating of Neutral Gas

We have considered above the change of the electron concentration and ion concentration N in the ionosphere, resulting from the change in the dissociative recombination coefficient when the plasma is heated in an alternating electric field of a radio wave. These changes of N are in general not very strong. In very strong electric fields, however, new ionization and recombination mechanisms can appear and can sharply increase the degree of plasma ionization. These will be briefly discussed in this section.

Change of Ionization Balance. In very strong fields, molecule ionization by accelerated electrons, and dissociative attachment that leads to

[11] It should be noted, however, that if the electric field were effectively to change the radiation temperature in the ionosphere, then the change in N could become essential.

the appearance of negative ions, play an essential role. The ionization balance equations for the concentrations N of the electrons and N^- of the negative ions then take the form

$$\frac{dN}{dt} = q_i + \nu_{ion}N - \nu_a N + \nu_d N^- - \alpha N(N + N^-),$$
$$\frac{dN^-}{dt} = \nu_a N - \nu_d N^- - \alpha_i N^-(N + N^-). \tag{2.251}$$

Here, as usual, q_i is the total intensity of the ionization produced by the external source and ν_{ion} is the frequency of the molecule ionization by the fast electrons [Eqs. (2.177), (2.178)]. We note that the exponential in Equation (2.177) contains a large quantity, so that the dependence of ν_{ion} on E_0, ω, and N_m is determined in the main by the exponential term.

Next, ν_a is the frequency of electron attachment to the molecules. Under the conditions of the lower ionosphere, an important role is played by the attachment of electrons to oxygen molecules in triple collisions, and also by dissociative attachment to the ozone and oxygen molecules

$$\nu_a = (k_1 N_{O_2} + k_2 N_{N_2}) N_{O_2} + k_3 N_{O_3} + k_a N_{O_2} \tag{2.251a}$$

Here k_1 and k_2 are the attachment coefficients for triple collisions. According to Phelps (1969),

$$k_1 = 1.4 \cdot 10^{-29} \exp\left(-\frac{600}{T_e}\right) \text{cm}^6/\text{s}, \qquad k_2 = 1.0 \cdot 10^{-31}\ \text{cm}^6/\text{s} \tag{2.251b}$$

The electron temperature is expressed here in K. The coefficient to ozone is $k_3 \simeq (1\text{–}10) \cdot 10^{-12}$ cm^3/s. The coefficient of dissociative attachment to oxygen k_a is given by Equation (2.177a) and (2.178a).

Next, ν_d is the electron detachment frequency. Important processes in the ionosphere are photodetachment, detachment in collisions with molecules, and associative detachment:

$$\nu_d = \nu_{ph} + (k_4 N_{O_2} + k_5 N_{N_2} + k_6 N'_{O_2}) + (k_7 N_O + k_8 N_N). \tag{2.251c}$$

Here ν_{ph} is the photodetachment frequency; under conditions of the daytime ionosphere, $\nu_{ph} \simeq 0.44$ s^{-1} (Whitten and Poppoff, 1965). The coefficients k_4 and k_5 for detachment in collisions with oxygen and nitrogen molecules, respectively, at lower-ionosphere temperatures $T \sim 200$–300 K, are small: $k_4 \sim 10^{-22}$–10^{-19} cm^3/s and $k_5 \sim 10^{-23}$–10^{-20} cm^3/s.

More significant here is apparently the detachment in collisions with excited oxygen, with $k_6 \approx 2 \cdot 10^{-10}$ cm^3/s (N'_{O_2} is the concentration of the excited $O_2('\Delta_g)$ molecules). The coefficients $k_7 = 2.5 \cdot 10^{-10}$ cm^3/s and $k_8 = 3 \cdot 10^{-10}$ cm^3/s describe associative detachment in collisions with oxygen and nitrogen atoms ($O_2^- + O \to O_3 + e$, $N + O_2^- \to NO_2 + e$). We note that we consider here detachment only for the O_2^- ions. The O^- ions vanish mainly through processes of associative detachment (Phelps, 1969).

Finally, α is the dissociative recombination coefficients:

$$\alpha = \alpha_1 n_{NO^+} + \alpha_2 n_{O_2^+} + \alpha_3 n_c. \qquad (2.251d)$$

In contrast to Equation (2.223), we took into account here also dissociative recombination of heavy ion clusters $\alpha_3 n_c$, with a coefficient $\alpha_3 \sim 10^{-5}$ cm^3/s and a relative cluster concentration $n_c = N_c^+/(N^- + N)$; it rises abruptly in the region of the D layer, at $z \leq 80$ km, where the ion clusters frequently predominate (Goldberg and Aikin, 1971; Bauer, 1973; Danilov and Simonov, 1975). The recombination coefficient of the positive and negative ions is $\alpha_i \sim 10^{-7}$ cm^3/s (Biondi, 1970; Bauer, 1973).

We emphasize that the complete system of balance equations must include the equations of ion kinetics, of the type considered in Section 2.5.1. We are forced here to confine ourselves to the approximate system of Equations (2.251), since the exact composition of the negative and positive ions in the D region is still not sufficiently well known, nor are the coefficients of the corresponding ionic reactions.

We proceed to the analysis of the stationary solution of Equations (2.251). From the second equation of Equation (2.251) it follows that

$$N^- = N\nu_a/[\nu_d + \alpha_i(N + N^-)].$$

Using the quasineutrality condition $N + N^- = N^+$, we express the concentrations of the electrons and of the negative ions in terms of the concentration of the positive ions N^+

$$N = N^+ \frac{\nu_d + \alpha_i N^+}{\nu_a + \nu_d + \alpha_i N^+}, \qquad N^- = N^+ \frac{\nu_a}{\nu_a + \nu_d + \alpha_i N^+}. \qquad (2.251e)$$

Substituting these expressions in the first equation of Equation (2.251), we have for N^+ the cubic equation

$$q_i(\nu_a + \nu_d + \alpha_i N^+) + \nu_{ion} N^+(\nu_d + \alpha_i N^+) - N^{+2}(\alpha_i \nu_a + \alpha\nu_d + \alpha\alpha_i N^+) = 0.$$

In the lower ionosphere, the following condition is usually well satisfied:

$$\alpha\alpha_i N^+ \ll \alpha_i \nu_a + \alpha \nu_d.$$

In this case we have

$$N^+ = \frac{\nu_{ion}\nu_d + \alpha_i q_i + [(\nu_{ion}\nu_d + \alpha_i q_i)^2 + 4q_i(\nu_a + \nu_d)(\alpha_i \nu_a + \alpha \nu_d - \alpha_i \nu_{ion})]^{1/2}}{2[\alpha_i(\nu_a - \nu_{ion}) + \alpha \nu_d]}. \tag{2.252}$$

We note that the term $\alpha_i q_i$ is frequently inessential. The expression for the electron concentration can then be represented in the form:

$$N \approx \frac{\nu_{ion} + \sqrt{\nu_{ion}^2 + 4q_i(1+\lambda)\left(\alpha_{ef} - \alpha_i \dfrac{\nu_{ion}}{\nu_d}\right)}}{2(1+\lambda)(\alpha_{ef} - \alpha_i \nu_{ion}/\nu_d)};$$

$$\lambda = \frac{N^-}{N} = \frac{\nu_a}{\nu_d + \alpha_i N^+}, \qquad \alpha_{ef} = \alpha + \alpha_i \lambda. \tag{2.252a}$$

In the absence of ionization by fast electrons, Equation (2.252a) goes over into the well-known expression

$$N = \sqrt{q_i/(1+\lambda)\alpha_{ef}}. \tag{2.252b}$$

Equations (2.251e) and (2.252) determine the stationary values of the electron and ion concentration. It is important that the ionization frequency ν_{ion}, the attachment frequency ν_a, and also the recombination coefficient α all depend strongly on the electric field amplitude. This is why the electron and ion concentrations can be altered by the action of radio waves. By way of example, Figure 16a shows plots of N/N_0 and N^-/N against E_0/E_p, determined from Equations (2.252) and (2.251e) for a height of 85 km in the day time ionosphere. We see that initially (at $E_0/E_p \lesssim 5$) the electron concentration increases. This is due to the decrease of the coefficient of dissociative recombination when the electrons are heated [the process considered in Sect. 2.5.5.—see Equations (2.248), (2.229), and (2.229a)]. We note that the concentration of negative ions is simultaneously changed by the variations of the attachment coefficient in triple collisions [Eq. (2.251b)]. At heights of 60–70 km, the role of the negative ions and of positive cluster ions is much greater, and more

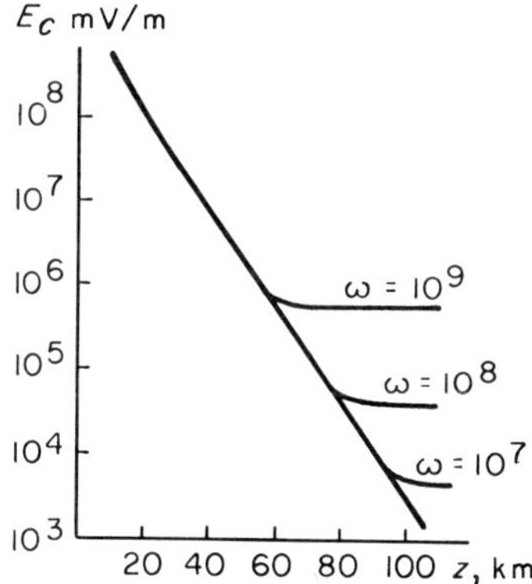

Fig. 16. Critical breakdown field in the upper atmosphere

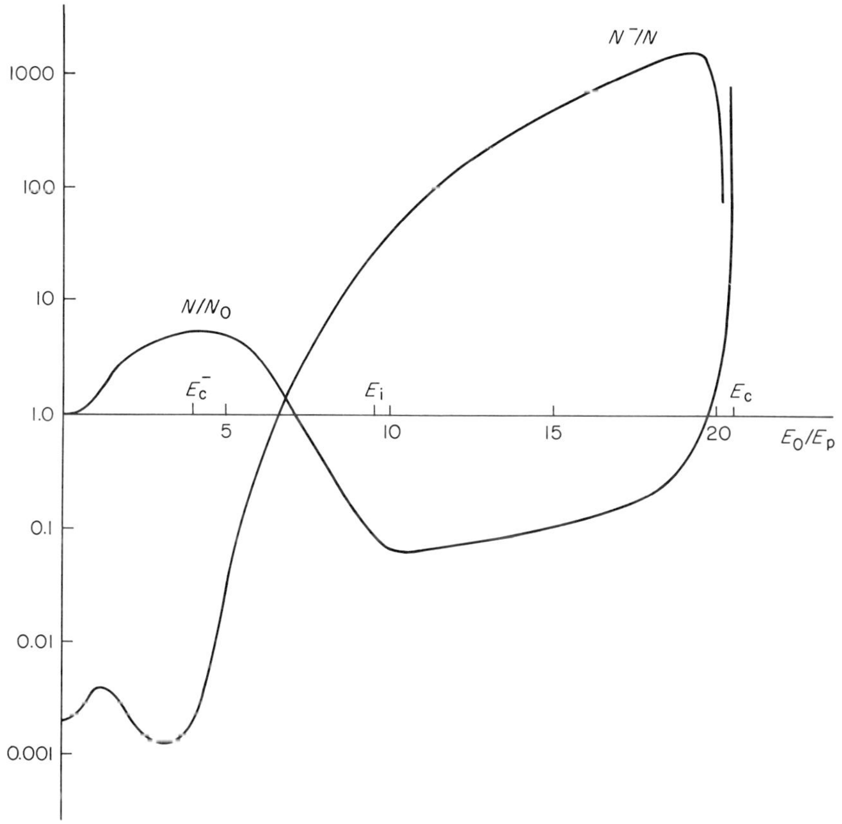

Fig. 16a. Changes induced in the electron density N/N_0 and the negative-ion density N^-/N of the daytime ionosphere by the field of a high-power radio wave. Height $z = 85$ km, unperturbed density $N_0 = 2.5 \cdot 10^3$ cm^{-3}, temperature $T = 185$°K, $\delta_0 = 4.8 \cdot 10^{-3}$. Effective wave frequency $\omega_{\mathrm{eff}} = \omega + \omega_H \cos\theta = 1.8 \cdot 10^7$ s^{-1}. Plasma field $E_p \simeq 0.7$ V/m. Dissociative attachment predominates at $E_0 > E_c^-$ and electron-accelerated ionization predominates at $E_0 > E_i$, where E_c is the critical breakdown field

complicated variations of the electron and ion concentrations are possible when the electrons are heated in fields $E_0 \sim E_p$.

At $E_0 > E_c^-$, dissociative attachment of the electrons to the oxygen molecules begins to predominate [Eq. (2.177a)]. The attachment frequency increases exponentially with increasing field amplitude, and this leads to a strong increase of the negative-ion concentration, and simultaneously to vanishing of the electrons, namely, the enhanced recombination [Eq. (2.252a)] causes the electron concentration to decrease by 1.5 or 2 orders of magnitude. The field of the strong radio wave suppresses, as it were, the lower ionosphere. The field E_c^- at which the strong increase of the negative-ion concentration begins is determined by the condition:

$$k_a(E_c^-) \approx \nu_{a0}/N_{O_2}$$

where $k_a(E_c^-)$ is the oxygen dissociation coefficient [Eq. (2.177a)], ν_{a0} is the adhesion frequency due to triple collisions and other processes [Eqs. (2.251a) and (2.251b)]. The field E_c^- is approximately equal to (3–5)E_p.

At $E_0 > E_i$, the decisive role is assumed by the additional ionization of the ionosphere by the fast electrons. The field $E_i \simeq 10E_p$ is defined by the condition

$$\nu_{ion}^2(E_i) \approx 4q_i(1 + \lambda)\alpha_{ef}$$

The ionization by an external source q_i loses its significance at $E_0 > E_i$. Therefore, for example, the change of q_i during the time of sudden perturbations should not affect the state of the plasma at $E_0 > E_i$. Owing to the additional ionization, the electron concentration ceases to decrease and begins to increase. This increase, however, is damped by the continuing increase of the dissociative-attachment frequency. At $E_0 \approx 20E_p$ the attachment coefficient stops growing [Eq. (2.178a)] and the ionization frequency becomes comparable with the attachment frequency. The electron concentration in Equation (2.252) then increases very abruptly, and even becomes infinite—in other words, Equations (2.251) have no stationary solution. This phenomenon—breakdown of the gas—sets in at $E_0 = E_c$, where E_c is the critical breakdown field. It is determined by the condition $\nu_{ion} = \nu_a + (\alpha/\alpha_i)\nu_d$. Taking Equation (2.177) for ν_{ion} into account, the expression for E_c assumes the form

$$E_0 = E_c, \qquad E_c = \frac{\sqrt{6m}}{ex_i}(A_i^2\omega^2 + B_i^2)^{1/2}K_i \tag{2.253}$$

where the constant x_i is defined by the relation

$$x_i^2 \exp(x_i) = \frac{2.6 \cdot 10^{-7}}{k_{am}}, \qquad x_i = 6.$$

We have allowed here for the fact that $v_a = k_{am} N_{O_2} \gg (\alpha/\alpha_i) v_d$ [see Eqs. (2.178a) and (2.251c)]. Equation (2.253) together with Equation (2.178), with corrections ΔB and ΔA included (p. 79), agrees well with laboratory measurement data (MacDonald, 1966). It is valid also in a constant electric field, provided the substitution $E_c \to \sqrt{2} E_{c0}$ is made. From Equations (2.253) and (2.178) we then have $E_{c0} = 0.72 \cdot 10^{-15} N_m$ V-cm^2, where N_m is the concentration of the air molecules.

The field E_c as a function of the height z in the atmosphere is shown in Figure 16. It is evident from the figure and from Equation (2.253) that the field E_c decreases rapidly with height (in proportion to N_m) until the field frequency ω becomes comparable with $v_0 = B_i/A_i \approx 1.7 \cdot 10^{-7} N_m$ cm^3/s. When this equality is reached, the field E_c becomes practically independent of N_m.

If $E_0 \geq E_c$, the electron concentration in the plasma increases rapidly. The foregoing, however, pertains only to a plasma situated in a given periodic electric field. If, however, the breakdown is produced by traveling radio waves, as is the case in the ionosphere, then allowance must be made for the fact that the increase of the electron concentration is accompanied by equally abrupt increase of the wave absorption. The wave intensity is rapidly decreased by the absorption, and this leads to a certain stabilization: the electron concentration at $E_0 \geq E_c$ does not increase too strongly. This is one of the examples of "self-action" of a wave in a plasma (for details see Sect. 3.2.3).

It is important that an appreciable perturbation of the plasma ionization appears already in fields weaker than E_c by one order of magnitude [see Fig. 16a]. An important role can be played here by excitation of instabilities that lead to focusing of the radio waves, to increased heating of the plasma, and to acceleration of the electrons (see Chap. 6). Perturbation of the ionization of the lower ionosphere by radio waves was observed in experiment (Gurevich and Shlyuger, 1975; Utlaut, 1975; Gurevich et al., 1976b).

Isothermal Ionization. It was assumed above that only electrons become heated. At low heights $z \lesssim 60$ km, an important role can be played by heating of the neutral gas (Gurevich, 1972c). Indeed, at high neutral-gas temperatures the isothermal ionization becomes significant. The degree of isothermal ionization, i.e., of the ionization of the gas under conditions of thermodynamic equilibrium, is determined by the Saha formula

$$N = N_{\mathrm{m}} \left(\frac{mT}{2\pi\hbar^2} \right)^{3/4} \exp(-\varepsilon_{\mathrm{i}}/2T). \tag{2.254}$$

Here T is the gas temperature and ε_{i} is the ionization energy. It is seen from Table 12 that the NO molecules have the smallest ionization potential. They play therefore the principal role at low temperatures. The number of NO molecules in air is given approximately by the formula (Nicolet, 1960):

$$N_{\mathrm{NO}} \approx \frac{N_{\mathrm{m}}}{50} \exp\left(-\frac{0.26}{T} \right) \tag{2.255}$$

where N_{m} is the total concentration of the molecules in the air, and T is the temperature in electron volts. Substituting Equation (2.254) in Equation (2.255) and recognizing that $N_{\mathrm{m}} = p_0/T$, where p_0 is the pressure of the unperturbed gas (this relation follows from the conditions of hydrodynamic equilibrium of the heated region in the air), we obtain:

$$N = \frac{\sqrt{p_0}}{7} T^{1/4} \left(\frac{m}{2\pi\hbar^2} \right)^{3/4} \exp\left(-\frac{4.76}{T} \right) \simeq \sqrt{p_0}\, T^{1/4} \exp\left(41 - \frac{4.76}{T} \right) \tag{2.256}$$

(the pressure p_0 is in atmospheres and the temperature T in electron volts). It follows therefore that already at $T \approx 0.25$–0.3 eV the electron density reaches 10^8–10^{11} cm^{-3} at $p_0 \sim 10^{-3}$–10^{-1} atm (i.e., at heights of 50–20 km).

The heated region of the atmosphere, naturally, cools down and loses energy. The heat losses are due to molecular thermal conductivity, radiation, and wind. The energy flux from the heated region due to thermal conductivity is

$$S_{\mathrm{m}} = \kappa_{\mathrm{m}} \nabla T \sim \frac{\kappa_{\mathrm{m}}(T - T_0)}{R_0}. \tag{2.257}$$

Here κ_{m} is the coefficient of molecular thermal conductivity (in air at temperatures of the order of 10^3 degrees we have $\kappa_{\mathrm{m}} \sim 10^4$ erg/cm-s-deg), R_0 is the characteristic dimension of the strongly heated region, T is the discharge temperature, and T_0 is the temperature of the unperturbed gas.

The radiation at low temperatures T and low pressures p_0, and also at not too large dimensions of the heated zone R_0, leaves freely the discharge region. The energy lost to radiation has in this case a volume character:

$$J = \sigma_f T^4 f \tag{2.257a}$$

Here J is the energy radiated by 1 cm^3 of gas per second, $\sigma_f = 5.67 \times 10^{-4}$ erg/cm^2 s-deg^4 is the Stefan–Boltzmann constant, and $f = 1/l(T, p_0)$, where l is the characteristic length. Numerical calculations (Avilova et al., 1970) show that for air at low temperatures ($\sim 10^3$ K) and low pressures ($p_0 \sim 0.1$–0.01 atm) we have $f \approx 2 \times 10^{-15} T^3 p_0^{3/2}$ cm^{-1}; here T is in degrees and p_0 in atmospheres.

Finally, the energy flux lost by the action of the wind is

$$S_w = N_m V_n T = \frac{T}{T_0} V_n P_0. \tag{2.257b}$$

Here V_n is the wind velocity normal to the discharge boundary in the unperturbed gas.

The high temperature of the gas can be maintained by ohmic losses of the energy of the powerful radio wave. Let P be the total power lost by the radio wave in the discharge region. Equating it to the total power loss [Eqs. (2.257), (2.257a)], we arrive at the relation

$$4\pi R_0 \kappa_m T + \frac{4\pi}{3} R_0^3 J = P, \tag{2.258}$$

which determines approximately the discharge dimension R_0. For simplicity we have assumed here the discharge to be spherical and neglected the influence of the winds. We note that if the wave causing the discharge is converging and if

$$R_0 \gg \Delta_s = c/4\pi\sigma, \tag{2.259}$$

where Δ_s is the depth of penetration of the field into the plasma, then it becomes in fact completely dissipated in the discharge. In this case P is the total power of the wave.

At low discharge temperatures, the electron concentration N is small and the absorption of the wave is negligible; at high temperatures, N is large and a strong reflection of the radio wave takes place. The discharge temperature T_d which is the most effective from the point of view of energy absorption is determined approximately by the relation

$$\frac{4\pi e^2 N(T_d)}{m\omega^2} \approx 1. \tag{2.260}$$

From this, taking into account the exponential dependence of N on T [Eq. (2.256)], we obtain

$$T_d \approx \frac{0.22 \text{ eV}}{1 - 0.093 \ln(\omega/10^9) + 0.023 \ln(p_0/10^{-2})}. \tag{2.261}$$

Here ω is in s^{-1} and p_0 is in atmospheres. It is this relation which leads to the characteristic temperatures $T_d \sim 0.2$–0.3 eV at $\omega \sim 10^8$–10^{10}.

Equations (2.258), (2.259), and (2.261) determine approximately the temperature, the dimension, and the conditions for the existence of the discharge. We note that when Equations (2.260) and (2.261) are satisfied we have $\Delta_s \approx c/v_e$. Equation (2.259) determines then the minimum dimension of the discharge

$$R_{0\,\min} \approx \alpha c/v_e, \tag{2.262}$$

and Equation (2.258) determines the minimum power of the discharge-producing wave

$$P_{\min} \approx \frac{4\pi\alpha c}{v_e}\left[\kappa_m T + \frac{\alpha^2}{3}\frac{c^2}{v_e^2} J\right]. \tag{2.263}$$

Here the value of the constant α is ~ 2–3. At $v_e \approx 10^9$ s^{-1}, $p_0 \approx 0.01$ atm, and $T \approx 3000°$ the minimum dimension is $R_{0\,\min} = 1$ m, and the minimum power is $P_{\min} \approx 5$ kW. At $v_e = 10^8$ s^{-1}, $p_0 = 0.001$ atm, and $T = 3000°$ we have $R_{0\,\min} = 10$ m and $P_{\min} \approx 200$ kW. The minimum power $P_{\min}$ increases rapidly with increasing height z. The minimum dimension of the discharge increases at the same time.

With increasing dissipated wave power, the dimension of the discharge increases [see Eq. (2.258)]. At small values $P \ll P_0$, where

$$P_0 = \frac{8\sqrt{3\pi}(\kappa_m T)^{3/2}}{J^{1/2}}, \qquad R_{01} \approx \left(3\kappa_m \frac{T}{J}\right)^{1/2} \tag{2.264}$$

the dimension R_0 increases linearly with P. In this case $R_0 \ll R_{01}$. At large values $P \gg P_0$, to the contrary, we have $R_0 > R_{01}$; then $R_0 \sim P^{1/3}$. In the former case, the decisive role is played by losses through thermal conductivity, and in the latter by losses through radiation. The values of P_0 and R_{01} increase with decreasing temperature and pressure: $P_0 \sim T^{-2} P_0^{-3/4}$ and $R_{01} \sim T^{-3} P_0^{-3/4}$. At $T = 3000°$ and $P_0 = 0.01$ atm we have $R_{01} \approx 2$ m and $P_0 \approx 15$ kW.

We now dwell on the conditions under which the assumed isothermal character of the discharge is valid. It was assumed primarily that the plasma is close to thermodynamic equilibrium. This is true if the electron temperature T_e is close to the temperature T of the neutral gas, i.e., if the field of the wave E_0 is weaker than the characteristic plasma field E_p. Recognizing that for a spherical wave $E_0 \approx (2P/cR_0^2)^{1/2}$, we find that the isothermy condition imposes the following bound on the wave power

$$P \lesssim \frac{3}{2} \frac{mcT\,\delta(\omega^2 + \nu_e^2)}{e^2} R_0^2 \tag{2.265}$$

The discharge dimension R_0 is determined here by Equation (2.258). We see that Equation (2.265) is violated at sufficiently large values of R_0 (i.e., at sufficiently large values of P).

We note that isothermy is violated also as a result of the emergence of the radiation from the discharge zone. The role of this process, as well as the influence of the heating of the electrons at appreciable deviations of T_e from T, can be investigated only by a detailed kinetic analysis. It should be noted, in addition, that we consider here in fact, not breakdown conditions, but a condition under which ionization already existing in the gas is maintained.

It is also important that we have neglected above the influence of the winds. As seen from Equations (2.257b) and (2.258), this is correct only if

$$V_n < \frac{\kappa_m T_0}{P_0 R_0}\left(1 + \frac{JR_0^2}{3\kappa_m T}\right).$$

It is easily seen that for heights 20–60 km this condition leads to very strong limitations on the wind velocity, $V_n \lesssim 1$ m/s. Under real conditions of the upper atmosphere the average wind velocity is larger by one order of magnitude. At such velocities V_n, the losses connected with the winds are very large and suppress the ionization. The estimates presented above are valid therefore only if the heated region moves in the atmosphere together with the wind, i.e., when the radio beam that heats the plasma follows the motion of the heated region. Note that a detailed theory of isothermal discharge was considered by Meerovich and Pitaevskii (1972), Meerovich (1973), and Gil'denburg and Golubev (1974).

The heating of the neutral component can lead not only to isothermal ionization, but also to other significant effects in the ionosphere (Gurevich, 1975). For example, sufficiently intense heating of the neutrals by the radiowave pushes the heated mass of the gas upward and causes it to "float." A region in which the electron temperature decreases with height ($dT/dz < 0$) can appear above the heated zone. Convective instability develops readily in this region. Nonstationary heating of the neutrals can lead also to generation of gravitational and acoustic waves, etc. We emphasize, however, that in the lower layers of the ionosphere the process of heating of the neutral gas by radio waves is usually very slow: essential changes of the neutral temperatures under the influence of a radio wave can occur only after a time $\Delta t \gtrsim 10^4$–10^6 s. It can apparently play a more essential role in the region of the F layer; the characteristic time of heating the neutrals in this case, when account is taken of the enhanced dissipation of the radio waves (see Chap. 6), is of the order 10^2 s (Grigor'ev, 1975).

3. Self-Action of Plane Radio Waves

3.1. Simplification of Initial Equations

The propagation of electromagnetic waves is described by Maxwell's equations:

$$\operatorname{rot} \boldsymbol{H} = \frac{4\pi}{c} \boldsymbol{j} + \frac{1}{c} \frac{\partial \boldsymbol{E}}{\partial t}, \tag{3.1}$$

$$\operatorname{div} \boldsymbol{D} = 4\pi\rho, \tag{3.2}$$

$$\operatorname{rot} \boldsymbol{E} = -\frac{1}{c} \frac{\partial \boldsymbol{H}}{\partial t}, \tag{3.3}$$

$$\operatorname{div} \boldsymbol{H} = 0, \tag{3.4}$$

$$\boldsymbol{D} = \hat{\varepsilon} \boldsymbol{E}, \qquad \boldsymbol{j} = \hat{\sigma} \boldsymbol{E} + \frac{\partial}{\partial t} \left(\frac{\hat{\varepsilon} - 1}{4\pi} \boldsymbol{E} \right). \tag{3.5}$$

Here $\hat{\sigma}$ and $\hat{\varepsilon}$ are the operators of the conductivity and of the dielectric constant of the plasma; the magnetic permeability of the non-ionized medium is assumed equal to unity. Applying the curl operation to Equation (3.3) and using Equations (3.1) and (3.5), we arrive at the following equation for the electric field of the wave:

$$\nabla^2 \boldsymbol{E} - \operatorname{grad} \operatorname{div} \boldsymbol{E} - \frac{4\pi}{c^2} \frac{\partial}{\partial t} (\hat{\sigma} \boldsymbol{E}) - \frac{1}{c^2} \frac{\partial^2}{\partial t^2} (\hat{\varepsilon} \boldsymbol{E}) = 0. \tag{3.6}$$

Equation (3.6) is nonlinear, since the tensors $\hat{\sigma}$ and $\hat{\varepsilon}$, generally speaking, depend on $\boldsymbol{E}$.

3.1.1. Nonlinear Wave Equation

We assume that the wave frequency ω is high enough so that the time $1/\omega$ during which an appreciable change takes place in the field is much less than the characteristic time $1/\delta\nu_e$ of establishment of the average energy (or temperature) of the electrons in the plasma:

$$\omega \gg \delta\nu_e. \tag{3.7}$$

In the ionosphere, as noted above, Equation (3.7) is usually well satisfied. In this case, as shown in Sections 2.1.2, 2.3.1, and 2.3.2, the effective electron temperature in the strong alternating electric field of the wave can differ appreciably from its unperturbed value. It is important, however that T_e is established at a level that is constant and independent of the time; the periodically varying corrections to this value are small, their amplitude is on the order of $\delta\nu_e/\omega$ [see Eq. (2.43)]. The conductivity and the dielectric constant of the plasma [Eqs. (2.44) and (2.151)] behave in exactly the same manner. This means that the solution of Equation (3.6) can be sought by successive approximations, expanding it in powers of the small parameter $\delta\nu_e/\omega$.

$$\boldsymbol{E} = \boldsymbol{E}^{(1)} + \boldsymbol{E}^{(2)} + \cdots, \qquad \hat{\sigma} = \hat{\sigma}^{(1)} + \hat{\sigma}^{(2)} + \cdots, \qquad \hat{\varepsilon} = \hat{\varepsilon}^{(1)} + \hat{\varepsilon}^{(2)} + \cdots.$$

Substituting this expansion in Equation (3.6), we assume that in first-order approximation the conductivity $\hat{\sigma}$ and the dielectric constant $\hat{\varepsilon}$ are constant in time. In this approximation, a wave arriving at the plasma boundary preserves its frequency ω, and Equation (3.6) reduces to the wave equation:

$$\begin{gathered} \nabla^2 \boldsymbol{E}^{(1)} - \operatorname{graddiv} \boldsymbol{E}^{(1)} + \frac{\omega^2}{c^2}\hat{\varepsilon}' \boldsymbol{E}^{(1)} = 0, \\ \hat{\varepsilon}' = \hat{\varepsilon}^{(1)} + i4\pi\hat{\sigma}^{(1)}/\omega. \end{gathered} \tag{3.8}$$

The expressions for the tensors $\hat{\varepsilon}^{(1)}$ and $\hat{\sigma}^{(1)}$ or for the tensor $\hat{\varepsilon}'$ were analyzed in the preceding chapter. The tensors depend on the form of the electron distribution function or on T_e, i.e., on the amplitude $|\boldsymbol{E}^{(1)}|$. Because of this, Equation (3.8) is a nonlinear equation and differs significantly, of course, from the usually considered linear wave equation.

In the next order of approximation we have

$$\begin{gathered} \nabla^2 \boldsymbol{E}^{(2)} - \operatorname{graddiv} \boldsymbol{E}^{(2)} - \frac{4\pi\hat{\sigma}^{(1)}}{c^2}\frac{\partial \boldsymbol{E}^{(2)}}{\partial t} - \frac{\hat{\varepsilon}^{(1)}}{c^2}\frac{\partial^2 \boldsymbol{E}^{(2)}}{\partial t^2} = \frac{4\pi}{c^2}\frac{\partial \boldsymbol{j}^{(2)}}{\partial t}, \\ \boldsymbol{j}^{(2)} = \hat{\sigma}^{(2)}\boldsymbol{E}^{(1)} + \frac{1}{4\pi}\frac{\partial}{\partial t}(\hat{\varepsilon}^{(2)}\boldsymbol{E}^{(1)}) = \frac{1}{4\pi}\frac{\partial}{\partial t}[\hat{\varepsilon}'^{(2)}\boldsymbol{E}^{(1)}]. \end{gathered} \tag{3.9}$$

Here the tensors $\hat{\sigma}^{(1,2)}$, $\hat{\varepsilon}^{(1,2)}$, and $\varepsilon'^{(2)}$ and the current $\boldsymbol{j}^{(2)}$ depend only on $\boldsymbol{E}^{(1)}$. Therefore Equation (3.9) for the electric field of the harmonic $\boldsymbol{E}^{(2)}$ is linear. The frequency of the wave $\boldsymbol{E}^{(2)}$ coincides with the frequency of the nonlinear correction to the current $\boldsymbol{j}^{(2)}$, i.e., it is equal to the sum or to the difference of the frequency of the fundamental wave ω and the frequencies of the alternating corrections to the conductivity and the dielectric constant of the plasma $\hat{\sigma}^{(2)}$ and $\hat{\varepsilon}^{(2)}$.

The equations for the higher harmonics can be obtained in similar fashion.

3.1.2. Nonlinear Geometrical Optics of a Plane Wave

The wave Equation (3.8) is quite complicated. It is known from the linear theory that it can be greatly simplified in the case when the medium is weakly inhomogeneous, i.e., when the properties of the medium vary little over the wave length. In this case, the geometrical-optics approximation is valid. The wave field in this approximation is represented in the form of a bundle of rays, characterized at each point by a definite intensity and propagation direction. An analogous approximation of geometrical optics can be constructed also for the nonlinear wave Equation (3.8).

By virtue of the presence of two independent states of polarization of the wave, it is frequently necessary to consider in geometrical optics a wave field consisting at each point of two or several rays. In the linear theory, these rays are independent (with the exception of special regions where geometrical optics is not valid). In the nonlinear theory, to the contrary, different rays are always interrelated.

We consider a plane-layered medium, i.e., we assume that the properties of the plasma vary only in one direction z [i.e., $\hat{\varepsilon} = \hat{\varepsilon}(z)$ and $\hat{\sigma} = \hat{\sigma}(z)$]. We assume further that the waves are plane, propagate in the z direction, and have one normal linear-polarization component. Then the conductivity and dielectric tensors of the plasma have each only one significant component, and the wave Equation (3.8) takes the form

$$\frac{d^2E}{dz^2} + \frac{\omega^2}{c^2}\,\varepsilon'(z, |\boldsymbol{E}|)E = 0. \tag{3.10}$$

Here $\varepsilon' = \varepsilon + i4\pi\sigma/\omega$. The boundary condition is

$$E\big|_{z=0} = E_0. \tag{3.11}$$

The solution of Equation (3.10) can be naturally represented as a sum of two waves propagating in opposite directions. We consider only one traveling wave. In other words, we seek the solution of Equation (3.10) in the form

$$E(z) = E_1(z) \exp \left\{ i \int k \, dz - \frac{\omega}{c} \int \kappa \, dz \right\}. \tag{3.12}$$

Here k is the wave vector

$$k = \frac{\omega}{c} n, \tag{3.13}$$

n and κ are the refractive index and the absorption coefficient of the wave

$$\begin{aligned} n &= \operatorname{Re} \sqrt{\varepsilon'(z)} = \left[\frac{\varepsilon}{2} + \sqrt{\left(\frac{\varepsilon}{2}\right)^2 + \left(\frac{2\pi\sigma}{\omega}\right)^2} \right]^{1/2}, \\ \kappa &= \operatorname{Im} \sqrt{\varepsilon'(z)} = \left[-\frac{\varepsilon}{2} + \sqrt{\left(\frac{\varepsilon}{2}\right)^2 + \left(\frac{2\pi\sigma}{\omega}\right)^2} \right]^{1/2}. \end{aligned} \tag{3.14}$$

In particular, if $\varepsilon \gg 4\pi\sigma/\omega$, then

$$n \simeq \sqrt{\varepsilon}, \qquad \kappa = 2\pi\sigma/\omega\sqrt{\varepsilon}. \tag{3.15}$$

Substituting Equation (3.12) in Equation (3.10), we arrive at the following equation for $E_1(z)$:

$$\frac{d^2 E_1}{dz^2} + 2i\sqrt{\varepsilon'} \frac{dE_1}{dz} + \frac{iE_1}{2\sqrt{\varepsilon'}} \frac{d\varepsilon'}{dz} = 0. \tag{3.16}$$

If

$$\left| \frac{\lambda}{4\pi} \frac{d\varepsilon'}{dz} \cdot (\varepsilon')^{-3/2} \right| \ll 1, \qquad \lambda = 2\pi/k, \tag{3.17}$$

then the term d^2E_1/dz^2 [Eq. (3.16)] can be neglected. It follows therefore that

$$E_1(z) = (\varepsilon')^{-1/4} E(0),$$

where $E(0)$ is the field at the plasma boundary [(Eq. 11)]. Thus, in geometrical-optics approximation we have

$$E(z) = \frac{E(0)}{(\varepsilon')^{1/4}} \exp\left(i \frac{\omega}{c} \int_0^z \sqrt{\varepsilon'} \, dz \right). \tag{3.18}$$

This relation is an integral equation in the nonlinear theory since ε' itself depends on the field amplitude $E(z)$. Changing in Equation (3.18) to $E_0(z) = |E(z)|$, we rewrite it in the form

$$E_0(z) = \frac{E_0(0)}{(n^2 + \kappa^2)^{1/4}} \exp\left(-\frac{\omega}{c}\int_0^z \kappa \, dz\right), \tag{3.19}$$

where $n = n[z, E_0(z)]$ and $\kappa = \kappa[z, E_0(z)]$. Far from the reflection point we have $n \gg \kappa$; in addition, the dependence of the refractive index on the field amplitude is usually slight, $n \neq n(E_0)$. In this case Equation (3.19) becomes simpler and is conveniently rewritten in differential form

$$\frac{dE_0}{dz} + \frac{E_0}{2n}\frac{dn}{dz} + \frac{\omega}{c}\kappa E_0 = 0. \tag{3.20}$$

Equations (3.19) and (3.20) are the equations of nonlinear geometrical optics for a plane wave propagating normal to the plasma layer. In contrast to the linear theory, the absorption coefficients and the refractive indices of the wave depend here on the wave amplitude E_0. We note that in a strong field, when $\varepsilon' = \varepsilon'(|\mathbf{E}|)$, to satisfy Equation (3.17) it is necessary to have

$$\kappa/n \ll 1. \tag{3.21}$$

3.2. Effect of Nonlinearity on the Amplitude and Phase of the Wave

We consider the propagation of a plane wave in a plasma layer. The wave propagates in the z direction (normal incidence on the layer). In the geometrical-optics approximation, the amplitude of the wave field in the plasma is described by Equation (3.20) or Equation (3.19).

3.2.1. Self-Action of a Weak Wave

We assume first that the wave arriving at the plasma boundary is weak $E_0^2 \ll E_p^2$ (Vilenskii, 1955; Vilenskii and Zykova, 1959). In this case the modifications it produces in the plasma are small. Accordingly, the non linear effects are also small, so that we can use the perturbation theory to solve Equation (3.20), assuming the first-order approximation to be linear:

$$E_0 = E_1(z) + E_2(z) + \cdots$$
$$E_1(z) = \frac{E_0(0)}{\sqrt{n_0(z)}} \exp\left(-\frac{\omega}{c}\int_0^z \kappa_0\, dz\right), \tag{3.22}$$

where $n_0(z)$ and $\kappa_0(z)$ are the linear refractive index and absorption coefficient of the medium. Substituting the expansion Equation (3.22) in Equation (3.20), we obtain

$$\frac{dE_2}{dz} + \frac{E_2}{2n_0}\frac{dn_0}{dz} + \frac{\omega}{c}\kappa_0 E_2 = -E_1 R,$$
$$R = \frac{\omega}{c}\left(\frac{d\kappa}{dT_e}\right)_{T_{e0}} \Delta T_e(E_1) + \frac{1}{2}\frac{d}{dz}\left[\left(\frac{dn}{dT_e}\right)_{T_{e0}} \frac{\Delta T_e(E_1)}{n_0}\right]. \tag{3.23}$$

We recognize here that the nonlinearity is due to heating of the plasma electrons in the wave field; ΔT_e is the perturbation of the effective electron temperature by the field, and T_{e0} is the electron temperature in the unperturbed plasma. The change of the electron effective temperature, more accurately, of the distribution function $f_0(v)$ (see Sect. 2.3), leads to a corresponding change of the concentration and frequency of the collisions of the electrons, and consequently of the absorption coefficient and refractive index of the wave. In the calculation of the modifications of n and κ we shall use below, for convenience, mainly the elementary theory.

Integrating Equation (3.23) with the boundary condition $E_2 \to 0$ as $z \to 0$, we have

$$E_2(z) = -E_1(z)\int_0^z R\, dz. \tag{3.24}$$

Thus the amplitude $E_0(z)$ of the wave field in the plasma, when account is taken of the nonlinearity, is given by the expression

$$E_0(z) = E_1(z)\left[1 - q - \frac{1}{2n_0}\left(\frac{dn}{dT_e}\right)_{T_{e0}} \Delta T_e(E_1)\right],$$
$$q = \int_0^z \Delta T_e(E_1)\frac{\omega}{c}\left(\frac{d\kappa}{dT_e}\right)_{T_{e0}} dz. \tag{3.25}$$

Here $E_1(z)$ is the wave-field amplitude in the linear approximation [Eq. (3.22)]. The coefficient q characterizes the influence of the nonlinearity on the field of the weak wave [inasmuch as dn/dT_e is proportional to

N_0, the last term in Equation (3.25) vanishes when the wave leaves the plasma, i.e., at $N_0 \to 0$].

Let us obtain concrete expressions for the coefficient q, and introduce certain simplifications for this purpose. We consider the region far from the wave reflection point, when the change of the refractive index can be neglected (the self-action in the region of wave reflection is considered in Sect. 5.3). In this case, at $n_0 \gg \kappa_0$, as follows from Equations (3.15) and (2.13), we have in the elementary-theory approximation

$$\kappa = \frac{2\pi e^2 N v_e}{m\omega n_0(\omega^2 + v_e^2)}, \tag{3.26}$$

where v_e is the effective electron collision frequency. Under the influence of the electric field of the wave, the electron temperature T_e rises; the result is a change in the collision frequency v_e and in the electron concentration N in the plasma. Accordingly

$$\left(\frac{d\kappa}{dT_e}\right)_{T_{e0}} = \left\{\frac{2\pi e^2 N(\omega^2 - v_e^2)}{m\omega(\omega^2 + v_e^2)^2 n_0}\left(\frac{dv_e}{dT_e}\right) + \frac{2\pi e^2 v_e}{m\omega(\omega^2 + v_e^2)n_0}\left(\frac{dN}{dT_e}\right)\right\}_{T_{e0}}. \tag{3.27}$$

We now recognize that in accordance with Equations (2.32a) and (3.22) we have

$$\Delta T_e(E_1) = \frac{e^2 E_0^2(0)}{3m\delta_0(\omega^2 + v_{e0}^2)n_0}\exp\left(-\frac{2\omega}{c}\int_0^z \kappa_0(z_1)\,dz_1\right), \tag{3.28}$$

where $\delta_0 = \delta(T_{e0})$ and $v_{e0} = v_e(T_{e0})$. We assume also that $v_{e0} \neq v_{e0}(z)$ and that $dN/dT_e = \gamma_1 N/T_{e0}$, where γ_1 is a constant (see Sect. 2.5.5). In this case the integration in Equation (3.25) can be carried out analytically. Substituting Equations (3.27) and (3.28) in Equation (3.25), and integrating, assuming $n_0 \neq n_0(z)$, we obtain;

$$q = \frac{1}{n_0}\left[\frac{E_0^2(0)}{E_p^2}\right]\left[\frac{T_{e0}}{2v_{e0}}\left(\frac{dv_e}{dT_e}\right)_{T_{e0}}\frac{\omega^2 - v_{e0}^2}{\omega^2 + v_{e0}^2} + \frac{\gamma_1}{2}\right][1 - \exp(-2K)], \tag{3.29}$$

$$K = \frac{\omega}{c}\int_0^z \kappa_0(z_1)\,dz_1. \tag{3.30}$$

We see therefore that the coefficient q is proportional to $[E_0(0)/E_p]^2$, i.e., the expansion [Eq. (3.22)] is indeed valid only for a weak field. As the wave penetrates into the plasma, q increases, tending to a constant value when the total absorption K of the wave in the plasma becomes large enough: $K \gtrsim 1$. The sign of q can be either positive (the wave is weakened by the nonlinearity) or negative (the wave is enhanced), depending on the ratio of the wave frequency ω to the collision frequency ν_{e0}, and also on the character of the dependence of ν_e on T_e and of N on T_e. We recall that in collisions with molecules we usually have $d\nu_e/dT_e > 0$, and in collisions with ions $d\nu_e/dT_e < 0$ (see Sect. 2.1.1), so that the character of the action of the nonlinearity on the wave field amplitude depends essentially on the degree of ionization of the plasma.

We note that in the foregoing calculation of the perturbations of T_e we used elementary field theory. When the kinetic corrections are taken into account (in the case $\nu_{ee} \gg \delta\nu_e$), additional factors K_σ appear in Equations (3.26) and (3.29) (Sect. 2.3.1).

The corrections to the phase of the wave are determined when account is taken of the influence of the nonlinearity on the refractive index of the wave:

$$\Delta\varphi = \frac{\omega}{c}\int_0^z \left[n(E_1^2) - n_0\right] dz - \frac{1}{4}\operatorname{Im}\left[\varepsilon'(E_1^2) - \varepsilon_0'\right]. \tag{3.31}$$

The last term takes into account the change of the phase by the presence of $(\varepsilon')^{1/4}$ in Equation (3.18). Substituting here Equation (3.14) for n, using Equations (2.13), (2.6), and (2.32a) for σ, ε, ν_e, and T_e, and using Equation (3.22) for $E_1^2(z)$ and then integrating with respect to z, we can calculate the nonlinear correction to the phase. In particular, under the same assumption as used to calculate q analytically in Equation (3.29), we obtain

$$\Delta\varphi = \left[\frac{E_0^2(0)}{E_p^2}\right]\left[\frac{\omega\nu_{e0}}{\omega^2 + \nu_{e0}^2}\frac{T_{e0}}{\nu_{e0}}\left(\frac{d\nu_e}{dT_e}\right)_{T_{e0}} - \frac{\gamma_1}{2}\frac{\omega}{\nu_{e0}}\right]\left[1 - \exp(-2K)\right]. \tag{3.32}$$

3.2.2. Self-Action of a Strong Wave

We proceed now to consider the self-action of strong waves, i.e., we do not limit the wave field amplitude (Gurevich, 1956). Equation (3.20), which describes the amplitude of the wave in the plasma, can be integrated numerically. The results of the corresponding calculations for the ionosphere will be discussed in Section 3.5. Here we consider simple models that make it possible to obtain analytic expressions for the field $E_0(z)$. We assume that the wave propagates far from the reflection point, so that $n = n_0 \neq n_0(z)$. The second term in Equation (3.20) can then be neglected, and this equation takes the form

$$\frac{dE_0}{dz} + \frac{\omega}{c}\kappa(z, E_0)E_0 = 0. \tag{3.33}$$

The coefficient of absorption of the wave is given by Equation (3.26) also in the case of a strong field. To separate here the nonlinearity, we rewrite it in the form

$$\kappa(z, E_0) = \kappa_0(z)\frac{N(T_e)}{N_0}\frac{\nu_e(T_e)}{\nu_{e0}}\frac{\omega^2/\nu_{e0}^2 + 1}{\dfrac{\omega^2}{\nu_{e0}^2} + \dfrac{\nu_e^2(T_e)}{\nu_{e0}^2}}. \tag{3.34}$$

Here $\kappa_0(z)$ is the wave absorption coefficient in the linear theory, and $\nu_e(T_e) = \nu_{em} + \nu_{ei}$ is the effective collision frequency. The dependence of ν_e and N on T_e and the dependence of the stationary electron temperature on the field amplitude E_0 are determined in Chapter 2. We assume that $\delta = \text{const}$ and that the collision frequency in the unperturbed plasma ν_{e0} and the temperatures T_{e0} and T do not depend on z, with $T_{e0} = T$. In this case Equation (3.33) can be integrated analytically.

Weakly Ionized Plasma. We consider first the case of a weakly ionized plasma, when the principal role is played by collisions with molecules as hard spheres. In this case the ratio $\nu_e(T_e)/\nu_{e0}$ equals $\sqrt{T_e/T}$ [see Eq. (2.6]. It is then convenient to use in Equation (3.34) in lieu of E_0 a new variable $\tau = \sqrt{T_e(E_0)/T}$. The dependence of the stationary temperature of the electrons T_e on the field E_0 is determined by Equation (2.33), which takes in terms of the new variables the form

$$\tau^2 - 1 = \left(\frac{E_0}{E_p}\right)^2 \frac{\omega^2 + \nu_{e0}^2}{\omega^2 + \nu_{e0}^2\tau^2}. \tag{3.35}$$

Differentiating Equation (3.35), we obtain

$$\frac{1}{E_0}\frac{dE_0}{dz} = \frac{d\tau}{dz}\frac{\tau(\omega^2 + 2\nu_{e0}^2\tau^2 - \nu_{e0}^2)}{(\tau^2 - 1)(\omega^2 + \nu_{e0}^2\tau^2)}. \tag{3.36}$$

Substituting this expression in Equation (3.33) taking Equation (3.34) into account, we obtain

$$\frac{d\tau}{dz}\left(\frac{1}{\tau^2 - 1} + \frac{2\nu_{e0}^2}{\omega^2 + \nu_{e0}^2}\right) + \frac{\omega}{c}\kappa_0(z)\frac{N(\tau^2)}{N_0} = 0. \tag{3.37}$$

In this section we consider the case when the electron concentration does not vary in the wave field, $N(\tau^2)/N_0 = 1$. Then the solution of Equation (3.37) takes the form

$$\frac{\tau - 1}{\tau + 1} \exp\left(\frac{4\nu_{e0}^2 \tau}{\omega^2 + \nu_{e0}^2}\right) = C \exp(-2K), \tag{3.38}$$

where K is the total linear absorption of the wave in the plasma, from the plasma boundary ($z = 0$) to the point z under consideration [see Eq. (3.30)]. On the plasma boundary ($z = 0$) the wave amplitude is $E_0(0)$. Consequently,

$$C = \frac{\tau_0 - 1}{\tau_0 + 1} \exp\left(\frac{4\nu_{e0}^2 \tau_0}{\omega^2 + \nu_{e0}^2}\right), \qquad \tau_0 = \sqrt{\frac{T_e[E_0(0)]}{T}}. \tag{3.39}$$

It is seen from Equation (3.38) that τ_0 is the maximum value of τ: with increasing z, more accurately, with increasing $K = \int_0^z \kappa_0 \, dz$, the value of τ decreases monotonically. In the interior of the plasma, at $K \gg 1$, the value of τ tends to unity, i.e., the wave becomes weak, as it should.

The solution [Eq. (3.38)] determines τ at each point z inside the plasma. Knowing $\tau(z)$, we can easily determine from Equation (3.35) the sought amplitude $E_0(z)$. It is convenient to represent it in the form

$$E_0(z) = E_0(0) \exp(-K) P \tag{3.40}$$

P is the factor that accounts for the result of the self-action of the wave in the plasma. The quantity $-\ln P$ is the nonlinear absorption of the wave. In a weak field $P = 1 - q$ and is close to unity (see Sect. 3.2.1). In the general case P depends on the wave amplitude at the plasma boundary, on the frequency, and on the depth of penetration of the wave into the plasma:

$$P = P\left(\frac{E_0(0)}{E_p}, \frac{\omega}{\nu_{e0}}, K\right). \tag{3.41}$$

The expression for the self-action factor has a simple form in the interior of the plasma, i.e., at $K \gg 1$. In fact, as is clear from Equation (3.35), $\tau(z)$ is close to unity in this case. From Equation (3.38) we find:

$$\tau(z) = 1 + 2 \exp\left(-\frac{4\nu_{e0}^2}{\omega^2 + \nu_{e0}^2}\right) C \exp(-2K). \tag{3.42}$$

Here $C = C(\tau_0)$ is determined by Equation (3.39). On the other hand, it follows from Equation (3.35) that as $\tau \to 1$ we have

$$E_0 = E_p\sqrt{2}\sqrt{\tau - 1}.$$

Substituting here $\tau - 1$ from Equation (3.42), we obtain

$$E_0 = 2E_p\sqrt{C}\exp\left[-\frac{2\nu_{e0}^2}{\omega^2 - \nu_{e0}^2} - K\right]. \tag{3.43}$$

Consequently, at $K \gg 1$ we have

$$P = 2\frac{E_p}{E_0(0)}\sqrt{\frac{\tau_0 - 1}{\tau_0 + 1}}\exp\left[\frac{2\nu_{e0}^2(\tau_0 - 1)}{\omega^2 + \nu_{e0}^2}\right]. \tag{3.44}$$

It is seen from Equation (3.44) that at high wave frequencies, $\omega^2 \gg 2\nu_{e0}^2(\tau_0 - 1)$, the self-action factor is smaller than unity. This is perfectly understandable, since at high wave frequencies the absorption coefficient increases with increasing T_e, i.e., with increasing amplitude E_0. In a very strong field at $E_0(0) \gg E_p$, i.e., $\tau_0 \gg 1$, the factor P decreases with increasing $E_0(0)$ in proportion to $E_p/E_0(0)$. In this case the amplitude of the wave passing into the interior of the plasma does not increase with increasing $E_0(0)$, but tends to a constant value independent of $E_0(0)$:

$$E_0(z) \to 2E_p \exp(-K). \tag{3.44a}$$

Something like saturation takes place, namely, the nonlinear absorption of the wave increases so strongly with increasing wave amplitude, that the amplitude of the wave passing into the interior of the plasma ceases to depend on its value on the plasma boundary (Fig. 17, curve 1).

When the inverse condition $\omega^2 \ll 2\nu_{e0}^2\tau_0$ is satisfied, the factor P in the interior of the plasma, on the contrary, increases rapidly with increasing $E_0(0)$. This is natural, since the absorption coefficient of a low frequency wave $\omega < \nu_{e0}$ decreases with increasing T_e. In a very strong field, the self-action factor

$$P = 2\frac{E_p}{E_0(0)}\exp\left[2\sqrt{\left(\frac{\nu_{e0}^2}{\omega^2 + \nu_{e0}^2}\right)^{3/2}\frac{E_0(0)}{E_p}}\right] \tag{3.45}$$

increases exponentially with increasing $E_0(0)$ (see Fig. 17, curve 2).

It follows from Equations (3.44) and (3.45) that at $\omega^2 \gg \nu_{e0}^2$ the self-action factor first decreases with increasing $E_0(0)/E_p$ and then increases.

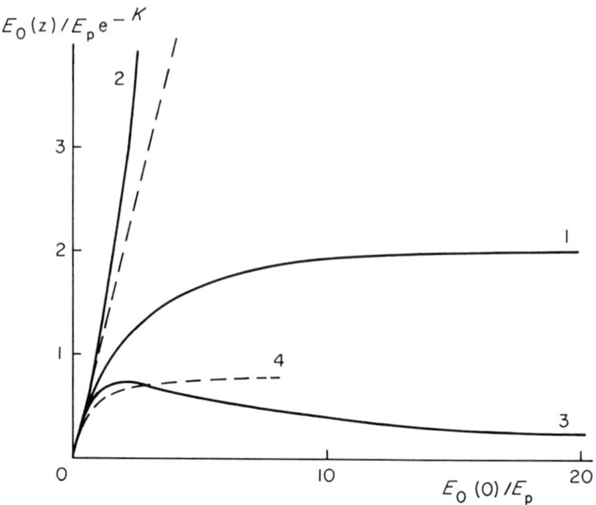

Fig. 17. Amplitude of wave field in plasma interior (curve 1: $\omega^2 \gg v_{e0}^2$; curve 2: $\omega^2 \ll v_{e0}^2$) and amplitude of reflected wave (curve 3, $\omega^2 \gg v_{e0}^2$). *Dashed line*: linear approximation (17); *dashed curve* 4: amplitude of field in the interior of the plasma at $\omega^2 \gg v_{e0}^2$ with allowance for the dependence of δ on T_e [Eq. (3.49b)]

The minimum value of P is

$$P_{\min} = \frac{2v_{e0}^3}{\omega(\omega^2 + v_{e0}^2)} \exp\left[\frac{2(\omega^2 - v_{e0}^2)}{\omega^2 + v_{e0}^2}\right]. \tag{3.46}$$

It is reached at

$$\left(\frac{E_0(0)}{E_p}\right)_{\min} = \frac{\omega\sqrt{\omega^4 - v_{e0}^4}}{v_{e0}^3}. \tag{3.47}$$

With increasing frequency, the value of $P_{\min}$ decreases in proportion to $(v_{e0}/\omega)^3$. The dependence of the self-action factor in the interior of the plasma (at $K \gg 1$) on the amplitude of the field on the plasma boundary and on the frequency of the wave is shown in Figures 18 and 19.

We have analyzed above the expression for the self-action field in the interior of the plasma ($K \gg 1$). It is valid also if the following condition is satisfied

$$K(z) > 1 + \frac{2v_{e0}^2(\tau_0 - 1)}{\omega^2 + v_{e0}^2}.$$

This condition determines the thickness of the characteristic plasma layer in which the wave is strong and exerts a strong influence on the medium.

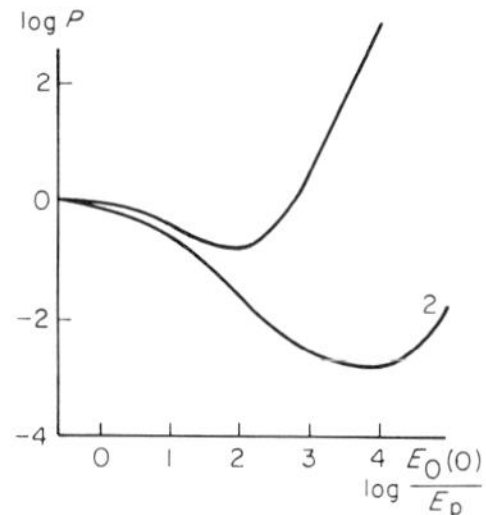

Fig. 18. Self-action factor. *1*: $\omega/\nu_{e0} = 4.2$; *2*: $\omega/\nu_{e0} = 20$

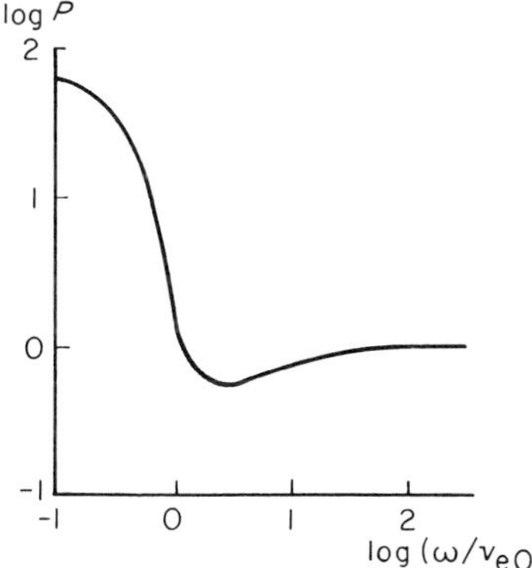

Fig. 19. Frequency dependence of the self-action factor; $E_0(0)/E_p = 20$

The change of P in the characteristic layer is determined by the general Equations (3.38) and (3.40). A simple limiting expression for P is obtained at high frequencies ($\omega^2 \gg 2\nu_{e0}^2\tau_0$):

$$P = \frac{2E_p}{E_0(0)}\sqrt{\frac{\tau_0 - 1}{\tau_0 + 1}}\left[1 - \frac{\tau_0 - 1}{\tau_0 + 1}\exp(-2K)\right]^{-1}, \qquad (3.48)$$

$$\tau_0 = \sqrt{1 + E_0^2(0)/E_p^2}.$$

We see therefore that the self-action factor at high frequencies decreases rapidly with increasing K. The stronger the field, the faster the decrease of P so that the thickness of the characteristic layer does not depend on $E_0(0)$ (it is determined by the condition $K \gtrsim 1$).

It was assumed above for simplicity that the frequency of the collisions of the electron with the neutral particles increases with increasing electron temperature like $\sqrt{T_e/T}$ (hard-sphere collisions). Actually, the dependence of the frequency of the collisions with the molecules on the temperature is more complicated (see Sects. 2.3.1 and 2.5.2). In particular, in air and in the lower ionosphere, the electron collision frequencies at not too high energies is proportional to $(T_e/T)^{5/6}$ [Eqs. (2.140) and (2.235)]. The calculation of the self-action effects in this case is perfectly analogous to that given above. Replacing E_0 in Equation (3.33), as before, by a new variable $\tau = \sqrt{T_e/T}$, defined by the relation

$$\tau^2 - 1 = \left(\frac{E_0}{E_p}\right)^2 \frac{\omega^2 + \nu_{e0}^2}{\omega^2 + \nu_{e0}^2 \tau^{10/3}},$$

we obtain in place of Equation (3.37) the following equation for $\tau(z)$:

$$\frac{d\tau}{dz}\left[\frac{8}{3}\nu_{e0}^2\tau^{2/3} + \frac{\tau^{2/3}}{\tau^2 - 1}\nu_{e0}^2 + \frac{\omega^2}{\tau^{2/3}(\tau^2 - 1)}\right] = -(\omega^2 + \nu_{e0}^2)\frac{\omega}{c}\kappa_0.$$

Integrating this expression, we obtain the function $\tau(z)$ in implicit form:

$$\sqrt{\frac{\tau - 1}{\tau + 1}} \exp\left[\frac{1}{\omega^2 + \nu_{e0}^2}\left(\frac{8}{5}\nu_{e0}^2\tau^{5/3} + \nu_{e0}^2 I_1(\tau) + \omega^2 I_2(\tau)\right)\right]$$
$$= \sqrt{\frac{\tau_0 - 1}{\tau_0 + 1}} \exp\left[\frac{1}{\omega^2 + \nu_{e0}^2}\left(\frac{8}{5}\nu_{e0}^2\tau_0^{5/3} + \nu_{e0}^2 I_1(\tau_0) + \omega^2 I_2(\tau_0)\right) - K(z)\right].$$

Here, as usual, τ_0 is the value τ on the plasma boundary [at $E_0 = E_0(0)$] $K(z)$ is the total absorption of the weak wave, and the functions $I_1(\tau)$ and $I_2(\tau)$ are defined by the relations

$$I_1(\tau) = \int_1^\tau \frac{x^{2/3} - 1}{x^2 - 1}\,dx, \qquad I_2(\tau) = \int_1^\tau \frac{x^{-2/3} - 1}{x^2 - 1}\,dx.$$

In particular

$$I_2(\tau) = \frac{3}{4}\ln\frac{3(\tau^{2/3} - \tau^{1/3} + 1)}{\tau^{2/3} + \tau^{1/3} + 1} - \frac{\sqrt{3}}{2}\left(\operatorname{arctg}\frac{\tau^{2/3} - 1}{\sqrt{3}\tau^{1/3}}\right).$$

The amplitude of the field of the wave in the plasma can be represented as before in Equation (3.40). For the self-action factor in the interior of the plasma at $K > 1$ we obtain:

$$P = 2\sqrt{\frac{\tau_0 - 1}{\tau_0 + 1}}\frac{E_p}{E_0(0)} \exp\left[\frac{\nu_{e0}^2}{\omega^2 + \nu_{e0}^2}\left(\frac{8}{5}(\tau_0^{5/3} - 1) + I_1(\tau_0)\right) + \frac{\omega^2}{\omega^2 + \nu_{e0}^2} I_2(\tau_0)\right]. \tag{3.49}$$

We see therefore that in a strong high-frequency field we have in the interior of the plasma

$$P = \frac{E_p}{E_0(0)} \cdot 2 \cdot 3^{3/4} \exp\left(-\sqrt{3}\pi/4\right), \tag{3.49a}$$

and, as before, a "saturation" effect takes place: as $E_0(0) \to \infty$ we have

$$E_0(z) \to E_p \cdot 2 \cdot 3^{3/4} \exp\left(-\sqrt{3}\pi/4 - K(z)\right).$$

The self-action factor of a low-frequency wave $\omega^2 \ll v_{e0}^2$ increases exponentially with increasing field amplitude:

$$P = 2\frac{E_p}{E_0(0)} \exp\left\{\frac{8}{5}(\tau_0^{5/3} - 1) + I_1(\tau_0)\right\}.$$

The expressions obtained here can also be easily generalized to the case when the fraction δ of the lost energy depends on T_e. We then obtain for the self-action factor of a high-frequency wave in the interior of the plasma

$$P = 2\sqrt{\frac{\tau_0 - 1}{\tau_0 + 1}}\frac{E_p}{E_0(0)} \exp\left[I_2(\tau_0) + B\right], \qquad B = \frac{1}{2}\int_1^{\tau_0} \frac{d\tau}{\tau^{5/3}} \frac{d\delta/d\tau}{\delta}. \tag{3.49b}$$

The amplitude of the field of a high-frequency wave $\omega^2 \gg v_{e0}^2$ in the interior of the plasma, under the conditions of a real ionosphere, i.e., with allowance for Equation (3.49b), is shown by the dashed curve 4 of Figure 17 (the dependence of δ on T_e was assumed in accordance with Fig. 10). The most characteristic feature of this curve is the abrupt turning on of the nonlinearity at $E_0(0) \sim E_p$, this being due to the decrease of $\delta(T_e)$ at low temperatures T_e, i.e., to the influence of the rotational effect (Sect. 2.3.2). In addition, when account is taken of the dependence of δ on T_e there is no saturation in the strict meaning of the word—the amplitude of the field in the interior of the plasma is not constant as in the case when δ is constant [Eqs. (3.44a) and (3.49a)], but increases slowly with increasing amplitude $E_0(0)$ of the incident wave.

We have considered above only the penetration of the wave into the interior of the plasma layer. If the wave is reflected in the plasma, then it passes again through the nonlinear region, i.e., it experiences a second nonlinear absorption. Then, at $K \gg 1$, the amplitude of the reflected wave is

$$E_r = E_0 \exp(-K)P^2. \tag{3.50}$$

Here K is the combined linear absorption of the incident and reflected waves. It is seen from Equations (3.44), (3.49), and (3.50) that owing to the nonlinear absorption in a strong high-frequency field the amplitude of the reflected wave not only fails to increase, but even decreases with increasing radiation power, $E_r \sim 1/E_0$ (Fig. 17, curve 3). The nonlinear abosprtion of the reflected wave at $E_0(0)/E_p \gg 1$, as seen from Figure 17, is very large (compare curve 3 with the dashed line representing the linear solution); it determines completely the amplitude of the reflected wave. We note that if $K < 1$, then the interaction of the incident and reflected waves has a more complicated character and Equation (3.50) is not valid in this case.

The phase corrections $\Delta\varphi$ necessitated by the self-action of the strong wave are calculated analogously. Substituting in Equation (3.31) the Equations (3.38)–(3.40) obtained for $E_0(z)$, we obtain

$$\Delta\varphi = \operatorname{arctg} \frac{\nu_{e0}\tau_0}{\omega} - \operatorname{arctg} \frac{\nu_{e0}\tau}{\omega}. \tag{3.51}$$

Here $\tau = \tau(z)$ is determined, as before, by Equation (3.38). As is clear from Equation (3.51), changes of the phase of a plane wave as the result of self-action are negligible at both high and low frequency limits. The largest value (of the order of $\pi/2$) of $\Delta\varphi$ is reached at $\omega^2 \gg \nu_{e0}^2$ in a sufficiently strong field, when $\omega^2 \ll \nu_{e0}^2\tau_0^2$.

Strongly Ionized Plasma. We consider now the case of a strongly ionized plasma, when the principal role is played by collisions of the electrons with the ions [Eqs. (2.7), (2.141) and (2.142)]. We assume for simplicity that the frequency of the collisions of the ions with the neutrals is high enough, $\nu_{im} \gg \delta_{ei}\nu_{ei}$. In this case, as is clear from Equations (2.56) and (2.210), the ion temperature is close to the temperature of the neutral particles, i.e., it does not depend on the amplitude of the field E_0. We note that only a high-frequency wave $\omega \gg \nu_{e0}$ can propagate in a strongly ionized isotropic plasma, since $\omega \gtrsim \omega_0 \gg \nu_{e0}$, where ω_0 is the Langmuir frequency of the plasma, and the relation $\omega_0 \gg \nu_{e0}$ is always satisfied by virtue of Equation (2.94). In this case the expression for the electron temperature takes the simple form of Equation (2.35):

$$\frac{T_e}{T} = 1 + \left(\frac{E_0}{E_p}\right)^2. \tag{3.52}$$

Taking into account now the concrete form of the frequency of the collisions with the ions [Eqs. (2.7) and (2.141)], and performing in Equation

(3.33) the same transformations as in the preceding section, we obtain the solution of Equation (3.33) in implicit form:

$$\frac{\tau - 1}{\tau + 1} \exp\left(\frac{2\tau^3}{3} + 2\tau\right) = \frac{\tau_0 - 1}{\tau_0 + 1} \exp\left(\frac{2\tau_0^3}{3} + 2\tau_0 - 2K\right). \tag{3.53}$$

$$\tau = \sqrt{1 + E_0^2/E_p^2}. \tag{3.54}$$

In the interior of the plasma at $K > (\tau_0^3 - 1)/3$ we obtain therefore

$$E_0 = E_0(0) \exp(-K)P \tag{3.55}$$

$$P = 2\frac{E_p}{E_0(0)}\sqrt{\frac{\tau_0 - 1}{\tau_0 + 1}} \exp\left[\frac{\tau_0^3 - 1}{3} + \tau_0 - 1\right]. \tag{3.56}$$

We see therefore that the self-action factor increases exceptionally rapidly with increasing field amplitude. The wave pierces, as it were, into the interior of the plasma. This is seen from Figure 20.

The coefficient of transmission of the wave through a plasma layer of "thickness" $K_S = (\omega/c) \int_S \kappa_0 \, dS$ (i.e., the ratio of the amplitude of the wave passing through the layer to the amplitude of the wave incident on

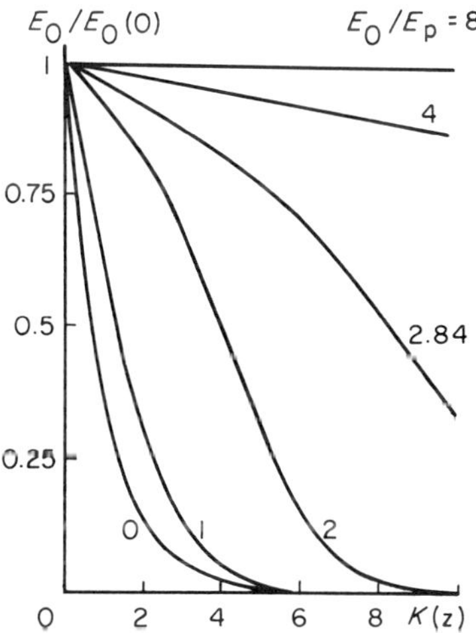

Fig. 20. Relative amplitude $E_0(z)/E_0(0)$ of wave field in plasma in the case of collisions with ions. The values of $E_0(0)/E_p$ are indicated in the *figure on curves*. The linear approximation corresponds to the curve $E_0(0)/E_p = 0$

the layer) rapidly approaches unity with increasing $E_0(0)/E_p$ (Fig. 21). If

$$E_0(0) > E_c \simeq 1.5\sqrt[3]{K_S}E_p \tag{3.57}$$

then a strong wave passes through the layer practically without damping. It is seen from Figure 21 that the effect of "bleaching of the plasma" by a strong wave is very abrupt, so that the field E_c defined by Equation (3.57) is so to speak the critical value for wave absorption in a layer of "thickness" K_S.

The corrections to the phase of the wave are small:

$$\Delta\varphi = -C\frac{\nu_{e0}}{\omega}\left(\frac{1}{\tau^2} - \frac{1}{\tau_0^2}\right), \qquad C = \left(\frac{9\pi^4}{16}\frac{\omega}{\nu_{e0}}\right)^{1/3}. \tag{3.58}$$

The coefficient C constitutes a correction to the elementary calculation obtained with the aid of the kinetic theory. In the case of collisions with ions we see that this coefficient is appreciable. We note that in collisions with ions the change of a phase has an opposite sign than in the case of collisions with molecules.

Change of Plasma Concentration. We have neglected above the change of the plasma concentration N as a result of the disturbance to the ionization-recombination balance. This is legitimate if the duration of the perturbing pulse is shorter than the electron lifetime τ_N (Table 13). In the

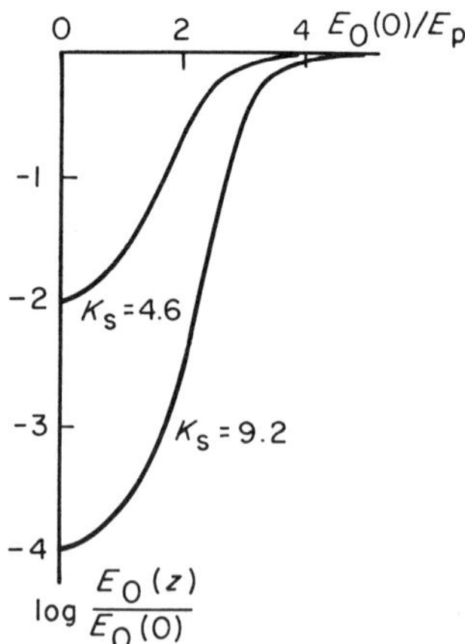

Fig. 21. Transmission coefficient of a wave through a plasma layer as a function of the field amplitude $E_0(0)$ at the layer boundary

opposite case, the changes of N can be appreciable. When account is taken of the $N(T_e)$ dependence, the self-action effects retain their character. Assuming the change of the concentration in Equation (3.37) to be given by Equations (2.248) and (2.249),

$$\frac{N(T_e)}{N_0} = \left(\frac{T_e}{T}\right)^{\gamma_1}, \tag{3.59}$$

we obtain

$$\frac{d\tau}{dz}\left[\frac{1}{\tau^2 - 1} + \frac{2v_{e0}^2}{\omega^2 + v_{e0}^2}\right] + \frac{\omega}{c}\,\kappa_0(z)\tau^{2\gamma_1} = 0. \tag{3.60}$$

The solution of this equation at $\gamma_1 = 0.5$ (see Table 14) takes the form

$$(\tau^2 - 1)\tau^{2a} = (\tau_0^2 - 1)\tau_0^{2a} \exp\left[-2K\right], \tag{3.61}$$

where

$$a = \frac{v_{e0}^2 - \omega^2}{v_{e0}^2 + \omega^2}, \quad \tau_0 = \sqrt{\frac{T_e[E_0(0)]}{T}}, \quad K = \frac{\omega}{e}\int_0^z \kappa_0\, dz. \tag{3.62}$$

The self-action factor P in the interior of the plasma is determined in this case by the expression

$$P = \frac{E_p}{E_0(0)}\sqrt{\tau_0^2 - 1\tau_0^a}. \tag{3.63}$$

It follows therefore that for a strong wave of high frequency $\omega^2 \gg v_{e0}^2$ the self-action factor is

$$P = \frac{E_p}{E_0(0)}. \tag{3.64}$$

In other words, in the case of a strong wave of high frequency, as before, "saturation" of the field in the plasma takes place: in the interior of the plasma ($K \gg 1$) the wave field amplitude does not exceed $E_p \exp(-K)$, regardless of the value of the field $E_0(0)$ at the plasma boundary. For a strong wave of low frequency we have $P = 1$. In this case the growth of the plasma concentration cancels out the decrease of the absorption coefficient of the wave as a result of the increase of the electron temperature. Because of this, the effects of self action in a strong low-frequency field $\omega^2 \ll v_{e0}^2$ turn out to be suppressed.

3.2.3. Self-Action of Waves in the Case of Artificial Ionization

In a sufficiently strong electric field $E \approx E_c$, an abrupt increase of the electron and ion concentration can be produced by the additional ionization of the plasma by fast electrons (see Sect. 2.5.6). This exerts a strong influence on the character of the self-action effects.

The concentration of the electrons in the plasma, with allowance for the additional ionization, is determined by the balance [Eq. (2.251)]. Under stationary conditions neglecting the negative ions we have approximately

$$q_i - \nu_r N + \nu_{ion} N = 0. \tag{3.65}$$

Here q_i is the number of electrons produced per second in a unit volume as a result of ionization by the external source (see Sect. 2.5.1), $\nu_r = k_a N_m$ is the recombination (attachment) frequency, ν_{ion} is the frequency of the additional ionization [Eq. (2.177)]. The frequency ν_{ion} is conveniently written in the form

$$\nu_{ion} = \nu_r \exp\left[Q\left(1 - \frac{E_c}{E}\right)\right]. \tag{3.66}$$

Here E is the amplitude of the wave field, $Q = \ln(\nu_{ion}^0/\nu_r)$ is a quantity on the order of 10 (see Sect. 2.5.6). We assume Q to be constant, i.e., we neglect the dependence of Q on E and z. Further, $E_c(z)$ is the critical breakdown field determined by Equation (2.253).

The field E_c in the lower ionosphere decreases rapidly with increasing height z at $B/A > \omega$ (see Fig. 16). We assume that $E_c(z)$ decreases more rapidly than the amplitude of the wave field $E(z)$,

$$\frac{1}{E_c}\frac{dE_c}{dz} \approx \left|\frac{1}{N_m}\frac{dN_m}{dz}\right| \frac{(B/A)^2}{\omega^2 + (B/A)^2} \gg \frac{\omega}{c}\kappa. \tag{3.67}$$

Then, if the wave field at small z is smaller than critical, then at a certain point z_0 it becomes comparable with the critical one, $E(z_0) = E_c(z_0)$. The plasma concentration in the vicinity of this point z_0 should increase appreciably.

Indeed, let the wave propagate in the plasma in the z direction. The field amplitude is again determined by Equations (3.33) and (3.34). The concentration of the plasma is determined in this case by Equation (3.65).

From Equations (3.65) and (3.66) follows a simple expression for the stationary concentration of the electrons and ions:

$$N = \frac{N_0(z)}{1 - \exp\left[Q\left(1 - \frac{E_c(z)}{E}\right)\right]}. \tag{3.68}$$

Here $N_0(z) = q_i/\nu_r$ is the plasma concentration in the absence of a perturbing field. At $E \ll E_c(z)$ the concentration is $N = N_0(z)$. At $E \approx E_c(z)$ the plasma concentration increases abruptly.

Equation (3.33), with allowance for Equations (3.34) and (3.68), takes the form

$$\frac{dE}{dz} = -\frac{\omega}{c}\kappa_0(z)\frac{\nu_e(T_e)}{\nu_{e0}}\frac{\omega^2 + \nu_{e0}^2}{\omega^2 + \nu_e^2(T_e)}\frac{E}{1 - \exp\left[Q\left(1 - \frac{E_c(z)}{E}\right)\right]}. \tag{3.69}$$

Here $T_e = T_e(E)$; it is defined by Equation (2.32). At $z \ll z_0$, the last factor in the right-hand side of Equation (3.69) is close to unity and Equation (3.69) coincides with Equation (3.33) and Equation (3.34) at $N(\tau)/N_0 = 1$, the solutions of which were considered in the preceding section. Near the point $z = z_0$ and beyond this point, the factor, to the contrary, plays the decisive role. It is natural to seek the solution of Equation (3.69) at $z \gtrsim z_0$ in the form of a series

$$E = E_c(z) + E_1 + \cdots. \tag{3.70}$$

Substituting the expansion [Eqs. (3.70)] in (3.69), we obtain the field $E_1(z)$:

$$E_1(z) = \frac{\omega}{c}\kappa(T_{ec})\frac{E_c^2}{Q\frac{dE_c}{dz}}; \qquad \kappa(T_{ec}) = \kappa_0\frac{\nu_e(T_{ec})}{\nu_{e0}}\frac{\omega^2 + \nu_{e0}^2}{\omega^2 + \nu_e^2(T_{ec})}. \tag{3.71}$$

Here $T_{ec} = T_e(E_c)$. By virtue of Equation (3.67) we have the field $|E_1| \ll |E_c|$. We note that $E_1 < 0$ (since $dE_c/dz < 0$), i.e., the stationary field E always remains somewhat lower than $E_c(z)$ also at $z \gtrsim z_0$. At sufficiently large $z \approx z_1$, however, owing to the decrease of $B \sim N_m(z)$ [Eq. (2.178)], the quantity B/A becomes comparable with ω. The derivative $|dE_c/dz|$ decreases very strongly [Eq. (3.67), see also Fig. 16] in this case, and the field E_1 becomes comparable with $E_c(z)$.

Substituting Equations (3.70) and (3.71) in Equation (3.68), we determine the change of the concentration

$$\frac{N}{N_0} = \begin{cases} \left(1 - \exp\left[Q\left(1 - \frac{E_c(z)}{E}\right)\right]\right)^{-1}, & \text{if} \quad z < z_0 \\ \left[-\frac{\omega}{c}\kappa_0 \frac{\nu(T_e)}{\nu_{e0}} \frac{\omega^2 + \nu_{e0}^2}{\omega^2 + \nu_e^2(T_e)} \frac{E_c}{dE_c/dz}\right]^{-1}, & \text{if} \quad z \geq z_0. \end{cases} \tag{3.72}$$

We note that the unperturbed electron concentration $N_0 = q_i/\nu_r$ in the lower ionosphere increases itself with the height z. The dependence of N on z in accordance with Equation (3.72), in the case of a linear growth of $N_0(z)$, is shown in Figure 22. We see that ahead of the point $z = z_0$ the ratio N/N_0 increases sharply. The characteristic dimension of the region of the sharp increase of N is

$$\Delta z \approx \left[\frac{\omega}{c}\kappa(T_{ec})Q\right]^{-1}. \tag{3.73}$$

The maximum value is

$$\left(\frac{N}{N_0}\right)_{\max} \approx -\frac{c}{\omega\kappa(T_{ec})}\frac{1}{E_c}\frac{dE_c}{dz}. \tag{3.74}$$

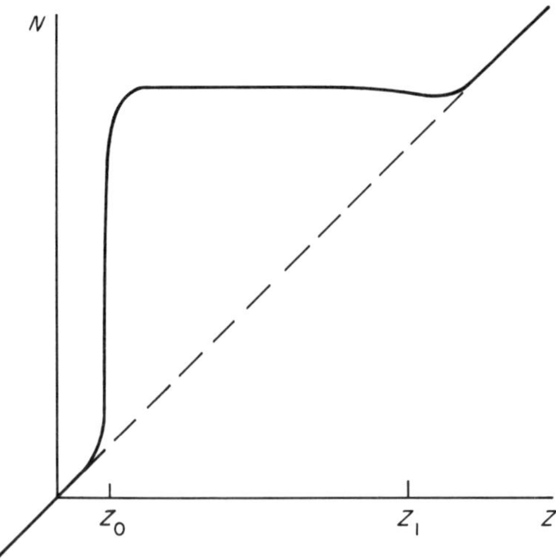

Fig. 22. Increase of plasma concentration in the vicinity of the critical point z_0. *Dashed line*: unperturbed concentration $N_0(z)$

At $z > z_0$, the quantity N remains practically constant up to the point $z \approx z_1$, where $B(z_1)/A = \omega$. The total width of the region of increased ionization is

$$z_1 - z_0 \approx H_0 \ln\left(\frac{E(z_0)}{E_{c\omega}}\right), \tag{3.75}$$

where $E(z_0)$ is the amplitude of the field of the wave at the point z_0, $E_{c\omega}$ is the critical field at $\omega \gg B/A$ [Eq. (2.253), Fig. 16], and H_0 is the distance over which there is a significant change in the field $E_c(z)$ which is proportional to molecule concentration in the plasma [it is assumed that $N_m = N_{m0} \exp(-z/H_0)$, where $H_0 = T/Mg$ is the height of the homogeneous atmosphere].

Thus, as z approaches z_0, i.e., when the amplitude of the plane wave field E_0 approaches the critical field E_c, the electron concentration in a weakly inhomogeneous plasma increases rapidly. At the same time, however, the wave absorption also increases sharply, so that the field E_0 always remains somewhat smaller than E_c. We note that in the region of increased ionization the plasma can become unstable—this is thermoionization instability (Gurevich and Shvartsburg, 1973). We note also that inasmuch as $E_c \gg E_p$, it follows that an important role even ahead of the point z_0 can be played by the self-action effects considered in the preceding section (Ginzburg, 1964). Calculations of the additional ionization for a converging wave are given by Lombardini (1965).

3.3. Change of Wave Modulation

We consider now the propagation of amplitude-modulated radio waves in a plasma. Assume that an amplitude-modulated low-frequency wave

$$E|_{z=0} = E_0(0)[1 + \mu_0 \cos \Omega t] \tag{3.76}$$

arrives at the plasma boundary $z = 0$. When the wave propagates in the plasma, it causes both constant and periodically varying (with low frequency Ω) perturbations of the conductivity and of the dielectric constant. The constant perturbations, as we have seen above, affect the total absorption and the phase of the wave. Owing to the variable perturbations of ε and σ, the wave modulation changes.

3.3.1. Weak Wave

We assume first that the wave is weak, $E_0^2 \ll E_p^2$. The amplitude of the wave field in the plasma can then be found by a perturbation method

(Hibberd, 1955, Vilenskii, 1955; Vilenskii and Zykova, 1959). Repeating the exposition given in Section 3.2.1 for the unmodulated wave, we have

$$E_0(z, t) = E_1(z, t) + E_2(z, t) + \cdots \tag{3.77}$$

$$E_1(z, t) = \frac{E_0(0)}{\sqrt{n_0(z)}}(1 + \mu_0 \cos \Omega t) \exp\left\{-\frac{\omega}{c}\int_0^z \kappa_0(z)\, dz\right\}. \tag{3.78}$$

Here $E_1(z, t)$, as before, is the amplitude of the wave field in the plasma in the linear approximation. When the nonlinearity is taken into account, the wave field is given by Equation (3.24) even in the presence of modulation. All that change are the perturbations ΔT_e of the electron temperature. From Equations (2.18), (2.136), and (3.78) it follows that ΔT_e is determined in our case by the equation

$$\begin{aligned} \frac{1}{\delta_0 \nu_{e0}} \frac{d\,\Delta T_e}{dt} + \Delta T_e &= \frac{e^2 E_1^2}{3m\,\delta_0(\omega^2 + \nu_{e0}^2)} \\ &= \frac{E_0^2(0)}{E_p^2}\frac{T_{e0}}{n_0}\left(1 + \frac{\mu_0^2}{2} + 2\mu_0 \cos \Omega t + \frac{\mu_0^2}{2}\cos 2\Omega t\right)\exp(-2K). \end{aligned} \tag{3.79}$$

Integrating this equation, we obtain the perturbations of the electron temperature

$$\Delta T_e = \Delta_0 T_e + \Delta_\Omega T_e,$$

$$\Delta_0 T_e = \left[\frac{E_0(0)}{E_p}\right]^2 \frac{T_{e0}}{n_0} \exp(-2K)\left(1 + \frac{\mu_0^2}{2}\right),$$

$$\begin{aligned} \Delta_\Omega T_e = 2\mu_0 \left[\frac{E_0(0)}{E_p}\right]^2 \frac{T_{e0}}{n_0} \times \Bigg\{ &\frac{1}{\Omega^2 + \delta_0^2 \nu_{e0}^2}[\delta_0^2 \nu_{e0}^2 \cos \Omega t + \Omega\, \delta_0 \nu_{e0} \sin \Omega t] \\ &+ \frac{\mu_0}{4(4\Omega^2 + \delta_0^2 \nu_{e0}^2)}[\delta_0^2 \nu_{e0}^2 \cos 2\Omega t + 2\Omega\, \delta_0 \nu_{e0} \sin 2\Omega t]\Bigg\} \exp(-2K) \end{aligned} \tag{3.80}$$

Substituting this expression in Equation (3.24), we obtain the field $E_2(z, t)$. The amplitude of the field of the modulated wave in the plasma, when account is taken of the nonlinearity [Equation (3.77)], is then represented in the form

$$E(z, t) = E_{\mu 0}(z)[1 + \mu_0(1 - \alpha_\Omega)\cos(\Omega t + \varphi_\Omega) - \mu_{2\Omega}\cos(2\Omega t + \varphi_{2\Omega}) - \mu_{3\Omega}\cos(3\Omega t + \varphi_{3\Omega})].$$

Here $E_{\mu 0}(z)$ is the part of the field that does not change with time; it is determined by Equations (3.25) and (3.29), provided we replace $[E_0(0)]^2$ in Equation (3.29) by

$$[E_0(0)]^2\left(1 + \frac{\mu_0^2}{2} + \frac{\mu_0^2\,\delta_0^2 v_{e0}^2}{\Omega^2 + \delta_0^2 v_{e0}^2}\right).$$

Further, α_Ω is the correction to the depth of modulation of the wave at the fundamental frequency Ω:

$$\alpha_\Omega = q\left[\frac{2\,\delta_0^2 v_{e0}^2(1 - \mu_0^2/2)}{\Omega^2 + \delta_0^2 v_{e0}^2} + \frac{\mu_0^2}{4}\,\frac{\delta_0^2 v_{e0}^2}{\delta_0^2 v_{e0}^2 + 4\Omega^2}\right], \tag{3.81}$$

φ_Ω is the correction to the phase of the modulation at the fundamental frequency:

$$\varphi_\Omega = q\mu_0\left[\frac{2\Omega^2}{\Omega^2 + \delta_0^2 v_{e0}^2} + \frac{\mu_0^2\Omega^2}{\delta_0^2 v_{e0}^2 + 4\Omega^2}\right]. \tag{3.82}$$

Here q is determined as before by Equation (3.29). The term $\sim\gamma_1$ in Equation (3.29), however, should be neglected here, since the changes of the concentration are characterized by the appreciable time τ_N, and at $\Omega\tau_N \gg 1$ they cannot affect the modulation of the wave. Furthermore, $\mu_{2\Omega}$, $\mu_{3\Omega}$, and $\varphi_{2\Omega}$, $\varphi_{3\Omega}$ are the coefficients and phases of the modulation at the frequencies 2Ω and 3Ω, and appear as the result of the self-action of the wave:

$$\mu_{2\Omega} = \frac{3}{2}\mu_0^2 q\,\delta_0 v_{e0}\sqrt{\frac{\delta_0^2 v_{e0}^2 + 25\Omega^2/9}{[\delta_0^2 v_{e0}^2 + \Omega^2][\delta_0^2 v_{e0}^2 + 4\Omega^2]}},$$

$$\mu_{3\Omega} = \frac{\mu_0^3 q}{4}\,\frac{\delta_0 v_{e0}}{\sqrt{\delta_0^2 v_{e0}^2 + 4\Omega^2}},$$

$$\varphi_{2\Omega} = \arctan\frac{\Omega}{\delta_0 v_{e0}}\,\frac{10\Omega^2 + 4v_{e0}^2}{9\Omega^2 + 3v_{e0}^2}, \qquad \varphi_{3\Omega} = \arctan\frac{2\Omega}{\delta_0 v_{e0}}.$$

We see therefore that changes of the modulation of the weak wave are proportional to q, i.e., they are proportional to $E_0^2(0)/E_p^2$. A distortion of the same order of magnitude is produced in the wave modulation by generation of overtones of the fundamental frequency of the modulation. We note that the distortions of the modulation are proportional also to the depth of the modulation of the fundamental wave $\mu_{2\Omega} \sim \mu_0^2$, $\mu_{3\Omega} \sim \mu_0^3$, therefore they are significant only at sufficiently large values of μ_0. The modulation at the fundamental frequency can either increase, $q < 0$, or decrease, $q > 0$. The former case (self-modulation of the wave), as seen from Equations (3.81) and (3.29), occurs in a weakly ionized plasma for low-frequency waves $\omega < \nu_{e0}$, and high-frequency waves $\omega > \nu_{e0}$ become demodulated in a weakly ionized plasma [see Eq. (3.29)].

3.3.2. Change of Amplitude Modulation of Strong Wave

Low Modulation Frequency. The self-action of the strong wave $E_0(0) \gtrsim E_p$ leads to a strong distortion of its modulation (Gurevich, 1958b). We consider first the quasi-stationary case, when the modulation frequency is so low that the amplitude of the wave in the plasma is altered significantly only during a time much longer than the electron temperature relaxation time:

$$\Omega \ll \delta_0 \nu_{e0}. \tag{3.83}$$

In that case, naturally, Equation (3.40) is valid for the intensity of the wave field in the plasma:

$$E(z, t) = E|_{z=0} \exp(-K)P, \tag{3.84}$$

where $E|_{z=0}$ is the amplitude at the plasma boundary [Eq. (3.76)]. This amplitude changes slowly with time at a frequency Ω. Accordingly, the amplitude $E(z, t)$ also changes. In the general case it is a complicated periodic function of the time. It can consequently be represented in the form

$$E = E_0(z)[1 + \mu_\Omega(z) \cos \Omega t + \mu_{2\Omega}(z) \cos 2\Omega t + \cdots]. \tag{3.85}$$

In order to gain a complete idea of the character of the modulation of the wave in the plasma, it is necessary to determine the amplitude of each harmonic $\mu_\Omega(z)$, $\mu_{2\Omega}(z)$ . . . , which in general is a complicated matter. We confine ourselves first to an analysis of the expression for the effective depth μ of modulation of the wave in the plasma, which is defined in natural manner as follows:

$$\mu(z) = \frac{E_{\max}(z) - E_{\min}(z)}{E_{\max}(z) + E_{\min}(z)} \tag{3.86}$$

Here $E_{\max} = E[E_0(0)(1 + \mu_0), z]$ and $E_{\min} = E[E_0(0)(1 - \mu_0), z]$ are the maximum and minimum values assumed by the wave field amplitude as it varies in time at the point z.

A simple expression is obtained from Equation (3.86) at $\mu_0 \ll 1$. Indeed, expanding $E(\mu_0)$ in Equation (3.86) in powers of μ_0 and confining ourselves to the first terms of the expansion, we have

$$\mu = \mu_0 \left[\frac{E_0(0)}{E} \frac{\partial E}{\partial E_0(0)} \right]_{\mu_0 = 0}. \tag{3.87}$$

It remains now to obtain the value of $\partial E/\partial E_0(0)$ at $\mu_0 = 0$, i.e., for an unmodulated wave. According to Equations (3.33) and (3.34) we have in a weakly ionized plasma at $N \neq N(T_e)$

$$f(E) = \int \frac{dE}{E} \frac{\omega^2 + v_{e0}^2 \tau^2}{\tau} = f[E_0(0)] - K(z), \tag{3.87a}$$

where $\tau(E)$ and $K(z)$ are defined by Equations (3.35) and (3.30). Hence

$$\frac{\partial E}{\partial E_0(0)} = \frac{\tau}{\tau_0} \frac{\omega^2 + v_{e0}^2 \tau_0^2}{\omega^2 + v_{e0}^2 \tau^2} \frac{E}{E_0(0)}, \tag{3.88}$$

where $\tau_0 = \tau[E_0(0)]$. Therefore

$$\mu = \mu_0 \frac{\tau}{\tau_0} \frac{\omega^2 + v_{e0}^2 \tau_0^2}{\omega^2 + v_{e0}^2 \tau^2}. \tag{3.89}$$

We note that although we have assumed above $v_e = v_{e0}\tau$, it follows from the derivation of Equation (3.89) that this formula is valid also for any form of $v_e(\tau)$:

$$\mu = \mu_0 \frac{v_e(\tau)}{v_e(\tau_0)} \frac{\omega^2 + v_e^2(\tau_0)}{\omega^2 + v_e^2(\tau)}. \tag{3.89a}$$

This expression is valid for an arbitrary dependence of δ on T_e.

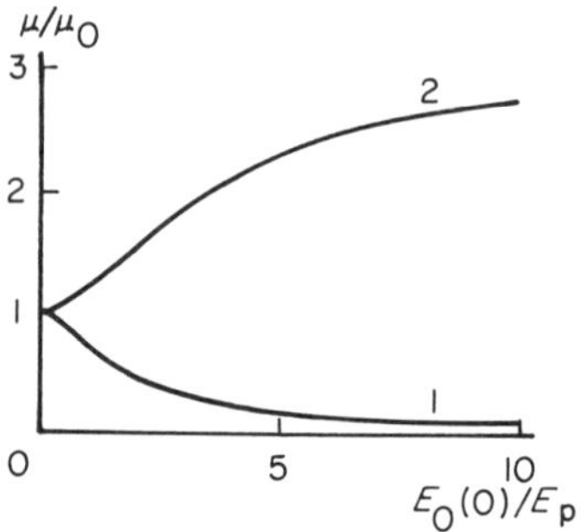

Fig. 23. Depth of wave modulation in plasma, *1*: $\omega \gg \nu_{e0}$; *2*: $\omega \ll \nu_{e0}$

Proceeding to an analysis of Equation (3.89), we note that, as is clear from Equation (3.35), τ is a monotonically increasing function of E. Therefore on the plasma boundary, where $E = E_0(0)$, τ is maximal and equal to τ_0. In the interior of the plasma, τ decreases monotonically to a value $\tau = 1$. From Equation (3.89) it follows therefore that at high frequency $\omega^2 \gg \nu_{e0}^2 \tau_0^2$ the depth of modulation of the wave in the plasma decreases because of the self-action—the wave becomes demodulated. In the interior of the plasma, where $K \gg 1$, we have

$$\mu = \mu_0/\tau_0 \tag{3.90}$$

We see therefore that demodulation of a strong high-frequency wave can be quite appreciable (even complete at $\tau_0 \gg 1$). At low frequency ($\omega^2 \ll \nu_{e0}^2$), to the contrary, μ increases. In the interior of the plasma we have

$$\mu = \mu_0 \tau_0. \tag{3.91}$$

Plots of μ/μ_0 against $E_0(0)/E_p$ and ω/ν_{e0} are shown in Figures 23 and 24. At $\omega > \nu_{e0}$ the modulation depth depends nonmonotonically on

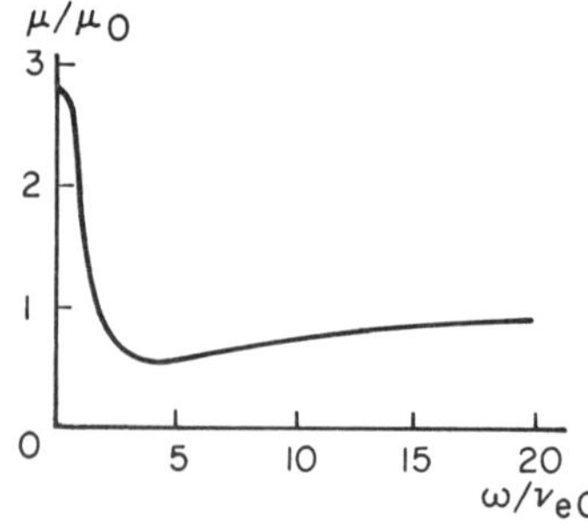

Fig. 24. Frequency dependence of depth of modulation

$E_0(0)/E_p$; the minimum value

$$\mu_{\min} = \mu_0 \frac{2\omega\nu_{e0}}{\omega^2 + \nu_{e0}^2}$$

is reached at $E_{0\,\min} = E_p\sqrt{\omega^2/\nu_{e0}^2 - 1}$ [an analogous dependence on $E_0(0)/E_p$ is possessed by the self-action factor; see Eq. (3.46)].

In the general case, when μ_0 is not small, the depth of modulation of the wave in the plasma is determined by Equation (3.86). The plot of μ against μ_0 constructed with the aid of Equation (3.86) is shown in Figure 25. It is seen from the figure that at $\mu \lesssim 0.5$ and $\mu_0 \lesssim 0.5$ the dependence of μ on μ_0 is close to linear. Consequently, under these conditions, Equation (3.89) is valid. It is also seen from the figure that the relative change of the depth of modulation (i.e., the deviation of μ/μ_0 from unity) is strongest at small μ_0 and is small at large values of μ_0. In particular, as $\mu_0 \to 1$ the ratio μ/μ_0 always stays close to 1.

The distortions of the modulation, to the contrary, are negligible at small μ_0 and very large (in strong fields) as $\mu_0 \to 1$. Consequently, in the last case, it is of interest to know not only the depth of modulation μ, but also the amplitudes of the individual harmonics of the modulation μ_Ω, $\mu_{2\Omega}, \mu_{3\Omega} \cdots$ [Eq. (3.85)]. They can be determined from a harmonic analysis of the plot of $E(z, t)$ [Eq. (3.84)], which represents the wave form of the signal modulation in the plasma. Characteristic curves showing the modulation wave forms of strong radio waves in a plasma at $\mu_0 = 1$ are drawn in Figure 26.

Arbitrary Modulation Frequency. We have considered above the variation of the wave modulation in the plasma in the case of a low modulation

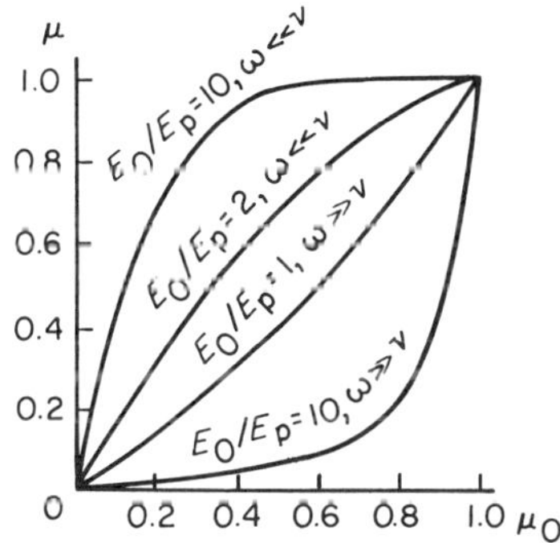

Fig. 25 Depth of modulation of wave in plasma as a function of the depth of modulation of the incident wave

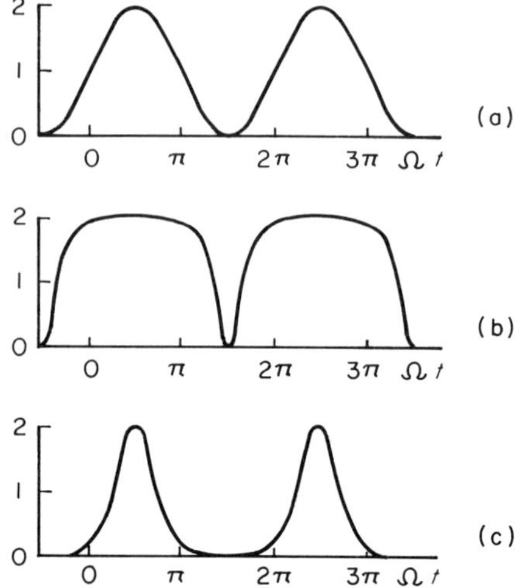

Fig. 26a–c. Distortion of modulation waveform; $E_0(0)/E_p = 10$, $\mu_0 = 1$. (a) Wave at plasma boundary; (b) Wave in the interior of plasma at $\omega \gg \nu_{e0}$; (c) Wave in interior of plasma at $\omega \ll \nu_{e0}$

frequency [Eq. (3.83)]. At an arbitrary frequency Ω it is necessary to solve simultaneously the nonstationary equation for the electron temperature in Equation (2.18) and Equation (3.33) for the wave field amplitude.

In a weakly ionized plasma ($\nu_e \sim \sqrt{T_e/T}$), these equations take the form

$$\frac{dE}{dz} + \frac{\omega}{c}\kappa_0(z)E\frac{\tau(\omega^2 + \nu_{e0}^2)}{\omega^2 + \nu_{e0}^2\tau^2} = 0, \tag{3.92}$$

$$\frac{d\tau}{dt} + \frac{\delta_0\nu_{e0}}{2}(\tau^2 - 1) = \frac{e^2E^2}{6mT}\frac{\nu_{e0}}{\omega^2 + \nu_{e0}^2\tau^2}. \tag{3.93}$$

Here, as usual, $\tau = \sqrt{T_e/T}$. On the plasma boundary we have

$$E|_{z=0} = E_0(0)[1 + \mu_0 \cos \Omega t]. \tag{3.94}$$

We consider the case of small $\mu_0 \ll 1$. The solution of Equations (3.92) and (3.93) can then be sought in the form

$$E = E(z)[1 + \mu(z)\cos(\Omega t - \varphi(z))] \tag{3.95}$$

(we neglect the term of order μ_0^2). Substituting Equation (3.95) in Equation (3.93) and integrating the latter with respect to t, we obtain

$$\tau = \tau(z) + \mu \frac{\delta_0 v_{e0}[\tau^2(z) - 1]}{\sqrt{\Omega^2 + [\delta' v_{e0}\tau(z)]^2}} \cos(\Omega t - \varphi - \varphi'), \tag{3.96}$$

where

$$\delta' = \delta_0 \left[1 + \frac{v_{e0}^2[\tau^2(z) - 1]}{\omega^2 + v_{e0}^2\tau^2(z)}\right], \qquad \operatorname{tg}\varphi' = \frac{\Omega}{\delta' v_{e0}\tau(z)}. \tag{3.97}$$

Here $\tau(z)$ and $E(z)$ are the values of τ and E for unmodulated waves, as defined by Equations (3.38) and (3.35).

Substituting Equations (3.96) and (3.95) in Equation (3.92) and separating the time variable, we arrive at the following system of equation, which determines the depth μ and the phase φ of the modulation:

$$\frac{d(\mu\cos\varphi)}{dz} + \frac{\omega}{c}\kappa_0\mu\cos(\varphi + \varphi')\frac{(\tau^2 - 1)\,\delta_0 v_{e0}}{\sqrt{\Omega^2 + (\delta' v_{e0}\tau)^2}}\frac{(\omega^2 - v_{e0}^2\tau^2)(\omega^2 + v_{e0}^2)}{(\omega^2 + v_{e0}^2\tau^2)^2} = 0 \tag{3.98}$$

$$\frac{d(\mu\sin\varphi)}{dz} + \frac{\omega}{c}\kappa_0\mu\sin(\varphi + \varphi')\frac{(\tau^2 - 1)\,\delta_0 v_{e0}}{\sqrt{\Omega^2 + (\delta' v_{e0}\tau)^2}}\frac{(\omega^2 - v_{e0}^2\tau^2)(\omega^2 + v_{e0}^2)}{(\omega^2 + v_{e0}^2\tau^2)^2} = 0. \tag{3.99}$$

The boundary conditions, as follows from Equation (3.94), are

$$\mu|_{z=0} = \mu_0, \qquad \varphi|_{z=0} = 0 \tag{3.100}$$

Using Equation (3.37), we now replace the variable z in Equations (3.98) and (3.99) by τ:

$$\frac{d}{d\tau}(\mu\cos\varphi) - \mu\cos(\varphi + \varphi')\frac{\delta_0 v_{e0}}{\sqrt{\Omega^2 + (\delta' v_{e0}\tau)^2}}\frac{(\omega^2 - v_{e0}^2\tau^2)(\omega^2 + 2v_{e0}^2\tau^2 - v_{e0}^2)}{(\omega^2 + v_{e0}^2\tau^2)^2}, \tag{3.101}$$

$$\frac{d}{d\tau}(\mu\sin\varphi) = \mu\sin(\varphi + \varphi')\frac{\delta_0 v_{e0}}{\sqrt{\Omega^2 + (\delta' v_{e0}\tau)^2}}\frac{(\omega^2 - v_{e0}^2\tau^2)(\omega^2 + 2v_{e0}^2\tau^2 - v_{e0}^2)}{(\omega^2 + v_{e0}^2\tau^2)^2}. \tag{3.102}$$

To separate now the sought functions μ and φ in Equations (3.101) and (3.102), we multiply Equation (3.101) by $\mu \cos \varphi$, and Equation (3.102) by $\mu \sin \varphi$, and adding the resultant expressions we arrive at the equation

$$\frac{d\mu}{d\tau} = \mu \frac{(\delta' v_{e0}\tau)^2}{\Omega^2 + (\delta' v_{e0}\tau)^2} \frac{\omega^2 - v_{e0}^2 \tau^2}{\tau(\omega^2 + v_{e0}^2 \tau^2)} \tag{3.103}$$

[we have used here also Equation (3.97) for φ' and δ']. The solution of this equation with the boundary condition [Eq. (3.100)] is

$$\mu(z) = \mu_0 \exp \left\{ \int_{\tau_0}^{\tau(z)} \frac{(\delta' v_{e0}\tau)^2}{\Omega^2 + (\delta' v_{e0}\tau)^2} \frac{\omega^2 - v_{e0}^2 \tau^2}{\tau(\omega^2 + v_{e0}^2 \tau^2)} \, d\tau \right\}. \tag{3.104}$$

Analogously, multiplying Equation (3.101) by $\sin \varphi$ and Equation (3.102) by $\cos \varphi$ and subtracting Equation (3.102) from Equation (3.101), we arrive at an equation for:

$$\varphi(z) = \int_{\tau_0}^{\tau(z)} \frac{\delta' v_{e0}\Omega}{\Omega^2 + (\delta' v_{e0}\tau)^2} \frac{\omega^2 - v_{e0}^2 \tau^2}{\omega^2 + v_{e0}^2 \tau^2} \, d\tau. \tag{3.105}$$

At $\Omega \ll \delta_0 v_{e0}$, the Equations (3.104) and (3.105) coincide with Equation (3.89), as they should. It is also clear from Equation (3.104) that with increasing frequency Ω the influence of the self-action of the wave in the plasma on its modulation always becomes weaker: $|\mu - \mu_0|$ decreases with increasing Ω. For example, at $\Omega \gg \delta' v_{e0}\tau_0$ the difference $|\mu - \mu_0|$ decreases like $(\delta' v_{e0}\tau_0/\Omega)^2$. Consequently, the expressions considered in the preceding section for μ show always the maximum possible change of the wave modulation as a result of its self-action in the plasma.

In the case of a wave of high frequency $\omega^2 \gg v_{e0}^2$ we obtain from Equations (3.104) and (3.105) a simple expression for the depth and phase of the modulation:

$$\mu = \mu_0 \sqrt{\frac{(\delta' v_{e0}\tau)^2 + \Omega^2}{(\delta' v_{e0}\tau_0)^2 + \Omega^2}}, \qquad \varphi = \operatorname{arctg} \frac{\Omega}{\delta' v_{e0}\tau_0} - \operatorname{arctg} \frac{\Omega}{\delta' v_{e0}\tau}. \tag{3.106}$$

Plots of μ/μ_0 and φ against $\Omega/\delta' v_{e0}$ are shown in Figure 27. It is seen from the figure that perturbations of the depth of modulation decrease monotonically with increasing Ω. The phase shift φ depends on Ω non-monotonically. It is small both at low modulation frequencies ($\Omega \ll \delta' v_{e0}$)

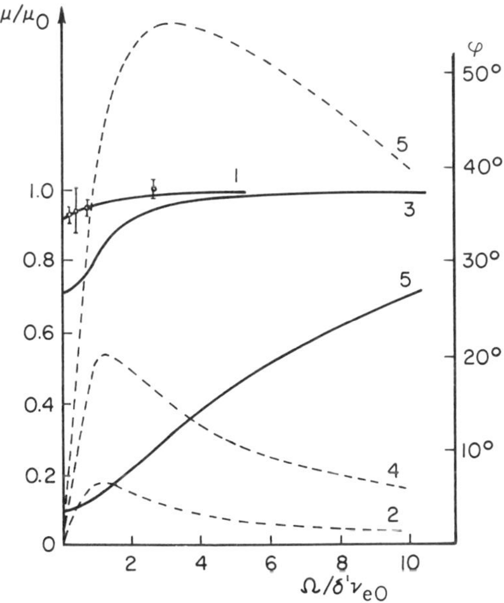

Fig. 27. Dependence of the modulation depth and phase (*dashed curve*) on the modulation frequency. Curves *1*, *2*, *3*, *4*, and *5* correspond to $E_0(0)/E_p = 0.4$, 0.75, 1.0, 1.73, and 10, respectively. *Points*: results of experiment (King, 1959)

and at high modulation frequencies ($\Omega \gg \delta' \nu_{e0}$). The maximum value

$$\varphi_{max} = \operatorname{arctg} \sqrt{\tau_0/\tau} - \operatorname{arctg} \sqrt{\tau/\tau_0}$$

is reached at $\Omega_{max} = \delta' \nu_{e0} \sqrt{\tau_0/\tau}$.

3.3.3. Phase Modulation

It was shown in Sections 3.2.1 and 3.2.2 that self-action changes both the amplitude and the phase ($\Delta\varphi$) of a plane wave in a plasma. Consequently, when an amplitude-modulated wave propagates in plasma, a phase modulation should result from the nonlinearity.

In the case of low modulation frequencies [Eq. (3.83)] $\Omega \ll \delta_0 \nu_{e0}$, the phase modulation can be determined directly from Equation (3.51), which shows the change $\Delta\varphi$ of the phase of the wave as a function of the amplitude of the field at the plasma boundary. It is merely necessary to take into account the fact that the field amplitude at the plasma boundary changes with time in accordance with Equation (3.76). Consequently the phase also changes periodically with time, with frequency Ω. At small μ_0, expanding $\Delta\varphi$ in Equation (3.51) in powers of μ_0 and confining ourselves to the first term of the expansion, we have:

$$\Delta\varphi = \Delta\varphi|_{\mu_0=0} + \mu_0 \left[E_0(0) \frac{\partial\, \Delta\varphi}{\partial E_0(0)} \right]_{\mu_0=0} \cos \Omega t + \cdots. \tag{3.107}$$

The first term of this expression gives a constant phase shift. Therefore the phase-modulation index is equal to

$$\beta = \mu_0 \left[E_0(0) \frac{\partial\, \Delta\varphi}{\partial E_0(0)} \right]_{\mu_0=0} \tag{3.108}$$

where the function $\Delta\varphi\,[E_0(0)]$ is determined by Equation (3.51).

The general Equation (3.108) becomes simpler in the interior of the plasma $[K(z) > 1,\ \tau(z) \to 1]$:

$$\beta = \mu_0 \frac{\tau_0^2 - 1}{\tau_0} \frac{\omega \nu_{e0}}{\omega^2 + 2\nu_{e0}^2\tau_0^2 - \nu_{e0}^2}. \tag{3.109}$$

It follows from Equation (3.109) that the phase-modulation index is small, both at high ($\omega^2 \gg \nu_{e0}^2\tau_0^2$) and at low frequencies ($\omega^2 \ll \nu_{e0}^2$). The maximum value of β as a function of ω is reached at $\omega_{max} = \nu_{e0}\sqrt{2\tau_0^2 - 1}$:

$$\beta_{max} = \mu_0 \frac{\tau_0^2 - 1}{2\tau_0\sqrt{2\tau_0^2 - 1}}. \tag{3.110}$$

The value of β_{max} increases with increasing $\tau_0 = \sqrt{T_e[E_0(0)]/T}$, i.e., with increasing field amplitude at the plasma boundary. At large $\tau_0 \gg 1$, the optimal value $\beta_{max} = \mu_0/2\sqrt{2}$ is reached.

3.3.4. Nonlinear Distortion of Pulse Waveform

Let us see how the nonlinearity deforms a rectangular radio pulse. It is known that the pulse is deformed also in the linear approximation, as a result of the wave velocity dispersion (Ginzburg, 1960). We shall consider, however, propagation far from the reflection point ($n \sim 1$), and assume that the pulse duration t_1 is large enough, $t_1 \gg [(\pi z/c)(d^2/d\omega^2)(\omega n(\omega))]^{1/2}$, so that the linear deformation due to the dispersion is of little significance.

Then the amplitude of the wave field in the plasma $E_0(z, t)$ at $n \gg \kappa$ is described by the equation

$$\frac{\partial E_0}{\partial t} + v_g \left(\frac{\partial E_0}{\partial z} + \frac{E_0}{2n} \frac{\partial n}{\partial z} + \frac{\omega}{c} \kappa E_0 \right) = 0. \tag{3.111}$$

Here $v_g = c/(d(n\omega)/d\omega)$ is the group velocity. Equation (3.111) follows from Equation (3.6) in the geometrical-optics approximation (cf. Sect. 3.1.2); under stationary conditions it coincides with Equation (3.20). On the plasma boundary we have $z = 0$:

$$E_0\big|_{z=0} = \begin{cases} 0, & \text{at} \quad t < 0 \text{ and at } t > t_1 \\ E_0(0), & \text{at} \quad 0 \le t \le t_1. \end{cases} \tag{3.112}$$

Here t_1 is the pulse duration. The absorption coefficient κ of the wave in the plasma depends on the electron temperature [Eq. (3.34)], which in turn is given by Equation (2.18) or Equation (2.136).

We consider first the case of a weak field. The influence of the nonlinearity on the wave-field amplitude can then be neglected in the first-order approximation $E_0 = E_1(z)$ [Eq. (3.22)], and the equation for the perturbation of the electron temperature ΔT_e takes the form

$$\frac{1}{\delta_0 \nu_{e0}} \frac{d(\Delta T_e)}{dt} + \Delta T_e = \begin{cases} 0, & \text{at} \quad t < t_0 \text{ and at } t > t_1 + t_0 \\ \dfrac{E_1^2(z)}{E_p^2} T_{e0}, & \text{at} \quad t_0 \le t \le t_1 + t_0. \end{cases} \tag{3.113}$$

Here t_0 is the instant of time when the leading edge of the pulse reaches the point z:

$$t_0 = \frac{1}{c} \int_0^z \frac{d}{d\omega} (n\omega) \, dz_1. \tag{3.114}$$

Integrating Equation (3.113), we obtain

$$\frac{\Delta T_e}{T_{e0}} = \begin{cases} 0, \quad \text{at} \quad t < t_0 \\ \dfrac{E_1^2(z)}{E_p^2} [1 - \exp\{-\delta_0 \nu_{e0}(t - t_0)\}], \ \text{at} \ t_0 \le t \le t_0 + t_1 \\ \dfrac{E_1^2(z)}{E_p^2} [1 - \exp(-\delta_0 \nu_{e0} t_1)] \exp[-\delta_0 \nu_{e0}(t - t_1 - t_0)]. \ \text{at} \ t > t_0 + t_1. \end{cases} \tag{3.115}$$

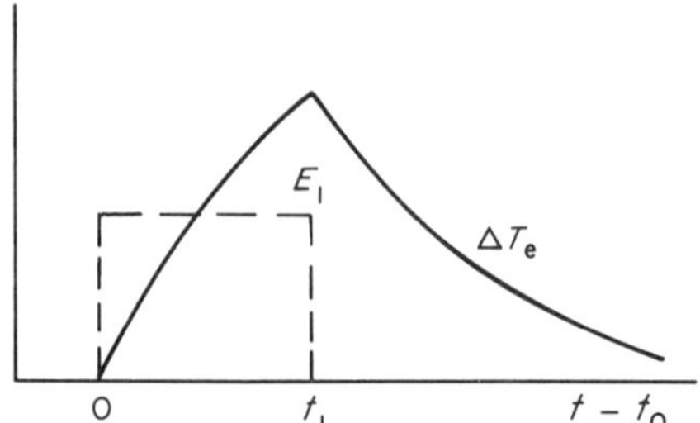

Fig. 28. Perturbation of electron temperature

The electron-temperature perturbation [Eq. (3.115)] is shown in Figure 28. From Equation (3.24) it follows now that

$$E(z, t) = \begin{cases} 0, & \text{at} \quad t < t_0 \text{ and at } t > t_0 + t_1 \\ E_1(z)[1 - q(1 - \exp[-\delta_0 \nu_{e0}(t - t_0)])], & \text{at} \quad t_0 \leq t \leq t_0 + t_1. \end{cases} \tag{3.116}$$

The quantity q is defined, as before, by Equations (3.25) and (3.29).

Equations (3.111), (3.34), and (2.18) were integrated numerically for a strong electric field. The results of the calculation for $\omega \gg \nu_e$ are shown in Figure 29. From the figure and from Equation (3.116) we see that during the time $t - t_0 \sim 1/\delta_0 \nu_{e0}$ the field amplitude in the plasma decreases (or increases at $\omega < \nu_e$). At $t - t_0 \gg 1/\delta_0 \nu_{e0}$ it assumes a stationary value [Eqs. (3.25) or (3.41)]. We see that a rectangular pulse of appreciable amplitude is strongly deformed by the nonlinearity. The physical cause

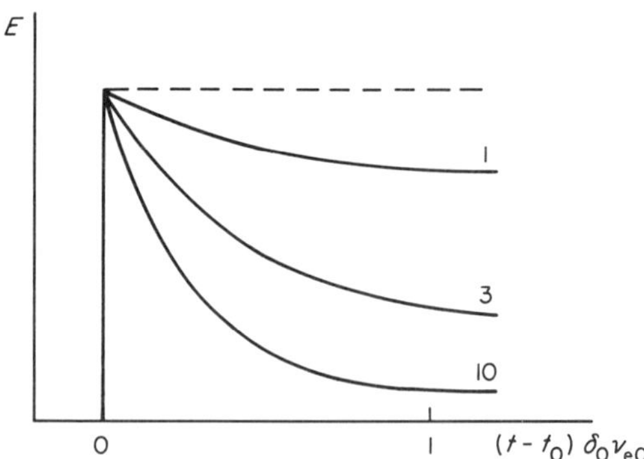

Fig. 29. Nonlinear distortion of the envelope of a rectangular pulse with $\omega \gg \nu_{e0}$; the values of $E_0(0)/E_p$ are indicated at the curves. *Dashed line*: linear approximation

of this deformation is quite understandable: the absorption of the wave of the plasma, when account is taken of the nonlinearity, depends on the electron temperature. The latter, however, is not established immediately, but after a time $\tau_T = (\delta_0 \nu_{e0})^{-1}$, during which the electrons become heated. This leads to a change in the absorption of the wave during the transient time τ_T, and consequently also to deformation of the pulse. In particular, the leading front of the pulse passes through the unheated unperturbed plasma. Its amplitude is therefore always determined only by the linear absorption.

3.4. Generation of Harmonic Waves and Nonlinear Detection

3.4.1. Frequency Tripling

The perturbations of the dielectric constant and of the conductivity by the electric field of a wave of frequency ω in a homogeneous plasma at $\omega \gg \delta\nu_e$ are in the main, constant in time. There are, however, alternating corrections of frequency 2ω to σ and ε, i.e., alternating corrections of frequency 3ω and ω to the current. Under Equation (3.7) these corrections are small, of the order of $\delta\nu_e/\omega$. A wave propagating in such a nonlinear medium preserves for the most part its frequency ω. However, owing to the presence of alternating current corrections of frequency 3ω, harmonic waves of frequency 3ω can be generated (Vilenskii, 1955; Ginzburg and Gurevich, 1960; Rosen, 1961; Sodha and Kaw, 1965; Ting-Wei Tang, 1966).

In the elementary theory the corrections $\boldsymbol{j}_{3\omega}^{(2)}$ of frequency 3ω to the current take the form of Equation (2.44):

$$\boldsymbol{j}_{3\omega}^{(2)} = \boldsymbol{j}_{0+}^{(2)} \exp(3i\omega t) + \boldsymbol{j}_{0-}^{(2)} \exp(-3i\omega t),$$

$$\boldsymbol{j}_{0+}^{(2)} = \frac{\delta}{8} \frac{E_0^2}{E_p^2} \frac{\overline{T}_e}{\nu_e} \left(\frac{\partial \nu_e}{\partial T_e} \right)_{T_e} \frac{3\omega^2 - 5\nu_e^2 - i\nu_e(\nu_e^2 - 7\omega^2)}{9\omega^2 + \nu_e^2} \boldsymbol{j}_0^{(1)}; \qquad \boldsymbol{j}_{0-}^{(2)} = \boldsymbol{j}_{0+}^{(2)*}. \tag{3.117}$$

Here $\boldsymbol{j}_{0+}^{(2)*}$ is the complex conjugate of $\boldsymbol{j}_{0+}^{(2)}$ and $\boldsymbol{j}_0^{(1)} = \sigma_0 \boldsymbol{E}$ is the amplitude of the conduction current at the fundamental frequency ω. A more rigorous kinetic calculation leads to the same form of the current $\boldsymbol{j}_{3\omega}^{(2)}$. Only the expressions for the amplitudes $\boldsymbol{j}_{0+}^{(2)}$ and $\boldsymbol{j}_{0-}^{(2)}$ are slightly changed.

The presence of the current $\boldsymbol{j}_{3\omega}^{(2)}$ in the plasma leads to generation of a corresponding harmonic wave $\boldsymbol{E}^{(2)}$ of frequency 3ω. The field of the wave $\boldsymbol{E}^{(2)}$ in the plasma is determined by Equation (3.9). Substituting Equation

(3.117) in Equation (3.9) and assuming that $\boldsymbol{E}^{(2)} = \boldsymbol{E}_{+}^{(2)} \exp(3i\omega t) + \boldsymbol{E}_{-}^{(2)} \exp(-3i\omega t)$ we arrive at the following equation for the amplitude $\boldsymbol{E}_{+}^{(2)}$:

$$\Delta \boldsymbol{E}_{+}^{(2)} - \text{graddiv}\, \boldsymbol{E}_{+}^{(2)} + \left[\frac{9\omega^2 \varepsilon_1}{c^2} + i\,\frac{12\pi\omega\sigma_1}{c^2}\right] \boldsymbol{E}_{+}^{(2)} = i\,\frac{12\pi\omega}{c^2}\,\boldsymbol{j}_{0+}^{(2)} \tag{3.118}$$

Here σ_1 and ε_1 are the conductivity and dielectric constant of the plasma at the frequency 3ω. The equation for the complex-conjugate amplitude $\boldsymbol{E}_{-}^{(2)}$ is similar in form.

We assume for simplicity that the plasma is isotropic and that the wave propagates in the z direction. Equation (3.118) then takes the form

$$\frac{d^2 E_{+}^{(2)}}{dz^2} + \left[\frac{9\omega^2 \varepsilon_1}{c^2} + i\,\frac{12\pi\omega\sigma_1}{c^2}\right] E_{+}^{(2)} = i\,\frac{12\pi\omega}{c^2}\,j_{0+}^{(2)}. \tag{3.119}$$

There is no harmonic of frequency 3ω at the plasma boundary $z = 0$. At large z, outside the interaction region, the wave $E_{+}^{(2)}$ propagates only in the z direction. The boundary conditions for Equation (3.119) are therefore

$$E_{+}^{(2)}\big|_{z=0} = 0, \qquad E_{+}^{(2)}\big|_{z\to\infty} = C \exp(ikz). \tag{3.120}$$

We assume for simplicity that σ_1 and ε_1 in the left-hand side of Equation (3.119) are independent of z. The solution of Equation (3.119) consists then of the solution of the inhomogeneous equation

$$E_{+}^{(2)} = \frac{6\pi\omega}{c^2 k_2}\left\{\exp(ik_2 z)\int_0^z j_{0+}^{(2)} \exp(-ik_2 z_1)\, dz_1 - \exp(-ik_2 z)\int_z^\infty j_{0+}^{(2)} \exp(ik_2 z_1)\, dz_1\right\} \tag{3.121}$$

and the homogeneous solution

$$E_0^{(2)} = C_1 \exp(ik_2 z) + C_2 \exp(-ik_2 z) \tag{3.122}$$

with arbitrary constants C_1 and C_2. Here k_2 is the complex wave vector of the wave of frequency 3ω:

$$k_2 = \frac{3\omega}{c}\sqrt{\varepsilon_1 + i\,\frac{4\pi\sigma_1}{3\omega}}. \tag{3.123}$$

Choosing the constants C_1 and C_2 such as to satisfy the boundary conditions in Equation (3.120), we get

$$E_{+}^{(2)} = \frac{6\pi\omega}{c^2 k_2} \left\{ \exp(ik_2 z) \left[\int_0^z j_{0+}^{(2)} \exp(-ik_2 z_1)\, dz_1 - \int_0^\infty j_{0+}^{(2)} \exp(ik_2 z_1)\, dz_1 \right] \right.$$

$$\left. + \exp(-ik_2 z) \int_z^\infty j_{0+}^{(2)} \exp(ik_2 z_1)\, dz_1 \right\}. \tag{3.124}$$

An analogous solution is obtained also for the amplitude $E_{-}^{(2)} = E_{+}^{(2)*}$.

It is seen from Equation (3.124) that the amplitude of the harmonic $E^{(2)}(z)$ increases effectively only over a short distance of the order of the wavelength near the plasma boundary. Its subsequent variation is oscillatory. At larger distances in the interior of the plasma, where the perturbations produced by the fundamental wave are already weak, we obtain for the harmonic amplitude $\boldsymbol{E}^{(2)}$

$$\left|\boldsymbol{E}^{(2)}\right|\Big|_{z\to\infty} = \frac{24\pi\omega}{c^2 k_2} \exp(-K_2) \left| \int_0^\infty j_0^{(2)} \sin(k_2 z)\, dz \right|. \tag{3.125}$$

Here $K_2(z)$ is the total absorption of the wave of frequency 3ω from the plasma boundary to the considered point z:

$$K_2(z) = \frac{3\omega}{c} \int_0^z \kappa(3\omega)\, dz.$$

The amplitude of the harmonic wave is always smaller than the amplitude of the fundamental if Equation (3.7) is satisfied, and is of the order of $\delta\nu_e/\omega$. With decreasing fundamental frequency ω, the amplitude of the harmonic increases. It can become appreciable in strong fields $E_0 \gtrsim E_p$ at $\omega \sim \delta\nu_e$. The amplitudes of higher harmonics of frequency $5\omega, 7\omega, \ldots,$ decrease like powers of the parameter $\delta\nu_e/\omega$.

We note that the nonlinear current $\boldsymbol{j}^{(2)}$ contains also corrections $\boldsymbol{j}_{\omega}^{(2)}$ at the fundamental frequency ω [see Eq. (2.44)], which determine the additional changes of the conductivity and the dielectric constant of the plasma. These changes are small in comparison with the main perturbations produced in σ and ε by the change of $\bar{T}_e$. They are, therefore, generally insignificant. The current $\boldsymbol{j}_{\omega}^{(2)}$, however, is of anisotropic character, and this can manifest itself in the propagation of elliptically polarized waves. Indeed, in the linear approximation, an elliptically polarized wave propagating in an isotropic plasma retains its previous polarization. The main nonlinear perturbations of $\varepsilon(\bar{T}_e)$ and $\sigma(\bar{T}_e)$ are scalar and therefore likewise have no effect on the wave polarization. The nonlinear corrections $\boldsymbol{j}_\omega^{(2)}$, however, are anisotropic. They can therefore alter the ratio of the phases of the polarization components, and this leads in final analysis to a change in the polarization ellipse (Gurevich and Shvartsburg, 1973).

We note also that harmonics of frequency 2ω can also be generated in an inhomogeneous and magnetoactive plasma (Ginzburg, 1958; Sodha and Kaw, 1966).

3.4.2. Nonlinear Detection

Assume that a constant electric current

$$\boldsymbol{j}_0 = \hat{\sigma}\boldsymbol{E} \tag{3.126}$$

flows in the plasma ($\boldsymbol{E}$ is the electric field produced by external sources). It is known that intense electric currents flow in the ionosphere at heights corresponding to the lower part of the E layer. Amplitude-modulated radio waves [Eq. (3.76)] produce periodically alternating perturbations, of low frequency Ω, in the temperature of the plasma electrons. The conductivity $\hat{\sigma}$ in Equation (3.126) is correspondingly altered. As a result, the current j_0 is modulated at the low frequency Ω (Kotik and Trakhtengerts, 1975).

The low-frequency perturbations $\Delta_\Omega T_e$ of the electron temperature were defined above and are given by Equation (3.80) in the case of a weak field and by Equations (3.96), (3.104), and (3.106) in the case of a strong field. The alternating corrections to the current [Eq. (3.126)] then take the form

$$\boldsymbol{j}_\Omega = \frac{\partial\hat{\sigma}}{\partial\nu_e}\frac{d\nu_e}{dT_e}\Delta_\Omega T_e\boldsymbol{E}, \tag{3.127}$$

where $\hat{\sigma}$ is the plasma conductivity tensor [Eq. (2.49)] for a low-frequency electric field. In particular, in a weakly ionized plasma (the lower ionosphere), we have

$$\begin{gathered}\frac{\partial\sigma_{zz}}{\partial\nu_e} = -\frac{e^2N}{m\nu_e^2}, \qquad \frac{\partial\sigma_{xx}}{\partial\nu_e} = \frac{\partial\sigma_{yy}}{\partial\nu_e} = \frac{e^2N(\omega_H^2 - \nu_e^2)}{m(\omega_H^2 + \nu_e^2)^2},\\ \frac{\partial\sigma_{xy}}{\partial\nu_e} = -\frac{\partial\sigma_{yx}}{\partial\nu_e} = \frac{2e^2N\nu_e\omega_H}{m(\omega_H^2 + \nu_e^2)^2}.\end{gathered} \tag{3.128}$$

We note that according to Equations (3.80) and (3.76) the current $\boldsymbol{j}_\Omega$ has components not only of frequency Ω, but also of its multiples 2Ω, 3Ω, etc.

When an amplitude-modulated radio wave propagates in an inhomogeneous plasma, corrections to the frequency-modulated current $\mathbf{j}_\Omega$ arise also in the absence of an external electric field (Ginzburg, 1958). The lower ionosphere is strongly inhomogeneous over a low-frequency wave length $2\pi c/\Omega$. Therefore the contribution of the inhomogeneous currents to $\mathbf{j}_\Omega$ is appreciable here and commensurate in a middle latitude ionosphere with Equation (3.127).

Owing to the presence of the alternating current $\mathbf{j}_\Omega$ [Eq. (3.127)], radio waves of frequency Ω, 2Ω, . . . , are generated in the plasma. The field of these waves in a homogeneous medium can be calculated in complete analogy with the preceding section. An essential feature of the ionosphere problem is that the low-frequency wave can propagate only in the earth-ionosphere wave guide (Wait, 1962; Al'pert et al., 1967; Al'pert, 1974); (or else in a channel along the earth's magnetic field). The excitation of the natural modes of the earth-ionosphere wave guide by the low-frequency current [Eq. (3.127)] was considered by Belustin et al. (1975).

3.5. Self-Action of Radio Waves in the Lower Ionosphere

The plasma field in the lower ionosphere is weak, $E_p \approx 40\ (\omega/10^6)$ mV/m. Field amplitudes E_0 that lead to noticeable nonlinear effects are therefore easy to attain here with radio waves of the medium-wave band $10^6 \lesssim \omega \lesssim 10^7$. If the radio transmitter power is high, E_0 becomes comparable with the plasma field E_p, and may even exceed it appreciably. Numerical estimates of the ratio E_0/E_p and of the nonlinear self-action effects in the case of vertical ($\psi = 0°$) and oblique ($\psi = 75°$) propagation of the wave in the lower ionosphere, are given in Table 15. Here ψ is the incidence angle, i.e., the angle between the wave vector $\mathbf{k}$ and the vertical, $\omega_{\text{ef}} = |\omega \pm \omega_H \cos \alpha|$ is the effective frequency (+ for the ordinary wave,—for the extraordinary wave, and α is the angle between $\mathbf{k}$ and the earth's magnetic field $\mathbf{H}$), W_0 is the effective radiation power, or the power of the equivalent dipole (see footnote 1 in Chapter 1), and P is the self-action factor [see Eq. (3.40)], which shows how the wave amplitude is altered by its self-action in the ionosphere. The ratio μ/μ_0 indicates the change in the depth of the amplitude modulation of the wave. The coefficients $P_r = P^2$ and $\mu_r/\mu_0 = (\mu/\mu_0)^2$ describe the same quantities for a wave reflected from the ionosphere, provided that this wave passes through the perturbed region.

It is seen from the table that the self-action of high-power radio waves can be quite appreciable in the lower ionosphere. For example, under night-time conditions, when the effective power of the transmitter at a frequency $\omega_{\text{ef}} = 10^6$ changes from $W_0^{(1)} = 10^2$ kW to $W_0^{(2)} = 10^4$ kW, the

Table 15. Self-action effects in the lower ionosphere

		$\psi = 0°$			$\psi = 75°$				
ω_{ef}, s^{-1}	W_0, kW	E_0/E_p	P	P_r	E_0/E_p	P	P_r	μ/μ_0	μ_r/μ_0
10^6 day-time	10^2 10^4	0.3 3.3	1.02 2.06	1.04 4.24	0.11 1.1	1.00 1.17	1.01 1.37	1.01 1.27	1.02 1.61
10^6 night-time	10^2 10^4	1.0 10	0.86 0.40	0.75 0.16	0.3 3.3	0.98 0.55	0.96 0.30	0.96 0.41	0.92 0.17
10^7	10^2 10^4 10^6	0.1 1.0 10	0.99 0.88 0.79	0.98 0.78 0.62	0.03 0.32 3.2	1.00 0.96 0.67	1.00 0.92 0.45	1.00 0.93 0.54	1.00 0.86 0.30
10^8	10^4 10^6 10^8	0.1 1.0 10	0.99 0.83 0.18	0.98 0.69 0.03	0.03 0.32 3.24	1.00 0.96 0.44	1.00 0.92 0.20	1.00 0.93 0.30	1.00 0.87 0.09

amplitude of the wave reflected from the ionosphere, as is clear from the table, is increased only 2.1 times:

$$\frac{E_0^{(2)} P_r[E_0^{(2)}]}{E_0^{(1)} P_r[E_0^{(1)}]} \approx 2.1.$$

If it were possible to neglect self-action-effects, then to obtain the same change in the reflected-wave amplitude the transmitter power would have to be increased not 100 times, but only four times. The amplitude of the wave reflected from the ionosphere may even decrease with increasing radiation power.

Figures 30–33 show some results of a numerical calculation of the self-action for plane radio waves of various frequencies ω propagating in the ionosphere (Gurevich, 1956). Longitudinal propagation of an extraordinary wave is considered. Curves 1 in the figures correspond to wave-field amplitude $E_0 = 50$ mV/m at the boundary of the ionosphere, curves 2 to $E_0 = 500$ mV/m, and curves 3 to $E_0 = 5000$ mV/m. The gyrofrequency ω_H is assumed equal to $9 \cdot 10^6$.

It is seen from the figures that for waves of resonance frequency $\omega = \omega_H$ the self-action factor increases monotonically as the wave propagates into the interior of the plasma [cf. Eq. (3.45)]. The wave, so to speak, pierces its way into the ionosphere. To be sure, owing to the very large absorption, the wave with $\omega = \omega_H$ attenuates rapidly at the beginning of the layer. Self-action effects help a strong wave to advance 5–7 km into the interior

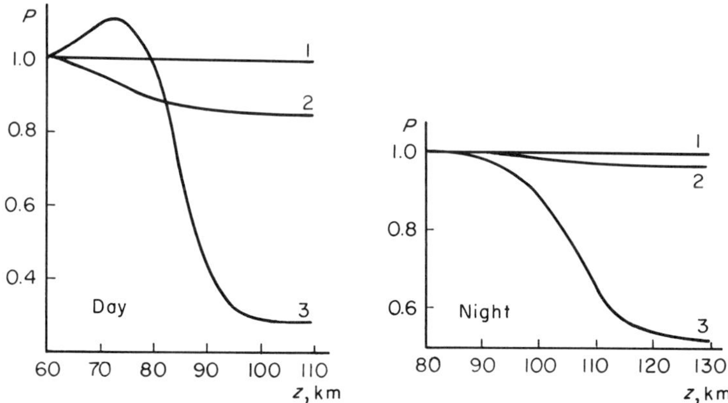

Fig. 30. Self-action factor in the ionosphere for a wave of frequency $\omega = 3 \cdot 10^7$

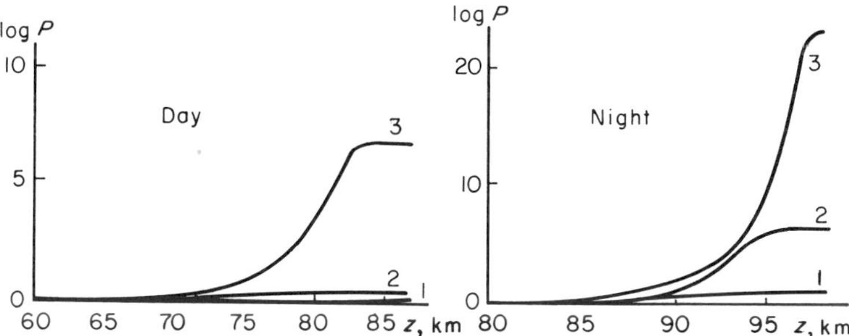

Fig. 31. Self-action factor in the ionosphere at gyroresonance $\omega = \omega_H = 9 \cdot 10^6$

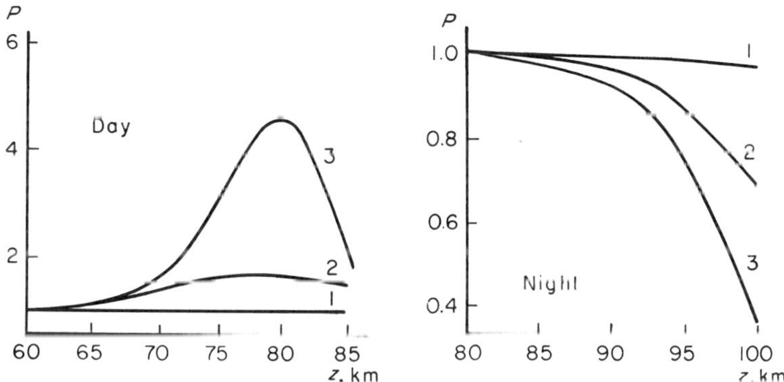

Fig. 32. Self-action factor in the ionosphere for a wave of frequency $\omega = 3 \cdot 10^6$

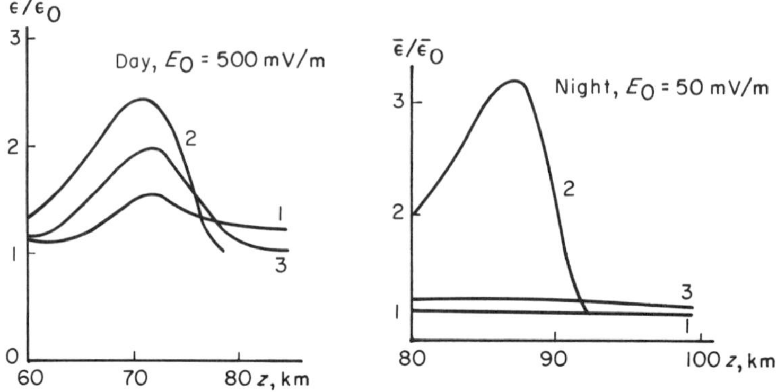

Fig. 33. Electron heating in the ionosphere. Curves *1* and *3* correspond to the frequencies $\omega = 3 \cdot 10^7$ and $\omega = 3 \cdot 10^6$, curve 2 corresponds to $\omega = \omega_H = 9 \cdot 10^6$

of the ionosphere. For waves at the nonresonant frequencies $\omega = 3 \cdot 10^6$ and $\omega = 3 \cdot 10^7$, the self-action factor is much closer to unity. It can either increase and decrease as the wave penetrates into the plasma.

Figure 33 shows the average energy (or the effective temperature) of the electrons, $\bar{\varepsilon}/\varepsilon_0 = T_e/T$, in the wave field. It is seen that at the gyrofrequency the plasma can be significantly heated at night-time even at $E_0 = 50$ mV/m. In this case the effect has a clearly pronounced resonant character, for there is practically no heating at the same value of the field amplitude at other frequencies. Under day-time conditions, and also at higher values of the wave field, the gyro resonance is less strongly pronounced.

Self-action effects in the ionosphere were investigated experimentally for both weak radio waves (King, 1959; Vilenskii, 1960–1966) and strong ones (Gurevich and Shlyuger, 1975). The experimental results are in sufficient agreement with the theory. Some of the measurement results are shown in Figures 27 and 34. Figure 34 shows a plot of the self-action factor P_r of the reflected wave and of its amplitude E_r against the radiation power (cf. Figs. 17 and 18). At the maximum radiation power $W_0 = W_{0m}$, the nonlinear absorption of the reflected-wave power reached 26 dB; its amplitude decreased with increasing radiation power at $E_0 \gtrsim E_p$.

Figure 34 reveals a rather abrupt transition boundary: at $E_0 \lesssim 0.7E_p$ the role of the nonlinear effect is still negligible, but it becomes decisive already at $E_0 > 1.5E_p$. This agrees with the theoretical concept that a strong nonlinearity is sharply "turned on" in a molecular plasma as a result of the singularities in the excitation of the rotational levels of the molecules by slow electrons (Sects. 2.2.3 and 2.3.2). Because of this circumstance it becomes possible to indicate a critical transmitter radiation

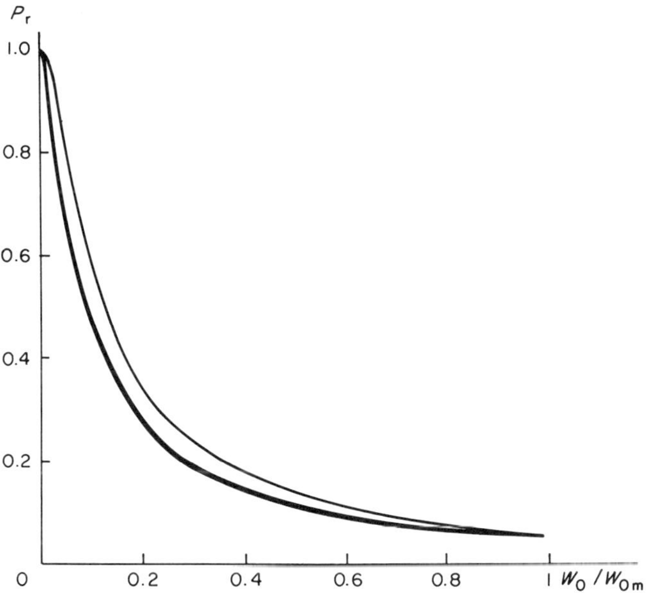

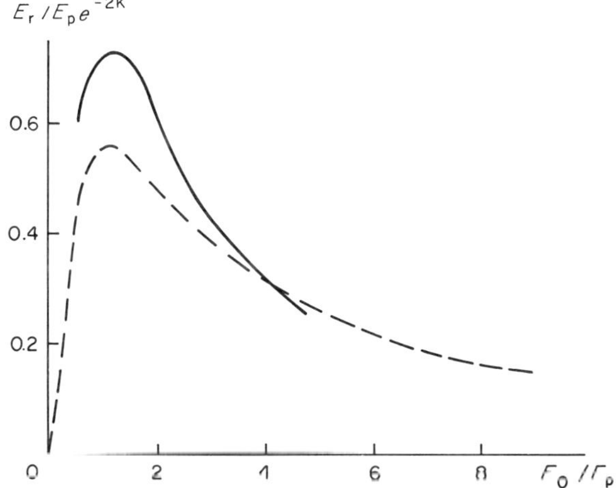

Fig. 34. Self-action coefficient P_r and reflected-wave amplitude E_r: $\omega = 8.48 \cdot 10^6$, $\alpha \simeq 20°$, ordinary wave, $\omega_{ef} \simeq 1.66 \cdot 10^7 W_{0m} \sim 10^6$ kW. *Thick curves*: averaged experimental results (Gurevich and Shlyuger, 1975). *Thin curve*: numerical calculation; *dashed curve*: calculation by Equations (3.49) and (3.50)

power W_{0c} (Ginzburg and Gurevich, 1960). At $W_0 > W_{0c}$ the nonlinear processes become decisive and the radiation energy goes mainly to heating of the ionosphere; the amplitude modulation of the wave is also strongly distorted in this case. For day time conditions in the short-wave band $\omega \geq \omega_H$ we have

$$W_{0c} \simeq \left[\frac{5(\omega_{ef}/10^7)^2}{\sin^2 \varphi_0 + 0.025}\right]. \tag{3.129}$$

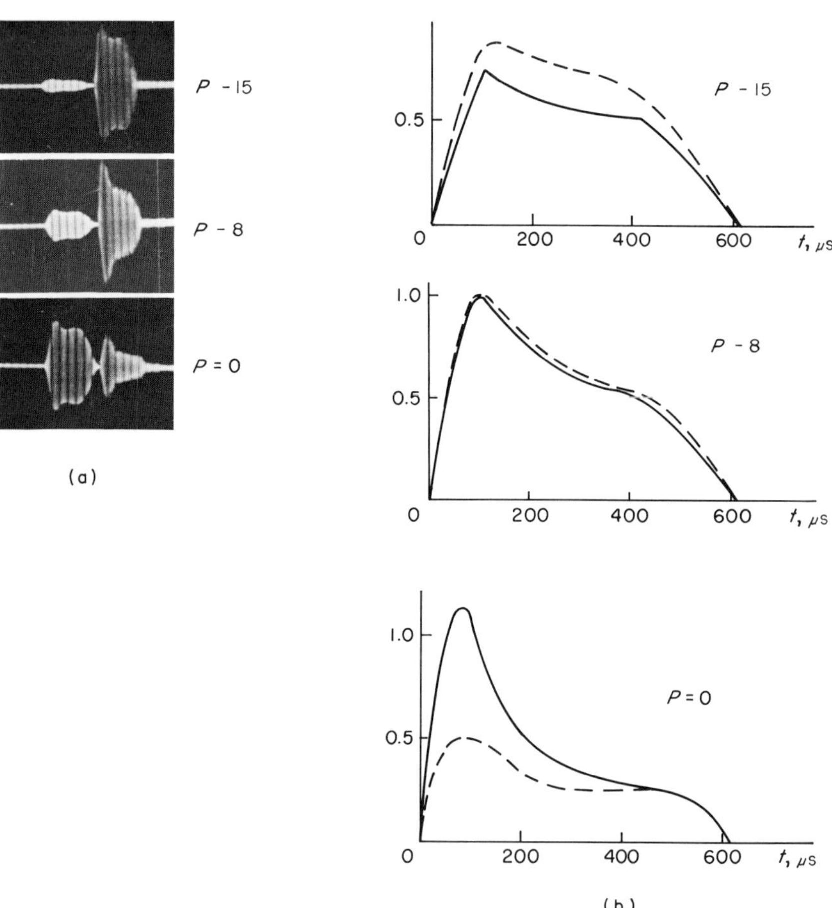

Fig. 35a, b. Nonlinear distortion of pulse waveform. (a) Experimental results. *Left*: radiated pulse, *right*: pulse reflected from the ionosphere (enlarged scale). *Black strips*: 100-μs time markers. *On the right* is indicated the relative radiation power in decibels, $P = -10 \log (W_0/W_{0m})$, where $W_{0m} \sim 10^6$ kW. It is seen that the pulse reflected from the ionosphere weakens with increasing radiation power. (b) Pulse envelope. *Dashed curves*: experiment; *solid curves*: theory (numerical calculation)

Here W_{0c} is the equivalent radiation power in MW, $\omega_{ef} = |\omega \pm \omega_H \cos\alpha|$, φ_0 is the elevation angle, i.e., the angle between $\boldsymbol{k}$ and the horizontal direction on earth for the maximum of the radiation. Thus, at $\varphi_0 = 15^\circ$ and at a wave frequency 5 MHz($\omega \approx 3.10^7$) we have $W_{0c} \approx$ 300 to 800 MW, but at 10 MHz we have already $W_{0c} \approx$ 1500–2000 MW. We emphasize that the foregoing calculation of the critical power took into account only effects connected with the heating of the lower ionosphere. Under certain conditions, the nonlinear effects in the upper ionosphere can lower the value of W_{0c} appreciably (see Sects. 6.2 and 6.3).

The nonlinear distortion of the waveform of a high-power radio pulse reflected from the ionosphere was also investigated experimentally. An example is shown in Figure 35. It is seen that the pulse envelope acquires characteristic "wings" (see Sect. 3.3.4 and Fig. 29). The calculation of the perturbation of the effective electron temperature by a radio pulse in the ionosphere at $W_0 = 10^6$ kW is illustrated in Figure 36. It is seen that in day time the electron temperature is increased by the wave field about twenty-fold, and at night by more than 40 to 50 times. The extraordinary and ordinary waves produce at gyroresonance $\omega = \omega_H$ approximately equal electron-temperature perturbations, i.e., there is practically no resonance effect at high radiation power. The perturbations due to the ordinary wave during the night time at large heights are preserved for a long time, owing to the growth of the characteristic temperature relaxation time τ_T (see Table 13). The extraordinary wave is strongly absorbed at night in the height region $z \approx 90$ km. The ordinary wave, to the contrary, is little absorbed at night, and the pulse is slightly deformed.

Getmantsev et al. (1974) have observed the detection effect of the ionosphere. The ionosphere was perturbed by a high-power high-frequency wave ($\omega = 3.6\ 10^7$, $W_{0m} \simeq 20$ MW) modulated in amplitude at a low-frequency $F = \Omega/2\pi = (1.7\text{–}8.2)$ kHz. The field E_Ω of the low-frequency wave resulting from the nonlinear detection in the ionosphere was observed at a distance of 200 km from the transmitter; it was much higher in the day than at night. The amplitude E_Ω increased linearly with increasing power of the high-frequency transmitter, as should be the case when $E_0^2 < E_p^2$, since $E_\Omega \sim j_\Omega \sim (E_0^2/E_p^2)$ [Eqs. (3.127) and (3.80)]. Figure 37 shows a typical plot of E_Ω against the frequency Ω. The decrease of E at high frequencies $\Omega > \delta_0 \nu_{e0}$ agrees qualitatively with Equations (3.80) and (3.127). The decrease of E_Ω at low frequencies, $\Omega/2\pi \lesssim 2$ kHz, is connected with peculiarities in the excitation of the Earth—ionosphere wave-guide (Belustin et al., 1975). The coefficient of the transformation of the energy of the high-frequency wave into a low-frequency one is $K \sim 10^{-10}$. We note that near the polar and equatorial electrojets the ionospheric electric fields and currents are much higher than in the

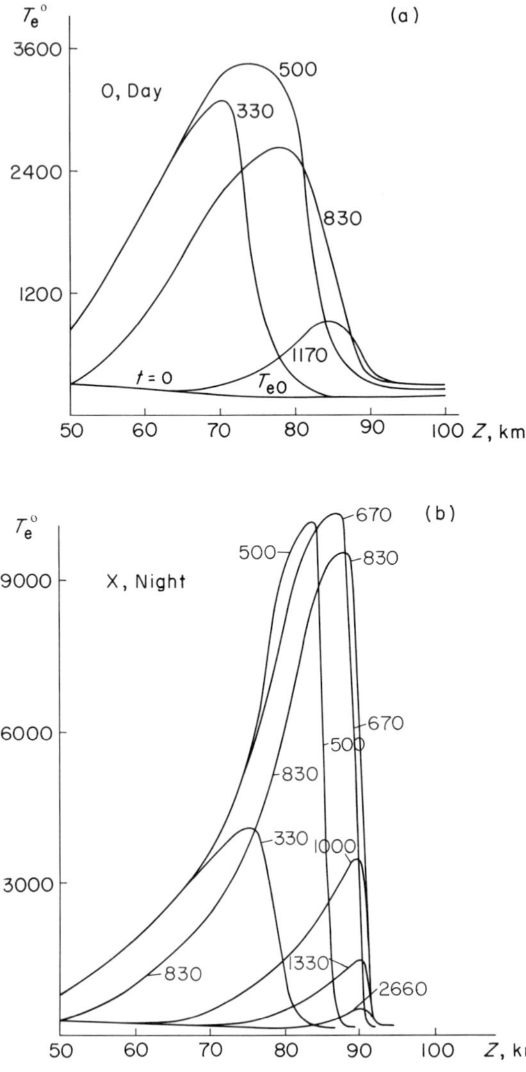

Fig. 36a–d. Effective electron temperature in the ionosphere. Modification of the ionosphere by a pulse of duration $t_1 \approx 500\ \mu$s (see Fig. 35), of frequency $\omega = \omega_H$, and power $W_0 = W_{0m} = 10^6$ kW. (a) O: ordinary wave, (b) X: extraordinary wave, T_{e0}: unperturbed electron temperature. The time in microseconds from the start of the pulse radiation on earth is marked next to the curves. In the case of the ordinary wave at night (c and d) the temperature relaxation proceeds slowly (d is a continuation of c); the secondary growth of T_e at t ~ 2000 μs at heights $z \sim 60$–80 km is the result of the action of the wave reflected from the ionosphere

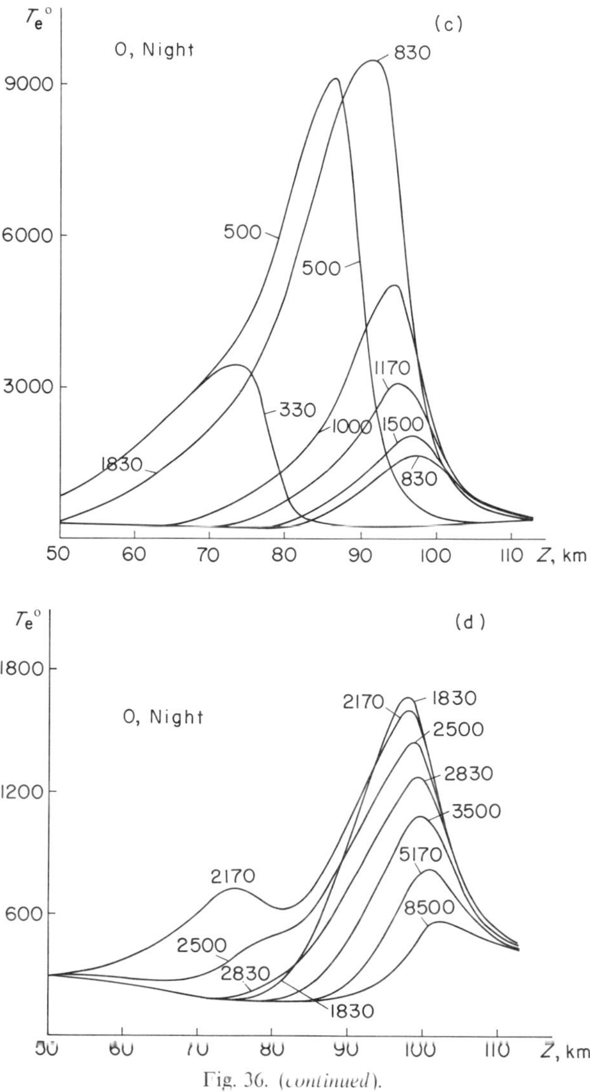

Fig. 36. (*continued*).

middle-latitude ionosphere. An appreciable increase of the nonlinear-detection effect was observed there (Kapustin et al., 1977). We note also that Equation (3.127) gives only a part of the current, of frequency Ω, which is produced in the ionosphere plasma. The point is that the stationary and quasi-stationary currents flowing in the ionosphere are nondivergent, i.e., they satisfy the condition div $\boldsymbol{j} = 0$ (for otherwise the charges would accumulate). This condition is not satisfied for the current (Eq. (3.127)]

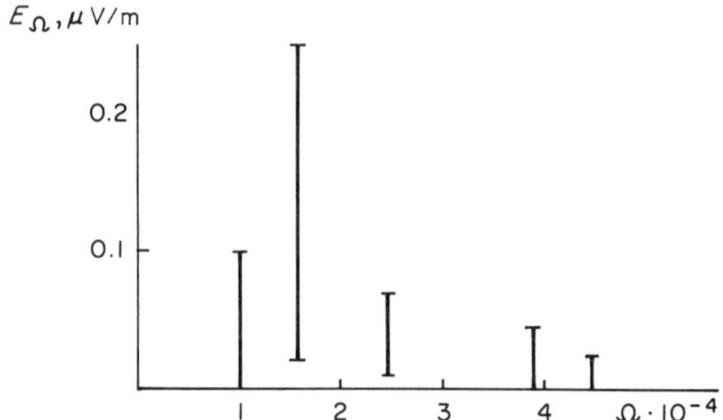

Fig. 37. Frequency dependence of the signal amplitude in the case of nonlinear detection

excited in a definite region heated by radio waves. This means that perturbations of the stationary electric field E'_Ω in Equation (3.126) arise, and additional current appears, such that div $(\boldsymbol{j}_\Omega + \boldsymbol{j}'_\Omega) = 0$. The combined current $\boldsymbol{j}_\Omega + \boldsymbol{j}'_\Omega$ is the source of the low-frequency radio waves.

An increase in the electron density in the lower ionosphere under the influence of a high-power wave was observed (see Sect. 2.5.6). However, this question still awaits a detailed experimental investigation.

It would be of considerable interest to investigate the nonlinear phenomena produced in the ionosphere by propagation of low-frequency radio waves. The plasma field E_p for these waves is weak: in the lower part of the E layer $E_p \approx 40$–60 mV/m, and in the D layer $E_p \approx 200$–500 mV/m. The low-frequency waves, however, propagate in the earth–ionosphere waveguide and cannot be described by the geometrical-optics approximation (Wait, 1962; Al'pert et al., 1967; Al'pert, 1974). The role of the nonlinear effects can be estimated in this case by using the results of calculations and measurements of the wave field amplitude E_0 at various distances R from the transmitting station. Thus, at $\omega = (1$–$2) \cdot 10^5$ and at $W = 1$ kW we have, according to Al'pert et al. (1967),

R, km	200	1000	2000	3000
E_0, mV/m	2	0.5	0.2	0.1

Since $E \sim \sqrt{W}$, the amplitude of the low-frequency electric field in the ionosphere becomes comparable with E_p at $W > 1000$ kW. At $W > 1000$ kW the nonlinearity can affect the absorption of the low-frequency waves substantially. Some self-action effects of radiowaves at $\omega \approx 10^5$ were observed by Molchanov et al. (1976).

One more feature of the nonlinear effects, in the case of very long waves, is connected with the fact that the frequency of these waves is low.

Whereas for medium and short radio waves the amplitudes of the harmonic waves produced in the plasma are small, by virtue of Equation (3.7): $\delta v_e \ll \omega$, they can be much larger for long waves. For ultralong waves, δv_e and ω are at any rate of the same order, so that self-action of strong ultralong waves of frequency $\omega \lesssim 10^3$–10^4 can lead to a noticeable distortion of their frequency spectrum. This can manifest itself, in particular, in the frequency spectrum of atmospherics, the integrated power of which is usually high: $W \sim 10^5$–10^6 kW.

4. Interaction of Plane Radio Waves

The modifications produced in a plasma by a high-power radio wave affect also other waves propagating in the perturbed region. This leads to interaction of the radio waves. The present chapter deals with the characteristic phenomena of radio-wave interaction in the lower ionosphere.

If the high-power wave is amplitude-modulated at a low frequency Ω, then the perturbations the wave produces in the plasma turn out to be modulated, as consequently also are other waves passing through the perturbed region. This phenomenon is called cross modulation. It is observed when radio waves propagate in the ionosphere and is of practical significance for radio broadcasting at medium wave lengths. A phenomenon close in character to cross modulation is also the interaction of the short pulses used for the experimental investigation of the ionosphere.

High-power unmodulated radio waves also modify the plasma; the result is a change in the absorption and in the phase of other waves propagating in the modified region. In addition to modifications that are constant in time, alternating modifications that vary with a frequency that is a multiple of that of the modifying wave are also produced in the plasma. Consequently the propagation of other radio waves in the medium should give rise to waves with combination frequencies.

4.1. Cross Modulation

4.1.1. Weak Waves

To calculate the effect of the cross modulation, it is first necessary to determine the magnitude of the low-frequency perturbations produced in the plasma by the high-power wave (the wave E_1) and then find how these perturbations affect another wave (wave E_2) propagating in the perturbed region. We consider first the case of an isotropic plasma and a relatively weak perturbing wave ($E_1 \ll E_p$), when the self-action of the wave is not significant (Bailey and Martin, 1934; Huxley and Ratcliffe, 1949; Ginzburg and Gurevich, 1960).

At the plasma boundary ($z = 0$) the amplitude of the field of the perturbing wave is

$$E_{10} = E_1(0)(1 + \mu_0 \cos \Omega t). \tag{4.1}$$

At a point z in the plasma we have in the geometrical-optics approximation

$$E_1(z) = \frac{E_1(0)}{\sqrt[4]{\varepsilon_1(z)}} \exp(-K_1)(1 + \mu_0 \cos \Omega t), \tag{4.2}$$

where $\varepsilon_1(z)$ is the dielectric constant for the wave E_1 at the point z, and

$$K_1(z) = \frac{\omega}{e} \int_0^z \kappa_1(z)\, dz$$

is the absorption of the wave in the layer.[12] The perturbations of the electron temperature T_e and of the collision frequency ν_e of a weakly-ionized plasma, which are due to the action of the field [Eq. (4.2)], and which vary periodically with frequency Ω, were calculated in Section 3.3.1 [Eq. (3.80)]:

$$\frac{\Delta_\Omega T_e}{T_{e0}} = \frac{2\mu_0 e^2 E_1^2(0)}{3m\, \delta_0 T_{e0}(\omega_1^2 + \nu_{e0}^2)\sqrt{\varepsilon_1(z)}} \exp[-2K_1(z)]$$
$$\cdot \left\{ \frac{\delta_0 \nu_{e0} \cos(\Omega t - \varphi_\Omega)}{\sqrt{\delta_0^2 \nu_{e0}^2 + \Omega^2}} + \frac{\mu_0 \delta_0 \nu_{e0} \cos(2\Omega t - \varphi_{2\Omega})}{4\sqrt{\delta_0^2 \nu_{e0}^2 + 4\Omega^2}} \right\}, \tag{4.3}$$

$$\Delta_\Omega \nu_e = \left(\frac{\partial \nu_e}{\partial T_e}\right)_{T_{e0}} \Delta_\Omega T_e, \qquad \operatorname{tg} \varphi_\Omega = \frac{\Omega}{\delta_0 \nu_{e0}}, \qquad \operatorname{tg} \varphi_{2\Omega} = \frac{2\Omega}{\delta_0 \nu_{e0}} \tag{4.4}$$

The field E_1 is weak, and therefore the perturbations $\Delta_\Omega T_e$ and $\Delta_\Omega \nu_e$ are small: $\Delta_\Omega T_e \ll T_{e0}$ and $\Delta_\Omega \nu_e \ll \nu_{e0}$.

The amplitude, at the point z, of any other wave, which we assume for simplicity to be unmodulated, is of course also determined by the expression

$$E_2(z) = \frac{E_2(0)}{\sqrt{\varepsilon_2(z)}} \exp[-K_2(z)], \qquad K_2(z) = \frac{\omega_2}{c} \int_0^z \kappa_2\, dz, \tag{4.5}$$

where $\varepsilon_2(z)$ is the dielectric constant and κ_2 is the absorption coefficient of the wave E_2.

---

[12] It is assumed here that the modulation frequency is low enough, $\Omega \ll c/\Delta z$, so that the phase shift of the modulation during the time of propagation of the wave in the perturbed region can be neglected. Assuming $\Delta z \sim 10$ km in the ionosphere, we see that this condition is satisfied only if $\Omega \ll 3 \cdot 10^4$.

The absorption coefficient depends on the collision frequency. Therefore when the wave E_2 propagates in the perturbed region of the plasma, a part of the absorption coefficient varies periodically with time. Separating this time-dependent part of κ_2,

$$\kappa_2 = \kappa_2(\nu_{e0}) + \left(\frac{\partial \kappa_2}{\partial \nu_e}\right)_{\nu_{e0}} \Delta_\Omega \nu_e \tag{4.6}$$

and substituting Equations (4.6) in Equations (4.5), we find that the amplitude of the wave passing through the perturbed layer of the plasma along a path S can be represented in the form

$$E_2 = E_2(0) \exp\left\{-\frac{\omega_2}{c}\int_S \kappa_2\, ds\right\}\left\{1 - \frac{\omega_2}{c}\int_S \left(\frac{\partial \kappa_2}{\partial \nu_e}\right)_{\nu_{e0}} \left(\frac{\partial \nu_e}{\partial T_e}\right)_{T_{e0}} \Delta_\Omega T_e\, ds\right\}, \tag{4.7}$$

where $\Delta_\Omega T_e$ is the alternating part of the electron-temperature perturbations in the plasma and is given by Equation (4.3).[13] We see therefore that after the wave E_2 passes through the perturbed layer it is amplitude-modulated at the frequencies Ω and 2Ω, i.e., it takes the form

$$E_2 = E_{20}\{1 - \mu_\Omega \cos(\Omega t - \varphi_\Omega) - \mu_{2\Omega} \cos(2\Omega t - \varphi_{2\Omega})\}. \tag{4.8}$$

The depth and phase of the cross modulation are then determined by the expressions

$$\mu_\Omega = \frac{\omega_2}{c}\int_S \frac{2\mu_0 e^2 E_1^2(0)}{3m\, \delta_0(\omega_1^2 + \nu_{e0}^2)} \left(\frac{\partial \nu_e}{\partial T_e}\right)_{T_{e0}} \left(\frac{\partial \kappa_2}{\partial \nu_e}\right)_{\nu_{e0}} \times \frac{\delta_0 \nu_{e0}}{\sqrt{\delta_0^2 \nu_{e0}^2 + \Omega^2}} \frac{\exp[-2K_1(s)]}{\sqrt{\varepsilon_1(z)}}\, ds \tag{4.9}$$

$$\mu_{2\Omega} = \frac{\omega_2}{4c}\int_S \frac{2\mu_0^2 e^2 E_1^2(0)}{3m\, \delta_0(\omega_1^2 + \nu_{e0}^2)} \left(\frac{\partial \nu_e}{\partial T_e}\right)_{T_{e0}} \left(\frac{\partial \kappa_2}{\partial \nu_e}\right)_{\nu_{e0}} \times \frac{\delta_0 \nu_{e0}}{\sqrt{\delta_0^2 \nu_{e0}^2 + 4\Omega^2}} \frac{\exp[-2K_1(s)]}{\sqrt{\varepsilon_1(z)}}\, ds \tag{4.10}$$

---

[13] We have put in Equation (4.7)

$$\exp\left\{-\frac{\omega_2}{c}\int_S \left(\frac{\partial \kappa_2}{\partial \nu_e}\right)_{\nu_{e0}} \left(\frac{\partial \nu_e}{\partial T_e}\right)_{T_{e0}} \Delta_\Omega T_e\, ds\right\} \simeq 1 - \frac{\omega_2}{c}\int_S \left(\frac{\partial \kappa_2}{\partial \nu_e}\right)_{\nu_{e0}} \left(\frac{\partial \nu_e}{\partial T_e}\right)_{T_{e0}} \Delta_\Omega T_e\, ds$$

Therefore Equation (4.7) is valid only if the cross-modulation depth is appreciably less than unity.

$$\operatorname{tg} \varphi_{\Omega} = \Omega/\delta_0 \nu_{e0}, \qquad \operatorname{tg} \varphi_{2\Omega} = 2\Omega/\delta_0 \nu_{e0}. \tag{4.11}$$

To obtain final expressions for the cross-modulation depth it is necessary also to integrate with respect to ds in Equations (4.9) and (4.10). We assume first that both waves E_1 and E_2 are normally incident on the plasma (i.e., that the normals to the wave fronts are parallel to the z axis, along which the plasma properties vary). The perturbed plasma layer then extends from the start of the plasma ($z = 0$) to the point $z = z_{01}$ at which the wave E_1 is reflected. We shall assume for simplicity that this point lies much lower than the point of reflection of the wave E_2, so that $\varepsilon_2(s) \simeq 1$ in the perturbed region. Then, recognizing that according to Equation (3.26)

$$\kappa_2(s) = \kappa_2(z) = \frac{2\pi e^2 N(z) \nu_{e0}}{m\omega_2(\omega_2^2 + \nu_{e0}^2)\sqrt{\varepsilon_2(z)}}$$

we obtain

$$\frac{\omega_2}{c}\left(\frac{\partial \kappa_2}{\partial \nu_e}\right)_{\nu_{e0}} \frac{1}{(\omega_1^2 + \nu_{e0}^2)\sqrt{\varepsilon_1(z)}} = \frac{\omega_2^2 - \nu_{e0}^2}{(\omega_2^2 + \nu_{e0}^2)^2} \frac{\omega_1}{c\nu_{e0}} \kappa_1(z), \tag{4.12}$$

where $\kappa_1(z)$ is the absorption coefficient of the wave E_1.

We now assume also that the plasma electron temperature T_{e0} and the collision frequency ν_{e0} in the perturbed region do not change with changing z (this is not a necessary assumption for high-frequency waves with $\omega_2^2 \gg \nu_{e0}^2$). Then, substituting Equations (4.12) in Equations (4.9) and recognizing that $(\omega_1/c)\kappa_1 = dK_1(z)/dz$ and that the wave can be represented with sufficient accuracy, up to its reflection point, as a sum of an incident wave and a reflected wave, we can easily integrate with respect to ds:

$$\begin{aligned}
\mu_\Omega &= \int_S \frac{2\mu_0 e^2 E_1^2(0)}{3m\,\delta_0} \frac{\delta_0 \nu_{e0}}{\sqrt{\delta_0^2 \nu_{e0}^2 + \Omega^2}} \frac{\omega_2}{c}\left(\frac{\partial \nu_e}{\partial T_e}\right)_{T_{e0}} \left(\frac{\partial \kappa_2}{\partial \nu_e}\right)_{\nu_{e0}} \frac{\exp[-2K_1(s)]\,ds}{(\omega_1^2 + \nu_{e0}^2)\sqrt{\varepsilon_1(z)}} \\
&= \frac{2\mu_0 e^2 E_1^2(0)}{3m\sqrt{\delta_0^2\nu_{e0}^2 + \Omega^2}} \left(\frac{\partial \nu_e}{\partial T_e}\right)_{T_{e0}} \frac{\omega_2^2 - \nu_{e0}^2}{(\omega_2^2 + \nu_{e0}^2)^2} \left[\int_0^{z_{01}} \frac{\omega_1}{c} \kappa_1(z) \exp[-2K_1(z)]\,dz \right. \\
&\quad \left. + \exp[-2K_1(z_{01})] \int_0^{z_{01}} \frac{\omega_1}{c} \kappa_1(z) \exp[-2K_1(z)]\,dz \right] \\
&= \frac{\mu_0 e^2 E_1^2(0)(\omega_2^2 - \nu_{e0}^2)}{3m(\omega_2^2 + \nu_{e0}^2)^2\sqrt{\delta_0^2\nu_{e0}^2 + \Omega^2}} \left(\frac{\partial \nu_e}{\partial T_e}\right)_{T_{e0}} [1 - \exp(-2K_1^0)]
\end{aligned} \tag{4.13}$$

It is assumed here that both waves propagate vertically;

$$K_1^0 = 2K_1(z_{01}) = 2\frac{\omega_1}{c} \int_0^{z_{01}} \kappa_1(z)\,dz$$

is the total absorption of the wave E_1 in the plasma. No allowance is made in Equation (4.13) for the fact that usually it is not only the incident wave, but also the reflected wave E_2 that passes through the modified region; this leads, naturally, to a doubling of the cross-modulation depth. The expression obtained for the depth of the cross modulation at the frequency 2Ω is similar to Equation (4.13); the cross-modulation depth at the second harmonic is always much lower than μ_Ω:

$$\mu_{2\Omega} = \frac{\mu_0}{4}\sqrt{\frac{\delta_0^2 \nu_{e0}^2 + \Omega^2}{\delta_0^2 \nu_{e0}^2 + 4\Omega^2}}\,\mu_\Omega. \tag{4.14}$$

Equations (4.13) and (4.14) can be easily generalized to include the case of oblique incidence of the waves E_1 and E_2 on the plasma, using the theorems concerning the connection between the absorption at oblique incidence with absorption at normal incidence (Ginzburg, 1960, Sect. 29). Thus, if the angle of incidence of the wave E_2 on the layer is equal to ψ_2, then μ_Ω and $\mu_{2\Omega}$ in Equations (4.13) and (4.14) must be multiplied by $\cos\psi_2$, and ω_2 must be replaced by $\omega_2 \cos\psi_2$. If the wave E_1 is incident at an angle ψ_1, then in the case when $\omega_1^2 \gg \nu_{e0}^2$ it suffices to multiply μ_Ω and $\mu_{2\Omega}$ by $\cos\psi_1$. Thus, for example, the expression for μ_Ω with allowance for the passage of the reflected beam E_2 through the perturbed region, with $\nu_e(T_e)$ defined in accordance with Equation (2.140), takes the form

$$\mu_\Omega = \frac{5\mu_0 e^2 E_1^2(0)}{9mT_{e0}} \frac{\omega_2^2 \cos^2\psi_2 - \nu_{e0}^2}{(\omega_2^2 \cos^2\psi_2 + \nu_{e0}^2)^2} \times \cos\psi_1 \cos\psi_2 \frac{\nu_{e0}}{\sqrt{\delta_0^2 \nu_{e0}^2 + \Omega^2}} [1 - \exp(-2K_1^0)]. \tag{4.15}$$

It is seen from Equations (4.15) and (4.13) that the depth of the cross modulation at the frequency Ω is proportional to the modulation depth of the perturbing wave and to its power. At $\omega_2^2 \cos\psi_2^2 \gg \nu_{e0}^2$, the depth of the cross modulation also increases with increasing angle of incidence of the wave E_2 on the plasma layer. This is valid, of course, only so long as the reflection point of the wave E_2 remains higher than the reflection point of the wave E_1; otherwise the dimensions of the interaction region also change appreciably with changing angle ψ_2. It is also seen from Equations (4.15) and (4.13) that the cross-modulation depth is maximal at low modulation frequencies $\Omega \ll \delta_0 \nu_{e0}$. At $\Omega \gg \delta_0 \nu_{e0}$ it decreases in proportion to $\delta_0 \nu_{e0}/\Omega$. The phase of the cross modulation [Eq. (4.11)] is small at small values of $\Omega/\delta_0 \nu_{e0}$ and increases to $\pi/2$ with increasing

$\Omega/\delta_0 \nu_{e0}$. Plots of the cross-modulation depth and phase against $\Omega/\delta_0 \nu_{e0}$ are shown in Figure 38.

The dependence of the depth of the cross modulation on the frequency ω_1 of the perturbing wave is determined by the factor

$$F(K_1^0) = 1 - \exp(-2K_1^0) \tag{4.16}$$

If the perturbing wave is appreciably absorbed in the interaction region, $K_1^0 \gtrsim 1$, then the factor F is always close to unity, and the depth of the cross modulation does not depend on the frequency ω_1. The cross modulation is, as it were, complete. If, to the contrary, the wave E_1 is only insignificantly absorbed in the plasma, $K_1^0 \ll 1$, then the depth of the cross modulation is proportional to its total absorption K_1^0. The dependence of the depth of modulation on the frequency ω_2 is given by

$$\mu_\Omega \sim \frac{\omega_2^2 \cos^2 \psi_2 - \nu_{e0}^2}{(\omega_2^2 \cos^2 \psi_2 + \nu_{e0}^2)^2}. \tag{4.17}$$

It was assumed above that the point of reflection of the wave E_2 lies much higher than the reflection point of the perturbing wave E_1. Equation (4.13) is valid also if this condition is not satisfied, but the form of the factor $F(K_1^0)$ [Eq. (4.16)] is altered somewhat. Consider, for example, the opposite case, when the reflection point of the wave E_2 lies much lower than that of the wave E_1. The total absorption of the wave E_2 in the plasma at $\omega_2^2 \gg \nu_{e0}^2$ (including the reflection point z_{02}) is given by the formula (Ginzburg, 1962, Sect. 17):

$$K_2 = \frac{2\omega_2}{c} \int_0^{z_{02}} \frac{2\pi\sigma_2}{\omega_2 \sqrt{\varepsilon_2}}\, dz = \frac{2\omega_2}{c} \left[\int_0^{z_{02}} \frac{2\pi\sigma_2}{\omega_2}\, dz + \int_0^{z_{02}} \frac{2\pi\sigma_2}{\omega_2} \left(\frac{1}{\sqrt{\varepsilon_2}} - 1 \right) dz \right]. \tag{4.18}$$

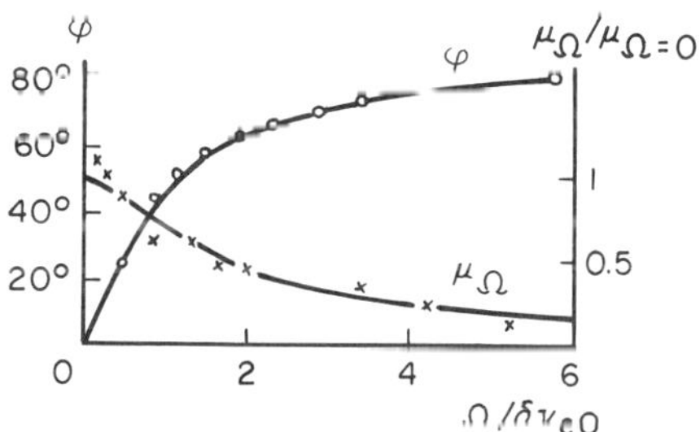

Fig. 38. Depth and phase of the cross modulation

Here z_{02} is the point of reflection of the wave E_2; σ_2 and ε_2 are the conductivity and the dielectric constant of the plasma for the wave E_2. The first term is identical with Equation (4.16). Therefore the same transformations of Equations (4.12) and (4.13) lead again to Equation (4.13) for the cross-modulation depths μ_Ω and $\mu_{2\Omega}$, but the factor $F(K_1^0) = 1 - \exp(-2K_1^0)$ is replaced by

$$F_1(K_1^0) = 1 - \exp(-2K_1^0) + f(K_1^0), \tag{4.19}$$

where

$$f(K_1^0) = \frac{2\omega_1}{c} \int_0^{z_{02}} \kappa_1 \exp(-2K_1) \left(\frac{1 - \sqrt{\varepsilon_2}}{\sqrt{\varepsilon_2}} \right) dz, \qquad K_1^0 = \frac{\omega}{c} \int_0^{\kappa_1} \kappa_1 \, dz. \tag{4.20}$$

It is important that K_1^0 is now the total absorption of the wave E_1 up to the point of reflection of the wave E_2. We neglect here the influence of the reflected beam E_1 on the cross modulation, for the sake of simplicity. The form of the function $f(K_1^0)$ depends somewhat on the character of the variation of the density $N(z)$ in the vicinity of the wave reflection point z_{02}. It is shown in Figure 39 for various dependences of N on z, namely $N = (z/z_{02})^n N_0$ with $n = 1$, $n = 3$, and $n \gg 1$. The same figure shows the factor $F_1(K_1^0)$ for $n = 1$. It is seen from the figure that at $K_1^0 \sim 1$ the factor $F_1(K_1^0)$ reaches in this case its maximum value $F_{1\,\max} \approx 1.6$. The height of the maximum changes somewhat, depending on the form of $N_0(z)$. The dashed curve in Figure 39 shows the factor $F = 1 - \exp(-2K_1^0)$, which, as shown, is the correct one to use at $z_{02} \gg z_{01}$. It is important that in general the difference between the limiting functions F and F_1 is small.

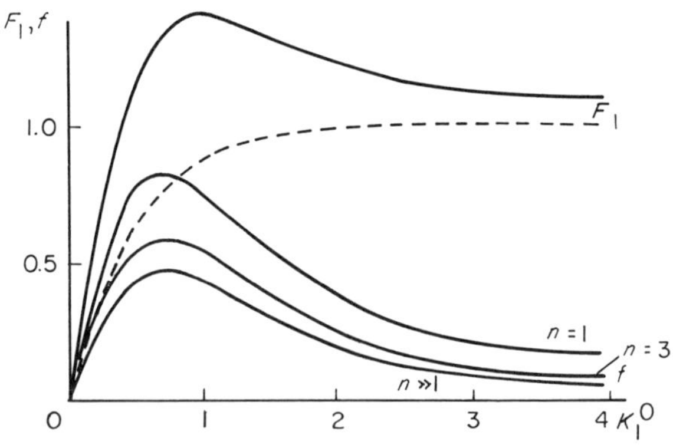

Fig. 39. The functions $f(K_1^0)$ and $F(K_1^0)$

We have considered here only amplitude cross modulation. The wave E_2 can also become phase-modulated (Vilenskii, 1962a).

4.1.2. Strong Perturbing Wave

It was assumed above that the perturbing wave is weak, $E_1^0(0) \ll E_p^2$. If this condition is not satisfied, then in the calculation of the cross-modulation effect it is necessary to take into account the self-action of the wave E_1, which changes its amplitude and the depth and phase of its modulation (Gurevich, 1958a). Just as before, we must first determine the perturbations produced in the plasma by the wave E_1, and then find how they influence the other wave E_2 propagating in the perturbed region.

The perturbations produced in the plasma by a strong amplitude-modulated wave E_1 were determined earlier in Section 3.3.2, where the self-action of the wave was taken into account. It is thus necessary to establish only how these perturbations influence the wave E_2.

We consider first the case of low modulation frequency $\Omega \ll \delta_0 \nu_{e0}$. Assuming that the wave E_2 is weak, we have

$$E_2(z) = \frac{E_2(0)}{\sqrt[4]{\varepsilon_2(z)}} \exp\left\{-\frac{\omega_2}{c}\int_S \kappa_2\, ds\right\} = \frac{E_2(0)}{\sqrt[4]{\varepsilon_2(z)}} \exp\left\{-\frac{\omega_2}{c}\int_S \kappa_{20}\, ds\right\} \cdot P_{12}. \tag{4.21}$$

Here κ_2 is the absorption coefficient [Eq. (3.26)] of the wave E_2 in the perturbed plasma, and κ_{20} is the same coefficient in the unperturbed plasma; P_{12} is a factor that shows how the perturbing wave E_1 influences the wave E_2. We recognize that according to Equation (3.34), in a weakly ionized plasma we obtain in the case of collisions with molecules [see Eq. (2.6)]

$$\kappa_2 = \kappa_{20} \frac{\tau(\omega_2^2 + \nu_{e0}^2)}{\omega_2^2 + \nu_{e0}^2 \tau^2}, \tag{4.22}$$

where $\tau = \sqrt{T_e(E_1)/T}$ is the perturbation of the electron temperature by the wave E_1, as defined by Equation (3.38). Consequently

$$P_{12} = \exp\left\{-\frac{\omega_2}{c}\int_S \kappa_{20}\left[\frac{\tau(\omega_2^2 + \nu_{e0}^2)}{\omega_2^2 + \nu_{e0}^2\tau^2} - 1\right]\right\} ds. \tag{4.23}$$

The factor P_{12} depends on the amplitude $E_1(0)$ of the wave E_1 at the plasma boundary. If the amplitude $E_1(0)$ is modulated at the low frequency Ω, then the factor P_{12} is likewise modulated. It is consequently this factor which determines the cross modulation of the wave E_2. Replacing in Equation (4.23), in accordance with Equation (3.26)

$$\frac{\omega_2}{c}\kappa_{20} \quad \text{by} \quad \frac{\omega_1}{c}\kappa_{10}\frac{\omega_1^2 + \nu_{e0}^2}{\omega_2 + \nu_{e0}^2}\sqrt{\frac{\varepsilon_1}{\varepsilon_2}}$$

and changing over to integration with respect to $d\tau$ [using Eq. (3.37) with $N(\tau)/N_0 = 1$], we obtain

$$P_{12} = \exp\left\{\overline{\sqrt{\frac{\varepsilon_1}{\varepsilon_2}}}\left[\frac{\omega_1^2 + \nu_{e0}^2}{\omega_2^2 + \nu_{e0}^2}\ln\frac{\tau + 1}{\tau_0 + 1}\right.\right.$$
$$\left.\left. + \left(1 - \frac{\omega_1^2 + \nu_{e0}^2}{2(\omega_2^2 + \nu_{e0}^2)}\right)\ln\frac{\omega_2^2 + \nu_{e0}^2\tau^2}{\omega_2^2 + \nu_{e0}^2\tau_0^2} + \frac{2\nu_{e0}^2(\tau_0 - \tau)}{\omega_2^2 + \nu_{e0}^2}\right]\right\}. \tag{4.24}$$

It is assumed here for simplicity that both waves propagate in a direction normal to the layer; $\tau_0 = \tau(E_{10})$ is the value of $\tau(E_1)$ at the plasma boundary $z = 0$. The field E_{10} is the amplitude of the perturbing wave at the plasma boundary; its time variation is given by Equation (4.1). Next, $\overline{\sqrt{\varepsilon_1/\varepsilon_2}}$ is the average ratio of the refractive indices in the interaction region.

This yields a simple expression for the cross-modulation depth μ_Ω if μ_0 is small. Indeed, just as in the derivation of Equations (3.87) and (3.89), we obtain:

$$\mu_\Omega = \mu_0\left[\frac{\partial P_{12}}{\partial E_1(0)}\cdot\frac{E_1(0)}{P_{12}}\right]_{\mu_0=0} = \mu_0\frac{\omega_1^2 + \nu_{e0}^2\tau_0^2}{\tau_0}\left\{\frac{\tau_0}{\omega_2^2 + \nu_{e0}^2\tau_0^2} - \frac{\tau}{\omega_2^2 + \nu_{e0}^2\tau^2}\right\}. \tag{4.25}$$

The ratio $\overline{\sqrt{\varepsilon_1/\varepsilon_2}}$ has been set here equal to unity.[14]

Equation (4.25) determines, first, the dependence of the depth of cross modulation μ_Ω on the amplitude (or on the power W_0) of the perturbing wave. A rigorously linear growth of μ_Ω as a function of the power W_0 of

[14] We take into account here wave interaction only in one direction. If there is also a reflected wave propagating in the perturbed region, then Equation (4.25) for the depth of the cross modulation must, naturally, be multiplied by 2.

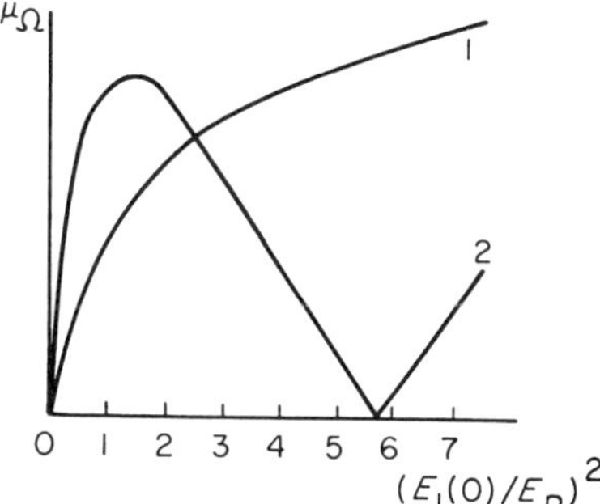

Fig. 40. Depth of cross modulation as a function of the power of the strong modifying wave $E_1^2(0)/E_p^2$

the transmitting station is obtained here only in the case of a weak perturbing field, $E_1^2(0) \ll E_p^2$. Indeed, for example, as is clear from Equation (4.25), at high frequencies $\omega_1^2 \gg v_{e0}^2\tau_0^2$ and $\omega_2^2 \gg v_{e0}^2\tau_0^2$ we have

$$\mu_\Omega = \mu_0 \frac{\omega_1^2}{\omega_2^2}\frac{\tau_0 - \tau}{\tau_0} = \mu_m \frac{\tau_0 - \tau}{\tau_0}. \tag{4.26}$$

At high frequencies ω_1 and ω_2, the cross-modulation depth [Eq. (4.26)] cannot exceed the value $\mu_m = \mu_0\omega_1^2/\omega_2^2$ at any power of the perturbing station (curve 1 of Fig. 40). If the frequency of the perturbing wave E_1 cannot be regarded as high, $\omega_1^2 \lesssim v_{e0}^2\tau_0^2$, then the growth of μ_Ω as a function of W_0 is also weaker than linear. The nonlinear dependence of μ_Ω on W_0 is particularly pronounced, however, under conditions when the frequency of the wave E_2 cannot be regarded as high, $v_{e0}^2 \lesssim \omega_2^2 \lesssim v_{e0}^2\tau_0^2$. In this case, as seen from curve 2 of Figure 40, μ_Ω can decrease with increasing W_0 and can even vanish. The principal role at $M_\Omega \to 0$ is assumed by cross modulation at the frequency 2Ω, so that the very character of the wave modulation undergoes strong changes.

This can be seen from Figure 41, which shows how the form of the cross modulation is altered under these conditions as a function of the power of the perturbing wave [here $\omega_2^2 = 2v_{e0}^2$, $\omega_1^2 = 2v_{e0}^2$, $K > 1$, $\Omega \ll \delta_0 v_{e0}$; $\mu_0 = \frac{1}{2}$; the calculation are performed using the Equation (4.24), the values of $E_1^2(0)/E_p^2$ are indicated in the figure]. We see that at $E_1^2(0)/E_p^2 = 6$ the cross-modulation frequency is in fact equal to 2Ω. The character of the distortions of the modulation in this case is similar to that of the distortions produced by "overmodulation" of the waves.

Equation (4.25) was derived under the condition that μ_0 is small ($\mu_0 \ll 1$). In the general case, when μ_0 is not small, the cross-modulation

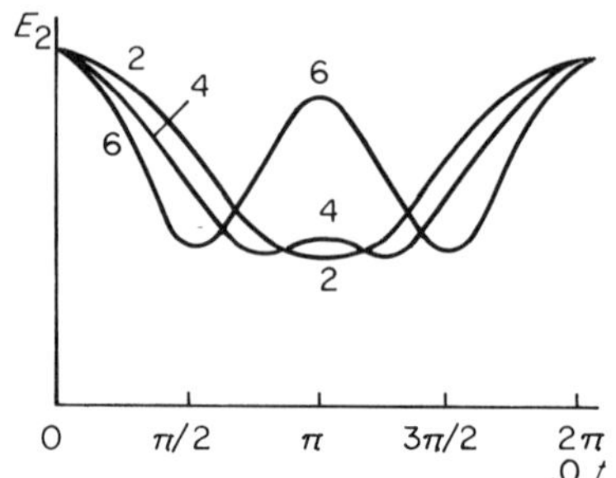

Fig. 41. Distortion of the modulation waveform. The curves are marked by the values of $E_1^2(\theta)/E_p^2$

depth is determined with the aid of the factor P_{12} [Eq. (3.24)]

$$\mu_\Omega = \frac{P_{12}[E_1(0)(1+\mu_0)] - P_{12}[E_1(0)(1-\mu_0)]}{P_{12}[E_1(0)(1+\mu_0)] + P_{12}[E_1(0)(1-\mu_0)]}. \tag{4.27}$$

The values of μ_Ω calculated from Equation (4.27) and from the simple Equation (4.25) differ insignificantly. Equation (4.25) can consequently be used with good accuracy at any value of μ_0. At large values of μ_0 and $E_1^2(0)/E_p^2$, however, the form of the cross modulation becomes strongly distorted (see Fig. 41).

It was assumed throughout the foregoing that the depth of the cross modulation is small, $\mu_\Omega \ll 1$, so that

$$\exp\{-a\cos(\Omega t - \varphi)\} \simeq 1 - a\cos(\Omega t - \varphi); \qquad \mu_\Omega = a. \tag{4.28}$$

This is true only if $a \ll 1$. Obviously, if $a \gtrsim 1$ the signal becomes distorted, i.e., the next higher harmonics also become appreciable. The depth of modulation of the fundamental tone is then given by

$$\mu_\Omega = 2\frac{\int_0^{2\pi}[\exp(-a\cos x)]\cos x\,dx}{\int_0^{2\pi}\exp(-a\cos x)\,dx} = 2\frac{I_1(a)}{I_0(a)} \tag{4.29}$$

where I_0 and I_1 are Bessel functions of imaginary argument and of zeroth and first order, respectively. At $a \gg 1$ the modulation depth μ_Ω tends to the value 2. Equally appreciable in this case also are the modulation depths of the second, third, . . . , n-th harmonics, up to $n \sim \sqrt{a}$. At the same time, the effective cross-modulation depth defined by Equation (4.27) is equal to unity. The signal takes the form of peaks similar to those shown in Figure 26.

We note, finally that we have considered above only the case of a perturbing station with a low modulation frequency, $\Omega \ll \delta_0 \nu_{e0}$. At an arbitrary frequency Ω the problem is complicated, but becomes simpler at small perturbing-wave modulation depths, $\mu_0 \lesssim \frac{1}{2}$. Equation (3.96) can then be used for the periodic perturbations of the electron temperature in the plasma, for arbitrary Ω. The resultant expressions for the depth and phase of cross modulation are simple in the case of high frequencies $(\omega_{1,2}^2 \gg \nu_{e0}^2)$:

$$\mu_\Omega = \mu_{\Omega \to 0} \cdot \frac{\delta_0 \nu_{e0} \tau_0}{\sqrt{\Omega^2 + \delta_0^2 \nu_{e0}^2 \tau_0^2}}; \qquad \operatorname{tg} \varphi_\Omega = \frac{\Omega}{\delta_0 \nu_{e0} \tau_0} \tag{4.30}$$

where $\mu_{\Omega \to 0}$ is the cross-modulation depth at the low frequency Ω considered above. These expressions are close in structure to Equations (4.11)

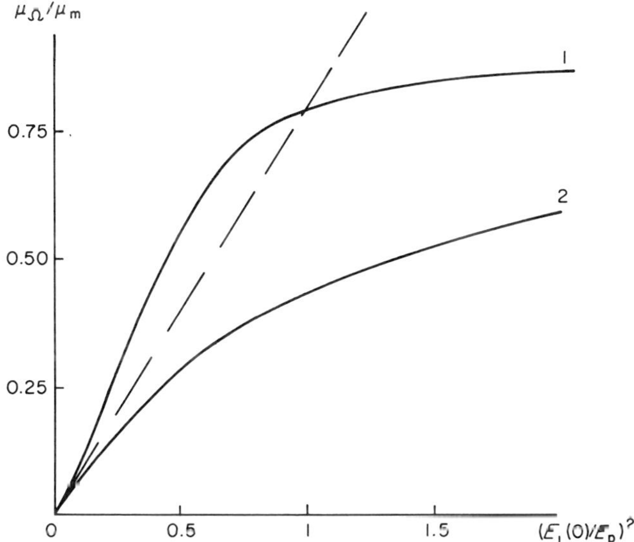

Fig. 42. Depth of cross modulation as a function of the power of the modifying wave with allowance for the actual functions $\delta(T_e)$ and $\nu_e(T_e)$ for the lower ionosphere

1. $\delta = \delta(T_e)$, $\quad \nu_e = \nu_{e0}(T_e/T)^{5/6}$
2. $\delta = \text{const.}$, $\quad \nu_e = \nu_{e0}(T_e/T)^{5/6}$

The maximum cross modulation depth is $\mu_m = \mu_0 \omega_1^2/\omega_2^2$. *Dashed line* shows the depth of the cross modulation in the case of a weak field and (4.13), which were obtained above for weak fields, except that ν_{e0} is replaced in them by $\nu_{e0}\tau_0$.

It is important to note that Equation (4.25) can be easily generalized to include the case of arbitrary dependences of ν_e and δ on T_e [cf. Eq. (3.89a)]:

$$\mu_\Omega = \mu_0 \frac{\omega_1^2 + \nu_e^2(\tau_0)}{\nu_e(\tau_0)} \left[\frac{\nu_e(\tau_0)}{\omega_2^2 + \nu_e^2(\tau_0)} - \frac{\nu_e(\tau)}{\omega_2^2 + \nu_e^2(\tau)} \right]. \qquad (4.25a)$$

Figure 42 (curve 1) shows the variation of μ_Ω in the interior of the plasma in accord with Equation (4.25a) for $\omega_1^2 \gg \nu_e^2$ and $\omega_2^2 \gg \nu_e^2$ with account taken of functions $\nu_e(T_e)$ [Eq. (2.235)] and $\delta(T_e)$ (Fig. 10) that are realistic for the lower ionosphere. An interesting distinguishing feature of this curve is that at $E_1^2/E_p^2 \lesssim 1$ the cross-modulation depth increases with increasing power more rapidly than in linear fashion (dashed straight line). The cause of this enhancement of the cross modulation is the decrease of $\delta(T_e)$ as a result of the rotation effect (see Sect. 2.3.2). At large values of $E_1^2(0)/E_p^2$ both curves of Figure 42 tend to the same value $\mu_m = \mu_0\omega_1^2/\omega_2^2$ [see Eq. (4.26)]. It is evident that the rotational effects accelerates noticeably the onset of saturation of μ_Ω.

4.1.3. Resonance Effects Near the Electron Gyrofrequency

We consider now the case when the perturbing-wave frequency ω_1 is close to the gyrofrequency ω_H (Bailey, 1937; Gurevich, 1958a; Hibberd, 1965). We assume for simplicity that the perturbing wave propagates longitudinally. It suffices then to consider only the extraordinary perturbing wave, since the perturbations produced by the ordinary wave have no resonance properties. The perturbations produced in the plasma by the extraordinary wave are given as before by Eq. (4.3), in which we must make the substitution

$$\frac{E_1^2(0)}{\omega_1^2 + \nu_{e0}^2} \rightarrow \frac{E_{1-}^2(0)}{(\omega_1 - \omega_H)^2 + \nu_{e0}^2}$$

where $E_{1-}(0)$ is the amplitude of the extraordinary perturbing wave at the plasma boundary (at $z = 0$). Similarly, on going from wave E_1 to wave E_{1-}, the absorption coefficient is changed:

$$\kappa_1(\nu_{e0}) \rightarrow \kappa_{1-}(\nu_{e0}) = \kappa_1(\nu_{e0}) \frac{\omega_1^2 + \nu_{e0}^2}{(\omega_1 - \omega_H)^2 + \nu_{e0}^2}.$$

Consequently the transformation [Eq. (4.12),] and hence also Equations (4.3) and (4.15) retain the same form if $E_1(0)$ is replaced in them by $E_{1-}(0)$ and K_1^0 is replaced by K_{1-}^0:

$$\mu_\Omega = \mu_0 \frac{5e^2 E_{1-}^2(0)}{9mT_{e0}} \frac{\omega_2^2 \cos^2 \psi_2 - v_{e0}^2}{(\omega_2^2 \cos^2 \psi_2 + v_{e0}^2)^2} \cos \psi_2 \cos \psi_1 \frac{v_{e0} F_1(K_{1-}^0)}{\sqrt{\delta_0^2 v_{e0}^2 + \Omega^2}} \tag{4.31}$$

where K_{1-}^0 is the total absorption of the wave E_{1-} up to the point of reflection of the wave E_2; the factor $F_1(K_{1-}^0)$ is defined by Equation (4.19) and is shown in Figure 39. It was assumed in Equation (4.31) for simplicity that the wave E_2 is transverse and ordinary.

It follows from Equation (4.31) that the depth of cross modulation at the gyrofrequency (i.e., at $\omega_1 \approx \omega_H$) does not exceed the cross-modulation depth at any other frequency ω_1 for which complete cross modulation is realized: in fact, if $K_{1-}^0 \gtrsim 1$, then $F_1 \approx 1$ and the cross-modulation depth [Eq. (4.31)] is practically independent of ω_1. The reason is that even though the extraordinary wave E_{1-} does indeed cause a strong perturbation of the plasma at $\omega_1 \approx \omega_H$ and $v_{e0} \ll \omega_H$, it does itself attenuate in a very thin layer (for example, at night the extraordinary wave attenuates in the ionosphere at $\omega_1 \approx \omega_H$ in a layer 1–5 km thick, see Fig. 36b). To the contrary, if the frequency ω_1 differs appreciably from the gyrofrequency, then the perturbed layer is considerably thicker despite the fact that the perturbations produced by it are much weaker than at $\omega_1 \approx \omega_H$. Therefore the integral ("complete") depth of the cross modulation is the same in these cases, provided only that the wave E_2 passes through the entire perturbed layer, i.e., if $K_{1-}^0 \gtrsim 1$.

At the same time, if the wave E_2 propagates only in a rather thin layer of the plasma (i.e., if the frequency ω_2 is low or its angle of incidence on the layer ψ_2 is large), then the "complete" cross modulation will be realized only for an E_{1-} wave having a frequency close to the gyrofrequency, for only such a wave is attenuated sufficiently in a thin plasma layer. Moreover, as is clear from Equation (4.31), the "complete" cross-modulation depth increases strongly with decreasing ω_2 or with increasing angle ψ_2, namely, $\mu_\Omega \sim 1/\omega_2^2 \cos \psi_2$. Therefore at low frequencies ω_2 or at large angles ψ_2 the cross-modulation depth has a clearly pronounced resonant peak when the frequency ω_1 is varied near the gyrofrequency.

This is seen from Figure 43, which shows a plot of μ_Ω against ω_1 at $\omega_1 \sim \omega_H$ and $\omega_2 \cos \psi_2 \sim v_{e0}$ in accordance with Equation (4.31). The resonance peak has either one or two humps. Two humps are clearly seen at $K_{1-}^0 \gtrsim 1$, where K_{1-}^0 is the total absorption of the E_1 wave at the

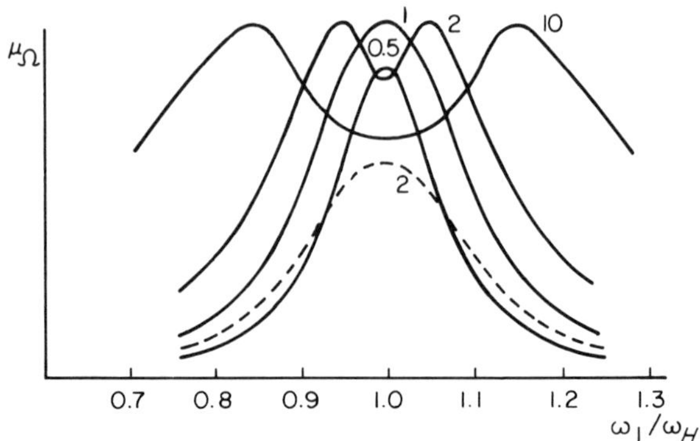

Fig. 43. Cross-modulation resonance near the gyrofrequency. The curves are marked by the values of K^0_{1-}

frequency $\omega_1 \approx \omega_H$ up to the point of reflection of the wave E_2. If the wave E_2 penetrates deep into the plasma, then $K^0_{1-} \gg 1$ and the resonance peak is washed out. The resonance can therefore become indistinguishable at very large values of K^0_{1-}, as was already mentioned earlier.

We note that the shape of the resonance curves is strongly influenced by the correction function $f(K^0_{1-})$ in the factor F_1 [Eq. (4.20)]; this correction is due to effects that occur in the region of reflection of the wave E_2. The resonance curve without allowance for $f(K^0_{1-})$ is shown dashed in Figure 43; it can have only one hump.

It was assumed above that the perturbing wave propagates longitudinally. In the case of non-longitudinal propagation, owing to the influence of the plasma polarization, the resonance effects take place already not at the gyrofrequency ω_H, but at the frequency ω_r given by

$$\omega_r = \left[\frac{\omega_H^2 + \omega_0^2}{2} \pm \sqrt{\frac{(\omega_H^2 + \omega_0^2)^2}{4} - \omega_H^2\omega_0^2 \cos^2\alpha}\right]^{1/2},$$

where α is the angle between the normal to the wave front and the direction of the magnetic field (Zheleznyakov, 1958). The frequency ω_r ranges from $\omega_r = \omega_H$ at $\alpha = 0$ to $\omega_r = \sqrt{\omega_0^2 + \omega_H^2}$ at $\alpha = \pi/2$. It must be borne in mind, however, that a wave of resonant frequency usually attenuates at the very beginning of the plasma layer. Therefore if the layer does not have a sharp boundary, then the perturbing wave is completely absorbed in the region of small ω_0. In this case the difference between the resonant cross modulation frequency in the case of non-longitudinal propagation, on the one hand, and the resonant frequency for longitudinal propagation, on the other, is negligible. For example, for cross modulation in the ionosphere, the indicated effect shifts the resonant frequency by at most 1–2%. The possibility of resonantly perturbing a plasma with a high-frequency wave amplitude-modulated at a frequency $\Omega = \omega_H$ or $\Omega = \omega_H/2$ was indicated by Cutolo (1964), Menzel (1965), and Gurevich and Shvartsburg (1968). Similar resonant perturbations can be produced in the ionosphere plasma by radio waves of frequency $\omega = \omega_H/2$ and $\omega = 2\omega_H$.

4.2. Interaction of Unmodulated Waves

4.2.1. Interaction of Short Pulses

A high-power unmodulated pulse E_1 of frequency ω_1 produces in a plasma perturbations that affect the propagation of another pulse E_2 of frequency ω_2. This is the cause of the interaction of radio pulses in the ionosphere.

Perturbations of the electron temperature at a point z, produced in an isotropic plasma by a high-power pulse E_1 are described by Equation (3.115) with the field $E_1(z)$ defined in accordance with Equation (3.22). The variation of the perturbation ΔT_e with the time t is shown in Figure 28. Naturally, the perturbations of the effective electron collision frequency behave in similar manner:

$$\Delta \nu_e = \left(\frac{\partial \nu_e}{\partial T_e}\right)_{T_{e0}} \Delta T_e$$

It is these perturbations which cause the cross modulation of the pulses. If the perturbing pulse is not strongly absorbed, then an appreciable perturbation can be produced also by a pulse reflected from the ionosphere. Then the perturbation ΔT_e consists of two parts, incident [Eqs. (3.115) and (3.22)] and reflected; the latter is also described by Equations (3.115) and (3.22), apart from a change in the quantity $K_1(z) = K_{12}$ and in the initial instant t_0 [here $K_1(z)$ is the total absorption of the wave E_1 up to the point z under consideration (Eq. (3.30))]. We shall assume for simplicity that the absorption K_{12} is appreciable ($K_{12} \gtrsim 1$), so that the influence of the reflected pulse can be neglected.

We consider now the second pulse E_2 at the frequency ω_2, propagating in the perturbed region. Its amplitude is

$$E_2 = E_{20}(z)\left[1 - \frac{\omega_2}{c}\int_0^z \left(\frac{\partial \kappa_2}{\partial \nu_e}\right)_{\nu_{e0}} \left(\frac{\partial \nu_e}{\partial T_e}\right)_{\nu_{e0}} \Delta T_e \, dz\right]. \tag{4.32}$$

Here $E_{20}(z)$ is the amplitude of the E_2 pulse if it has propagated in the unperturbed plasma, and κ_2 is the absorption coefficient of the E_2 wave. The second term in the square brackets of Equation (4.32) gives the perturbation produced in the E_2 wave by its interaction with the E_1 wave, and ΔT_e is defined in Equation (3.115).

We assume for simplicity that both pulses, E_1 and E_2, propagate along the z direction. The pulse E_2 starts at the instant t_2 and its duration is assumed to be negligibly small. Depending on the relation between t_2, t_{10} (the start of the pulse E_1) and t_1 (the duration of E_1), three cases can be distinguished:

$$\text{I) } t_2 < t_{10}; \quad \text{II) } t_2 > t_{10} + t_1; \quad \text{III) } t_{10} < t_2 < t_{10} + t_1. \tag{4.33}$$

In the first case the perturbing pulse follows the pulse E_2. It can act only on the E_2 wave reflected from the ionosphere. The two pulses thus travel in opposition to each other in the interaction region. Let z_1 be the point at which the pulse E_2 encounters the leading front of the pulse E_1. It is here that the interaction sets in. Furthermore let z_0 be the point at which the pulses cease to interact directly and the so-called aftereffect begins: the pulse E_2 propagates in a medium in which the perturbation due to the pulse E_1 has not yet relaxed fully. The total perturbation of the pulse E_2 consists then of the perturbation due to the direct interaction and the perturbation due to the after effect. Substituting ΔT_e from Equations (3.115) in Equations (4.32), recognizing that the pulses propagate in opposite directions, and assuming that in the interaction region $n_1 \approx n_2 \approx 1$, i.e., $t - t_0 = 2(z_1 - z)/c$ and $t_1 = 2(z_1 - z_0)/c$, we get

$$\begin{aligned}
\mu = \frac{E_{20} - E_2}{E_{20} + E_2} &= \frac{\omega_2}{2c} \int_0^{z_1} \left(\frac{\partial \kappa_2}{\partial \nu_e}\right)_{\nu_{e0}} \left(\frac{\partial \nu_e}{\partial T_e}\right)_{T_{e0}} \Delta T_e \, dz \\
&= \int_{z_0}^{z_1} D \exp\left[-2K_1(z)\right] \left[1 - \exp\left(-\frac{2\,\delta\nu_{e0}}{c}(z_1 - z)\right)\right] dz \\
&\quad + \int_0^{z_0} D \exp\left[-2K_1(z)\right] \left[1 - \exp(-\delta\nu_{e0} t_1)\right] \exp\left[-\frac{2\,\delta\nu_{e0}}{c}(z_0 - z)\right] dz
\end{aligned} \tag{4.34}$$

$$\begin{aligned}
D &= \frac{E_1^2(0)}{2n_1 E_p^2} \frac{\omega_2}{c} \frac{T_{e0}}{\nu_{e0}} \left(\frac{\partial \nu_e}{\partial T_e}\right)_{T_{e0}} \left(\nu_{e0} \frac{\partial \kappa_2}{\partial \nu_{e0}}\right) \\
&\simeq \frac{E_1^2(0)}{2E_p^2} \frac{\omega_2}{c} \frac{\omega_2^2 - \nu_{e0}^2}{\omega_2^2 + \nu_{e0}^2} \frac{T_{e0}}{\nu_{e0}} \left(\frac{\partial \nu_e}{\partial T_e}\right)_{T_{e0}} \kappa_2 .
\end{aligned}$$

This expression determines the change of the amplitude of the pulse E_2 due to its interaction with the pulse E_1, i.e., the depth of the cross modulation of the E_2 wave. We note that in air (i.e., in the lower ionosphere) we have according to Equations (2.140) and (2.235)

$$\frac{T_{e0}}{\nu_{e0}}\left(\frac{\partial \nu_e}{\partial T_e}\right)_{T_{e0}} = \frac{5}{6}.$$

In the second case, the pulse 2 follows the perturbing pulse E_1. Only the incident wave E_2 is then cross modulated, and the entire interaction is by way of an aftereffect. We obtain for the cross-modulation depth

$$\mu = \int_0^{z_2} D \exp\left[-2K_1(z)\right]\left[1 - \exp\left(-\delta\nu_{e0}t_1\right)\right] \times \exp\left\{-\delta\nu_{e0}\left[t_2 - (t_{10} + t_1)\right]\right\} dz. \tag{4.35}$$

Here $t_2 - (t_{10} + t_1)$ is the time by which the pulse E_2 is delayed relative to the rear edge of the pulse E_1; z_2 is the point of reflection of the wave E_2 (or E_1); Under strong-interaction conditions, $K_1(z_2) \gtrsim 1$, we can assume $z_2 \to \infty$. If we neglect the variation of $\delta\nu_{e0}$, ν_{e0}, and T_e with the height z, then Equation (4.35) can be greatly simplified. Indeed, by using Equations (4.12) and (4.13), we can integrate in Equation (4.35) with respect to dz. We then obtain

$$\mu = D_0\left[1 - \exp\left(-\delta\nu_{e0}t_1\right)\right] \exp\left\{-\delta\nu_{e0}(t_2 - t_{10} - t_1)\right\},$$
$$D_0 = \frac{e^2 E_1^2(0)}{12m\,\delta\nu_{e0}} \frac{\omega_2^2 - \nu_{e0}^2}{(\omega_2^2 + \nu_{e0}^2)^2}\left(\frac{\partial \nu_e}{\partial T_e}\right)_{T_{e0}}. \tag{4.36}$$

Finally, in the third case the perturbing pulse interacts with both the incident and the reflected E_2 waves. Accordingly, the cross-modulation depth is

$$\begin{aligned}\mu = \mu_1 + \mu_2 = &\int_0^{z_1} D \exp\left[-2K_1(z)\right]\left[1 - \exp\left(-\delta\nu_{e0}(t_2 - t_{10})\right)\right] dz \\ &+ \int_0^{z_2} D \exp\left[-2K_1(z)\right]\left[1 - \exp\left(-\delta\nu_{e0}t_1\right)\right] \\ &\times \exp\left[-\delta\nu_{e0}\left(t_2 + 2\tau - t_{10} - t_1 - \frac{z}{c}\right)\right] dz\end{aligned} \tag{4.37}$$

Here τ is the group propagation time of the pulse E_2 from the transmission point to the reflection point ($\tau > z_2/c$); it is assumed that $t_2 + \tau > t_{10} + t_1$.

The first term in Equation (4.37) can be recast in the form of Equation (4.36):

$$\mu_1 = D_0[1 - \exp(-\delta\nu_{e0}(t_2 - t_{10}))]. \tag{4.38}$$

In addition to amplitude cross modulation, phase cross modulation is also produced when short pulses interact. The change of the phase of the E_2 wave under the influence of the E_1 wave is obviously given by the expression

$$\mu_\varphi = \frac{\omega_2}{2c}\int_0^z \left(\frac{\partial n_2}{\partial \nu_e}\right)_{\nu_{e0}} \left(\frac{\partial \nu_e}{\partial T_e}\right)_{T_{e0}} \Delta T_e \, dz,$$

where n_2 is the refractive index of the E_2 wave. Recognizing that

$$\left(\frac{\partial n_2}{\partial \nu_e}\right)_{\nu_{e0}} = \frac{\nu_{e0}\omega_0^2}{n_2(\omega_2^2 + \nu_{e0}^2)^2} = \frac{\omega_2}{\omega_2^2 + \nu_{e0}^2}\kappa_2 = \frac{\omega_2\nu_{e0}}{\omega_2^2 - \nu_{e0}^2}\frac{\partial \kappa_2}{\partial \nu_{e0}}$$

we find that the same Equations (4.34)–(4.38) hold for the phase-modulation depth μ_φ, except that D and D_0 must be replaced by

$$D_\varphi = D\frac{\omega_2\nu_{e0}}{\omega_2^2 - \nu_{e0}^2}, \qquad D_{\varphi 0} = D_0\frac{\omega_2\nu_{e0}}{\omega_2^2 - \nu_{e0}^2}. \tag{4.39}$$

We see therefore that at $\nu_{e0} \sim \omega_2$ the interaction of short pulses leads to an appreciable phase cross modulation. This cross modulation is small at $\nu_{e0} \ll \omega_2$ or $\nu_{e0} \gg \omega_2$ (cf. Sect. 3.3.3). If a magnetic field is present in the plasma, then the polarization of both waves plays an important role in the interaction of short pulses, just as for ordinary cross modulation.

An important advantage of interaction between pulses is that it takes place in a relatively small region. By varying the delay time t_2–t_{10} it is possible to shift the point where the pulses meet, i.e., the region of their effective interaction. Local sounding of the plasma is therefore made possible by an analysis of the character of the interaction (Anderson et al., 1953; Fejer, 1955; Smith, 1966).

4.2.2. Change in the Absorption of a Wave Propagating in a Perturbed Plasma Region

A high-power wave E_1 of frequency ω_1 produces, in the main, plasma perturbations that are constant in time. Changes occur, in particular, in the electron temperature. This change of T_e affects the absorption of other waves that propagate in the perturbed region.

Assume that the perturbing wave E_1 propagates in a direction normal to the plasma layer. In this case the wave field is described by Equations (3.19) and (3.20). These equations are solved for a weakly ionized plasma in Section 3.2.2. The electron temperature in the perturbed region is $T_e = T\tau^2(z)$, where $\tau(z)$ is implicitly defined by Equation (3.38). We consider now another wave, of frequency ω_2, propagating likewise in the direction z normal to the layer. We assume this wave to be weak and neglect its self-action. The amplitude of the field of the E_2 wave is

$$E_2(z) = \frac{E_2(0)}{\sqrt{n_2(z)}} \exp\left\{-\frac{\omega_2}{c}\int \kappa_2(z, T_e)\,dz\right\}$$
$$= \frac{E_2(0)}{\sqrt{n_2(z)}} \exp\left\{-\frac{\omega_2}{c}\int \kappa_{20}(z)\,dz\right\} P_{12}.$$

Here κ_{20} is the absorption coefficient of the E_2 wave in the unperturbed plasma, $\kappa_2(z, T_e)$ is the same coefficient in a plasma perturbed by the E_1 wave, and P_{12} is an interaction factor that shows how the amplitude of E_2 is altered by E_1. In a weakly ionized plasma, the factor P_{12} is given by Equation (4.24).

In the interior of the plasma we have $\tau(z) \to 1$ (see Sect. 3.2.2). In this case the wave interaction is the strongest and most complete. The factor P_{12} for the case of complete interaction is equal to

$$P_{12} = \exp\left\{\left(\frac{\varepsilon_1}{\varepsilon_2}\right)^{1/2}\left[-\frac{\omega_1^2 + \nu_{e0}^2}{\omega_2^2 + \nu_{e0}^2}\ln\frac{\tau_0 + 1}{2}\right.\right.$$
$$\left.\left.-\left(1 - \frac{\omega_1^2 + \nu_{e0}^2}{2(\omega_2^2 + \nu_{e0}^2)}\right)\ln\frac{\omega_2^2 + \nu_{e0}^2\tau_0^2}{\omega_2^2 + \nu_{e0}^2} + \frac{2\nu_{e0}^2(\tau_0 - 1)}{\omega_2^2 + \nu_{e0}^2}\right]\right\}. \tag{4.40}$$

If the E_1 wave is powerful enough, then the factor P_{12} can assume a significant role. Thus, in the case of a high-frequency E_2 wave ($\omega_2^2 \gg \nu_{e0}^2\tau_0^2$) we have

$$P_{12} \simeq \left(\frac{\tau_0 + 1}{2}\right)^{-(\omega_1^2/\omega_2^2)\sqrt{\varepsilon_1/\varepsilon_2}}$$

The interaction factor can be much smaller than unity in this case. The powerful E_1 wave suppresses the E_2 wave, as it were (at $\omega_1^2 \gtrsim \omega_2^2 \gg \nu_{e0}^2\tau_0$). The reason is that the absorption of the high-frequency E_2 wave increases in the plasma region heated by the strong E_1 wave. On the contrary, for a low-frequency E_2 wave, the absorption in the heated region decreases and the plasma can become transparent. In particular, at $\omega_2^2 \ll \nu_{e0}^2$ and $\omega_1^2 \ll \nu_{e0}^2$ we have

$$P_{12} = \left[\frac{\tau_0(\tau_0 + 1)}{2}\right]^{-\sqrt{\varepsilon_1/\varepsilon_2}} \exp\{2\sqrt{\varepsilon_1/\varepsilon_2}(\tau_0 - 1)\}. \tag{4.41}$$

The interaction factor P_{12} is of the same order of magnitude as the self-action factor P. The values of P_{12} for the ionosphere show therefore that interaction effect can play a noticeable role for waves of sufficiently high power (cf. Table 15).

4.2.3. Generation of Waves with Combination Frequencies

A high-power unmodulated wave produces in the plasma not only constant perturbations, but also perturbations that vary with time at a frequency $2\omega_1$. Wave interaction in a plasma therefore generates waves with the combination frequencies $\omega_2 \pm 2\omega_1$ (Vilenskii, 1954).

The propagation of the E_2 wave in a plasma is described by the equation [see Eq. (3.6)]

$$\frac{\partial^2 E_2}{\partial z^2} - \frac{1}{c^2}\frac{\partial^2 E_2}{\partial t^2} - \frac{4\pi}{c^2}\frac{\partial j_2}{\partial t} = 0. \tag{4.42}$$

It is assumed here, as usual, that the E_2 wave propagates in the z direction.

The action of a wave E_1 of frequency ω_1 on the plasma produces time-varying perturbations of T_e of frequency $2\omega_1$ [Eq. (2.43).] In view of the smallness of the variable corrections, we solve Equation (4.42) by successive approximations: $E_2 = E_2^{(0)} + E_2^{(1)} + \cdots$. In first-order approximation, we neglect the variable corrections to σ and ε. The wave then retains its frequency ω_2, and its field is given by

$$E_2^{(0)} = E_2(0) \exp[i(k_{20}z + \omega_2 t)]. \tag{4.43}$$

We consider here only one wave traveling in the direction of the z axis. The plasma is assumed for simplicity to be homogeneous, and k_{20} is the complex wave vector of the E_2 wave:

$$k_{20} = \frac{\omega_2}{c}\sqrt{\varepsilon(\omega_2) + i\frac{4\pi\sigma(\omega_2)}{\omega_2}}. \tag{4.44}$$

Substituting Equation (4.43) in Equation (4.42) and taking the plasma perturbations of frequency $2\omega_1$ into account, we obtain in the next higher approximation [see Eq. (3.9)]:

$$\frac{\partial^2 E_2^{(1)}}{\partial z^2} - \frac{1}{c^2}\frac{\partial^2 E_2^{(1)}}{\partial t^2} - \frac{4\pi\sigma}{c^2}\frac{\partial E_2^{(1)}}{\partial t} - \frac{\varepsilon - 1}{c^2}\frac{\partial^2 E_2^{(1)}}{\partial t^2} - \frac{4\pi}{c^2}\frac{\partial j^{(2)}}{\partial t} = 0. \quad (4.45)$$

Here $j^{(2)}$ is the correction that must be introduced into the current to allow for the presence of perturbations of frequency $2\omega_1$ in the medium [cf. Eq. (3.117)]:

$$j^{(2)} = j_+^{(2)} \exp\left[i(\omega_2 + 2\omega_1)t\right] + j_-^{(2)} \exp\left[i(\omega_2 - 2\omega_1)t\right]. \quad (4.46)$$

We see therefore that the sideband waves have the frequencies $\omega_2 + 2\omega_1$ and $\omega_2 - 2\omega_1$.

The solution of Equation (4.45) with the boundary conditions

$$E_2^{(1)}|_{z=0} = 0, \qquad E_2^{(1)}|_{z\to\infty} = C\exp(ikz).$$

is of the form

$$E_2^{(1)} = E_{21}^+ \exp\left[i(\omega_2 + 2\omega_1)t\right] + E_{21}^- \exp\left[i(\omega_2 - 2\omega_1)t\right]$$

where

$$E_{21}^{\pm} = -\frac{2\pi(\omega_2 \pm 2\omega_1)}{k^{\pm}c^2}\Bigg\{\exp(ik_1^{\pm}z)\int_z^{\infty}\left[\exp(-ik_1^{\pm}z_1)\right]j_{\pm}^{(2)}\,dz_1$$

$$+\exp(-ik_1^{\pm}z)\left[\int_0^z\left[\exp(ik_1^{\pm}z_1)\right]j_{\pm}^{(2)}\,dz_1 - \int_0^{\infty}\left[\exp(-ik_1^{\pm}z_1)\right]j_{\pm}^{(2)}\,dz_1\right]\Bigg\},$$

$$k_1^{\pm} = \frac{\omega_2 \pm 2\omega_1}{c^2}\sqrt{\varepsilon(\omega_2 \pm 2\omega_1) + \frac{4\pi i\sigma(\omega_2 \pm 2\omega_1)}{\omega_2 \pm 2\omega_1}} \quad (4.47)$$

$k_1^{\pm}$ is the complex wave number for the wave of "sideband" frequency $\omega_2 \pm 2\omega_1$.

For the amplitude of the "sideband" waves passing through the interaction region in the plasma we obtain from Equation (4.47) the expression

$$|E_{21}^{\pm}| = \frac{4\pi|\omega_2 \pm 2\omega_1|}{c^2|k_1^{\pm}|}\left|\int_0^{\infty} j_{\pm}^{(2)}\sin(k_1^{\pm}z)\,dz\right|.$$

The ratio of the "sideband" amplitude to the fundamental E_2 wave amplitude is of the order of

$$\eta = \frac{|E_{21}^{\pm}|}{|E_2(0)|} \simeq 10^{-2} \frac{e^4 E_1^2(0) N \nu_{e0}}{m^2 \omega_1^3 (\omega_2 \pm 2\omega_1)^2} \exp(K_0 - K_1^{\pm}), \qquad (4.48)$$

where K_0 and $K_1^{\pm}$ are the total-absorption coefficients of the fundamental and "sideband" waves.

If the perturbing transmitter operates at $W_0 = 100$ kW and at frequencies $\omega_1 \approx \omega_2 \sim 10^6 - 10^7$, then we have for the ionosphere $\eta \sim 5 \cdot 10^{-6}$–$10^{-8}$. Under the resonance conditions $\omega_2 = 2\omega_1$, the variable corrections increase, and the amplitude of the sideband waves increases correspondingly. At $\omega_2 = 2\omega_1 = 4 \cdot 10^6$ we have $\eta \simeq 3 \cdot 10^{-5}$; at $\omega_2 = 2\omega_1 = 2 \cdot 10^6$ we have $\eta \simeq 10^{-5}$. The sideband amplitudes increase with decreasing wave frequencies ω_1 and ω_2. They can become appreciable for strong waves at $\omega_1 \sim \omega_2 \lesssim \nu_{e0}$. In this case, however, the frequencies ω_1 and ω_2 are low and the waves propagate in the earth-ionosphere waveguide (see Sects. 3.4.2 and 3.5). We emphasize that we have considered above only a homogeneous plasma; sum and difference frequencies $\omega_1 \pm \omega_2$ can be generated in an inhomogeneous plasma (Ginzburg and Gurevich, 1960; Sodha and Kaw, 1966). Special notice should be taken of the generation at the infralow frequency $\Omega = \omega_2 - \omega_1 \sim (1$–$10)$ kHz in the lower ionosphere. This phenomenon is closely related to the ionosphere detection effect (Sect. 3.4.2).

4.3. Radio Wave Interaction in the Lower Ionosphere

4.3.1. Cross Modulation

Cross modulation of radio waves in the ionosphere has been well investigated experimentally. The effect is usually observed at night in the medium wave length band ($\lambda \sim 200$–2000 m), and the wave interaction takes place in the lower part of the E layer, at heights 80–90 km. The largest cross-modulation depth $\mu_{\Omega\,\max}$ is reached (according to average data) under the conditions $200 \lesssim \Omega \lesssim 500$, $\omega_1 \sim 10^6$, $\omega_2 \sim 5 \cdot 10^6$, $\cos\psi_2 \sim 0.25$, $\cos\psi_1 \sim 0.7$, and $R_1 \simeq 140$ km (R_1 is the distance from the perturbing station to the interaction region). In this case we have $\mu_{\Omega\,\max} \simeq 0.05$ for a transmitter power $W_0 \simeq 100$ kW and for $\mu_0 = 1$ (Huxley and Ratcliffe, 1949). Calculation by Equation (4.15) yields, under the same conditions, $\mu_{\Omega\,\max} \approx (0.04$–$0.05)$, which is in sufficient agreement with experiment.

The observed dependence of the cross-modulation phase φ_Ω on the modulation frequency Ω is shown in Figure 38 (Huxley et al., 1948;

Ratcliffe and Shaw, 1948). The experimental data are in good agreement with the theory. The value of $\delta_0 \nu_{0e}$ measured in these experiments is in the range $10^3 \lesssim \delta_e \nu_{0e} \lesssim 2 \cdot 10^3$, corresponding to a wave-interaction region $z \approx 80$–90 km. The dependence of the cross-modulation depth on the modulation frequency, shown in Figure 38, is also in good agreement with the conclusion of the theory, but the deviations from the theoretical curve [Eq. (4.15)] are nevertheless larger. It is noted, first that μ_Ω decreases more rapidly at large Ω (Huxley, 1952), and second, that the depth of

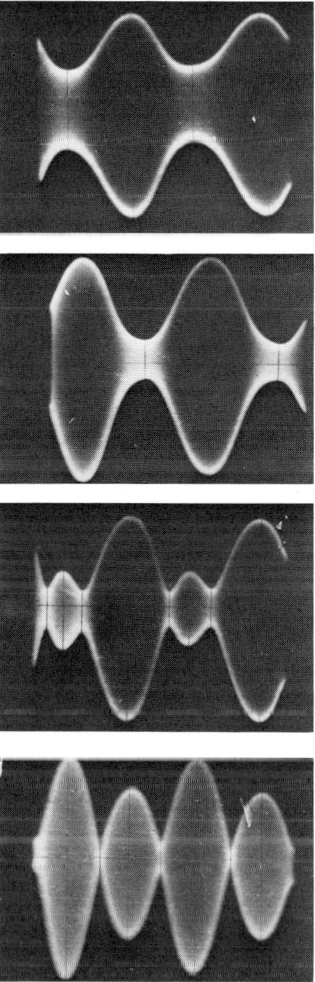

Fig. 44. Distortion of the cross-modulation waveform: $\omega_1 = 8.2 \cdot 10^6 \approx \omega_H$; $\omega_2 = 1.63 \cdot 10^6$; $\mu_0 = 0.80$; $\Omega = 6.3 \cdot 10^2$

modulation decreases to some extent as $\Omega \to 0$ (Huxley et al., 1948; Bell, 1951). The first anomaly seems to have a natural explanation: if $\Omega > c/\Delta z$ (where Δz is the width of the interaction region), the variation of the modulation phase in the interaction becomes significant and leads in fact to a lowering of μ_Ω. The cause of the second (low-frequency) anomaly is not clear.

The experimentally obtained dependence of μ_Ω on μ_0 is always linear in a weak field (Huxley et al., 1948; Ratcliffe and Shaw, 1948) in agreement with Equations (4.13) and (4.15) of the theory.

The cross-modulation depth μ_Ω increases linearly with increasing power W_0 of the transmitting station, at not too high radiation powers (Huxley et al., 1948; Ratcliffe and Shaw, 1948; Huxley, 1952), in accordance with the theory (Sect. 4.1.1). At high powers ($W_0 \gtrsim 10^3$ kW), the rate of growth of μ_Ω with increasing W_0 slows down, and the form of the modulation becomes appreciably distorted (Shlyuger, 1974; Gurevich and Shlyuger, 1975); this agrees with the theory of cross modulation of strong radio waves (see Sect. 4.1.2 and Figs. 40–42). An example is shown in Figure 44. It is seen that the cross-modulation depth not only reaches large values, but a doubling of the modulation frequency, similar to "over-modulation," sets in (see Fig. 41).

The dependence of the depth of the cross modulation on the frequency of the modifying wave ω_1 in the gyroresonance region $\omega_1 \approx \omega_H$ is shown in Figure 45 (Bailey et al., 1952). It is in satisfactory agreement with the

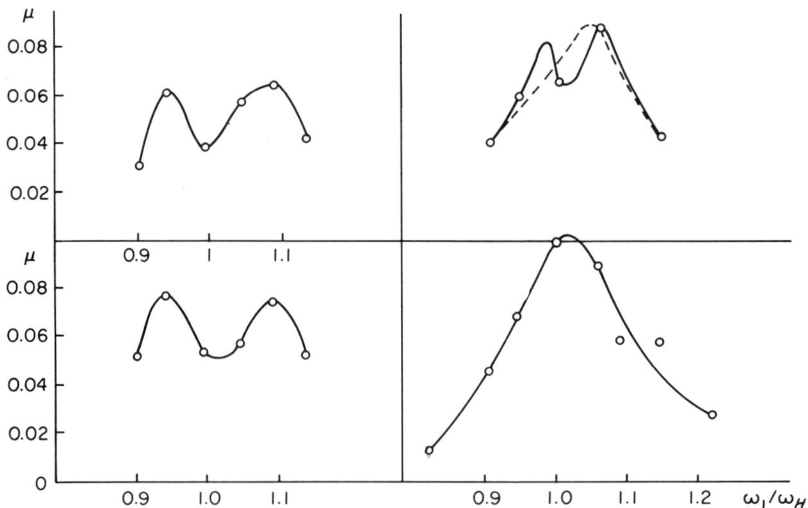

Fig. 45. Cross-modulation resonance near the gyrofrequency (Bailey et al., 1952)

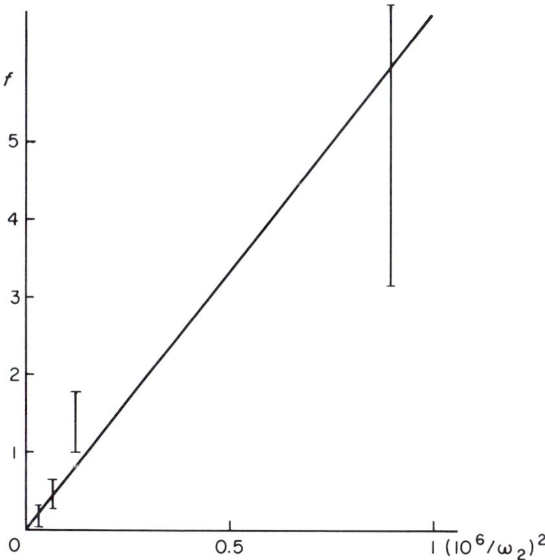

Fig. 46. Maximum depth of cross modulation at gyroresonance $\omega_1 = \omega_H$; $f = (\mu_{\Omega\,\max}/\mu_0) \cdot (\cos \psi_2/W_0\,[\mathrm{kW}]) \cdot 10^3$

results of the theory (Fig. 43). Figure 46 shows the maximum cross-modulation depth at gyroresonance $\omega_1 \simeq \omega_H$ as a function of the frequency of the E_2 wave as obtained in a number of studies (Cutolo, 1950; Bell, 1951; Bailey et al., 1952; Huxley, 1952). The solid line was calculated in accordance with the experimental conditions from Equation (4.31).

4.3.2. Fejer's Method

Fejer (1955) used the interaction of short pulses, which was considered in Sect. 4.2.1, to investigate the ionosphere.

In Fejer's method, twice as many weak pulses E_2 than strong pulses E_1 are emitted in the same time interval. The time between pulses is long enough for complete relaxation of the perturbation. Therefore, whereas one of the E_2 pulses interacts with the E_1 field, the second passes through the undisturbed ionosphere. Comparing the amplitudes and phases of the perturbed and unperturbed pulses, we can determine the effect of the interaction with high accuracy, up to $\mu \sim 10^{-5}$, where μ is the pulse cross-modulation depth (see Sect. 4.2.1). Variation of the pulse delay time changes the meeting height of the pulses. This makes it possible to plot the characteristics of the amplitude and phase interaction of the pulses in the ionosphere.

An amplitude characteristic of this type is shown in Figure 47 (Smith, 1966). The ordinates represent the cross-modulation depth μ in relative units, and the abscissas the delay time $t_2 - t_{10}$ [Eq. (4.33)]. Consequently, the cross-modulation depth rises and falls twice with changing delay time $t_2 - t_{10}$ (the sizes of the maxima can differ appreciably in this case).

Let us compare the amplitude characteristic of Figure 47 with Equations (4.34)–(4.38) which describe the interaction of the pulses. The left-hand part of the curve corresponds to large values of $t_2 - t_{10}$ and describes, according to Equation (4.33), the case II. As seen directly from Equation (4.36), the depth of the cross modulation μ increases exponentially with decreasing $t_2 - t_{10}$ [all the terms except the last one in Equation (4.36) are independent of the delay time].

As $t_2 - t_{10}$ decreases, we go over into region III. An essential role is played here by the interaction with both the forward and reflected E_2 pulses. The principal role is assumed initially by the interaction with the forward E_2 pulse [the term μ_1 in Equation (4.37)]. It is seen from Equation (4.38) that μ_1 decreases with decreasing delay time $t_2 - t_{10}$, in agreement with Figure 47. The interaction with the reflected pulse, to the contrary, is enhanced with decreasing $t_2 - t_{10}$. At $t_2 - t_{10} < 0$, the interaction with the forward pulse ceases completely and we go over to region I. What is essential here is that the region of the strongest interaction—the region of the maximum of the function D in Equation (4.34)—coincides in rough approximation with the maximum of the product $N\nu_{e0}$, i.e., with heights z_{max} on the order of 80–90 km at night and 60–70 km in day time in the ionosphere. At the start of region I, the pulses meet at $z > z_{max}$. With

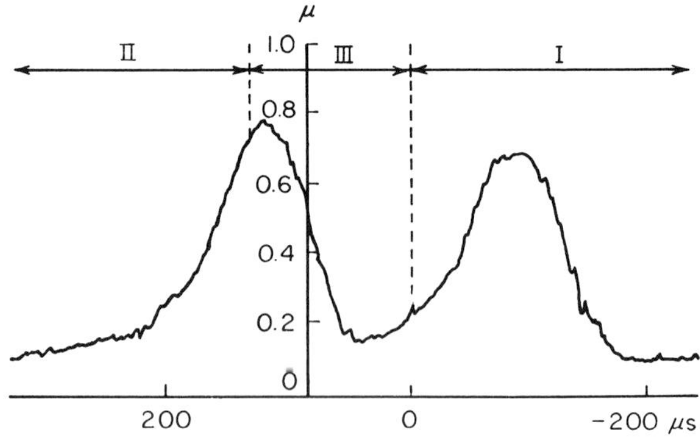

Fig. 47. Amplitude characteristic of pulse interaction

decreasing $t_2 - t_{10}$ the meeting point drops and μ increases. In the region of the second maximum of the function μ, the meeting point is just near z_{max}. With further decrease of $t_2 - t_{10}$ the pulse interaction region gradually leaves the ionosphere and the cross-modulation depth decreases again.

The dependence of the phase cross modulation on the delay time is similar in form. The variation of the functions μ and μ_φ with $t_2 - t_{10}$ depends, as we have seen above, not only on δv_{e0} but also on the height distributions of the collision frequency $v_{e0}(z)$ and of the concentration $N(z)$ in the lower ionosphere. A numerical reduction of the observation plots [a comparison of Equations (4.34)–(4.38) with the experimental data] makes it possible to determine the height profiles of the collision frequency and of the electron density in the lower ionosphere. This is one of the effective methods of investigating the ionosphere D-layer (Bailey and Goldstein, 1958; Bjelland et al., 1959; Barrington and Thrane, 1962; Weisbrok et al., 1964; Georges, 1966; Smith et al., 1966).

We note that the pulse cross-modulation depth is usually small, $\mu \sim 10^{-3}$–10^{-5}. It increases strongly, however, when pulses of very high power act on the ionosphere: cross modulation depths up to 90% were observed by Gurevich and Shlyuger (1975). The dependence of the cross-modulation depth on the power of the modifying pulse is in this case patently nonlinear in accordance with the theory. Experiment confirms well the effects of saturation and rotational enhancement of the cross modulation of the pulses at $\omega_1^2 \gg v_e^2$ and $\omega_2^2 \gg v_e^2$ (see Fig. 48).

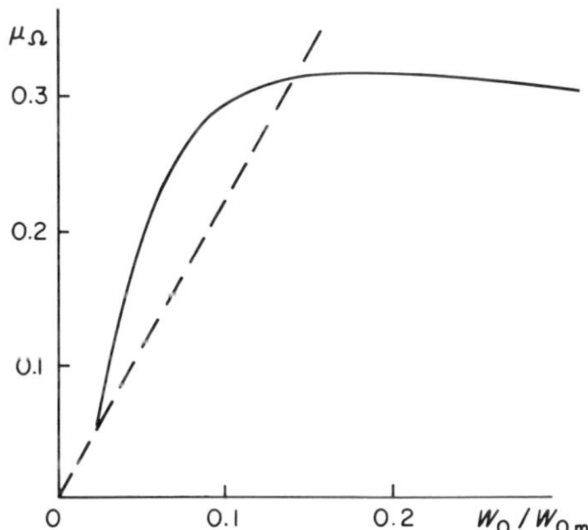

Fig. 48. Depth of cross modulation of high-power pulses with $\omega_1 = 8.48 \cdot 10^6$, $\omega_2 = 1.32 \cdot 10^7$, and $W_{0m} \sim 10^6$ kW. *Solid curve*: experiment; *dashed*: linear increase of μ_Ω with W_0. Rotation amplification and saturation of the cross modulation is seen (see Sect. 4.1.2 and Fig. 42).

4.3.3. Nonstationary Process in the Interaction of Strong Radio Waves

Gurevich et al. (1975) investigated the effect of a high-power pulsed wave of frequency $\omega_1 = 8.48 \cdot 10^6$ on cw waves E_2. Typical waveforms of the received E_2-wave signals are shown in Figure 49. The modifying E_1 pulse was turned on at the instant marked by the arrow on the figure. It is seen that the E_2 wave is appreciably modified by the wave E_1. The E_2 wave amplitude can be both decreased—suppressed by the E_1 wave (Fig. 49a)—and, to the contrary, enhanced (Fig. 49b). The observed effects agree with the theory (Sect. 4.2.2). The suppression, as seen, can be very strong: the nonlinear damping of the E_2-wave power reached 30 dB.

A characteristic feature of the presented curve is the rapid growth of the disturbance and its rather slow decrease: the relaxation time is longer by more than one order than the rise time of the disturbance. The cause of the rapid rise lies in the fact that the electron collision frequency increases intensely in the lower ionosphere with increasing electron energy [Eq. (2.140)]. This leads to a very rapid heating of the electron gas in the strong electric field (see Sect. 2.3.2). Indeed, from the elementary-theory Equations (2.18) amd (2.140), in the approximation that neglects the

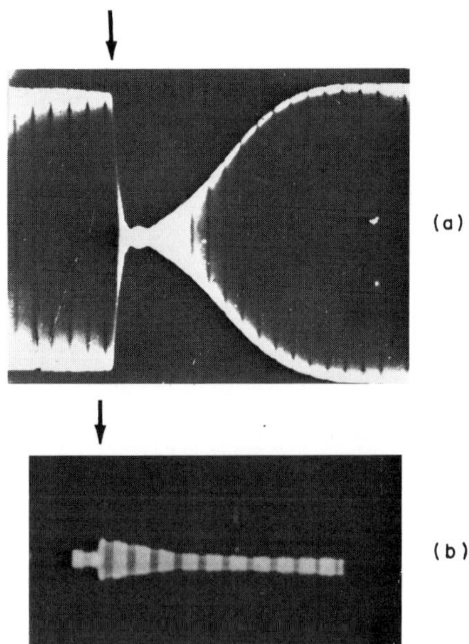

Fig. 49. Perturbation of cw radio wave by a high-power pulse. $\omega_1 = 8.48 \cdot 10^6$, $\omega_2 = 2.35 \cdot 10^6$. *Dark strips*: 1 millisecond time marker. High-power pulse duration ~0.5 ms

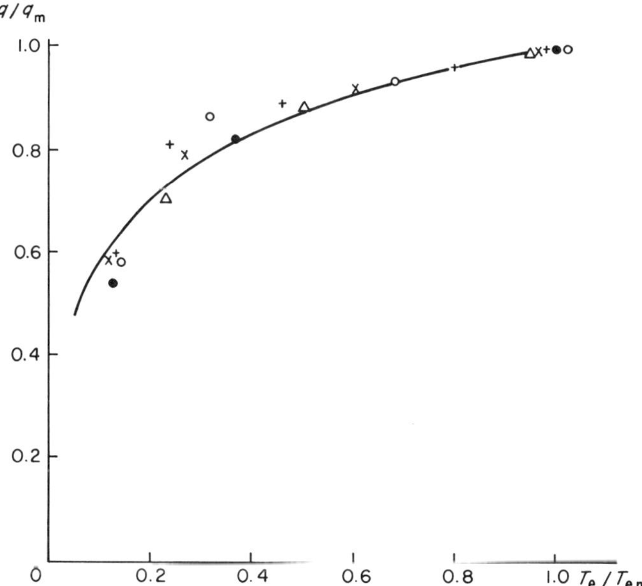

Fig. 50. Increase of the perturbation of the wave E_2 with time. Here $q(t) = |d[\ln(E_{20}/E_2)]/dt|$, where E_{20} and $E_2(t)$ are the unperturbed and perturbed amplitudes of the wave E_2, q_m is the maximum value of q, and T_{em} is the electron temperature at $q = q_m$. *Plot*: theory [Eq. (4.49)], *points*: experiment

relaxation terms, it follows that

$$\frac{T_e}{T} = (1 + Gt)^6, \qquad G = e^2 E_0^2 \nu_{e0}/3m\omega_{ef}^2 T. \tag{4.49}$$

We see therefore that when the heating is appreciable ($Gt \gtrsim 1$), the electron temperature increases rapidly with time. Equation (4.49) is in good agreement with experiment (Fig. 50). The dependence of ν_e on T_e, obtained from the data of ionosphere experiments, turned out to be close to Equation (2.140).

The relaxation of the disturbance is determined mainly by electron collisions accompanied by excitation of rotational levels (Sect. 2.3.2). The electron energy losses in the rotation region are low, and this stretches out the relaxation [Eq. (2.188).] An experimental investigation of the relaxation can yield the function of the rotational losses in the ionosphere; this function is in agreement with the theory Equation (2.123) (see Fig. 51).

Cross modulation of a strong modifying wave ($E_0 \gg E_p$) of frequency $\omega_1 \sim (3\text{–}4) \cdot 10^7$ was investigated by Utlaut and Violette (1974). They

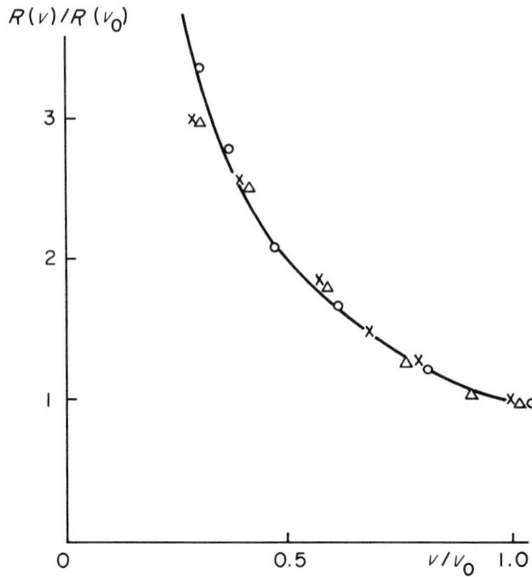

Fig. 51. Rotational-loss function $R(v)$; $v_0 = \sqrt{2\varepsilon_r/m}$ is the limit of the rotation region ($\varepsilon_r \approx 0.4$ eV). *Curve*: theory [Eq. (2.123)]; *points*: experiment

observed an interesting stretching of the relaxation of the disturbance: the additional absorption of the sounding wave E_2 remained noticeable even 10 minutes after the modifying transmitter was turned off.

5. Self-Action and Interaction of Radio Waves in an Inhomogeneous Plasma

We have considered so far the homogeneous problem, i.e., we have neglected transport processes such as diffusion, thermal diffusion, and heat conduction. This is true, in the main, in the lower ionosphere, where the electron and ion mean free paths are small. In the upper ionosphere, in the region of the F layer and above, the transport processes, to the contrary, are quite important. They influence the modification of the plasma by the radio waves and therefore to a considerable degree determine the character of the thermal nonlinear effects.

5.1. Inhomogeneous Electric Field in a Plasma

5.1.1. Fundamental Equations

Consider an alternating electric field with variable amplitude in a plasma. The electric field heats the plasma. The distribution of the electron and ion temperatures in such a field become inhomogeneous. This gives rise to a pressure gradient that produces a particle flux and leads in final analysis to an inhomogeneous distribution of the electron and ion concentrations. To describe these processes it is natural to use the macroscopic equations

$$\frac{\partial N_{\mathrm{e}}}{\partial t} + \operatorname{div} \boldsymbol{j}_{\mathrm{e}} = q, \tag{5.1}$$

$$\frac{\partial N_{\mathrm{i}}}{\partial t} + \operatorname{div} \boldsymbol{j}_{\mathrm{i}} = q, \tag{5.2}$$

$$N_{\mathrm{e}} \frac{\partial T_{\mathrm{e}}}{\partial t} + \boldsymbol{j}_{\mathrm{e}} \nabla T_{\mathrm{e}} + (\gamma_{\mathrm{e}} - 1) N_{\mathrm{e}} T_{\mathrm{e}} \operatorname{div} (\boldsymbol{j}_{\mathrm{e}}/N_{\mathrm{e}}) + \operatorname{div} \boldsymbol{q}_{\mathrm{e}} = R_{\mathrm{e}} \tag{5.3}$$

$$N_{i} \frac{\partial T_{i}}{\partial t} + \boldsymbol{j}_{\mathrm{i}} \nabla T_{\mathrm{i}} + (\gamma_{\mathrm{i}} - 1) N_{\mathrm{i}} T_{\mathrm{i}} \operatorname{div} (\boldsymbol{j}_{\mathrm{i}}/N_{\mathrm{i}}) + \operatorname{div} \boldsymbol{q}_{\mathrm{i}} = R_{\mathrm{i}}. \tag{5.4}$$

Equations (5.1)–(5.4), just as Equations (2.17) and (2.51) or Equations (2.135) and (2.136), which were considered in the preceding sections, are in fact balance equations that express the conservation laws for the energy and for the number of particles (electrons and ions). Here $q = q_i - q_r$ is the number of electron-ion pairs produced by ionization (q_i) and lost to volume recombination (q_r) per cm^3 and per second; $\frac{3}{2}R_e$ and $\frac{3}{2}R_i$ are the energies acquired by the electrons and ions per cm^3 and per second. In the presence of inhomogeneities, particle currents and energy fluxes are also produced. In Equations (5.1)–(5.4), $\boldsymbol{j}_e$ and $\boldsymbol{j}_i$ are the electron and ion flux densities, $\boldsymbol{g}_e$ and $\boldsymbol{g}_i$ are the heat fluxes, and γ_e and γ_i are the specific-heat ratios (c_p/c_v).

It is important that inhomogeneities in a plasma are bound to be accompanied by electric and magnetic fields. Therefore Equations (5.1)–(5.4) should be supplemented by Maxwell's field equations. Assume that the pressure of the electron-ion component of the plasma is much lower than the energy density of the external magnetic field:

$$(N_e T_e + N_i T_i) \ll H^2/8\pi, \tag{5.5}$$

and that the perturbation propagation velocity is less than the Alfven velocity. The perturbations of the magnetic field can then be neglected. The electric field $\boldsymbol{E}_s$ due to the presence of the inhomogeneity in a plasma is determined in this case by the Poisson equation

$$\nabla \boldsymbol{E}_s = 4\pi e(N_i - N_e). \tag{5.6}$$

Equations (5.1)–(5.4) are valid if the inhomogeneity is weak:

$$\begin{aligned} l\left|\frac{d}{dx}(E_0^2, N_e, N_i, T_e, T_i)\right| &\ll (E_0^2, N_e, N_i, T_e, T_i), \\ \nu^{-1}\left|\frac{d}{dt}(E_0^2, N_e, N_i, T_e, T_i)\right| &\ll (E_0^2, N_e, N_i, T_e, T_i). \end{aligned} \tag{5.7}$$

i.e., if the amplitude E_0 of the alternating electric field in the plasma, as well as the concentrations and temperatures of the electrons and the ions, change little over the mean free path l for the collisions of the electrons and ions with the molecules, or over their mean free path time ν^{-1}. For example, if the amplitude E_0 were to change appreciably over a path length, $l|dE_0^2/dx| \gg E_0^2$, then an important role would be played by the striction force, which is not taken into account in the equations. On the other hand, under Equation (5.7) this force is negligible, since it is small in comparison with the pressure gradient $\frac{d}{dx}(N_e T_e + N_i T_i)$. We note that in order to neglect the striction effects it sufficies to have l equal to the mean free path for arbitrary electron collisions (with molecules as well as with ions).

We note also that in the presence of a strong magnetic field (at $\Omega_H > \nu_{im}$) in a direction perpendicular to H, the first condition of Equation (5.7) can be made much less stringent by replacing in it the mean free path l by the ion Larmor radius ρ_{Hi}. It is assumed, in addition, that $N_e \ll N_m$ and $N_i \ll N_m$.

Under Equation (5.7), the directional electron and ion velocities are small, and they are subject to negligible inertial effects. The electron and ion flux densities are then proportional to the concentration and temperature gradients. They can therefore be represented in the form

$$\boldsymbol{j}_e = -\frac{\hat{\sigma}'_e}{e}\boldsymbol{E}_s - \hat{D}_e \nabla N_e - \hat{D}_{Te}\frac{N_e}{T_e}\nabla T_e - \hat{D}_{ei}\nabla N_i - D_{Tei}\frac{N_i}{T_i}\nabla T_i, \quad (5.8)$$

$$\boldsymbol{j}_i = \frac{\hat{\sigma}'_i}{e}\boldsymbol{E}_s - \hat{D}_i \nabla N_i - \hat{D}_{Ti}\frac{N_i}{T_i}\nabla T_i - \hat{D}_{ie}\nabla N_e - \hat{D}_{Tie}\frac{N_e}{T_e}\nabla T_e, \quad (5.9)$$

$$\boldsymbol{g}_e = -\hat{\kappa}_e \nabla T_e - \hat{\kappa}_{ei}\nabla T_i - \hat{\beta}_e \boldsymbol{j}_e - \hat{\beta}_{ei}\boldsymbol{j}_i, \quad (5.10)$$

$$\boldsymbol{g}_i = -\hat{\kappa}_i \nabla T_i - \hat{\kappa}_{ie}\nabla T_e - \hat{\beta}_i \boldsymbol{j}_i - \hat{\beta}_{ie}\boldsymbol{j}_e. \quad (5.11)$$

Here $\hat{\sigma}'_e$, $\hat{D}_e$, and $\hat{D}_{Te}$ are the conductivity, diffusion, and thermal-diffusion tensors for the electrons; $\hat{\sigma}'_i$, $\hat{D}_i$, and $\hat{D}_{Ti}$ are the same tensors for the ions; $\hat{D}_{ei}$, $\hat{D}_{Tei}$, $\hat{D}_{ie}$, and $\hat{D}_{Tie}$ are the corresponding mutual-diffusion and thermal diffusion tensors; $\hat{\kappa}_e$, $\hat{\kappa}_i$ and $\hat{\kappa}_{ei}$, $\hat{\kappa}_{ie}$ are the thermal conductivity and mutual thermal conductivity tensors for the electrons and ions; $\hat{\beta}_e$, $\hat{\beta}_i$, $\hat{\beta}_{ei}$ and $\hat{\beta}_{ie}$ are the thermal-force tensors.

The presence, say, of an ion concentration gradient produces an ion current that drags the electrons with it as a result of the collisions. It is this circumstance which is reflected by the terms representing the mutual diffusion, the mutual thermal diffusion, and the mutual thermal conductivity.[15] The mutual diffusion tensors, however, are not independent but

[15] It must be borne in mind that, under the conditions of real interest to us, the neutral-molecule concentration in the plasma exceeds greatly the charged-particle concentration, $N_m \gg N_e$ and $N_m \gg N_i$. This is true in practically the entire ionosphere. The condition that the total pressure be constant, which must be satisfied if the diffusion processes considered here are to take place, is always ensured by the small perturbations of the molecule concentration, so that the motion of the molecules does not affect the mutual diffusion of the electrons and ions.

We note also that an important feature of a low-temperature plasma is that the cross section for the collisions between the charged particles is larger by several orders of magnitude than the cross section for the collisions with the neutrals. The interaction of the charged particles with one another, and particularly their mutual dragging, is therefore essential even at comparatively low degrees of ionization.

are connected with $\hat{\sigma}'_e$ and $\hat{D}_e$ or with $\hat{\sigma}'_i$ and $\hat{D}'_i$ by the relations

$$\hat{D}_{ei} = \frac{T_i}{T_e}\hat{D}_e - \frac{T_i}{e^2 N_e}\hat{\sigma}'_e, \qquad \hat{D}_{ie} = \frac{T_e}{T_i}\hat{D}_i - \frac{T_e}{e^2 N_i}\hat{\sigma}'_i. \tag{5.12}$$

Equations (5.12) will be derived in the next section.

The transport coefficients satisfy the Onsager symmetry relations (see Landau and Lifshitz, 1963, Sect. 122). Thus, in the absence of a magnetic field we have at $T_e = T_i = T$

$$\frac{\partial D_e}{\partial T} = \frac{1}{T}\frac{\partial}{\partial N_e}(N_e D_{Te} + N_i D_{Tei}), \qquad \frac{\partial D_{ei}}{\partial T} = \frac{1}{T}\frac{\partial}{\partial N_i}(N_e D_{Te} + N_i D_{Tei}),$$

$$\frac{\partial D_e}{\partial N_i} = \frac{\partial D_{ei}}{\partial N_e}, \qquad \frac{\partial D_i}{\partial N_e} = \frac{\partial D_{ie}}{\partial N_i}, \tag{5.13}$$

$$\frac{\partial D_i}{\partial T} = \frac{1}{T}\frac{\partial}{\partial N_i}(N_i D_{Ti} + N_e D_{Tie}), \qquad \frac{\partial D_{ie}}{\partial T} = \frac{1}{T}\frac{\partial}{\partial N_e}(N_i D_{Ti} + N_e D_{Tie}).$$

At a low degree of plasma ionization, when the decisive role is played by collisions with the neutral particles, the mutual-diffusion, thermal-diffusion, and thermal-conductivity tensors of the electrons and ions are negligible. Equations (5.8)–(5.11) then simplify to:

$$\boldsymbol{j}_e = -\frac{\hat{\sigma}'_e}{e}\boldsymbol{E}_s - \hat{D}_e \nabla N_e - \hat{D}_{Te}\frac{N_e}{T_e}\nabla T_e, \tag{5.14}$$

$$\boldsymbol{j}_i = \frac{\sigma'_i}{e}\boldsymbol{E}_s - \hat{D}_i \nabla N_i - \hat{D}_{Ti}\frac{N_i}{T_i}\nabla T_i, \tag{5.15}$$

$$\boldsymbol{g}_e = -\hat{\kappa}_e \nabla T_e - \hat{\beta}_e \boldsymbol{j}_e, \tag{5.16}$$

$$\boldsymbol{g}_i = -\hat{\kappa}_i \nabla T_i - \hat{\beta}_i \boldsymbol{j}_i. \tag{5.17}$$

Equations (5.12) are then equivalent to the Einstein relations:

$$\hat{\sigma}'_e = \frac{e^2 N_e}{T_e}\hat{D}_e, \qquad \hat{\sigma}'_i = \frac{e^2 N_i}{T_i}\hat{D}_i, \tag{5.18}$$

while the Onsager Equations (5.13) take the form

$$\frac{\partial D_e}{\partial T_e} = \frac{D_{Te}}{T_e}, \qquad \frac{\partial D_i}{\partial T_i} = \frac{D_{Ti}}{T_i}. \tag{5.18a}$$

The expressions for the transport tensors and a rigorous derivation of Equations (5.8)–(5.11) or Equations (5.14)–(5.17) are given in the kinetic theory (see Sect. 5.2). We note that when the conditions in Equation (5.7) are satisfied we can neglect the second term in the left-hand side of Equations (5.3) and (5.4) ($\boldsymbol{j}_e \nabla T_e$, $\boldsymbol{j}_i \nabla T_i$), as well as disregard the heating of the plasma by the field $\boldsymbol{E}_s$. The system of Equations (5.1)–(5.4) and Equations (5.8)–(5.11) then becomes quasi-linear.

Also note that we have assumed above that the electron and ion gas is at rest relative to the constant magnetic field. Under ionospheric conditions the charged particles usually drift relative to $\boldsymbol{H}$, owing to the motion of the neutral gas $\boldsymbol{v}_m$ or to the external electric field $\boldsymbol{E}$. The electron and ion drift velocities $\boldsymbol{v}_{e0}$ and $\boldsymbol{v}_{i0}$ are determined by the following expressions:

$$\boldsymbol{v}_{e0} = \boldsymbol{v}_m - \frac{\hat{\sigma}'_e}{eN_e}\left\{\boldsymbol{E} + \frac{1}{c}\left[\boldsymbol{v}_m \times \boldsymbol{H}\right]\right\}, \qquad \boldsymbol{v}_{i0} = \boldsymbol{v}_m + \frac{\hat{\sigma}'_i}{eN_i}\left\{\boldsymbol{E} + \frac{1}{c}\left[\boldsymbol{v}_m \times \boldsymbol{H}\right]\right\}.$$

The perturbation of the plasma by the inhomogeneous electric field in the presence of a drift is described as before by Equations (5.1)–(5.4). All that need be taken into account is that the drift gives rise to corrections to Equations (5.8)–(5.11) for the fluxes, namely, terms $(N\boldsymbol{v}_0)_{ei}$ are added to the electron and ion current densities, and terms $(N\boldsymbol{v}_0 T)_{ei}$ to the energy fluxes.

Simplification of the Initial Equations. We proceed now to a simplification of the system of Equations (5.1)–(5.4), Equation (5.6), and Equations (5.8)–(5.11). The first important simplification is due to the possibility of using the quasi-neutrality condition. The point is that at relatively large inhomogeneity scales, with characteristic dimensions that are large compared with the Debye radius $D = (T_e/4\pi e^2 N)^{1/2}$, i.e., when the condition

$$D|\nabla N| \ll N \tag{5.19}$$

is satisfied, the plasma is quasineutral. In other words, the electron and ion concentrations in the plasma are approximately equal:

$$N_e \simeq N_i. \tag{5.20}$$

In this case the continuity equations for the electrons and ions [Eqs. (5.1) and (5.2)] can be replaced by a single equation for their common concentration $N_e = N_i = N$:

$$\frac{\partial N}{\partial t} + \operatorname{div} \boldsymbol{j} = q \tag{5.21}$$

This obviously is possible only if the following additional condition is satisfied:

$$\operatorname{div} \boldsymbol{j}_e = \operatorname{div} \boldsymbol{j}_i = \operatorname{div} \boldsymbol{j} \tag{5.22}$$

Equation (5.22) can be satisfied by choosing the longitudinal electric field $\boldsymbol{E}_s$. Consequently, the additional condition of Equation (5.22) must now be regarded as an equation that determines the electric field $\boldsymbol{E}_s$ and thus replaces the Poisson Equation (5.6). The latter, at a given field $\boldsymbol{E}_s$, determines the amount by which the ion and electron concentrations must differ in order that the field be produced. By virtue of Equation (5.19), this difference is only a small fraction of N, on the order of $(D|\nabla N|/N)^2$. The quasineutrality condition [Eq. (5.20)] is consequently satisfied with the same accuracy. In the ionosphere, the Debye radius is $D \sim 0.1$ to 1 cm. It is always small in comparison with the characteristic dimensions of the inhomogeneities.

The second simplification is connected with the fact that the heat flux is determined for the most part by the thermal conductivity $\kappa \nabla T$. This is also connected with the quasineutrality condition. Indeed, it follows from Equation (5.22) that the ion and electron diffusion currents $\boldsymbol{j}_i$ and $\boldsymbol{j}_e$ are of the same order, although the electron diffusion coefficient D_e is usually much larger than the ion diffusion coefficient. The reason for this lies in the action of the electric field. The electron inhomogeneity, owing to the larger electron mobility, tends to spread out more rapidly than that of the ions. This, however, would disturb the quasineutrality. The result is an electric field that forces the electrons and ions to diffuse at equal rates. The quasineutrality conditions thus determine the character of the diffusion process in the plasma, which proceeds, generally speaking, at a rate on the order of the ion diffusion rate (ambipolar diffusion; Schottky, 1924). In the case of diffusion transverse to a strong magnetic field, the same process impedes the ion motion.

The thermal conductivity is not restricted by the quasineutrality condition and frequently proceeds much faster. In this case the principal terms in Equations (5.3), (5.4) and in Equations (5.10) and (5.11) for the energy flux are those due to the thermal conductivity $\nabla(\kappa_e \nabla T_e)$, whereas the thermal force $\nabla(\beta T)$ can be neglected in comparison.[16] In particular, in an isotropic plasma (as well as in an anisotropic plasma, if heat transport along the magnetic field is the only important factor), if Equation (5.7) is satisfied, the electron thermal conductivity plays the decisive role if

[16] It is also important that the right-hand sides of Equations (5.1) and (5.2) are stationary or vary slowly with time. This variation is characterized by the electron lifetime, which is usually longer than, say, the temperature relaxation time $1/\delta\nu_e$.

$$\frac{M\nu_{im}}{m(\nu_{ei}+\nu_{em})} \gg 1. \tag{5.22a}$$

The term proportional to $(\gamma_e - 1)$ in Equation (5.3) is then also negligible. In the ionosphere, Equation (5.22a) is satisfied at heights $\lesssim 500$ km.

Thus, when Equations (5.5), (5.7), (5.9), and (5.22a) are satisfied, the distributions of the electron and ion densities, of their temperatures, and of the longitudinal electric field in an inhomogeneous plasma are described by the equations

$$\frac{\partial N}{\partial t} + \operatorname{div} \boldsymbol{j} = q, \tag{5.23}$$

$$\operatorname{div} \boldsymbol{j}_e = \operatorname{div} \boldsymbol{j}_i = \operatorname{div} \boldsymbol{j}, \tag{5.24}$$

$$\boldsymbol{j}_e = -\frac{\hat{\sigma}_e}{e}\boldsymbol{E}_s - \hat{D}_{ee}\,\nabla N - \hat{D}_{Te}\frac{N}{T_e}\nabla T_e - \hat{D}_{Tei}\frac{N}{T_i}\nabla T_i, \tag{5.25}$$

$$\boldsymbol{j}_i = \frac{\sigma_i}{e}\boldsymbol{E}_s - \hat{D}_{ii}\,\nabla N - \hat{D}_{Ti}\frac{N}{T_i}\nabla T_i - \hat{D}_{Tie}\frac{N}{T_e}\nabla T_e. \tag{5.26}$$

$$\frac{\partial T_e}{\partial t} = -\frac{1}{N}\nabla(\boldsymbol{g}_{e\perp} - \kappa_{e\parallel}\nabla_{\parallel} T_e) - (\gamma_e - 1)T_e \operatorname{div}(\boldsymbol{j}_e/N) - \frac{2}{3N}(\boldsymbol{E}\hat{\sigma}_{e\omega}\boldsymbol{E})$$
$$- \delta_{ei}\nu_{ei}(T_e - T_i) - \delta_{em}\nu_{em}(T_e - T) - \frac{2}{3N}Q, \tag{5.27}$$

$$\frac{\partial T_i}{\partial t} = -\frac{1}{N}\nabla \boldsymbol{g}_i - (\gamma_i - 1)T_i \operatorname{div}(\boldsymbol{j}_i/N) + \frac{2}{3N}(\boldsymbol{E}\hat{\sigma}_{i\omega}\boldsymbol{E})$$
$$- \delta_{ei}\nu_{ei}(T_i - T_e) - \nu_{im}(T_i - T_m). \tag{5.28}$$

The expressions for $\boldsymbol{j}_e$ and $\boldsymbol{j}_i$ take into account the fact that $N_e = N_i = N$. Here

$$\hat{D}_{ee} = \hat{D}_e\left(1 + \frac{T_i}{T_e}\right) - \frac{T_i}{e^2 N}\hat{\sigma}_e, \tag{5.29}$$

$$\hat{D}_{ii} = \hat{D}_i\left(1 + \frac{T_e}{T_i}\right) - \frac{T_e}{e^2 N}\hat{\sigma}_i. \tag{5.30}$$

In these Equations the tensors $\hat{D}_{ei}$ and $\hat{D}_{ie}$ are eliminated with the aid of Equation (5.12). Further, $\boldsymbol{E}$ is an alternating electric field of frequency ω, which heats the plasma and gives rise to its inhomogeneity, and $\hat{\sigma}_\omega$ is the plasma conductivity at the frequency ω. The right-hand sides of Equations (5.27) and (5.28) were in fact already considered in Sections 2.1 and 2.5; Q/N is the average energy acquired by the plasma electron per unit time from other heat sources (for example, the photoelectron energy that heats the ionosphere); δ_{im} has been set equal to unity [see Eq. (2.218)].

In the absence of a magnetic field we can neglect in Equation (5.28) the heating of the ions by an alternating electric field and ion thermal conductivity. If a magnetic field is present in the plasma, the same holds true for an alternating electric field of sufficiently high frequency [Eq. (2.50)], and also for inhomogeneities of sufficiently large dimension $R_\perp$. We note that when the condition

$$\frac{\kappa_{\text{i}\perp}}{NR_\perp^2(\delta_{\text{ei}}\nu_{\text{ei}} + \nu_{\text{im}})} \ll 1 \qquad \frac{R_{\|}}{R_\perp} < \frac{\Omega_H}{\nu_{\text{im}}} \tag{5.31}$$

is satisfied the transverse transport processes are of no importance at all. Equation (5.28) for the ion temperature then coincides with Equations (2.53) and (2.210). In Equation (5.27) for the electron temperature we can neglect the flux $\boldsymbol{g}_{\text{e}\perp}$ and the term proportional to $(\gamma_\text{e} - 1)$.

Electric Field. The equations can be further simplified because the stationary (or quasistationary) potential electric field $\boldsymbol{E}_\text{s}$ can be determined directly and eliminated. This field is defined by Equation (5.24). We assume first that the currents $\boldsymbol{j}_\text{e}$ and $\boldsymbol{j}_\text{i}$ [Eqs. (5.25) and (5.26)] can be represented in gradient form:

$$\boldsymbol{j}_\text{e} = \nabla\psi_\text{e}; \qquad \boldsymbol{j}_\text{i} = \nabla\psi_\text{i}. \tag{5.32}$$

It follows from Equation (5.32) that

$$\operatorname{rot} \boldsymbol{j}_\text{e} = \operatorname{rot} \boldsymbol{j}_\text{i} = 0. \tag{5.33}$$

The solution of Equation (5.24) under the boundary conditions $\boldsymbol{j}_\text{e} = \boldsymbol{j}_\text{i} = 0$ as $r \to \infty$, with Equation (5.33) taken into account, is obviously of the form

$$\boldsymbol{j}_\text{e} = \boldsymbol{j}_\text{i}. \tag{5.34}$$

Substituting in this equation, Equations (5.25) and (5.26) for the currents $\boldsymbol{j}_\text{e}$ and $\boldsymbol{j}_\text{i}$, we determine the electric field $\boldsymbol{E}_\text{s}$. Thus, at $\boldsymbol{H} = 0$ we have

$$\boldsymbol{E}_\text{s} = -\frac{e}{\sigma_\text{e} + \sigma_\text{i}}\left[(D_{\text{ee}} - D_{\text{ii}})\,\nabla N + \frac{N}{T_\text{e}}(D_{T\text{e}} - D_{T\text{ie}})\,\nabla T_\text{e} + \frac{N}{T_\text{i}}(D_{T\text{ei}} - D_{T\text{i}})\,\nabla T_\text{i}\right]. \tag{5.35}$$

Eliminating now with the aid of Equation (5.35) the electric field from Equation (5.23), we rewrite the latter expression in the form

$$\frac{\partial N}{\partial t} - \nabla\left[D_{\mathrm{a}}\nabla N + \frac{N}{T_{\mathrm{e}}}D_{T\mathrm{ea}}\nabla T_{\mathrm{e}} + \frac{N}{T_{\mathrm{i}}}D_{T\mathrm{ia}}\nabla T_{\mathrm{i}}\right] = q, \qquad (5.36)$$

$$D_{\mathrm{a}} = \frac{\sigma_{\mathrm{e}}D_{\mathrm{ii}} + \sigma_{\mathrm{i}}D_{\mathrm{ee}}}{\sigma_{\mathrm{e}} + \sigma_{\mathrm{i}}}, \qquad D_{T\mathrm{ea}} = \frac{\sigma_{\mathrm{e}}D_{T\mathrm{ie}} + \sigma_{\mathrm{i}}D_{T\mathrm{e}}}{\sigma_{\mathrm{e}} + \sigma_{\mathrm{i}}},$$

$$D_{T\mathrm{ia}} = \frac{\sigma_{\mathrm{e}}D_{T\mathrm{i}} + \sigma_{\mathrm{i}}D_{T\mathrm{ei}}}{\sigma_{\mathrm{e}} + \sigma_{\mathrm{i}}}, \qquad D_{T\mathrm{ea}} = k_{T\mathrm{ea}}D_{\mathrm{a}}, \qquad D_{T\mathrm{ia}} = k_{T\mathrm{ia}}D_{\mathrm{a}}. \qquad (5.37)$$

Here D_{a} is the coefficient of ambipolar diffusion of the plasma with allowance for the quasineutrality conditions [Eqs. (5.19) and (5.20); Schottky, 1924]. Correspondingly, $D_{T\mathrm{ea}}$ and $D_{T\mathrm{ia}}$ are the electron and ion coefficients of ambipolar thermal diffusion, while $k_{T\mathrm{ea}}$ and $k_{T\mathrm{ia}}$ are the ambipolar thermal-diffusion ratios.

Equations (5.36), (5.27) and (2.53) comprise, when Equation (5.32) is satisfied, a closed system of equations describing the distribution of the electron and ion densities and their temperatures in an inhomogeneous and nonstationary quasi neutral plasma at $\boldsymbol{H} = 0$. All the possible simplifications have been made here.

In the absence of a magnetic field ($\boldsymbol{H} = 0$), the conditions in Equation (5.32) are frequently satisfied. For example, at $T_{\mathrm{e}} = T_{\mathrm{i}} = T$ the currents $\boldsymbol{j}_{\mathrm{e}}$ and $\boldsymbol{j}_{\mathrm{i}}$ can be represented in the form

$$\boldsymbol{j}_{\mathrm{e}} = -\frac{\sigma_{\mathrm{e}}}{e}\boldsymbol{E}_{\mathrm{s}} + \nabla f_{\mathrm{e}}(N_{\mathrm{e}}, N_{\mathrm{i}}, T), \qquad \boldsymbol{j}_{\mathrm{i}} = \frac{\sigma_{\mathrm{i}}}{e}\boldsymbol{E}_{\mathrm{s}} + \nabla f_{\mathrm{i}}(N_{\mathrm{e}}, N_{\mathrm{i}}, T). \quad (5.38)$$

Indeed, it is seen from Equations (5.8) and (5.9) that the kinetic coefficients

$$D_{\mathrm{e}} = \frac{\partial f_{\mathrm{e}}}{\partial N_{\mathrm{e}}}, \qquad D_{\mathrm{ei}} = \frac{\partial f_{\mathrm{e}}}{\partial N_{\mathrm{i}}}, \qquad \frac{N_{\mathrm{e}}D_{T\mathrm{e}} + N_{\mathrm{i}}D_{T\mathrm{i}}}{T} = \frac{\partial f_{\mathrm{e}}}{\partial T} \text{ etc}$$

and the conditions of the type $\partial^2 f_{\mathrm{e}}/\partial N_{\mathrm{e}}\,\partial T = \partial^2 f_{\mathrm{e}}/\partial T\,\partial N_{\mathrm{e}}$, which are necessary if the currents are to be expressed in the form of Equation (5.38), are precisely the Onsager relations [Eq. (5.13)]. Examining now rot $\boldsymbol{j}_{\mathrm{e}}$ and rot $\boldsymbol{j}_{\mathrm{i}}$, we find that Equations (5.32) are always satisfied if $\nabla N_{\mathrm{e}} \| \nabla T_{\mathrm{e}} \| \nabla T_{\mathrm{i}}$.

We note also that Equations (5.32) are satisfied for one-dimensional inhomogeneities (when all the quantities depend on only one variable), for spherically and cylindrically symmetrical inhomogeneities, etc. On the contrary, in the presence of a magnetic field, Equations (5.32) and hence also the Equation (5.34), which makes it possible to eliminate the field $\boldsymbol{E}_s$ directly from the equations, are valid only in special cases, namely, if the inhomogeneity is strictly along or strictly across the magnetic field, or if the inhomogeneity is one-dimensional (all the quantities depend only on one variable). In the general case, on the other hand, the eddy currents are not equal to zero and Equation 5.24 is a second-order differential equation in the potential $\varphi_s(\boldsymbol{E}_s = -\nabla\varphi_s)$. The fact that eddy currents can be produced is an important distinguishing feature of ambipolar plasma diffusion in a magnetic field (Gurevich and Tsedilina, 1967).

5.1.2. Distribution of Density and Temperatures in Plasma

We now consider the plasma perturbations due to a stationary alternating electric field $E_0(r)$ of variable amplitude. In the absence of a magnetic field and under stationary conditions, Equations (5.36), (5.27), and (2.53) take the form

$$-\nabla_\xi\left[\gamma\left(\nabla_\xi N + \frac{N}{T_e} k_{Tea}\,\nabla_\xi T_e + \frac{N}{T_i} k_{Tia}\,\nabla_\xi T_i\right)\right] = \tau_N(q_i - q_r) \quad (5.39)$$

$$-\frac{N_0}{N}\nabla_\xi(\alpha\,\nabla_\xi T_e) = \left(\frac{E_0(r)}{E_p}\right)^2 T_{e0} - \frac{\delta_{ei}\nu_{ei}}{\delta\nu_e}(T_e - T_i)$$
$$- \frac{\delta_{em}\nu_{em}}{\delta\nu_e}(T_e - T) + \frac{2}{3}\frac{Q}{N\,\delta\nu_e}, \quad (5.40)$$

$$\nu_{im}(T_i - T) - \delta_{ei}\nu_{ei}(T_e - T_i) = 0. \quad (5.41)$$

We have changed over here to the dimensionless variable

$$\xi = r/R_0, \quad (5.42)$$

where R_0 is the characteristic dimension of the inhomogeneity produced in the plasma by the alternating field E_0; $\delta\nu_e = (\delta_{ei}\nu_{ei} + \delta_{em}\nu_{em})_{Te0}$; T_{e0} and T_{i0} are the temperatures of the electrons and ions in the absence of the field E_0; τ_N is the average electron lifetime determined by the recombination processes [Eq. (2.250)]:

$$\tau_N = \left(\frac{dq_r}{dN}\right)^{-1}$$

q_r is the number of electrons that vanish per cm^3 and per second as a result of volume recombination. Further, γ and α are characteristic dimensionless parameters:

$$\gamma = \frac{D_a \tau_N}{R_0^2} = \left(\frac{L_N}{R_0}\right)^2, \qquad L_N = \sqrt{D_a \tau_N}, \tag{5.43}$$

$$\alpha = \frac{\kappa_e}{N_0\, \delta v_e R_0^2} = \left(\frac{L_T}{R_0}\right)^2, \qquad L_T = \sqrt{\frac{\kappa_e}{N_0\, \delta v_e}}. \tag{5.44}$$

L_N and L_T are characteristic lengths in space which determine the role of the density and temperature inhomogeneity.

We can write down one more characteristic dimensionless ratio:

$$\eta = \frac{\alpha}{\gamma} = \left(\frac{L_T}{L_N}\right)^2. \tag{5.45}$$

We proceed now to an analysis of the solutions of the system of Equations (5.39)–(5.41) and consider the following cases:

1. *Large-scale inhomogeneity*:

$$R_0 \gg L_N,\ R_0 \gg L_T, \quad \text{or} \quad \gamma \ll 1,\ \alpha \ll 1. \tag{5.46}$$

The solution of Equations (5.39)–(5.41) is obtained by expansion in powers of the parameters α and γ. In first-order approximation, the left-hand sides of Equations (5.39) and (5.40) can be neglected, and Equations (5.39)–(5.41) reduce to the equations of a uniform plasma, which were considered in the preceding sections. Thus, the system in Equations (5.39)–(5.41), under the conditions in Equation (5.46), is locally homogeneous. In the next-order approximation, it is easy to obtain from Equations (5.39)–(5.41) the correction to the locally homogeneous solution; it is small (of the order of α or γ).

2. *Medium-scale inhomogeneity*:

$$L_T \ll R_0 \ll L_N, \quad \text{or} \quad \gamma \gg 1,\ \alpha \ll 1. \tag{5.47}$$

This case is possible only at $\eta \ll 1$ [Eq. (5.45)]. The plasma temperature turns out to be locally homogeneous as before. The density distribution, to the contrary, is determined by the transport processes. Expanding the solution of Equation (5.39) in powers of $1/\gamma$, we find that in first-order approximation

$$\nabla N + \frac{N}{T_e} k_{Tea} \nabla T_e + \frac{N}{T_i} k_{Tia} \nabla T_i = 0, \tag{5.48}$$

where, according to Equation (5.37)

$$k_{Tea} = \frac{\sigma_e D_{Tie} + \sigma_i D_{Te}}{\sigma_e D_{ii} + \sigma_i D_{ee}}, \qquad k_{Tia} = \frac{\sigma_e D_{Ti} + \sigma_i D_{Tei}}{\sigma_e D_{ii} + \sigma_i D_{ee}}. \tag{5.49}$$

This equation determines the plasma distribution under hydrodynamic equilibrium ($\boldsymbol{j} = 0$).

In particular, in a fully ionized plasma $k_{Tea} = T_e/(T_e + T_i)$, $k_{Tia} = T_i/(T_e + T_i)$, and the hydrodynamic-equilibrium condition [Eq. (5.48)] takes the form [see Eq. (5.107)]:

$$N = N_0 \frac{(T_{e0} + T_{i0})}{(T_e + T_i)}. \tag{5.50}$$

In other words, in a fully ionized plasma the condition for hydrodynamic equilibrium is the constancy of the total electron and ion pressure, $p = N(T_e + T_i) = \text{const}$.

At a low degree of plasma ionization, when the collisions of the electrons with the ions are inessential, the mutual thermal diffusion can be neglected in Equation (5.49): $D_{Tei} \to 0$ and $D_{Tie} \to 0$. By using Einstein's relations [Eq. (5.18)] we can then rewrite Equation (5.48) in the form

$$\nabla N + \frac{N k_{Te}}{T_e + T_i} \nabla T_e + \frac{N k_{Ti}}{T_e + T_i} \nabla T_i = 0; \qquad k_{Te} = \frac{D_{Te}}{D_e}, \; k_{T_i} = \frac{D_{T_i}}{D_i}. \tag{5.51}$$

If the electron and ion thermal-diffusion coefficients are equal,

$$k_{Te} = k_{Ti} = k, \tag{5.52}$$

then the hydrodynamic equilibrium-equation take the form (Bass et al., 1971)

$$N = N_0 \left(\frac{T_{e0} + T_{i0}}{T_e + T_i} \right)^k. \tag{5.53}$$

Thus, the hydrodynamic-equilibrium conditions lead to constancy of the total pressure of the ionized component $N(T_e + T_i)$ only in a fully ionized plasma. In the general case, however, the ion and electron pressure is not constant. The excess pressure is balanced by the action of the thermal force due to the thermal-diffusion transport.

It follows therefore that in the case of medium-scale inhomogeneity [Eq. (5.47)] the plasma concentration is determined in first-order approximation by the hydrodynamic-equilibrium [Eq. (5.48)] and the temperature is locally homogeneous. The next higher approximation leads to corrections of the order of α and γ^{-1}.

3. *Small-scale inhomogeneity*:

$$R_0 \ll L_N, \qquad R_0 \ll L_T. \tag{5.54}$$

In this case the plasma concentration is determined as before by the hydrodynamic-equilibrium condition, and the electron temperature is determined by the heat-conduction Equation (5.40). In the principal region modified by the alternating field $E_0(r)$, the electron energy lost in collisions is negligible by virtue of Equation (5.54), and Equation (5.40) takes the simple form (it is expressed here in terms of the dimensional variables r):

$$\nabla(\kappa_e \nabla T_e) = A\,\delta(r), \tag{5.55}$$

where

$$A = -N T_{e0}\,\delta\nu_e \int \frac{E_0^2(r)}{E_p^2}\,dr, \tag{5.56}$$

Let us take into account the fact that $\kappa_e = \kappa_e(T_e)$. In the general case $\kappa_e(T_e) = \kappa_e[T_e, N(T_e), T_i(T_e)]$, where $N(T_e)$ and $T_i(T_e)$ are determined by Equations (5.41) and (5.48). Introducing therefore $\lambda = \int \kappa_e(T_e)\,dT_e$, we rewrite Equation (5.55) in the form $\Delta\lambda = A\delta(r)$ and obtain its solution in the form

$$\lambda = \int_{T_{e0}}^{T_e} \kappa_e\,dT_e = -\frac{A}{4\pi r}. \tag{5.57}$$

This expression is valid at $r < L_T$; at large values of r, the temperature perturbation decreases exponentially [see Eq. (5.118)].

A perfectly analogous situation arises also in a plasma situated in a constant magnetic field $\boldsymbol{H}$. The same lengths L_T [Eq. (5.44)] and L_N [Eq. (5.43)] determine here the conditions for the establishment of a locally homogeneous or hydrodynamically balanced distribution of the plasma. The reason is that the diffusion and the thermal conductivity in the plasma along the magnetic field do not depend on the field and are determined by the same expressions as in an isotropic plasma.

The transverse components of the thermal-conductivity tensors of the electrons and ions decrease rapidly with increasing H. Therefore in a strongly magnetized plasma, if $\omega_H \gg \nu_e$ and if Equation (5.31) is satisfied, the transverse thermal conductivity of the electrons and the term proportional to $(\gamma_e - 1)$ are negligible, and Equation (5.27) for the stationary electron temperature takes the form:

$$-\frac{1}{N}\frac{\partial}{\partial x_{\|}}\left(\kappa_e \frac{\partial T_e}{\partial x_{\|}}\right) = \frac{e^2 E_0^2(r)\nu_e \varphi_p}{3m\omega^2} - \delta_{ei}\nu_{ei}(T_e - T_i) - \delta_{em}\nu_{em}(T_e - T) + \frac{2}{3}\frac{Q}{N}. \tag{5.58}$$

Here $x_{\|}$ is the coordinate along $\boldsymbol{H}$, and φ_p is the polarization factor [Eqs. (2.42) and (2.244)]. If Equation (5.31) is satisfied, the ion temperature is determined as before by Equation (5.41).

The Equations (5.23) and (5.24) for the plasma concentration N and for the electric field under stationary conditions take the form

$$\operatorname{div} \boldsymbol{j}_e = \operatorname{div} \boldsymbol{j}_i = q_i - q_r. \tag{5.59}$$

The currents $\boldsymbol{j}_e$ and $\boldsymbol{j}_i$ are defined in accordance with Equations (5.25) and (5.26). In the general case of an anisotropic plasma it is impossible to eliminate the field $\boldsymbol{E}_s$ and to reduce these equations to the ambipolar-diffusion Equations (5.36) and (5.39).

The locally-homogeneous distribution N, which is valid if Equations (5.46) are satisfied, is determined as before by the ionization balance equation $q_i = q_r$. The hydrodynamic-equilibrium distribution, which is valid under Equation (5.47), is determined by the equation

$$\operatorname{div} \boldsymbol{j}_e = \operatorname{div} \boldsymbol{j}_i = 0. \tag{5.60}$$

We consider an important particular case when the transverse dimension of the field inhomogeneity $R_{\perp}$ is large enough

$$\frac{R_{\perp}^2 D_{i\|}}{(L_T^2 + R_{\|}^2) D_{i\perp}} \gg 1. \tag{5.61}$$

Here $R_{\|}$ is the longitudinal dimension of the inhomogeneity, and $D_{i\|}$ and $D_{i\perp}$ are the coefficients of the longitudinal and transverse ion diffusion.

In this case the transverse diffusion is negligible. The equilibrium condition [Eq. (5.60)] is then rewritten as

$$\frac{\partial}{\partial x_{\|}} j_{e\|} = \frac{\partial}{\partial x_{\|}} j_{i\|}; \qquad j_{e\|} = j_{i\|}. \tag{5.62}$$

Here $j_{e\|}$ and $j_{i\|}$ are the longitudinal current components. Equation (5.62) is identical with Equation (5.34) in an isotropic plasma. The electric field is therefore determined by Equation (5.35), and the equilibrium conditions [Eq. (5.62)] take the simple form of Equation (5.48). Thus, in Equation (5.61), the hydrodynamic equilibrium in a plasma situated in a magnetic field is determined by the same expressions as in an isotropic plasma.

5.2. Kinetics of Inhomogeneous Plasma

5.2.1. Kinetic Coefficients. Elementary Theory

For an approximate calculation of the conductivity, diffusion, and thermal-diffusion tensors in an inhomogeneous plasma, we can use the elementary theory (see Sect. 2.1). In the elementary theory the directional velocity of an "average" electron in a homogeneous plasma is described by Equation (2.8). We recognize that in an inhomogeneous plasma the electrons are acted upon also by a force due to the presence of the electron-pressure gradient ∇p_e. In a weakly ionized plasma in the absence of a magnetic field, Equation (2.8) takes then the form

$$m \frac{d\boldsymbol{v}_e}{dt} = -e\boldsymbol{E} - m\nu_e \boldsymbol{v}_e - \frac{1}{N} \nabla p_e. \tag{5.63}$$

Under stationary conditions ($d\boldsymbol{v}_e/dt = 0$), the velocity $\boldsymbol{v}_e$ is equal to:

$$\boldsymbol{v}_e = -\frac{e}{m\nu_e} \boldsymbol{E} - \frac{1}{m\nu_e N} \nabla p_e.$$

Since $\boldsymbol{j}_e = N\boldsymbol{v}_e$ and $p_e = NT_e$, we obtain from this

$$\boldsymbol{j}_e = -\frac{eN}{m\nu_e} \boldsymbol{E} - \frac{T_e}{m\nu_e} \nabla N - \frac{N}{m\nu_e} \nabla T_e.$$

Comparing this expression with Equation (5.14), we arrive at the following expressions for the transport coefficients:

$$\sigma_e = \frac{e^2 N}{m\nu_e}, \quad D_e = \frac{T_e}{m\nu_e}, \quad D_{Te} = \frac{T_e}{m\nu_e}. \tag{5.64}$$

Here ν_e is the frequency of the electron-molecule collisions.

It is seen from Equation (5.64) that the symmetry relations [Eq. (5.18a)] for the coefficients D_e and D_{Te} are satisfied only when ν_e does not depend of T_e. This is perfectly understandable, since the Onsager relations are the results of exact kinetics, while the elementary theory is approximate and is valid only accurate to a factor on the order of unity. Therefore at $\nu_e = \nu_e(T_e)$ it is more correct to express D_{Te} in the form

$$D_{Te} = k_{Te} \frac{T_e}{m\nu_e} = k_{Te} D_e. \tag{5.65}$$

Here k_{Te} is the thermal-diffusion ratio; in elementary theory, $k_{Te} = 1$. From Equation (5.18a) it follows that

$$k_{Te} = 1 - \frac{T_e}{\nu_e} \frac{d\nu_e}{dT_e}. \tag{5.66}$$

A rigorous calculation with the aid of kinetic theory confirms this result (see Sect. 5.2.2).

In the presence of a magnetic field, the equation

$$-e\boldsymbol{E} - \frac{e}{c}[\boldsymbol{v}_e \times \boldsymbol{H}] - m\nu_e \boldsymbol{v}_e - \frac{1}{N}\nabla p_e = 0,$$

yields the stationary velocity of the electrons [cf. Eq. (2.15)]:

$$\boldsymbol{v}_e = \frac{e\nu_e}{m(\omega_H^2 + \nu_e^2)} \left\{ -\boldsymbol{E} - \frac{\omega_H^2}{\nu_e^2} \frac{\boldsymbol{H}(\boldsymbol{E} \cdot \boldsymbol{H})}{H^2} + \frac{\omega_H}{\nu_e} \frac{[\boldsymbol{E} \times \boldsymbol{H}]}{H} \right\}$$
$$+ \frac{\nu_e}{mN(\omega_H^2 + \nu_e^2)} \left\{ -\nabla p_e - \frac{\omega_H^2}{\nu_e^2} \frac{\boldsymbol{H}(\boldsymbol{H} \cdot \nabla p_e)}{H^2} + \frac{\omega_H}{\nu_e} \frac{[\nabla p_e \times \boldsymbol{H}]}{H} \right\}$$

The flux is $\boldsymbol{j}_e = N\boldsymbol{v}_e$. Comparing this expression for the flux with Equation (5.14) we obtain the components of the tensors $\hat{\sigma}'_e$, $\hat{D}_e$, and $\hat{D}_{Te}$. The longitudinal components coincide with Equation (5.64). The transverse components of the conductivity tensor are

$$\sigma_{exx} = \sigma_{eyy} = \frac{e^2 N \nu_e}{m(\omega_H^2 + \nu_e^2)}, \qquad \sigma_{exy} = -\sigma_{eyx} = -\frac{e^2 N \omega_H}{m(\omega_H^2 + \nu_e^2)}. \tag{5.67}$$

The components of the tensor $\hat{D}_e$ are connected with the corresponding components of the tensor $\hat{\sigma}'_e$ by the Einstein relations [Eq. (5.18)]. The tensor $\hat{D}_{Te}$ coincides with $\hat{D}_e$ in the elementary theory.

The very same expressions are obtained also for the ion-transport tensors $\hat{\sigma}'_i$, $\hat{D}_i$, and $\hat{D}_{Ti}$. All that need be done is to replace m, $-e$, T_e, ν_{em}, and ω_H in Equations (5.64) and (5.67) by the corresponding quantities M, e, T_i, ν_{im}, and Ω_H for the ions.

At an arbitrary degree of plasma ionization both the collisions with the neutral molecules and the collisions with the ions must be taken into account. Under stationary conditions, with allowance for the pressure gradients ∇p_e and ∇p_i, Equations (2.46) for the electron and ion velocities $\boldsymbol{v}_e$ and $\boldsymbol{v}_i$ take the form

$$-e\boldsymbol{E} - \frac{e}{c}[\boldsymbol{v}_e \times \boldsymbol{H}] - m\nu_{ei}(\boldsymbol{v}_e - \boldsymbol{v}_i) - m\nu_{em}\boldsymbol{v}_e - \frac{1}{N}\nabla p_e = 0,$$

$$e\boldsymbol{E} + \frac{e}{c}[\boldsymbol{v}_i \times \boldsymbol{H}] - m\nu_{ei}(\boldsymbol{v}_i - \boldsymbol{v}_e) - M\nu_{im}\boldsymbol{v}_i - \frac{1}{N}\nabla p_i = 0.$$

Solving them, we obtain the conductivity, diffusion, and thermal-diffusion tensors. The electron and ion conductivity tensors are given as before by Equation (2.49). For diffusion tensors [Eqs. (5.29) and (5.30)] we have, putting, $m\nu_{em}/M\nu_{im} \to 0$, as in Eq. (2.49),

$$(D_{ee})_{xx} = (D_{ee})_{yy} = \frac{B_1 T_e + B_2 T_i}{mA},$$

$$(D_{ee})_{xy} = -(D_{ee})_{yx} = -\frac{\omega_H(B_3 T_e + B_4 T_i)}{mA} \tag{5.68}$$

$$(D_{ee})_{zz} = \frac{T_e + B_4(T_e + T_i)}{m\nu_e}; \qquad (D_{ii})_{zz} = \frac{\nu_{ei} T_e + \nu_e T_i}{M\nu_{im}\nu_e},$$

$$(D_{ii})_{xx} = (D_{ii})_{yy} = \frac{B_7 T_e + B_6 T_i}{MA\nu_{im}},$$

$$(D_{ii})_{xy} = -(D_{ii})_{yx} = -\frac{\Omega_H(T_e \nu_{ei} - B_5 T_i)}{m\nu_{im} A},$$

where

$$B_1 = \nu_e\left(1 + \frac{\Omega_H^2}{\nu_{im}^2}\right) + \frac{m\nu_{ei}^2}{M\nu_{im}}, \qquad B_2 = \nu_{ei}\left(\frac{\Omega_H^2}{\nu_{im}^2} + \frac{m\nu_{ei}}{M\nu_{im}}\right),$$

$$B_3 = 1 + \frac{\Omega_H^2}{\nu_{im}^2} + 2\frac{m\nu_{ei}}{M\nu_{im}}, \qquad B_4 = \frac{m\nu_{ei}}{M\nu_{im}}, \qquad B_5 = \frac{m(\omega_H^2 + \nu_e^2 - \nu_{ei}^2)}{M\nu_{im}},$$

$$B_6 = \nu_e^2 + \omega_H^2(1 + B_4), \qquad B_7 = \nu_e\nu_{ei} + B_4\omega_H^2,$$

$$A = \nu_e^2 + \omega_H^2 B_3, \qquad \nu_e = \nu_{ei} + \nu_{em}.$$

The thermal-diffusion and mutual thermal-diffusion tensors for the electrons and ions are given by

$$\left.\begin{aligned}
&(D_{Te})_{xx} = (D_{Te})_{yy} = \frac{B_1 T_e}{mA}, \qquad (D_{Te})_{xy} = -(D_{Te})_{yx} = -\frac{\omega_H B_3 T_e}{mA},\\
&(D_{Te})_{zz} = \frac{(1 + B_4)T_e}{m\nu_e}, \qquad (D_{Tei})_{zz} = \frac{B_4 T_i}{m\nu_e},\\
&(D_{Tei})_{xx} = (D_{Tei})_{yy} - \frac{B_2 T_i}{mA}, \qquad (D_{Tei})_{xy} = -(D_{Tei})_{yx} = \frac{\omega_H B_4 T_i}{mA},\\
&(D_{Ti})_{xx} = (D_{Ti})_{yy} = \frac{B_6 T_i}{MA\nu_{im}}, \qquad (D_{Ti})_{xy} = -(D_{Ti})_{yx} = \frac{\Omega_H T_i B_5}{mA\nu_{im}},\\
&(D_{Ti})_{zz} = \frac{T_i}{M\nu_{im}}, \qquad (D_{Tie})_{zz} = \frac{T_e}{M\nu_{im}}\frac{\nu_{ei}}{\nu_e},\\
&(D_{Tie})_{xx} = (D_{Tie})_{yy} = \frac{B_7 T_e}{MA\nu_{im}}, \qquad (D_{Tie})_{yx} = -\frac{\Omega_H T_e \nu_{ei}}{mA\nu_{im}}.
\end{aligned}\right\} \quad (5.69)$$

The thermal-conductivity tensors take in the elementary approximation the form

$$\left.\begin{aligned}
&(\kappa_e)_{xx} = (\kappa_e)_{yy} = \frac{NT_e\nu_e}{m(\omega_H^2 + \nu_e^2)}, \qquad (\kappa_e)_{zz} = \frac{NT_e}{m\nu_e},\\
&(\kappa_e)_{xy} = -(\kappa_e)_{yx} = \frac{NT_e}{m}\frac{\omega_H}{\omega_H^2 + \nu_e^2}, \qquad (\kappa_i)_{zz} = \frac{NT_i}{M\nu_i},\\
&(\kappa_i)_{xx} = (\kappa_i)_{yy} = \frac{NT_i\nu_i}{M(\Omega_H^2 + \nu_i^2)}, \qquad (\kappa_i)_{xy} = -(\kappa_i)_{yx} = \frac{NT_i\Omega_H}{M(\Omega_H^2 + \nu_i^2)},\\
&\nu_i = \nu_{im} + \nu_{ii}, \qquad \nu_e = \nu_{ei} + \nu_{em}.
\end{aligned}\right\} \quad (5.70)$$

5.2.2. Kinetic Theory

Consider the electron distribution function in a plasma situated in an alternating inhomogeneous electric field $\boldsymbol{E}$. Assume that the inhomogeneity is weak, i.e., the field amplitude changes little over the electron or ion mean free path and during the free path time [see Eq. (5.7)]

$$l \ll \frac{E_0^2}{|\nabla E_0^2|}, \qquad \nu^{-1} \ll \frac{E_0^2}{|\partial E_0^2/\partial t|}. \tag{5.71}$$

In this case, Equations (2.73) are satisfied, and to describe the electron distribution function we can use the kinetic Equations (2.131) and (2.132). We assume first that

$$\nu(v) = \nu_{\mathrm{m}}(v) \gg \nu_{\mathrm{ee}}, \tag{5.72}$$

i.e., the degree of ionization of the plasma is low, so that the decisive influence on the distribution function $\boldsymbol{f}_1$ is exerted by the collisions with the neutrals. In this case Equation (2.132) for the function $\boldsymbol{f}_1$, takes the form

$$\frac{\partial \boldsymbol{f}_1}{\partial t} + \nu(v)\boldsymbol{f}_1 - \frac{e}{mc}[\boldsymbol{H} \times \boldsymbol{f}_1] = \frac{e\boldsymbol{E}}{m}\frac{\partial f_0}{\partial v} - v\,\nabla_{\boldsymbol{r}} f_0 + \frac{e\boldsymbol{E}_{\mathrm{s}}}{m}\frac{\partial f_0}{\partial v}. \tag{5.73}$$

Allowance is made here for the fact that an electric field $\boldsymbol{E}_{\mathrm{s}}$ always appears in the plasma in the presence of an inhomogeneity (see Sect. 5.1). It has the time variation as the inhomogeneity itself. By virtue of Equations (5.71) and (2.28), the field $\boldsymbol{E}_{\mathrm{s}}$ and the function f_0 in Equation (5.73) are quasistationary. Taking this into account, we easily obtain a solution of Equation (5.73). For example, in the absence of a magnetic field we have

$$\begin{gathered} \boldsymbol{f}_1 = \boldsymbol{f}_{1\omega} + \boldsymbol{f}_{1\mathrm{s}} = -\boldsymbol{u}_{\omega}\frac{\partial f_0}{\partial v} - \frac{v}{\nu}\nabla_{\boldsymbol{r}} f_0 + \frac{e\boldsymbol{E}_{\mathrm{s}}}{m\nu}\frac{\partial f_0}{\partial v}, \\ \boldsymbol{u}_{\omega} = -\frac{e\boldsymbol{E}}{m(-i\omega + \nu)}. \end{gathered} \tag{5.74}$$

We substitute this expression for $\boldsymbol{f}_1$ in Equation (2.131) and average, as usual, with respect to t [$1/\delta\nu \gg \Delta t \gg 1/\omega$; allowance is made here for the fact that $\omega \gg \delta\nu$ in accordance with Eq. (2.28)]. We obtain

$$\frac{\partial f_0}{\partial t} + \frac{v}{3}\nabla_{\boldsymbol{r}}\left(-\frac{v}{\nu}\nabla_{\boldsymbol{r}} f_0 + \frac{e\boldsymbol{E}_{\mathrm{s}}}{m\nu}\frac{\partial f_0}{\partial v}\right) + \frac{e}{3mv^2}\frac{\partial}{\partial v}\left(v^2\,\langle \boldsymbol{E}\boldsymbol{u}_{\omega}\rangle_t\,\frac{\partial f_0}{\partial v}\right) + S_0(f_0) = 0. \tag{5.75}$$

We have taken into account here the fact that by virtue of Equation (5.71) the contribution of the direct current to the plasma heating ($\boldsymbol{E}_s \boldsymbol{f}_{1s}$) can be neglected compared with $\boldsymbol{E}\boldsymbol{f}_{1\omega}$ (see Sect. 5.1.1).

The form of the function f_0 at

$$\nu_{ee} \gg \delta \nu_e \tag{5.76}$$

is determined by the interelectron collisions (Sect. 2.3.1). Using in this case an expansion similar to Equation (2.133), we easily verify that in the zero-order approximation f_0 is the Maxwellian function Equation (2.134)

$$f_{00} = N_e \left(\frac{m}{2\pi T_e}\right)^{3/2} \exp\left(-\frac{mv^2}{2T_e}\right). \tag{5.77}$$

From the conditions that the equations have a solution in the next approximation, we obtain, as in Section 2.3.1, equations for the electron density N_e and the temperature T_e. The same equations can be obtained by substituting the distribution function Equation (5.77) in Equation (5.75), multiplying Equation (5.75) in succession by $4\pi v^2$ and $2\pi m v^4$, and integrating it with respect to dv from 0 to ∞. The first of these equations becomes

$$\frac{\partial N_e}{\partial t} + \operatorname{div} \boldsymbol{j}_e = q_i - q_r. \tag{5.78}$$

Here q_i and q_r are the number of ionization and recombination acts in a unit volume per second, and $\boldsymbol{j}_e$ is the electron flux density

$$\boldsymbol{j}_e = \frac{4\pi}{3} \int_0^\infty v^3 \boldsymbol{f}_{1s}\, dv = -\frac{4\pi}{3} \nabla_{\boldsymbol{r}} \left(\int_0^\infty \frac{v^4}{\nu(v)} f_0\, dv\right) + \frac{4\pi}{3} \frac{e\boldsymbol{E}_s}{m} \int_0^\infty \frac{v^3}{\nu(v)} \frac{\partial f_0}{\partial v}\, dv. \tag{5.79}$$

Recognizing here that f_0 is Maxwellian [Eq. (5.77)], we rewrite this relation in the form

$$\boldsymbol{j}_e = -\frac{T_e}{m} A \nabla N_e - \frac{N_e}{m} \frac{d}{dT_e}(T_e A) \nabla T_e - \frac{eN_e}{m} A \boldsymbol{E}_s,$$
$$A(T_e) = \frac{8}{3\sqrt{\pi}} \int_0^\infty \frac{z^4 e^{-z^2}\, dz}{\nu(z\sqrt{2T_e/m})}, \qquad z = v/v_{Te} = v\sqrt{m/2T_e}. \tag{5.80}$$

On the other hand, from Equation (5.14) we have

$$\boldsymbol{j}_e = -D_e \nabla N_e - \frac{N_e}{T_e} D_{Te} \nabla T_e - \frac{\sigma_e}{e} \boldsymbol{E}_s, \tag{5.81}$$

where D_e, D_{Te}, and σ_e are the coefficients of the diffusion, thermal diffusion, and conductivity. Consequently,

$$D_e = \frac{T_e}{m} A, \qquad D_{Te} = \frac{T_e}{m}\left(A + T_e \frac{dA}{dT_e}\right), \qquad \sigma_e = \frac{e^2 N_e}{m} A. \tag{5.82}$$

The diffusion and conductivity coefficients are connected by Einstein's relation [Eq. (5.18)]. They satisfy the Onsager relations [Eq. (5.18a)].

The actual form of the transport coefficients depend on the electron collision frequency $\nu(v)$. It is natural to represent these coefficients in a form analogous to Equation (2.150):

$$D_e = \frac{T_e}{m\nu_e} K_D, \qquad \sigma_e = \frac{e^2 N_e}{m\nu_e} K_\sigma(0). \tag{5.83}$$

Here ν_e is the effective electron collision frequency [Eq. (2.137)], while the coefficients K_D and K_σ account for the difference between the exact kinetic calculation and the elementary Equation (5.64). In our case

$$K_D = K_\sigma(0) = \frac{8}{3\sqrt{\pi}} \nu_e \int_0^\infty \frac{z^4 e^{-z^2}\, dz}{\nu(z\sqrt{2T_e/m})} = A\nu_e. \tag{5.84}$$

Here $K_\sigma(\omega/\nu_e)$ is defined by Equation (2.151). In particular, for the elastic-sphere collision model [Eq. (2.78)] we have $\nu(v) \sim v$, i.e. $\nu(z) \sim z$ and

$$K_D - K_\sigma(0) - \frac{32}{9\pi} \simeq 1.13 \tag{5.85}$$

(see Table 5). The thermal-diffusion coefficient is in this case one-half the diffusion coefficient.

It is convenient to introduce the thermal-diffusion ratio (Sect. 5.1):

$$k_{Te} = \frac{D_{Te}}{D_e} = 1 + \frac{T_e}{A} \frac{dA}{dT_e}. \tag{5.86}$$

In elementary theory $k_{Te} = 1$ [see Eq. (5.64)]. Actually, as seen from Equation (5.86), k_{Te} can differ greatly from unity. It depends essentially on the form of the function $\nu(v)$. In particular, if $\nu \sim v^{\alpha}$, *then*

$$k_{Te} = 1 - \frac{\alpha}{2}. \tag{5.87}$$

For example, $k_{Te} = 1$ if ν is constant (i.e., $\alpha = 0$), $k_{Te} = \frac{1}{2}$ at $\nu \sim v$ Equation (2.78), $k_{Te} = \frac{1}{6}$ at $\nu \sim v^{5/3}$ [air, Eq. (2.140)], and $k_{Te} = 0$ at $\nu \sim v^2$. Equations (5.86) and (5.80) for k_{Te} coincide with Equation (5.66).

The second equation is obtained by substituting the function f_{00} Equation (5.77) in Equation (5.75), multiplying it in $2\pi m v^4$, and integrating with respect to v from 0 to ∞. It takes the form

$$N_e \frac{\partial T_e}{\partial t} + (\boldsymbol{j}_e \nabla) T_e + \nabla \boldsymbol{g}_e = -\frac{2}{3} \langle \boldsymbol{E} \boldsymbol{j}_{\omega} \rangle_t - N_e \delta(T_e) \nu_e(T_e)(T_e - T) + \frac{2}{3} Q. \tag{5.88}$$

In the derivation of this equation we took into account the fact that in accordance with Equation (5.78) we have

$$\frac{\partial}{\partial t}(N_e T_e) = N_e \frac{\partial T_e}{\partial t} + (\boldsymbol{j}_e \nabla) T_e - [\nabla(T_e \boldsymbol{j}_e) - T_e(q_i - q_r)] \tag{5.89}$$

The first term in the square brackets is included in the flux $\boldsymbol{g}_e$, and the second in Q. Comparing Equation (5.88) with Equation (5.3), we see that $\gamma_e = 1$ in a weakly ionized plasma.

The heat flux $\boldsymbol{g}_e$ in Equation (5.88) is defined by the expression

$$\boldsymbol{g}_e = -\frac{4\pi m}{9} \nabla_{\boldsymbol{r}} \left(\int_0^{\infty} \frac{v^6}{\nu(v)} f_{00}\, dv \right) - \frac{4\pi e}{9 T_e} \boldsymbol{E}_s \left(\int_0^{\infty} \frac{v^6}{\nu(v)} f_{00}\, dv \right) - T_e \boldsymbol{j}_e. \tag{5.90}$$

Recognizing here that the function f_{00} is Maxwellian [Eq. (5.77)], we rewrite Equation (5.90) in the form

$$\begin{gathered} \boldsymbol{g}_e = -\frac{N_e}{m} \nabla(T_e^2 B) + \frac{B T_e^2}{m} \left(-\nabla N_e - \frac{e N_e \boldsymbol{E}_s}{T_e} \right) - T_e \boldsymbol{j}_e, \\ B = \frac{16}{9\sqrt{\pi}} \int_0^{\infty} \frac{z^6 e^{-z^2}\, dz}{\nu(z\sqrt{2T_e/m})}. \end{gathered} \tag{5.91}$$

Using Equations (5.80) and (5.81), we can express here the next to the last term of Equation (5.91) with the aid of $\boldsymbol{j}_e$ and ∇T_e:

$$-\nabla N_e - \frac{eN_e}{T_e}\boldsymbol{E}_s = \frac{\boldsymbol{j}_e}{D_e} + \frac{N_e}{T_e} k_{Te} \nabla T_e, \tag{5.92}$$

where k_{T_e} is the thermal-diffusion ratio [Eq. (5.86)]. Substituting Equation (5.92) in Equation (5.91) we obtain ultimately

$$\boldsymbol{g}_e = -\kappa_e \nabla T_e - \beta_e \boldsymbol{j}_e, \tag{5.93}$$

$$\kappa_e = \frac{N_e T_e}{m}\left[B(2 - k_{Te}) + T_e \frac{dB}{dT_e}\right], \qquad \beta_e = T_e\left(1 - \frac{B}{A}\right). \tag{5.94}$$

The integrals A and B and the coefficients k_{T_e} are defined by Equations (5.80), (5.91), and (5.86).

The thermal conductivity coefficient κ_e is also conveniently represented in the form

$$\kappa_e = \frac{N_e T_e}{m v_e} K_\kappa. \tag{5.95}$$

The coefficient K_κ accounts for the difference between the kinetic calculation of κ_e and the calculation by the elementary Equation (5.70). For the elastic-sphere collision model Equation (2.78) we have

$$K_\kappa = 128/27\pi \approx 1.51. \tag{5.96}$$

At a constant collision frequency $v \neq v(v)$ we have $K_\kappa = 5/3$, and for $v \sim v^{5/3}$ we have $K_\kappa = 224/81\sqrt{3} \approx 1.60$.

For the sake of simplicity, we have neglected above the influence of the constant magnetic field. The problem can be solved quite analogously also when $\boldsymbol{H}$ is taken into account. We use Equation (5.73) to obtain the function $\boldsymbol{f}_1$:

$$\boldsymbol{f}_1 = -\boldsymbol{u}_\omega \frac{\partial f_{00}}{\partial v} + \frac{v}{v^2 + \omega_H^2}\left\{-v\left(\nabla_{r\perp} + \frac{e}{mcv}[\boldsymbol{H} \times \nabla_r]\right)\right.$$
$$\left. + \frac{e}{m}\left(\boldsymbol{E}_s + \frac{e}{mcv}[\boldsymbol{H} \times \boldsymbol{E}_s]\right)\frac{\partial}{\partial v}\right\} f_{00} - \frac{v}{v}\nabla_{r\parallel} f_{00}. \tag{5.97}$$

Here $\boldsymbol{u}_\omega$ is the electron velocity produced by the alternating field $\boldsymbol{E}$ [see Eq. (2.15)].

Substituting $\boldsymbol{f}_1$ in Equation (2.131), we multiply the latter in succession by $4\pi v^2$ and $2\pi m v^4$ and integrate with respect to v from 0 to ∞ with allowance for the fact that f_{00} is a Maxwellian function [Eq. (5.77)]. We then obtain again Equations (5.78) and (5.88) for the electron density and temperature. The fluxes $\boldsymbol{j}_e$ and $\boldsymbol{g}_e$ are defined as before by Equations (5.81) and (5.93), except that the coefficients D_e, σ_e, D_{Te}, κ_e, and β_e are now replaced by the tensors $\hat{D}_e$, $\hat{\sigma}'_e$, $\hat{D}_{Te}$, $\hat{\kappa}_e$, and $\hat{\beta}_e$. The longitudinal tensor components, naturally, coincide with the already considered coefficients D_e, σ_e, ... The complete tensor $\hat{\sigma}'_e = \hat{\sigma}_e - (i\omega/4\pi)(\hat{\varepsilon} - \hat{1})$ is defined by Equations (2.152) and (2.153) with $\omega = 0$. It differs from the "elementary" tensor $\hat{\sigma}'_e$ [Eq. (2.16)] by the coefficients $K_\sigma(x)$ and $K_\varepsilon(x)$, which are listed in Table 5 (here $x = \omega_H/\nu_e$). The diffusion tensor $\hat{D}_e$ is connected with the tensor $\hat{\sigma}'_e$ by the Einstein relation [Eq. (5.18)]. The transverse components of the diffusion, thermal-diffusion, thermal-force, and thermal-conductivity tensors take the form

$$(D_e)_{xx} = (D_e)_{yy} = \frac{T_e}{m} A_H, \qquad (D_e)_{xy} = -(D_e)_{yx} = -\frac{T_e \omega_H}{m} A'_H, \tag{5.98a}$$

$$\begin{aligned} (D_{Te})_{xx} = (D_{Te})_{yy} &= \frac{T_e}{m}\left[A_H + T_e \frac{dA_H}{dT_e}\right], \\ (D_{Te})_{xy} = -(D_{Te})_{yx} &= -\frac{T_e \omega_H}{m}\left[A'_H + T_e \frac{dA'_H}{dT_e}\right], \end{aligned} \tag{5.98b}$$

$$\begin{aligned} \beta_{xx} = \beta_{yy} &= T_e\left(1 - \frac{B_H}{A_H} - \omega_H^2 \frac{A'_H}{A_H} \frac{B'_H A_H - B_H A'_H}{A_H^2 + \omega_H^2 A_H'^2}\right), \\ \beta_{xy} = -\beta_{yx} &= T_e \omega_H \frac{B'_H A_H - B_H A'_H}{A_H^2 + \omega_H^2 A_H'^2}, \end{aligned} \tag{5.98c}$$

$$\begin{aligned} \kappa_{xx} = \kappa_{yy} = \frac{N T_e}{m}\Bigg\{ & B_H + T_e \frac{dB_H}{dT_e} - \frac{T_e B_H}{A_H}\frac{dA_H}{dT_e} \\ & + \omega_H^2 \frac{B'_H A_H - B_H A'_H}{A_H^2 + \omega_H^2 A_H'^2}\left(T_e \frac{dA'_H}{dT_e} - \frac{A'_H}{A_H} T_e \frac{dA_H}{dT_e}\right)\Bigg\}, \end{aligned}$$

$$\begin{aligned} \kappa_{xy} = -\kappa_{yx} = -\frac{N T_e}{m}\omega_H \Bigg\{ & B'_H + T_e \frac{dB'_H}{dT_e} - B_H \frac{A'_H}{A_H} - \frac{B_H T_e}{A_H^2 + \omega_H^2 A_H'^2} \\ & \times \left(A_H \frac{dA'_H}{dT_e} - A'_H \frac{dA_H}{dT_e}\right) - \frac{T_e B'_H}{2(A_H^2 + \omega_H^2 A_H'^2)} \frac{d}{dT_e}(A_H^2 + \omega_H^2 A_H'^2)\Bigg\}. \end{aligned}$$

$$A_H = \frac{8}{3\sqrt{\pi}} \int_0^\infty \frac{z^4 \nu(z\sqrt{2T_e/m}) e^{-z^2}}{\omega_H^2 + \nu^2(z\sqrt{2T_e/m})} dz = \frac{K_\varepsilon(x)}{\nu_e(x^2+1)},$$

$$A'_H = \frac{8}{3\sqrt{\pi}} \int_0^\infty \frac{z^4 e^{-z^2} dz}{\omega_H^2 + \nu^2(z\sqrt{2T_e/m})} = \frac{K_\sigma(x)}{\nu_e^2(x^2+1)}, \qquad x = \frac{\omega_H}{\nu_e}.$$

$$B_H = \frac{16}{9\sqrt{\pi}} \int_0^\infty \frac{z^6 \nu(z\sqrt{2T_e/m}) e^{-z^2}}{\nu^2(z\sqrt{2T_e/m}) + \omega_H^2} dz,$$

$$B'_H = \frac{16}{9\sqrt{\pi}} \int_0^\infty \frac{z^6 e^{-z^2} dz}{\omega_H^2 + \nu^2(z\sqrt{2T_e/m})}. \tag{5.98d}$$

If the collision frequency does not depend on the velocity, $\nu \neq \nu(v)$, then the functions A_H, A'_H, B_H, and B'_H take on the values

$$A_H = \frac{\nu_e}{\omega_H^2 + \nu_e^2}, \qquad A'_H = \frac{1}{\omega_H^2 + \nu_e^2},$$

$$B_H = \frac{5}{3}\frac{\nu_e}{\omega_H^2 + \nu_e^2}, \qquad B'_H = \frac{5}{3}\frac{1}{\omega_H^2 + \nu_e^2}.$$

In this case the components of the tensors $\hat{\beta}$ and $\hat{\kappa}$ are of the simple form

$$\beta_{xx} = \beta_{yy} = \beta_{zz} = -\tfrac{2}{3}T_e, \qquad \beta_{xy} = -\beta_{yx} = 0,$$

$$\kappa_{xx} = \kappa_{yy} = \frac{5}{3}\frac{NT_e}{m}\frac{\nu_e}{\omega_H^2 + \nu_e^2}, \qquad \kappa_{zz} = \frac{5}{3}\frac{NT_e}{m\nu_e},$$

$$\kappa_{xy} = -\kappa_{yx} = -\frac{5}{3}\frac{NT_e}{m}\frac{\omega_H}{\omega_H^2 + \nu_e^2}$$

Equations (5.78) and (5.88) were derived under the assumption that the electrons have a Maxwellian distribution function. In a weakly ionized plasma in a strong alternating electric field, if Equation (5.76) is not satisfied, the electron distribution can differ substantially from a Maxwellian (see Sects. 2.3.2 and 2.3.3). Strictly speaking, Equations (5.1)–(5.4) are no longer valid in this case. However, the replacement of the exact electron distribution function by a Maxwellian usually has little effect on the average quantities (Sect. 2.3.2).

The transport coefficients for the ions are obtained in similar fashion.

In the foregoing determination of the function f_1 we have neglected the collisions between the charged particles, which is permissible under

Equation (5.72). This condition is not satisfied at a sufficiently high degree of plasma ionization. It is then necessary to take into account in Equation (2.132) for the electrons and for the ions the integral terms that describe the interaction of the charged particles. The integral term in Equation (2.132) includes in this case the ion distribution function $\boldsymbol{f}_{1\mathrm{i}}$, while the equation for the ions includes the electron distribution function $\boldsymbol{f}_{1\mathrm{e}}$. The equations must therefore be solved simultaneously. The corresponding solution is obtained by expanding the functions $\boldsymbol{f}_{1\mathrm{i}}$ and $\boldsymbol{f}_{1\mathrm{e}}$ in series of Sonine or Laguerre polynomials or else by Grad's method, as in the theory of an inhomogeneous gas mixture (Chapman and Cowling, 1952; Shkarofsky et al., 1966; Hochstim and Massel, 1969; Silin, 1972). Substituting $\boldsymbol{f}_{1\mathrm{i}}$ and $\boldsymbol{f}_{1\mathrm{e}}$ in the equations for $\boldsymbol{f}_{0\mathrm{e}}$ [Eq. (2.131)] and $\boldsymbol{f}_{0\mathrm{i}}$ [Eq. (2.197)], and recognizing that the functions $f_{0\mathrm{e}}$ and $f_{0\mathrm{i}}$ are Maxwellian, we obtain in the usual manner the equations for N_e, N_i, T_e, and T_i, which take as before the forms (5.78), (5.88), and (5.1)–(5.4). The fluxes $\boldsymbol{j}_\mathrm{e}$, $\boldsymbol{j}_\mathrm{i}$, $\boldsymbol{g}_\mathrm{e}$, and $\boldsymbol{g}_\mathrm{i}$ are now determined by the general Equations (5.8)–(5.11).

Without presenting the corresponding solutions here, we show only that in the general case the kinetic coefficients are connected by Equations (5.12), and the heat fluxes can always be represented by Equations (5.10) and (5.11). Indeed, the right-hand side in an equation such as Equation (2.132) for the functions $\boldsymbol{f}_{1\mathrm{e}}$ or $\boldsymbol{f}_{1\mathrm{i}}$ can be rewritten as

$$\boldsymbol{J}_\mathrm{e} = -v\,\nabla_{\boldsymbol{r}} f_{0\mathrm{e}} + \frac{e\boldsymbol{E}}{m}\frac{\partial f_{0\mathrm{e}}}{\partial v}, \qquad \boldsymbol{J}_\mathrm{i} = -v\,\nabla_{\boldsymbol{r}} f_{0\mathrm{i}} - \frac{e\boldsymbol{E}}{M}\frac{\partial f_{0\mathrm{i}}}{\partial v}. \tag{5.99}$$

The functions $f_{0\mathrm{e}}$ and $f_{0\mathrm{i}}$ are Maxwellian [Eq. (5.77)]. Consequently

$$\begin{aligned}\nabla_{\boldsymbol{r}} f_{0\mathrm{e}} &= \left(\frac{m}{2\pi T_\mathrm{e}}\right)^{3/2} \exp\left(-mv^2/2T_\mathrm{e}\right) \nabla_{\boldsymbol{r}} N_\mathrm{e} \\ &\quad + N_\mathrm{e}\frac{\partial}{\partial T_\mathrm{e}}\left[\left(\frac{m}{2\pi T_\mathrm{e}}\right)^{3/2} \exp\left(-mv^2/2T_\mathrm{e}\right)\right] \nabla_{\boldsymbol{r}} T_\mathrm{e}, \\ \frac{\partial f_{0\mathrm{e}}}{\partial v} &= -\frac{mv}{T_\mathrm{e}} N_\mathrm{e}\left(\frac{m}{2\pi T_\mathrm{e}}\right)^{3/2} \exp\left(-mv^2/2T_\mathrm{e}\right)\end{aligned}$$

and Equation (5.99) can be represented in the form

$$\begin{aligned}\boldsymbol{J}_\mathrm{e} &= -v\exp\left(-\frac{mv^2}{2T_\mathrm{e}}\right)\left(\frac{m}{2\pi T_\mathrm{e}}\right)^{3/2}\left[\nabla N_\mathrm{e} + \frac{eN_\mathrm{e}}{T_\mathrm{e}}\boldsymbol{E}\right] - N_\mathrm{e}\frac{d}{dT_\mathrm{e}}\left(\frac{f_{0\mathrm{e}}}{N_\mathrm{e}}\right)\nabla T_\mathrm{e}, \\ \boldsymbol{J}_\mathrm{i} &= -v\exp\left(-\frac{Mv^2}{2T_\mathrm{i}}\right)\left(\frac{M}{2\pi T_\mathrm{i}}\right)^{3/2}\left[\nabla N_\mathrm{i} - \frac{eN_\mathrm{i}}{T_\mathrm{i}}\boldsymbol{E}\right] - N_\mathrm{i}\frac{d}{dT_\mathrm{i}}\left(\frac{f_{0\mathrm{i}}}{N_\mathrm{i}}\right)\nabla T_\mathrm{i}.\end{aligned} \tag{5.100}$$

It follows therefore that the density gradients ∇N_e and ∇N_i, as well as the quasistationary field $\boldsymbol{E}_s$, enter in the expressions for $\boldsymbol{f}_{1e}$ and $\boldsymbol{f}_{1i}$ only as the combinations

$$\nabla N_e + \frac{eN_e}{T_e}\boldsymbol{E}_s, \qquad \nabla N_i - \frac{eN_i}{T_i}\boldsymbol{E}_s.$$

The same combinations of ∇N_e and $\boldsymbol{E}_s$ enter consequently also in the expressions for the fluxes:

$$\boldsymbol{j}_e = \frac{4\pi}{3}\int_0^\infty \boldsymbol{f}_{1e}v^3\,dv = -\hat{D}_e\left(\nabla N_e + \frac{eN_e}{T_e}\boldsymbol{E}_s\right) - \hat{D}_{ei}\left(\nabla N_i - \frac{eN_i}{T_i}\boldsymbol{E}_s\right) + O(\nabla T_e, \nabla T_i). \tag{5.101}$$

$$\boldsymbol{j}_i = -\hat{D}_i\left(\nabla N_i - \frac{eN_i}{T_i}\boldsymbol{E}_s\right) - \hat{D}_{ie}\left(\nabla N_e + \frac{eN_e}{T_e}\boldsymbol{E}_s\right) + O(\nabla T_i, \nabla T_e). \tag{5.102}$$

On the other hand, the fluxes $\boldsymbol{j}_e$ and $\boldsymbol{j}_i$ are determined by the general Equations (5.8) and (5.9). Comparing them with Equations (5.101) and (5.102), we establish relation in Equation (5.12).

The same combinations of ∇N_e and $\boldsymbol{E}_s$ enter also in the expressions for the heat fluxes:

$$\boldsymbol{g}_e = -\hat{\gamma}_{ee}\left(\nabla N_e + \frac{eN_e}{T_e}\boldsymbol{E}_s\right) - \hat{\gamma}_{ei}\left(\nabla N_i - \frac{eN_i}{T_e}\boldsymbol{E}_s\right) + O(\nabla T_e, \nabla T_i), \tag{5.103}$$

$$\boldsymbol{g}_i = -\hat{\gamma}_{ii}\left(\nabla N_i - \frac{eN_i}{T_i}\boldsymbol{E}_s\right) - \hat{\gamma}_{ie}\left(\nabla N_e + \frac{eN_e}{T_e}\boldsymbol{E}_s\right) + O(\nabla T_e, \nabla T_i). \tag{5.104}$$

We determine then from Equations (5.101) and (5.102)

$$\begin{aligned}\nabla N_e + \frac{eN_e}{T_e}\boldsymbol{E}_s &= \hat{\alpha}_{1e}\boldsymbol{j}_e + \hat{\alpha}_{2e}\boldsymbol{j}_i + O(\nabla T_e, \nabla T_i),\\ \nabla N_i - \frac{eN_i}{T_i}\boldsymbol{E}_s &= \hat{\alpha}_{1i}\boldsymbol{j}_i + \hat{\alpha}_{2i}\boldsymbol{j}_e + O(\nabla T_e, \nabla T_i).\end{aligned} \tag{5.105}$$

Substituting Equations (5.105) in Equations (5.103) and (5.104) for $\boldsymbol{g}_e$ and $\boldsymbol{g}_i$, we arrive at Equations (5.10) and (5.11) for the heat fluxes.

We note that if the functions f_0 are not Maxwellian, then Equation (5.100), and consequently also Equations (5.12) for the kinetic coefficients and Equations (5.10) and (5.11) for $\boldsymbol{g}_e$ and $\boldsymbol{g}_i$ are, strictly speaking, not valid.

5.2.3. Fully Ionized Plasma

The case of a fully ionized plasma is singled out because Equation (5.7) is not satisfied here for the direction along the magnetic field. As seen from Equation (5.68) the longitudinal-diffusion coefficients $(D_{ee})_{zz}$ and $(D_{ii})_{zz}$ tend to infinity as $\nu_{im} \to 0$. In other words, longitudinal ambipolar diffusion is impossible in a fully ionized plasma. The reason is that diffusion calls for a constant total pressure, $\nabla p = 0$ (Chapman and Cowling, 1952). If the plasma is partially ionized, then this condition is ensured by the perturbation of the neutral molecules: $\nabla p = \nabla(p_e + p_i) + \nabla p_m = 0$. In a fully ionized plasma $\nabla p_m = 0$. Consequently $\nabla(p_e + p_i) = T_e \nabla N_e + T_i \nabla N_i = 0$. Recognizing that $\nabla N_e = \nabla N_i = \nabla N$ by virtue of the quasineutrality, we see that the condition $\nabla p = 0$ is satisfied only if $\nabla N = 0$.

In a fully ionized plasma $\gamma_e = \gamma_i = 5/3$. In place of Equations (5.25) and (5.26) it is more convenient here to use the equations

$$
\begin{aligned}
&-\nabla(N_e T_e) - eN_e\left(\boldsymbol{E} + \frac{1}{cN_e}[\boldsymbol{j}_e \times \boldsymbol{H}]\right) + \boldsymbol{R} = 0 \\
&-\nabla(N_i T_i) + eN_i\left(\boldsymbol{E} + \frac{1}{cN_i}[\boldsymbol{j}_i \times \boldsymbol{H}]\right) - \boldsymbol{R} = 0.
\end{aligned}
\tag{5.106}
$$

Here $\boldsymbol{R}$ is the electron and ion friction force. Equations (5.106) are degenerate: adding them, we arrive at the relation (at $N_e = N_i$):

$$
\frac{\partial}{\partial z}(N_e T_e + N_i T_i) = 0. \tag{5.107}
$$

The z axis is directed parallel to $\boldsymbol{H}$. This condition should be satisfied independently of the values of $\boldsymbol{j}_e$ and $\boldsymbol{j}_i$.

In a strongly magnetized plasma $\omega_H \gg \omega_{ei}$ (as, in particular, is the ionosphere in the F-layer region), the expressions for the friction force and for the transport coefficient take the form (Braginskii, 1965)

$$
\boldsymbol{R} = \frac{1}{m\nu_{ei}}\left(\frac{\boldsymbol{j}_{\parallel}}{1.96} + \boldsymbol{j}_{\perp}\right) - 0.71 N\, \nabla_{\parallel} T_e - \frac{3}{2}\frac{N\nu_{ei}}{\omega_H}\frac{[\boldsymbol{H} \times \nabla T_e]}{H}. \tag{5.108}
$$

The parallel and perpendicular symbols in the subscripts denote the components along and across the magnetic field, while ν_{ei} is the effective electron-ion collision frequency [Eq. (2.141)]. Furthermore,

$$
\begin{gathered}
\boldsymbol{g}_{\mathrm{e}} = 0.47 T_{\mathrm{e}} \boldsymbol{j}_{\|} + \frac{T_{\mathrm{e}} \nu_{\mathrm{ei}}}{\omega_H H} |\boldsymbol{H} \times \boldsymbol{j}_{\perp}| - \kappa_{\mathrm{e}\|} \nabla_{\|} T_{\mathrm{e}} - \kappa_{\mathrm{e}\perp} \nabla_{\perp} T_{\mathrm{e}} \\
- \frac{5}{3} \frac{N T_{\mathrm{e}}}{m H \omega_H} |\boldsymbol{H} \times \nabla T_{\mathrm{e}}], \quad \boldsymbol{j} = \boldsymbol{j}_{\mathrm{e}} - \boldsymbol{j}_{\mathrm{i}} \\
\kappa_{\mathrm{e}\|} = 2.08 \frac{N T_{\mathrm{e}}}{m \nu_{\mathrm{ei}}}, \qquad \kappa_{\mathrm{e}\perp} = 3.1 \frac{N T_{\mathrm{e}} \nu_{\mathrm{ei}}}{m \omega_H^2}.
\end{gathered} \tag{5.109}
$$

The ion heat flux is

$$
\begin{gathered}
\boldsymbol{g}_{\mathrm{i}} = -\kappa_{\mathrm{i}\|} \nabla T_{\mathrm{i}} - \kappa_{\mathrm{i}\perp} \nabla_{\perp} T_{\mathrm{i}} + \frac{5}{3} \frac{N T_{\mathrm{i}}}{M \Omega_H H} |\boldsymbol{H} \times \nabla T_{\mathrm{i}}], \\
\kappa_{\mathrm{i}\|} = 2.6 \frac{N T_{\mathrm{i}}}{M \nu_{\mathrm{ii}}}, \qquad \kappa_{\mathrm{i}\perp} = \frac{4}{3} \frac{N T_{\mathrm{i}} \nu_{\mathrm{ii}}}{M \Omega_H^2}.
\end{gathered} \tag{5.110}
$$

Here ω_H and Ω_H are the gyromagnetic frequencies for the electrons and ions, and ν_{ii} is the frequency of the collisions between the ions [Eq. (5.19)].

5.3. Modification of the F Region of the Ionosphere by Radio Waves

The perturbations of the electron temperature and of the concentration of a homogeneous ionospheric plasma under the influence of an alternating electric field of radio waves was considered in Sect. 2.5. We consider here the perturbations of an inhomogeneous plasma with allowance for the transport processes, an important factor in the region of the F layer.

5.3.1. Modification of the Electron Temperature and of the Plasma Concentration

We consider the modification of the ionosphere plasma in an inhomogeneous alternating electric field, with allowance for the transport processes, namely the diffusion, the thermal diffusion, and the thermal conductivity. We assume that the wave is weak, $E_0^2 \ll E_p^2$, so that the perturbations caused by the wave are small: $\Delta N \ll N_0$ and $\Delta T_e \ll \Delta T_{e0}$.

Then, linearizing Equations (5.36) and (5.27), we get

$$\frac{\partial\, \Delta N}{\partial t} - D_{\mathrm{a}} \frac{\partial^2\, \Delta N}{\partial x_{\parallel}^2} - D_{T\mathrm{ea}} \frac{N_0}{T_{\mathrm{e}0}} \frac{\partial^2\, \Delta T_{\mathrm{e}}}{\partial x_{\parallel}^2} = -\left(\frac{\partial q_{\mathrm{r}}}{\partial N}\right) \Delta N - \left(\frac{\partial q_{\mathrm{r}}}{\partial T_{\mathrm{e}}}\right) \Delta T_{\mathrm{e}}, \tag{5.111a}$$

$$\frac{\partial\, \Delta T_{\mathrm{e}}}{\partial t} - k_{\mathrm{e}} \frac{\partial^2\, \Delta T_{\mathrm{e}}}{\partial x_{\parallel}^2} = -\,\delta\nu_{\mathrm{e}}\, \Delta T_{\mathrm{e}} \varphi_{\mathrm{T}} + T_{\mathrm{e}} \left(\frac{E_0}{E_{\mathrm{p}}}\right)^2 \varphi_{\mathrm{p}}\, \delta\nu_{\mathrm{e}}. \tag{5.111b}$$

We have taken it into account here that under the conditions $\nu_{\mathrm{im}} \gg \delta\nu_{\mathrm{e}}$, which are well satisfied in the ionosphere up to heights $z \sim 600$ km, the heating of the ions is negligible and can be neglected. Next, $x_{\parallel}$ is the coordinate in the $\boldsymbol{H}$ direction [Eq. (5.58)], $\delta\nu_{\mathrm{e}} = \delta_{\mathrm{ei}}\nu_{\mathrm{ei}} + \delta_{\mathrm{em}}\nu_{\mathrm{em}}$; φ_{p} is the polarization factor [Eq. (2.42)], and φ_{T} is the non-isothermy factor [Eq. (2.246)]. We recognize also that the transverse dimension of the inhomogeneity is large enough [Eq. (5.61)], so that only the longitudinal thermal conductivity and diffusion are significant. The longitudinal electric field is eliminated from Equations (5.23)–(5.26) with the aid of Equations (5.62), (5.34), and (5.35). We point out that, as follows from Equations (5.37), (5.68), and (5.69):

$$D_{\mathrm{a}} = (T_{\mathrm{e}} + T_{\mathrm{i}})/M\nu_{\mathrm{im}}, \qquad D_{T\mathrm{ea}} = T_{\mathrm{e}}/M\nu_{\mathrm{im}}, \qquad \kappa_{\mathrm{e}}/N = T_{\mathrm{e}}/m(\nu_{\mathrm{em}} + \nu_{\mathrm{ei}}). \tag{5.111c}$$

The kinetic corrections to Equation (5.111c) are given by Equations (5.95), (5.96), and (5.109).

We consider a stationary perturbation of the plasma (Gurevich, 1971). Equations (5.111) are then rewritten in

$$L_N^2 \frac{d^2\, \Delta N}{dx_{\parallel}^2} + k_T L_N^2 \frac{N_0}{T_{\mathrm{e}0}} \frac{d^2\, \Delta T_{\mathrm{e}}}{dx_{\parallel}^2} = \Delta N - \gamma_1 \frac{N_0}{T_{\mathrm{e}0}} \Delta T_{\mathrm{e}}, \tag{5.112a}$$

$$L_T^2 \frac{d^2\, \Delta T_{\mathrm{e}}}{dx_{\parallel}^2} = \Delta T_{\mathrm{e}} - T_{\mathrm{e}0} \left(\frac{E_0}{E_{\mathrm{p}}}\right)^2 \varphi; \qquad \varphi = \varphi_{\mathrm{p}}/\varphi_{\mathrm{T}}; \qquad L_T^2 = k_{\mathrm{e}}/\delta\nu_{\mathrm{e}}\varphi_T \tag{5.112b}$$

The characteristic lengths L_N and L_T are defined here in accordance with Equations (5.43) and (5.44). For the ionosphere they are listed in Table 16. The thermal-diffusion ratio $k_T = k_{T\mathrm{ea}} = D_{T\mathrm{ea}}/D_{\mathrm{a}}$ is determined from Equations (5.111c) and (5.49). The ionization-equilibrium shift coefficient

Table 16. Temperature and density inhomogeneity characteristics

	z, km	D_a, cm^2/s	$k_e = \kappa_e/N_e$, cm^2/s	L_T, cm	L_N, cm	η
Day-time	100	$1.2 \cdot 10^5$	$9 \cdot 10^8$	$2.8 \cdot 10^3$	$2.3 \cdot 10^3$	1.38
	120	$3.2 \cdot 10^6$	$1.2 \cdot 10^{10}$	$3.5 \cdot 10^4$	$1.2 \cdot 10^4$	8.50
	150	$7.6 \cdot 10^7$	$1.4 \cdot 10^{11}$	$4.1 \cdot 10^5$	$5.2 \cdot 10^4$	62.2
	200	$1.7 \cdot 10^9$	$5.6 \cdot 10^{11}$	$1.7 \cdot 10^6$	$2.3 \cdot 10^5$	55.4
	250	$6 \cdot 10^9$	$6.1 \cdot 10^{11}$	$2.5 \cdot 10^6$	$9.2 \cdot 10^5$	7.42
	300	$1.6 \cdot 10^{10}$	$6.1 \cdot 10^{11}$	$2.8 \cdot 10^6$	$2.8 \cdot 10^6$	1.0
	400	$7.2 \cdot 10^{10}$	$1 \cdot 10^{12}$	$4.9 \cdot 10^6$	$1.7 \cdot 10^7$	0.083
	500	$2.5 \cdot 10^{11}$	$2.2 \cdot 10^{12}$	$1 \cdot 10^7$	$9.8 \cdot 10^7$	0.0104
	600	$1.3 \cdot 10^{12}$	$9.5 \cdot 10^{12}$	$2.2 \cdot 10^7$	$6 \cdot 10^8$	0.0013
Night-time	100	$1.5 \cdot 10^5$	$7.5 \cdot 10^8$	$2.5 \cdot 10^3$	$2.0 \cdot 10^4$	0.015
	120	$2.8 \cdot 10^6$	$1.0 \cdot 10^{10}$	$3 \cdot 10^4$	$6.1 \cdot 10^4$	0.24
	150	$6 \cdot 10^7$	$1.3 \cdot 10^{11}$	$4.1 \cdot 10^5$	$3.1 \cdot 10^5$	1.75
	200	$1.5 \cdot 10^9$	$1.2 \cdot 10^{12}$	$3.1 \cdot 10^6$	$3 \cdot 10^5$	27
	250	$6.1 \cdot 10^9$	$3.7 \cdot 10^{12}$	$1.03 \cdot 10^7$	$1.6 \cdot 10^6$	41.5
	300	$2.4 \cdot 10^{10}$	$1.5 \cdot 10^{12}$	$9.9 \cdot 10^6$	$7.2 \cdot 10^6$	1.89
	400	$1.6 \cdot 10^{11}$	$7.0 \cdot 10^{11}$	$6.0 \cdot 10^6$	$9 \cdot 10^7$	0.0045
	500	$6.5 \cdot 10^{11}$	$1.4 \cdot 10^{12}$	$1.1 \cdot 10^7$	$8 \cdot 10^8$	$1.5 \cdot 10^{-4}$

$$\gamma_1 = \frac{T_{e0}(\partial q_r/\partial T_e)}{N_0(\partial q_r/\partial N)} \tag{5.113}$$

is determined by Equation (2.249). Neglecting transport processes, Equations (5.112) coincide with Equations (2.249) and (2.246) for a uniform plasma.

We note that in the derivation of Equation (5.112) it was assumed that N_0 and T_{e0} do not depend on $x_\parallel$. Under real ionosphere condition, N_0 and T_{e0} depend on the height z, i.e., on $x_\parallel \cos\alpha$, where α is the angle between the vertical and the direction of the magnetic field. Equations (5.112) are therefore valid only under the condition that the inhomogeneity ΔN or ΔT_e is smaller in dimension than $L/\cos\alpha$, where L is the characteristic dimension over which $N_0(z)$ and $T_{e0}(z)$ change.

The solution of Equation (5.112b) under the boundary conditions $\Delta T_e \to 0$ as $x_\parallel \to \infty$ is written in the form

$$\Delta T_e = \frac{T_{e0}}{2L_T}\int_{-\infty}^{\infty} Q(\boldsymbol{r}_\perp, x'_\parallel)\exp\left(-\frac{|x_\parallel - x'_\parallel|}{L_T}\right)dx'_\parallel\,; \qquad Q = \frac{E_0^2(\boldsymbol{r}_\perp, x'_\parallel)}{E_p^2}\varphi. \tag{5.114}$$

The solution of Equation (5.112a) for the concentration perturbations is similar in form:

$$\begin{aligned}\Delta N &= \frac{1}{2L_N}\int_{-\infty}^{\infty}\psi(x'_{\parallel})\exp\left(-\frac{|x_{\parallel}-x'_{\parallel}|}{L_N}\right)dx'_{\parallel}\\ &= N_0\frac{\gamma_1}{2}\frac{L_N L_T}{L_N^2-L_T^2}\int_{-\infty}^{\infty}Q(x'_{\parallel})\,dx'_{\parallel}\left[\frac{\exp(-|x_{\parallel}-x'_{\parallel}|/L_N)}{L_T}\right.\\ &\quad\left.-\frac{\exp(-|x_{\parallel}-x'_{\parallel}|/L_T)}{L_N}\right]-N_0\frac{k_T}{2}\frac{L_N^2}{L_N^2-L_T^2}\int_{-\infty}^{\infty}Q(x'_{\parallel})\,dx'_{\parallel}\\ &\quad\times\left[\frac{\exp(-|x_{\parallel}-x'_{\parallel}|/L_T)}{L_T}-\frac{\exp(-|x_{\parallel}-x'_{\parallel}|/L_N)}{L_N}\right].\end{aligned} \tag{5.115}$$

Here

$$\psi=\gamma_1\frac{N_0}{T_{e0}}\Delta T_e+k_T L_N^2\frac{N_0}{T_{e0}}\frac{d^2\,\Delta T_e}{dx_{\parallel}^2}=\frac{N_0}{T_{e0}}\left(\gamma_1+k_T\frac{L_N^2}{L_T^2}\right)\Delta T_e-k_T N_0\frac{L_N^2 Q}{L_T^2}. \tag{5.116}$$

Equations (5.114)–(5.116) determine the perturbations ΔT_e and ΔN for an arbitrary dependence of the modifying-field amplitude E_0 on x. If the field E_0 depends little on $x_{\parallel}$, i.e., if it varies little over the characteristic lengths L_T and L_N,

$$\frac{L_T}{E_0^2}\left|\frac{dE_0^2}{dx_{\parallel}}\right|\ll 1,\qquad \frac{L_N}{E_0^2}\left|\frac{dE_0^2}{dx_{\parallel}}\right|\ll 1 \tag{5.117}$$

then we can take E_0^2 outside the integral signs in Equations (5.114)–(5.116), and these expressions for ΔT_e and ΔN now go over into Equations (2.246) and (2.249). If, on the contrary, the length L_T is large, so that a condition opposite to Equation (5.117) is satisfied, then the field E_0^2 in Equation (5.114) has a sharp maximum at a certain point x_0, and Equation (5.114) becomes [cf. Eq. (5.56)]:

$$\Delta T_e=\frac{T_{e0}\tilde{Q}}{2L_T}\exp\left(-\frac{|x_{\parallel}-x_0|}{L_T}\right),\qquad \tilde{Q}=\int Q(x_{\parallel})\,dx_{\parallel}. \tag{5.118}$$

We see that the maximum value of ΔT_e decreases in comparison with Equation (5.88) in proportion to $\Delta x_{\parallel}/2L_T$, where $\Delta x_{\parallel}$ is the characteristic width of the distribution of $E_0^2(x_{\|})$. Under these conditions we obtain from Equation (5.115) for the concentration perturbation

$$\Delta N = -\frac{N_0 k_T}{2}\frac{L_N^2 \tilde{Q}}{L_N^2 - L_T^2}\left(\frac{\exp\left[-|x_{\|} - x_0|/L_T\right]}{L_T} - \frac{\exp\left[-|x_{\|} - x_0|/L_N\right]}{L_N}\right)$$
$$+\frac{N_0 \gamma_1}{2}\frac{L_N L_T \tilde{Q}}{L_N^2 - L_T^2}\left(\frac{\exp\left[-|x_{\|} - x_0|/L_N\right]}{L_T} - \frac{\exp\left[-|x_{\|} - x_0|/L_N\right]}{L_N}\right) \quad (5.118a)$$

If $L_N^2 \gg L_T^2$, then Equations (5.118a) and (5.118) lead to the hydrodynamic-equilibrium distribution (Sect. 5.1.2)

$$\Delta N = -k_T \frac{N_0}{T_{e0}} \Delta T_e. \quad (5.119)$$

In the opposite case $L_I^2 \gg L_N^2$ we have

$$\Delta N = \frac{N_0 \tilde{Q}}{2L_T}\left[\gamma_1 L_T \exp\left(-\frac{|x_{\|} - x_0|}{L_T}\right) - k_T L_N \exp\left(-\frac{|x_{\|} - x_0|}{L_N}\right)\right] \quad (5.120)$$

It is seen that the concentration perturbations are small in this case, smaller than in the case of hydrodynamic equilibrium [Eq. (5.119)], and are proportional to $\Delta x_{\|} L_N / L_T^2$ at $\gamma_1 \simeq 0$.

It is seen from Table 16 that at $z \lesssim 200$ km the dimensions L_T and L_N are small. Equation (5.117) is satisfied here and the local homogeneous approximation considered in Sect. 2.5 is valid. At $z \sim 200$—300 km we have $L_T \gtrsim L_N$ and the concentration perturbation is usually small [Eq. (5.120)]. At $z > 300$ km, the dimension L_N increases very rapidly and exceeds the characteristic dimensions of the region modified by the radio waves. A hydrodynamic-equilibrium distribution of the concentration [Eq. (5.119)] is then established in the modified zone [see also Eqs. (5.50) and (5.53)].

The concentration perturbations on the axis of a Gaussian beam $E^2 = E_0^2(y, z) \exp(-x_{\|}^2/a^2)$, calculated from Equations (5.114)–(5.116), are shown in Figure 52. It is seen from the curves that at heights lower than 180–200 km the electron density is increased by the field of the wave, as follows from Equation (2.249). At heights 200 km $\lesssim z \lesssim$ (250–300) km, the electron-density perturbations are negligible. In the region above the F-layer maximum, these perturbations increase again. The concentration

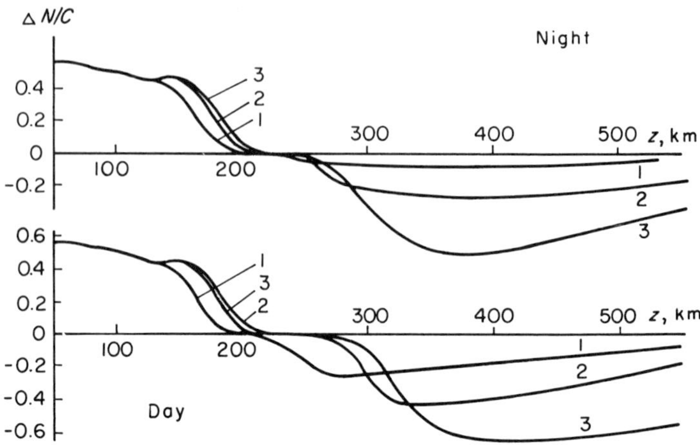

Fig. 52. Concentration perturbation $\Delta N/C$ on the beam axis. $C = N_0 \varphi E_0^2/E_p^2$. Curves *1*, *2*, and *3* correspond to an effective Gaussian-beam width $a = 10$, 30, and 100 km

N decreases here under the influence of the field of the wave.[17] This is the region where the hydrodynamic-equilibrium distribution of the plasma [Eq. (5.119)] is established.

In a plane perpendicular to the z axis, at $a > L_N$ and $a > L_T$ (i.e., at small heights), the perturbations ΔN and ΔT_e are replicas of the distribution of the field $\boldsymbol{E}$. At $L_T \gtrsim a$, the perturbations are elongated in the direction of the magnetic field $\boldsymbol{H}$. Figure 53 shows a form of the perturbed region in the xy plane, which is perpendicular to the propagation direction z of an axially-symmetrical beam, at different values of the ratio L_T/a. It is seen that at small values of L_T/a the perturbed region, just as the beam, is axially symmetrical. The transport processes are negligible in this case and the structure of the perturbed region coincides with the structure of the beam; the temperature and concentration distributions are then locally homogeneous. On the other hand, if the ratio $L_T/a \gg 1$, then a strong anisotropy in the distribution of the perturbation sets in. The perturbed region becomes elongated in the direction of the magnetic field.

We point out that the case when the radio waves propagate in the direction of the magnetic field $\boldsymbol{H}$ is special. The thermal conductivity is then of little effect, so that the electron temperature perturbations are always determined by the local Equations (2.246). The concentration per-

[17] We emphasize that the quantity $C \sim Q \sim E_0^2(y, z)$ is itself a strongly varying function of the height z. In particular, it increases strongly in the wave-reflection region (since $E_0^2 \sim 1/\sqrt{\varepsilon_0}$), and this increases the perturbation appreciably in this region (see Sect. 5.3.2).

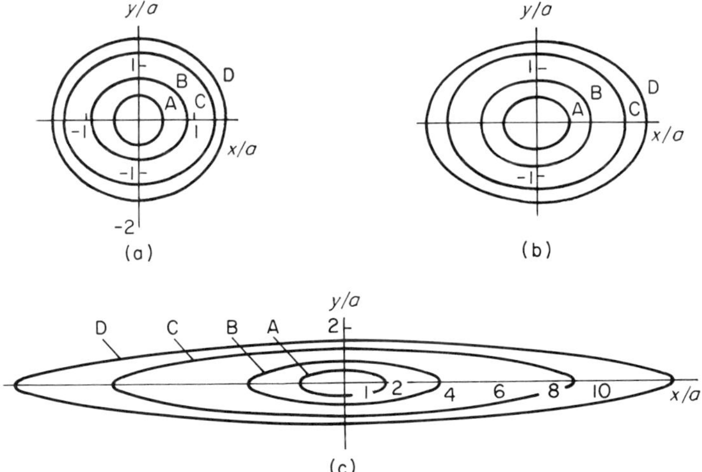

Fig. 53a–c. Concentration-perturbation curves N = const. in a plane perpendicular to the beam axis

$$\text{(a)}\ \frac{L_T}{a} = 0.2; \qquad \text{(b)}\ \frac{L_T}{a} = 0.5; \qquad \text{(c)}\ \frac{L_T}{a} = 5.$$

Curves A, B, C, and *D* correspond to $\Delta N/\Delta N(0) = 0.8$, 0.5, 0.2, and 0.1, respectively. Here $\Delta N(0)$ is the perturbation on the beam axis

turbations at $z \lesssim 250$–300 km are also determined by the local ionization balance [Eq. (2.249)]. In the region of the maximum of the layer and above it, diffusion assumes an essential role. However, the dimension of the inhomogeneity $\Delta N(z)$ is here the same as in the main plasma, $N_0(z)$, so that a rigorous solution can be obtained only by numerical means.

We have considered above only the case of a weak electric field. It is easy to estimate also, however, the character of the effects produced in the case of a strong plasma perturbation (Vas'kov and Gurevich, 1976a). Indeed, let us take into account the fact that if $\nu_{ei} \gg \nu_{em}$ the electronic thermal-conductivity coefficient [Eqs. (5.70) and (5.95)] increases rapidly with increasing electron temperature: $\kappa_e \sim T_e/\nu_{ei} \sim T_e^{5/2}$. Accordingly, the characteristic length $L_T \sim T_e^{5/4}$ also increases at a rapid rate [(see Eq. (5.44)]. In a strong field $E_0^2 \gtrsim E_p^2$ this causes the growth of the electron temperature to slow down, as follows directly from Equations (5.114) and (5.118): $T_e \sim E_0^2/L_T$ The electron-density perturbations increase even more slowly. Indeed, in the region below the maximum of the F layer we have $\gamma_1 \approx 0$ and $L_T \gtrsim L_N$ (see Tables 14 and 16). With increasing electron temperature, L_T increases, so that in a strong field we have $L_T \gg L_N$. In this case, as seen from Equation (5.120) we have $\Delta N \sim E_0^2/L_T^2$.

Let us consider, by way of example, the region below the maximum of the F layer. The perturbations of the concentration in this region, as will be shown below, are always small, $\Delta N \ll N_0$. In the thermal-conductivity Equations (5.27) or (5.58) we can therefore neglect the change of N, so that this equation takes the form

$$-\frac{d}{dx_{\parallel}}\left(k_{\mathrm{e}}\frac{dT_{\mathrm{e}}}{dx_{\parallel}}\right) = -\delta\nu_{\mathrm{e}}(T_{\mathrm{e}} - T) + \frac{2Q}{3N_0} + \frac{e^2 E_0^2(x_{\parallel})\nu_{\mathrm{e}}\varphi_{\mathrm{p}}}{3m\omega^2}. \quad (5.121)$$

Assume that the perturbing field $E_0^2(x_{\parallel})$ is concentrated in the vicinity of a certain point x_0 (more accurately, that the dimension of the region near x_0 occupied by the field is smaller than L_T). We can then put $E_0^2(x_{\parallel}) \simeq A\delta(x_{\parallel} - x_0)$, where $A = \int E_0^2(x_{\parallel})\, dx_{\parallel}$. Equation (5.121) can be easily integrated here in general form. Indeed, introducing the function $\psi = \int k_{\mathrm{e}}\, dT_{\mathrm{e}}$, we obtain the following solution of the homogeneous Equation (5.121):

$$\left(\frac{d\psi}{dx_{\parallel}}\right)^2 = \Phi(T_{\mathrm{e}}), \qquad \Phi(T_{\mathrm{e}}) = 2\int k_{\mathrm{e}}\left[\delta\nu_{\mathrm{e}}(T_{\mathrm{e}} - T) - \frac{2Q}{3N_0}\right] dT_{\mathrm{e}},$$

$$x_{\parallel} - x_0 = \pm\int_{T_{\mathrm{e}}}^{T_{\mathrm{eM}}} \frac{k_{\mathrm{e}}\, dT_{\mathrm{e}}}{\sqrt{\Phi(T_{\mathrm{e}})}}. \quad (5.121\mathrm{a})$$

Here T_{eM} is the maximum value of T_{e}. It is reached at the heating point x_0. Integrating Equation (5.121) with respect to $dx_{\parallel}$ in the vicinity x_0, we find that the derivative $dT_{\mathrm{e}}/dx_{\parallel}$ experiences at the point x_0 a discontinuity equal to

$$k_{\mathrm{e}}\left(\frac{dT_{\mathrm{e}}}{dx_{\parallel}}\right)_{x_0 - 0} - k_{\mathrm{e}}\left(\frac{dT_{\mathrm{e}}}{dx_{\parallel}}\right)_{x_0 + 0} = \delta_0 T_{\mathrm{e}0}\tilde{Q}\nu_{\mathrm{e}}, \qquad \tilde{Q} = \frac{A\varphi_{\mathrm{p}}}{E_{\mathrm{p}}^2}, \qquad \delta_0 = \frac{(\delta\nu_{\mathrm{e}})_{T_{\mathrm{e}0}}}{\nu_{\mathrm{e}0}}. \quad (5.121\mathrm{b})$$

Here $T_{\mathrm{e}0}$ is the unperturbed electron temperature, defined by the condition $(\delta\nu_{\mathrm{e}})_{T_{\mathrm{e}0}}(T_{\mathrm{e}} - T) = 2Q/3N_0$ [see Eq. (2.241)]. Recognizing, on the other hand, that Equation (5.121a) leads to $k_{\mathrm{e}}\; dT_{\mathrm{e}}/dx_{\parallel} = d\psi/dx = \pm\sqrt{\Phi(T_{\mathrm{e}})}$, we rewrite Equation (5.121b) in the form

$$2\sqrt{2}\left\{\int_{T_{\mathrm{e}0}}^{T_{\mathrm{eM}}} k_{\mathrm{e}}[\delta\nu_{\mathrm{e}}(T_{\mathrm{e}} - T) - 2Q/3N_0]\, dT_{\mathrm{e}}\right\}^{1/2} = \delta_0 T_{\mathrm{e}0}\tilde{Q}\nu_{\mathrm{e}}(T_{\mathrm{eM}}). \quad (5.121\mathrm{c})$$

This relation determines the maximum electron temperature T_{eM}. In the region of the F-layer maximum we have $\delta_{em}\nu_{em} \gg \delta_{ei}\nu_{ei}$ and $\nu_{ei} \gg \nu_{em}$ (see Tables 6 and 8). We then have Equation (5.109) $k_e = k_{e0}(T_e/T_{e0})^{5/2}$, $k_{e0} = 2.08T_{e0}/m\nu_{e0}$, $\nu_e \simeq \nu_{ei}$, and $\delta\nu_e \simeq \delta_{em}\nu_{em} \simeq$ const. The dependence of T_{em} on $\tilde{Q}/2L_{T0}$ for these conditions is shown in Figure 54. Here $L_{T0} = (k_{e0}/\delta_{em}\nu_{em})^{1/2}$ is the length L_T in the unperturbed plasma (at $T_e = T_{e0}$). At large values of T_{eM} it follows from Equation (5.121c) that

$$T_{eM} \simeq T_{e0}(3\tilde{Q}/4L_{T0})^{4/15}, \qquad \tilde{Q} = \frac{\varphi_p}{E_p^2}\int E_0^2(x_{\|})\,dx_{\|}.$$

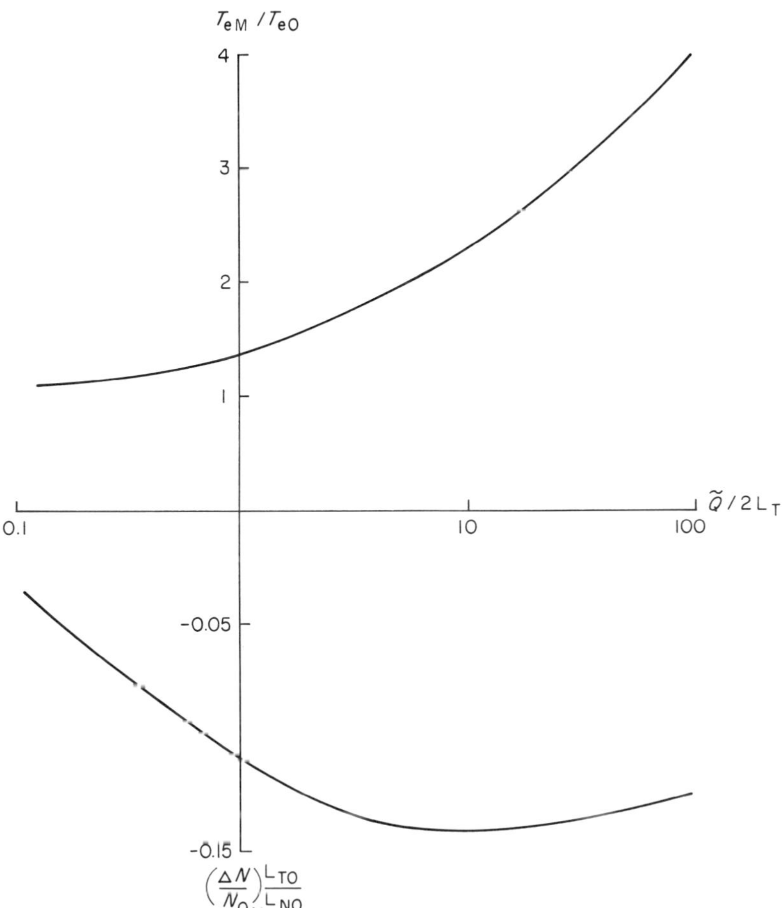

Fig. 54. Maximal temperature and electron-density perturbations as functions of the integrated wave power $\tilde{Q} = \varphi_p \int E_0^2(x_{\|})\,dx_{\|}/E_p^2$

It is seen that the electron temperature increases quite slowly with increasing perturbing field. This slowing down of the growth of T_e is due to the increase of the thermal-conductivity length $L_T(T_e)$, as was indeed noted above.

We now determine the values of the concentration perturbations. Assuming that $\Delta N \ll N_0$, we linearize Equation (5.36) with respect to ΔN. We have [Eqs. (5.43) and (5.49)]:

$$-\frac{d}{dx_{\|}}\left(L_N^2 \frac{d\,\Delta N}{dx_{\|}}\right) + \Delta N = \frac{d}{dx_{\|}}\left(k_{T_{e0}} L_N^2 \frac{N_0}{T_e}\frac{dT_e}{dx_{\|}}\right),$$

$$k_{T_{e0}} = T_e/(T_e + T), \qquad L_N^2 = \tau_N(T_e + T)/M\nu_{im}. \tag{5.121d}$$

We take into account the fact that the derivative $dT_e/dx_{\|}$ has a discontinuity [Eq. (5.121b)] at the point x_0. It follows from Equations (5.121d) that the derivative $d\,\Delta N/dx_{\|}$ is also discontinuous at this point:

$$\left(\frac{d\,\Delta N}{dx_{\|}}\right)_{x_0+0} - \left(\frac{d\,\Delta N}{dx_{\|}}\right)_{x_0-0} = \frac{N_0}{T_e + T}\left\{\left(\frac{dT_e}{dx_{\|}}\right)_{x_0-0} - \left(\frac{dT_e}{dx_{\|}}\right)_{x_0+0}\right\}$$

$$= \frac{N_0\,\delta_0 T_{e0}\tilde{Q}\nu_e}{k_e(T_{eM} + T)}. \tag{5.121e}$$

The solution of Equations (5.121d) and (5.121e) shows that the perturbation of the concentration in the vicinity of the point x_0 is negative. It has a maximum value at $x_{\|} = x_0$. In this case $(\Delta N/N_0)_M$ is determined from the expression

$$\left(\frac{\Delta N}{N_0}\right)_M \simeq -\frac{\tilde{Q}}{2L_{T0}}\frac{L_{N0}}{L_{T0}}\frac{T_{e0}^5}{T_{eM}^4(T_{eM} + T)^{1/2}(T_{e0} + T)^{1/2}}. \tag{5.121f}$$

The dependence of $(\Delta N/N_0)_M$ on $\tilde{Q}/2L_{T0}$ is shown in Figure 54. It is seen that the concentration perturbations are indeed always small (we recall that $L_{N0}/L_{T0} < 1$). With increasing amplitude of the perturbing field, they reach a maximum, and then even decrease slightly. The maximum value of $|\Delta N/N_0|$ under the conditions of the F-layer (at $z \sim 200$–300 km) is of the order of (1–10)%.

We emphasize that inasmuch as the plasma concentration in the F layer decreases in the plasma region modified by the wave [Eq. (5.119)], a sufficiently powerful radio wave pierces, so to speak, the F layer and forms a region of decreased concentration—a "hole" in the uniformly-layered distribution of the plasma density (Farley, 1963; Gurevich, 1965b). The strongest influence is exerted in this case by radio waves of frequency close to the critical frequency of the layer. To be sure, the concentration perturbations are usually small [Eq. (5.121f)]. They can, however, increase significantly when the maximum of the F layer rises to heights $z \sim 400$ km, i.e., the region in which the length L_N increases abruptly. In the actual calculations of these phenomena it is necessary to take into account the absorption of the waves and the defocusing of the beam in the reflection region, and in the case of sufficiently powerful and narrow beams it is also necessary to allow for their self-action (Utlaut, 1970; Meltz and Lelevier, 1970; Thomson, 1970; Meltz et al., 1974). An important role is played also by excitation of plasma instability (see Chap. 6).

5.3.2. Radio Wave Reflection Region

Enhancement of Field in Reflection Region. We consider the region of reflection of radio waves of frequency $\omega > \omega_H$ incident normally on the ionosphere (Ginzburg, 1960). The earth's magnetic field is directed at an angle α to the vertical z. The ordinary wave is reflected at a point z_0^+ defined by the condition $\varepsilon_+(z_0^+) = 0$, i.e.,

$$4\pi e^2 N(z_0^+)/m = \omega^2, \tag{5.122a}$$

and the extraordinary wave is reflected at a point z_0^-:

$$4\pi e^2 N(z_0^-)/m = \omega^2 - \omega\omega_H. \tag{5.122b}$$

The electric fields of the ordinary and extraordinary waves are polarized in the reflection region along $\boldsymbol{H}$ and perpendicular to the $\boldsymbol{H}$ plane, respectively. Accordingly, the polarization factor φ_p for the ordinary and extraordinary waves [Eq. (2.42)] is equal to

$$\varphi_\mathrm{p}^+ = 1, \qquad \varphi_\mathrm{p}^- = \frac{\omega^2}{(\omega - \omega_H)^2}\left[1 - \frac{\omega_H^2(3\omega^2 - \omega_H^2)\sin^2\alpha}{\omega^2(\omega + \omega_H)^2}\right]. \tag{5.123}$$

A standing wave is produced in the reflection region. The wave-field amplitude E_0 increases on approaching the reflection point $\varepsilon \to 0$. In the geometrical-optics approximation we have

$$E_0^2(z) = \frac{4E_0^2(0)\varphi_H}{\sqrt{\varepsilon(z)}}\sin^2\left(\frac{\omega\varphi_H}{c}\int_z^{z_0}\sqrt{\varepsilon}\,dz + \frac{\pi}{4}\right), \qquad \varepsilon_+ = 1 - \omega_0^2/\omega^2, \tag{5.124}$$

$$\varepsilon_- = 1 - \omega_0^2/\omega(\omega - \omega_H), \qquad \varphi_H^+ = (\sin\alpha)^{-1}, \qquad \varphi_H^- = [(\cos^2\alpha + 1)/2]^{-1/2}.$$

Here $E_0(0)$ is the incident-wave amplitude at the plasma-layer boundary (the absoprtion is disregarded), and φ_H is the magnetic polarization factor, which has different values for the ordinary and extraordinary waves. From Equation (5.124) it is seen that the factor φ_H^+ increases at high latitudes, where the direction of the magnetic field is close to vertical, $\alpha \to 0$. The reason is that the group velocity v_g of the ordinary wave is orthogonal to $\boldsymbol{H}$ in the reflection region. Therefore the component of v_g in the vertical z direction decreases in proportion to $\sin \alpha$, and it is this which causes the field amplitude to increase. We note that Equation (5.124) for φ_H^+ is valid only at $\alpha \gtrsim 10°$, i.e., up to latitudes $\theta \sim 70°$ (see Ginzburg, 1960, Sec. 28).

Geometrical optics is not valid in the immediate vicinity of the reflection point, where the field distribution is described by an Airy function. The maximum field amplitude is reached in the principal (first) maximum of the standing wave at a distance $z_0 - z_1$ from the reflection point:

$$z_0^+ - z_1^+ = 1.02(c^2 \sin^2 \alpha/\omega^2\mu)^{1/3}, \qquad \mu = \frac{1}{N_0}\left(\frac{dN}{dz}\right)_{z_0}$$
$$E_{\max}^+ = E(z_1^+) = 1.90\left(\frac{\omega}{c\mu}\right)^{1/6} (\sin \alpha)^{-2/3} E_0(0). \tag{5.125}$$

Here μ is the relative plasma concentration gradient at the reflection point z_0.

Plasma Perturbation in the Reflection Region. Owing to growth of the wave-field amplitude, the plasma perturbation becomes amplified in the reflection region. The resultant change in the electron density influences in turn the distribution of the wave field. In a sufficiently weak field, however, this effect of wave self-action can be neglected. The plasma perturbation is then described, as before, by Equations (5.114) and (5.115), where the field amplitude $E_0(z)$ is defined in accordance with Equation (5.124).

We assume that the dimensions L_T and L_N in the F region of the ionosphere are large enough (see Table 16), so that the condition

$$L_T \cos \alpha, L_N \cos \alpha \gg \lambda \tag{5.126}$$

is usually well satisfied ($\lambda = \lambda_0/\sqrt{\varepsilon}$ is the radio wavelength). In this case the ΔN and ΔT_e oscillations [Eq. (5.124)] in the standing wave are smoothed out by thermal conductivity and diffusion. The field $E_0^2(z)$ in the integrals of Equations (5.114) and (5.115) can then be averaged over a dimension on the order of the wavelength, i.e., it can be assumed equal to

$$\overline{E_0^2(z)} = \begin{cases} 2\,\dfrac{E_0^2(0)\varphi_H}{\sqrt{\varepsilon}}, & \text{if} \quad z \leq z_0 \\ 0, & \text{if} \quad z > z_0 \end{cases} \tag{5.127}$$

where z_0 is the reflection point. We assume further that the layer in the reflection region is linear:

$$\varepsilon_0(z) = \mu(z_0 - z), \qquad \mu = \frac{1}{N_0}\left(\frac{dN}{dz}\right)_{z_0}. \tag{5.128}$$

Substituting Equation (5.127) in Equations (5.114) and (5.115) and integrating, we obtain

$$\begin{aligned} \Delta T_e &= q_0 T_{e0} \Psi(t), \\ \Delta N &= -k_T q_0 N_0 \frac{L_N^2}{L_N^2 - L_T^2}\left[\Psi(t) - \sqrt{\frac{L_T}{L_N}}\,\Psi\left(\frac{L_T}{L_N}t\right)\right] \end{aligned} \tag{5.129}$$

where

$$t = \frac{z_0/\cos\alpha - x_\parallel}{L_T}, \qquad q_0 = \frac{\sqrt{\pi}\varphi\varphi_H}{\sqrt{\mu L_T \cos\alpha}}\frac{E_0^2(0)}{E_p^2}, \qquad x_\parallel = \frac{z}{\cos\alpha} \tag{5.130}$$

$$\begin{aligned} \Psi(t) &= \begin{cases} [1 - \Phi(\sqrt{t})]\exp(t) + f(\sqrt{t})\exp(-t), & \text{if } t > 0 \\ \exp(t) & \text{if } t < 0 \end{cases} \\ \Phi(\sqrt{t}) &= \frac{2}{\sqrt{\pi}}\int_0^{\sqrt{t}} \exp(-y^2)\,dy, \qquad f(\sqrt{t}) = \frac{2}{\sqrt{\pi}}\int_0^{\sqrt{t}} \exp(y^2)\,dy. \end{aligned} \tag{5.131}$$

We have neglected here for simplicity the term proportional to γ_1, since $\gamma_1 - 0$ in the F-layer region (see Table 14). The perturbations of the plasma concentration and of the temperature are illustrated in Figure 55. The perturbations are maximal near the wave reflection point z_0, and the plasma is crowded out of this region, $\Delta N < 0$. Far from the reflection point, however, the concentration increases, $\Delta N > 0$. If $L_T > L_N$, the perturbations ΔN have the same structure but their absolute values are smaller.

Shift of Reflection Point. Near the wave-reflection point, the plasma-concentration perturbations are maximal. As is clear from Equation

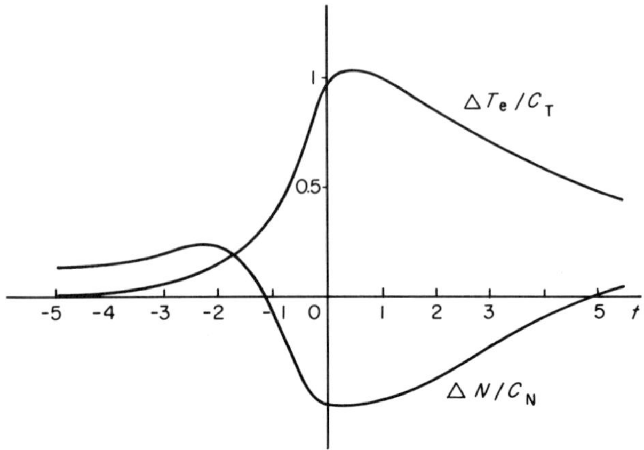

Fig. 55. Plasma perturbation in the wave-reflection region. $L_N = 4L_T$, $C_T = T_{e0}q_0$, $C_N = \frac{16}{15}k_T N_0 q_0$, $t = (z_0 - z)/L_T \cos\alpha$

(5.129), we have here $\Delta N < 0$, i.e., the plasma is crowded out by the wave. The wave reflection point in the F region of the ionosphere is then shifted upward, as if the wave were to pierce its way into the interior of the plasma (Gurevich, 1965b).

Consider this effect for the case of a weak field, $E_0^2 \ll E_p^2$. Equation (5.112b) for the electron temperature, with allowance for Equation (5.127), takes the form

$$L_T^2 \cos^2\alpha \frac{d^2 \Delta T_e}{dz^2} - \Delta T_e = -\frac{2\varphi\varphi_H E_0^2(0) T_{e0}}{E_p^2 \sqrt{\varepsilon}}, \qquad z < z_0 \quad (5.132)$$

When the concentration perturbation is taken into account, the plasma in the reflection region has a dielectric constant

$$\varepsilon = \mu(z_0 - z) - \frac{\Delta N}{N_0}, \qquad \frac{\Delta N}{N_0} = -k_T \frac{\Delta T_e}{T_{e0}}. \quad (5.133)$$

Here $N_0 = N(z_0)$ is the plasma concentration at the reflection point [see Eq. (5.122)]; the layer is assumed to be linear [Eq. (5.128)]. In addition, it is assumed for simplicity that $L_N \gg L_T$, so that the concentration perturbations ΔN are connected with the electron-temperature perturbations ΔT_e by the simple relation Equation (5.119). Equations (5.132) and (5.133) describe the ΔN and ΔT_e perturbations in the wave-reflection region with allowance for the self-action of the wave.

We change over in Equations (5.132) and (5.133) to the dimensionless variables:

$$x = \beta\varepsilon = \beta\left[\mu(z_0 - z) + k_T \frac{\Delta T_e}{T_{e0}}\right], \tag{5.134}$$

$$t = \frac{z_1 - z}{L_T \cos\alpha}, \qquad \beta = (k_T^{1/2}\sqrt{\varphi\varphi_H E_0/E_p})^{-4/3}. \tag{5.135}$$

Here z_0 is the point of reflection of the wave in the unperturbed medium in Equation (5.122), z_1 is the point of reflection of the wave with allowance for the perturbation in Equation (5.133), i.e., the point at which $\varepsilon(z_1) = 0$. Equation (5.132) is then rewritten in the form

$$\frac{d^2x}{dt^2} = x + x_0 - \alpha_1 t - \frac{2}{\sqrt{x}}. \tag{5.136}$$

where α_1 is the characteristic dimensionless parameter

$$\alpha_1 = \beta\mu L_T \cos\alpha. \tag{5.137}$$

The value of α_1 depends on the relation between the plasma concentration gradient, the characteristic length L_T, and the wave field amplitude, while x_0 determines the shift, due to nonlinearity, of the wave reflection point:

$$x_0 = (z_1 - z_0)\mu\beta = \alpha_1 t_1. \tag{5.138}$$

The boundary conditions for Equation (5.136) are

$$x_{t\to 0} = 0, \qquad \left.\frac{dx}{dt}\right|_{t\to\infty} = \alpha_1, \qquad \left.\frac{dx}{dt}\right|_{t\to 0} = \alpha_1 + x_0. \tag{5.139}$$

The first of these conditions follows from the definition of the quantity $x \sim \varepsilon(z)$ (the dielectric constant vanishes at $z = z_1$, i.e., at $t - 0$). The two other conditions follow from the requirement that the solutions of the heat-conduction Equation (5.132) remain bounded as $z \to \pm\infty$. Equation (5.136) with the three boundary conditions [Eq. (5.139)] determines not only the course of the function $x(t)$, but also the constant x_0, i.e., the shift of the wave reflection point.

We consider first the case when the electron density gradient in the unperturbed plasma is large, $\alpha_1 \gg 1$. In this case it is natural to seek the solution of Equation (5.136) in the form $x = \alpha_1 t + x_1$. From Equation

(5.136) we then obtain

$$\frac{d^2x_1}{dt^2} = x_1 + x_0 - \frac{2}{\sqrt{\alpha_1 t + x_1}}. \tag{5.140}$$

We solve this equation by successive approximations. In the first approximation, recognizing that $\alpha_1 \gg 1$, we neglect the value of x_1 under the square root in comparison with $\alpha_1 t$. Taking the boundary conditions in Equation (5.139) into account, we obtain

$$x_1 = \sqrt{\frac{\pi}{\alpha_1}}\{-1 + \exp(t)[1 - \Phi(\sqrt{t})] + [\exp(-t)]f(\sqrt{t})\}. \tag{5.141}$$

This expression agrees with Equations (5.129) and (5.131). In other words, at $\alpha_1 \gg 1$ the self-action of the wave has little effect on the plasma perturbations. It follows then from Equation (5.139) that

$$x_0 = \left.\frac{dx_1}{dt}\right|_{t\to 0} = \sqrt{\pi/\alpha_1}. \tag{5.142}$$

Thus, the shift of the wave reflection point in case $\alpha_1 \gg 1$ under consideration is determined by the expression

$$z_1 - z_0 = k_T\varphi\varphi_H \frac{E_0^2}{E_p^2}\left[\frac{\pi}{\mu^3 L_T \cos\alpha}\right]^{1/2}. \tag{5.143}$$

Figure 56 shows the dielectric constant of the plasma as a function of the distance z with allowance for the nonlinearity, $x(z) = \beta\varepsilon(z)$. The

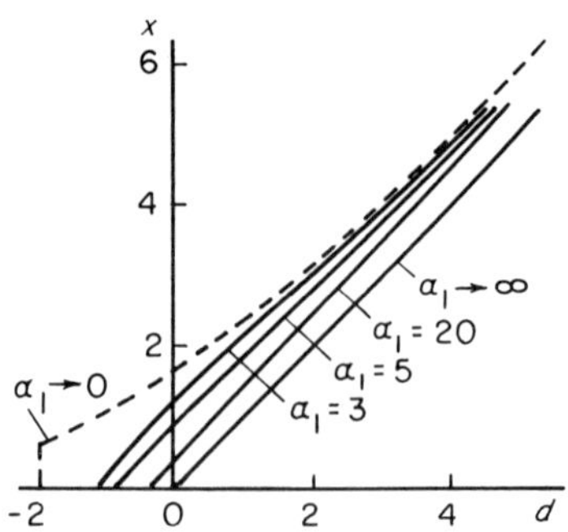

Fig. 56. Perturbation of dielectric constant in the reflection region $x = \beta\varepsilon(z)$, $d = \beta\varepsilon_0(z) = \beta\mu(z_0 - z)$; $d = 0$ is the wave-reflection point in the unperturbed plasma

wave reflection point z_1 is determined by the condition $x(z_1) = 0$. It is seen from the figure that the shift of the reflection point increases with decreasing α_1, i.e., with decreasing concentration gradient $(dN/dz)_0$.

In the opposite case, when the concentration gradient is small,

$$\alpha_1 \ll 1 \tag{5.144}$$

we can neglect the term $\alpha_1 t$ in Equation (5.136), which is then easily integrated. We have:

$$\frac{dx}{dt} = [(x + x_0)^2 - 8\sqrt{x} + C]^{1/2}. \tag{5.145}$$

The first and third boundary conditions [Eq. (5.139)] lead at $\alpha_1 = 0$ to the relation $dx/dt = x_0$ at $x = 0$. It follows therefore that $C = 0$.

The second condition of Equation (5.139) shows that at $\alpha_1 = 0$ and as $t \to \infty$ the derivative dx/dt tends to zero, i.e., x tends to a constant value x_∞. Consequently, as $x \to x_\infty$, the right-hand side of Equation (5.145) should vanish in proportion to $x - x_\infty$. This leads to two algebraic relations for the constants x_0 and x_∞:

$$(x_0 + x_\infty)^2 - 8x_\infty^{1/2} = 0, \tag{5.146}$$

$$x_\infty^{3/2} + x_0 x_\infty^{1/2} - 2 = 0. \tag{5.147}$$

Determining x_0 from Equation (5.147) and substituting it in Equation (5.146), we obtain x_∞, and then also x_0:

$$x_\infty = 2^{-2/3}, \qquad x_0 = 3 \cdot 2^{-2/3}. \tag{5.148}$$

The shift of the wave-reflection point is maximal as $\alpha_1 \to 0$ (see the dashed curve on Fig. 56) and is equal to

$$z_1 - z_0 = \frac{3}{2^{2/3}} \left(k_T^{1/2} \sqrt{\varphi \varphi_H} \frac{E_0}{E_p} \right)^{4/3} \frac{N(z_0)}{(dN/dz)_0}. \tag{5.149}$$

As $(dN/dz)_0 \to 0$, the wave reflection point is shifted by a considerable distance. This case is realized, in particular, if the frequency of the wave propagating in the plasma layer is close to the critical frequency ω_{c0} of the layer, where

$$\omega_{c0} = \sqrt{4\pi e^2 N_{\max}/m}; \tag{5.150}$$

ω_{c0} is the Langmuir frequency for the maximum electron density N_{max} in the layer. In the linear approximation, the condition for the passage of the wave through the plasma layer takes naturally the form $\omega \geq \omega_{c0}$. When the nonlinearity is taken into account, this condition becomes

$$\omega \geq \omega_c = \omega_{c0}\left[1 - \frac{3}{2^{5/3}\beta}\right]. \tag{5.151}$$

Thus, the frequency of the waves that can pass through a plasma layer is decreased by the nonlinearity by an amount $\Delta\omega$, where

$$\Delta\omega = \frac{3\omega_{c0}}{2^{5/3}}\left(k_T^{1/2}\sqrt{\varphi\varphi_H}\,\frac{E_0}{E_p}\right)^{4/3}. \tag{5.152}$$

With increasing field amplitude, the frequency shift increases in proportion to $E_0^{4/3}$. The foregoing, of course, pertains only to the case $\gamma_1 \to 0$, when the plasma concentration decreases with increasing electron temperature (Sect. 5.3.1).

From Equations (5.149) and (5.134) follow expressions for the plasma perturbation in the weak-wave reflection region, $E_0^2 \ll E_p^2$, as $\alpha_1 \to 0$:

$$\Delta N = (z_1 - z_0)\left(\frac{dN}{dz}\right)_0 = \frac{3N_0}{2^{2/3}}\left(k_T^{1/2}\sqrt{\varphi\varphi_H}\,\frac{E_0}{E_p}\right)^{4/3},$$

$$\Delta T_e = -\frac{\Delta N}{k_T N_0}\,T_{e0}. \tag{5.153}$$

The perturbation increases noticeably in this case in comparison with the perturbation at $\alpha_1 \gg 1$, when ΔN and ΔT_e are proportional to $(E_0/E_p)^2$ [see Eqs. (5.129)–(5.131)].

We emphasize that we have considered in this section only the case $L_N \gg L_T$. A more realistic condition for the ionosphere in the region below the maximum of the F layer, as seen from Table 16, is $L_T \gtrsim L_N$. The concentration perturbations are much weaker in this case than the temperature perturbations [see Eq. (5.129)]; the shift of the reflection point [Eqs. (5.143), and (5.149)] is then smaller.

We note, on the other hand, that we have taken into account here only the usual ohmic heating of the plasma. In the region of the reflection of the ordinary wave, an important role can be played also by collisionless dissipation mechanisms, which increase appreciably the absorption of the wave energy (see Sects. 6.2 and 6.3). The structure of the perturbed region in the vicinity of the point of reflection of the wave in the ionosphere

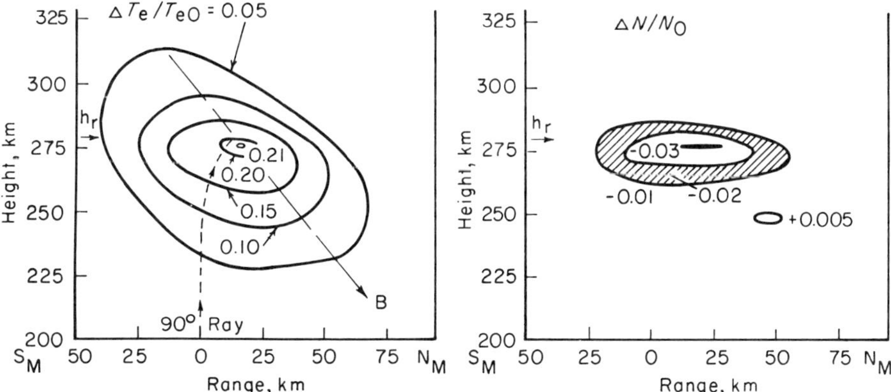

Fig. 57. Structure of the temperature and plasma-concentration perturbations in the magnetic-meridian plane (Meltz et al., 1974). The perturbing wave is ordinary, $W_0 \approx 2 \cdot 10^4$ kW ($P = 100$ kW, $G = 200$), $\omega = 3.75 \cdot 10^7$, $\omega_c = 3.92 \cdot 10^7$

(numerical calculation with the anomalous heating taken into account; see Meltz, Holway, and Tomijanovich, 1974) is shown in Figure 57. It agrees fully with Equation (5.129) at $L_T \sim 30$ km (see Fig. 54). We see that the concentration perturbations are small, since $L_N < L_T$ ($L_N \sim 10$–15 km). Nonlinear passage through the ionosphere F layer [Eqs. (5.151) and (5.152)] is therefore possible only for waves with frequency close enough to ω_{c0}.

5.3.3. Growth and Relaxation of the Perturbations

We consider now nonstationary processes. The nonstationary solution of the heat conduction Equation (5.111b) at a given source $Q(x, t)$ is

$$\Delta T_e(x_{\parallel}, t) = T_{e0} \int_{-\infty}^{\infty} \frac{dx'}{L_T} \int_{-\infty}^{t} \frac{dt'\, Q(x', t')}{\sqrt{4\pi\tau_T(t - t')}} \exp\left[-\frac{t - t'}{\tau_T} - \frac{(x_{\parallel} - x')^2}{4k_e(t - t')}\right]$$

$$\tau_T = (\delta\nu_e\varphi_T)^{-1} = \left[(\delta_{ei}\nu_{ei} + \delta_{em}\nu_{em})\varphi_T\right]^{-1}. \tag{5.154}$$

Analogously, by solving the diffusion Equation (5.111a), we can obtain the following expression for the concentration perturbations (Vas'kov and Gurevich, 1975a):

$$\Delta N(x_{\parallel}, t) = N_0 \int_{-\infty}^{t} \frac{dt''}{\tau_T} \int_{-\infty}^{\infty} \frac{dx''}{L_T} G(x_{\parallel} - x'', t - t'')Q(x'', t''), \tag{5.155}$$

where the Green's function is

$$G(x, t - t'') = \int_{t''}^{t} \frac{dt'}{\tau_N} \frac{L_T}{\sqrt{\pi b}} \left[\gamma_1 + \frac{4L_N^2 k_T}{b} \left(\frac{x^2}{b} - \frac{1}{2} \right) \right] \times \exp\left[-\frac{t - t'}{\tau_N} - \frac{t' - t''}{\tau_T} - \frac{x^2}{b} \right] \tag{5.156}$$

$$b = 4D_a(t - t') + 4k_e(t' - t''), \qquad k_e = \kappa_e/N.$$

Equations (5.154)–(5.156) describe in the general case nonstationary processes in a weak field. If the effective dimensions of the region modified by the field are large [Eq. (5.117)]:

$$R_{\parallel} \gg L_T, L_N \tag{5.157}$$

then $Q(x', t')$ can be taken outside the sign of integration with respect to dx' in Equation (5.154) at the point $x' = x$. Then

$$\Delta T_e(x_{\parallel}, t) = T_{e0} \int_0^t \frac{dt'}{\tau_T} Q(x_{\parallel}, t') \exp\left(\frac{-t + t'}{\tau_T} \right).$$

In the buildup stage, assuming that the modifying field $E_0^2(x)$ is turned on at the instant $t = 0$, we obtain from this

$$\Delta T_e(x_{\parallel}, t) = \Delta T_e(x_{\parallel}) (1 - \exp(-t/\tau_T)). \tag{5.158}$$

Here $\Delta T_e(x_{\parallel})$ is the stationary perturbation of the electron temperature [Eqs. (5.114) and (2.246)]. For the relaxation of the perturbation we obtain analogously

$$\Delta T_e(x_{\parallel}, t) = \Delta T_e(x_{\parallel}) \exp(-t/\tau_T). \tag{5.159}$$

We see therefore that in Equation (5.157) the electron temperature builds up and relaxes, as in a uniform plasma, within a characteristic time τ_T (see Table 13). Similar formulas describe the buildup and the relaxation of the concentration perturbations, except that τ_T is replaced by τ_N (it is assumed that $\tau_N \gg \tau_T$, see Table 13). Thus, in Equations (5.157) the nonstationary processes have the same character as in a uniform plasma; the transport processes are inessential in this case. To the contrary, under conditions that are the converse of Equation (5.157), when the perturbation region is small, the principal role in the initial period

$$t < \tau_N, \tau_T \frac{k_e}{D_a} \tag{5.160}$$

is played by the thermal conductivity and by diffusion. Green's function [Eq. (5.156)] takes in this case the simple form

$$G(x, t) = -\frac{k_T L_T}{2\sqrt{\pi t}} \frac{D_a}{k_e - D_a} \left[\frac{1}{\sqrt{D_a}} \exp\left(-\frac{x^2}{4D_a t} \right) - \frac{1}{\sqrt{k_e}} \exp\left(-\frac{x^2}{4k_e t} \right) \right]. \tag{5.161}$$

From Equations (5.155) and (5.161), recognizing that in the case of a small perturbation region we have

$$Q(x_\parallel) = \tilde{Q}\delta(x_\parallel - x_0), \tag{5.162}$$

where $\tilde{Q}$ is defined by Equation (5.118), we get

$$\Delta N = -\frac{N_0 k_T \tilde{Q}}{\sqrt{\pi} \tau_T} \frac{\sqrt{D_a t}}{k_e - D_a} \left[\psi\left(\frac{|x_\parallel - x_0|}{2\sqrt{D_a t}} \right) - \sqrt{\frac{D_a}{k_e}} \psi\left(\frac{|x_\parallel - x_0|}{2\sqrt{k_e t}} \right) \right] \tag{5.163}$$

$$\psi(z) = \exp(-z^2) - \sqrt{\pi} z[1 - \Phi(z)].$$

It is seen therefore that the density perturbation during the initial period builds up with time like $\sqrt{t}$. Near the field interaction region ($x_\parallel \sim x_0$) this perturbation is negative, but at a distance

$$|x_\parallel - x_0| > \sqrt{2D_a t \ln(k_e/D_a)}, \qquad k_e \gg D_a \tag{5.164}$$

it becomes positive. This is understandable; during the initial period [Eq. (5.160)], before recombination becomes essential, the total number of particles in the plasma is conserved and the crowding out of the plasma from the field interaction region is accompanied by an increased plasma concentration outside this region. A perturbation wave propagates, as it were, from the point x_0. We emphasize that transport processes play a decisive role only in the initial stage [Eq. (5.160)]. The final establishment of the stationary state is always determined by the characteristic times τ_T and τ_N.

Figure 58 shows typical density profiles produced when the perturbation develops in the vicinity of the wave reflection point. The perturbation is assumed to be local [Eq. (5.162)] and concentrated at the reflection point. It is seen that the shift of the reflection point increases with time in proportion to $\sqrt{t}$, in accordance with Equation (5.163). The unperturbed linear layer is shown in the figure by the thin straight line. It is seen that the crowding out of the plasma from the region near the reflection point is accompanied by an increase of the concentration in other regions. The character of the perturbation depends significantly here on the value of the parameter

$$K = \frac{k_T \tilde{Q}}{4\sqrt{\pi}\mu L_T^2 \cos\alpha(1 - D_a/k_e)}, \tag{5.165}$$

which is proportional to the integrated perturbation intensity [Eq. (5.118)]. In the case $K > 1$ a concentration maximum N_{max} is produced below the reflection point, and this maximum can exceed the value of N at the reflection point (dashed curve in Fig. 58). As a result, the reflection point should drop sharply, and this can bring about relaxation oscillations in the plasma during the buildup time. The relaxation oscillations produced under conditions when the concentration builds up in the wave reflection region were considered by Gurevich (1972a).

The modification of the F region of the ionosphere by high-power radio waves has been under intensive study in recent years. Figure 59 shows the modification of the electron temperature in the ionosphere, as reported by Gordon and Carlson (1974). It is seen that the maximum modification takes place in the vicinity of the wave reflection point $z_0 \sim$ 300–320 km. An estimate by Equations (5.129) and (5.153) for the conditions of these experiments ($W_0 = 2 \cdot 10^4$ kW, $T_{e0} = 900°$, $\omega = 3.2 \cdot 10^7$)

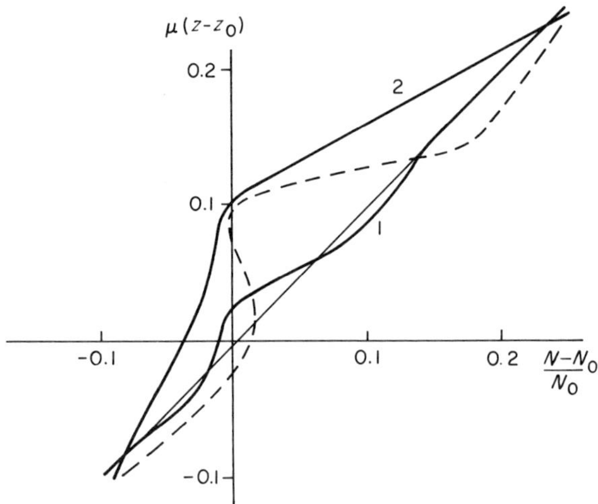

Fig. 58. Concentration distribution in the wave-reflection region. The wave is incident upward. The reflection point is $N = N_0 = m\omega^2/4\pi e^2$. Curves *1* and *2*: weak perturbation (parameter $K = 0.33$) at different instants of time $t_2 = 15t_1$. *Dashed curve*: strong perturbation, $K = 2.13$ for the instant $t = t_1$. *Thin line*: unperturbed distribution (linear layer)

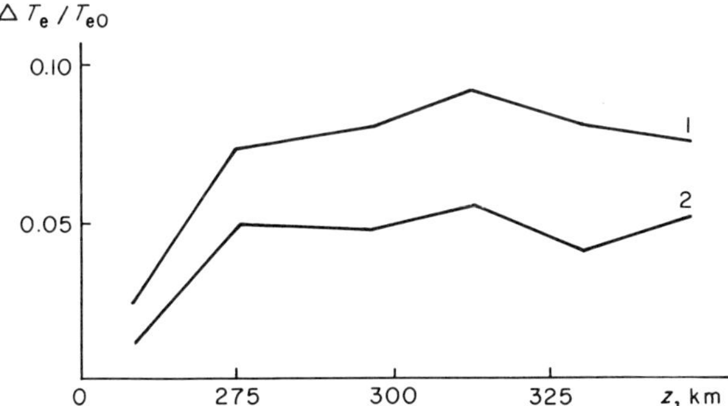

Fig. 59. Perturbation of the electron temperature in the ionosphere (Gordon and Carlson, 1974). $\omega = 3.2 \cdot 10^7$, $\omega_c = (3.7 - 4.0) \cdot 10^7$. Reflection point $z_0 \approx 300 - 320$ km, maximum of layer $z_m \approx 350 - 360$ km. Curve 1 – power $W_0 \approx 2 \cdot 10^4$ kW; curve *2* – power $0.5 W_0$

yields a maximum value $\Delta T_e \sim (0.1\text{–}0.2) T_{e0}$, which agrees with the experimental data. The variation of the modification in height is also in qualitative agreement with Equations (5.129) and (5.141) (length $L_T \sim 30$ km). A nonlinear dependence of the perturbation on the power W_0 is noted in the reflection region and agrees qualitatively with Equation (5.153): $\Delta T_e \sim W_0^{2/3}$. We note that according to (Biondi et al., 1970), when more power is applied to the plasma ($W_0 \approx 5 \cdot 10^4$ kW), the change in the electron temperature reached $0.3 T_{e0}$, which also indicates a slow growth of ΔT_e with increasing W_0 (see Sect. 5.3.1).

Experiment has also revealed a lowering of the electron density in the wave reflection region ($\Delta N \lesssim -0.1\ N_0$) (Utlaut and Violette, 1972, 1974; Gordon and Carlson, 1974), a shift of the reflection point (Utlaut and Violette, 1972), and a decrease of the density in the region of the F-layer maximum (piercing a "hole" through the ionosphere) (Fialer, 1974).

The buildup and the relaxation of the electron temperature agrees in the main with the theory (see Fig. 60). The time during which the electron density changes is of the order of several minutes, i.e., of the order of τ_N (see Table 13). Following a prolonged action on the ionosphere in the hours past midnight, the density perturbations can take as much as several hours to relax (Utlaut and Cohen, 1971). This is understandable, because at night the lengths L_N and L_T are large and the perturbations propagate in the region above the F-layer maximum, where the relaxation time τ_N is appreciable (see Table 13).

We stress that the phenomena considered here are masked and are modified to a considerable degree as a result of the intense development of instability in the perturbed region of the F layer, and this instability

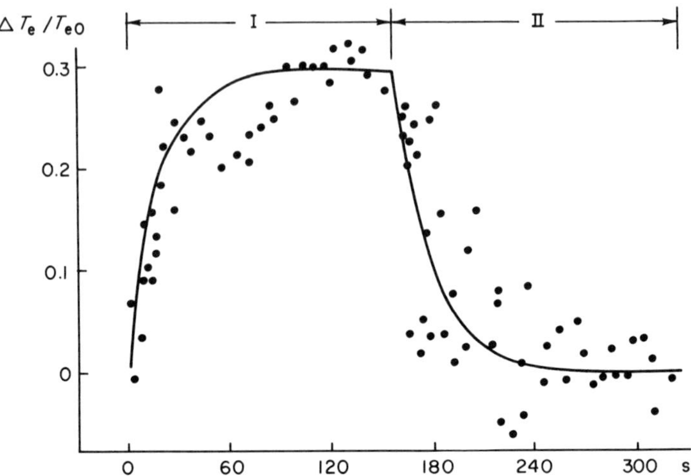

Fig. 60. Establishment and relaxation of electron temperature. I: Transmitter on; II: off. *Points*: results of experiment (Gordon and Carlson, 1974); *curve*: calculation by Equations (5.158) and (5.159). The time is $\tau_T \approx 20$ s under the experimental conditions ($N_0 = 3.2 \cdot 10^5$; $T_{e0} = 880°$, $z_0 \approx 300$ km)

leads to a complicated turbulent structure of the modified zone (see Chap. 6).

5.4. Focusing and Defocusing of Radio Wave Beams

In the preceding chapters we have considered plane radio waves. Unusual nonlinear effects are produced also when radio beams propagate in a plasma. These effects are of considerable interest because it is strongly directional beams which can provide radio waves of maximum intensity.

In the linear approximation, a beam can always be represented as an aggregate of plane waves propagating in different directions. Nonlinearity leads to violation of the superposition principle. This results in new singularities in the propagation of the radio-wave beams. Indeed, owing to the nonlinearities, the refractive index of the wave is modified in the beam propagation region. This leads to a bending of the ray trajectories. They can converge, in which case the entire beam is contracted or focused; this is called self-focusing of the beam. The inverse process, nonlinear divergence of the beam or defocusing, is also possible. Both can be realized in the ionosphere. It is important that in narrow beams even slight distortions of the ray trajectories lead to strong changes in the field distribution, so that nonlinear refraction plays a particularly important role in their case.

5.4.1. Nonlinear Geometrical Optics

The propagation of a radio wave of frequency ω in a plasma, with self-action taken into account, is described by the nonlinear wave Equation (3.8)

$$\Delta \boldsymbol{E} - \text{graddiv}\, \boldsymbol{E} + \frac{\omega^2}{c^2}\, \hat{\varepsilon}'(\boldsymbol{E})\boldsymbol{E} = 0. \tag{5.166}$$

We assume for simplicity that the plasma is isotropic.

In the ionosphere, this is valid at $\omega^2 \gg \omega_H^2$. In this case the complex dielectric tensor $\hat{\varepsilon}'$ is a scalar, and can be represented in the form

$$\varepsilon' = \varepsilon_0(s) + \varepsilon_{\text{n}}(\boldsymbol{r}, \boldsymbol{E}) + i\varepsilon_1$$

Here s is the wave propagation direction, ε_0 and $\varepsilon_1 = 4\pi\sigma_0/\omega$ are the real and imaginary parts of the dielectric constant in the inhomogeneous medium, while ε_{n} is the nonlinear increment to ε_0 and its actual form depends on the physical mechanism of the nonlinearity.

Assume now that ε is not very greatly altered over the wave length by the inhomogeneity of the plasma and by the nonlinear effect. In this case Equation (5.166) can be simplified by using Van der Pol's method. The idea of the method is to separate the rapidly and slowly varying parts of the field of the wave propagating in the s direction; more accurately, this field is represented in the form

$$\boldsymbol{E} = \frac{1}{2}\left[\boldsymbol{e}E(\boldsymbol{r}) \exp\left(i \int_0^s k\, ds\right) + \text{c.c.}\right] \tag{5.167}$$

Here $k = \dfrac{\omega}{c}\sqrt{\varepsilon_0(s)}$ is the modulus of the wave vector as determined in the linear approximation and e is the unit polarization vector. When the field is represented in the form of Equation (5.167), it is implied that the field amplitude $E(\boldsymbol{r})$ varies much more slowly than the phase

$$\lambda|\partial E/\partial s| \ll E.$$

The amplitude $E(\boldsymbol{r})$ is altered because of the absorption of the wave and because of the slight difference between the wave field and the planar field produced as a result of nonlinearity, diffraction, and inhomogeneity of the medium. We assume first for simplicity that the polarization is constant, $\boldsymbol{e} \neq \boldsymbol{e}(\boldsymbol{r})$.

Substituting Equation (5.167) in Equation (5.166) we obtain for $E(\mathbf{r})$ the scalar equation

$$\frac{\partial^2 E}{\partial s^2} + 2ik\frac{\partial E}{\partial s} + iE\frac{dk}{ds} + \Delta_\perp E + \frac{\omega^2}{c^2}(\varepsilon_n + i\varepsilon_1)E = 0, \tag{5.168}$$

where $\Delta_\perp$ is the Laplace operator in the ρ plane perpendicular to the beam propagation direction s.

We consider here the case of weak nonlinearity, $\varepsilon_n \ll \varepsilon_0$, which is essential only for narrow wave beams. We shall therefore deal only with narrow beams, i.e., beams with effective width a much smaller than the characteristic longitudinal beam dimension along which the amplitude is significantly altered by the nonlinearity or by the inhomogeneity of the plasma [Eq. (3.17)]. In this case $\partial^2 E/\partial s^2$ in Equation (5.168) is much smaller than $\Delta_\perp E$, so that it can be neglected. In place of Equation (5.168) we then obtain

$$\Delta_\perp E + 2ik\frac{\partial E}{\partial s} + iE\frac{dk}{ds} + \frac{\omega^2}{c^2}(\varepsilon_n + i\varepsilon_1)E = 0. \tag{5.169}$$

This equation is usually called a nonlinear parabolic equation. We seek its solution in the form

$$E = E_0 \exp(ikS). \tag{5.170}$$

The function $S(s, \boldsymbol{\rho})$ defines the phase difference between two points in a plane perpendicular to the beam propagation direction s; one of the points lies on the beam axis, while the other is located at a distance $\boldsymbol{\rho}$ from the axis. The quantity kS is the increment that must be added to the eikonal to account for the deformation of the wave front in a nonlinear and inhomogeneous medium. Substituting Equations (5.170) in Equations (5.169) and separating the real and imaginery parts, we obtain

$$\frac{\partial S}{\partial s} + \frac{1}{2}(\nabla_\perp S)^2 = -\frac{S}{k}\frac{dk}{ds} + \frac{\varepsilon_n}{2\varepsilon_0} + \frac{1}{2k^2 E_0}\Delta_\perp E_0, \tag{5.171a}$$

$$\frac{\partial E_0^2}{\partial s} + \frac{E_0^2}{k}\frac{dk}{ds} + (\nabla_\perp E_0^2)(\nabla_\perp S) + E_0^2\,\Delta_\perp S + \frac{\omega^2}{c^2}\frac{\varepsilon_1}{\varepsilon_0}E_0^2 = 0. \tag{5.171b}$$

Equation (5.171a) is the eikonal equation in a weakly nonlinear and weakly inhomogeneous medium with account taken of the diffraction spreading of the beam. It describes the deformation of the wave front and the change of the phase velocity of the wave in the beam. Applying the operation $\nabla_\perp$ to Equation (5.171a), we rewrite the latter in the form

$$\frac{\partial \boldsymbol{u}}{\partial s} + (\boldsymbol{u}\nabla_\perp)\boldsymbol{u} = -\frac{\boldsymbol{u}}{k}\frac{dk}{ds} + \frac{1}{2\varepsilon_0}\nabla_\perp\varepsilon_n + \frac{1}{2}\nabla_\perp\left(\frac{1}{k^2 E_0}\Delta_\perp E_0\right). \quad (5.172)$$

Here $\boldsymbol{u} = \nabla_\perp S$ is a vector lying in a plane perpendicular to the beam propagation direction. It gives the projection of the ray propagation direction on this plane. If all the vectors $\boldsymbol{u}$ are directed towards the beam axis, then the rays converge, i.e., the beam becomes focused. The beam diverges, however, if the vectors $\boldsymbol{u}$ are directed away from the axis. The first term in the right-hand side of Equation (5.172) describes the bending of the ray trajectories—the refraction due to the inhomogeneity of the medium. The second term describes the refraction due to the nonlinearities. The third takes approximate account of the diffraction effects.

In the limiting case of the plane wave ($\Delta_\perp E_0 = 0$) and in the absence of nonlinearity ($\varepsilon_n = 0$) or inhomogeneity ($dk/ds = 0$), the solution of Equation (5.172) nautrally takes the form $\boldsymbol{u} = \text{const} = \boldsymbol{u}_0$. In an inhomogeneous medium ($dk/ds \neq 0$), Equation (5.172) leads to the plane-wave refraction law which is well known from linear geometrical optics. In the absence of nonlinearity ($\varepsilon_n = 0$) and of inhomogeneity ($dk/ds = 0$) we obtain from Equation (5.172) the parabolic equation used in the approximate theory of diffraction (Leontovich, 1944).

Equation (5.171b) describes the change of the field amplitude in the beam. Its physical meaning becomes clear if it is rewritten in the form of a balance equation for the energy flux. Indeed, we recall that the group velocity $\boldsymbol{v}_g$ of the waves and the absorption coefficient are given in the linear theory by the expressions:

$$\boldsymbol{v}_g = \frac{\partial\omega}{\partial\boldsymbol{k}} = c\sqrt{\varepsilon_0}\frac{\boldsymbol{k}}{k} = \frac{c^2}{\omega}\boldsymbol{k}, \qquad \kappa = \frac{1}{2}\frac{\varepsilon_1}{\sqrt{\varepsilon_0}}.$$

The energy balance equation of the wave in the linear theory takes the form

$$\operatorname{div}(\boldsymbol{v}_g E_0^2) + 2\frac{\omega}{c}\kappa \boldsymbol{v}_g E_0^2 = 0. \quad (5.173)$$

Substituting here the expressions for $\boldsymbol{v}_g$ and κ we can easily verify that the linear energy-balance Equation (5.173) is identically equal to Equation (5.171b). Thus, in the approximation considered here the nonlinearity has no effect whatever on the energy balance equation [Eqs. (5.171b) and (5.173)].

The scalar Equation (5.171b) for the field amplitude was derived under the assumption that the wave is linearly polarized with a polarization vector $\boldsymbol{e} \neq \boldsymbol{e}(\boldsymbol{r})$. This is obviously valid for a two-dimensional beam $\boldsymbol{E} = \boldsymbol{E}(x, s)$ if the field is polarized in the direction of the y axis (TE wave). Indeed, in this case $\varepsilon(s, E^2) = \varepsilon(s, x)$ and the polarization direction cannot vary as the wave propagates. A similar situation obtains also in an axially-symmetrical beam $\boldsymbol{E} = \boldsymbol{E}(\boldsymbol{\rho}, s)$ polarized along $\boldsymbol{e}_\rho$. To be sure, in this case $\varepsilon = \varepsilon(\rho, s)$ and generally speaking a field component should appear in the s direction. In a narrow beam, however, at $|\boldsymbol{u}| \ll 1$, we always have $E_s \ll E_\rho$, so that $E^2 \approx E_\rho^2$. Therefore the scalar Equation (5.171b) is valid also for narrow axially symmetrical beams. In the general case, however, vector equations must be used for a wave with elliptic polarization in an isotropic medium, as well as for waves in an anisotropic medium.

If the beam width is large in comparison with the wave length, then we can neglect in Equations (5.171a) and (5.172) the term

$$\frac{1}{2} \nabla_\perp \left(\frac{1}{k^2 E_0} \Delta_\perp E_0 \right),$$

which describes the diffraction effects. Indeed, the term $\Delta_\perp E_0 / 2k^2 E_0$ is of the order of $\frac{1}{2}(k^2 a^2)^{-1}$, where a is the effective width of the beam. Comparing it with the nonlinear term $\varepsilon_n / 2\varepsilon_0$, we verify that if

$$\left(a \frac{\omega}{c} \right)^2 \varepsilon_n = \frac{4\pi^2 a^2 \varepsilon_n}{\lambda^2 \varepsilon_0} \gg 1, \tag{5.174}$$

then the diffraction term in Equations (5.171a) and (5.172) can be neglected. Without this term, Equations (5.171b) and (5.172) describe the propagation of a radio beam in the geometrical-optics approximation. In other words, the beam can be regarded in this case as made up of rays that have at each point of the medium a definite intensity I and a definite ray direction on a plane perpendicular to the beam axis. Equation (5.171b) then represents, as before, the energy or ray-intensity flux balance, while Equation (5.172) describes the change in the directions of the rays that make up the beam.

In a homogeneous ($dk/ds = 0$) non-absorbing ($\varepsilon_1 = 0$) weakly nonlinear medium, the geometrical-optics equations become

$$\frac{\partial I}{\partial s} + \nabla_\perp(\boldsymbol{u} I) = 0, \qquad \frac{\partial \boldsymbol{u}}{\partial s} + (\boldsymbol{u}\nabla_\perp)\boldsymbol{u} + \beta\, \nabla_\perp I = 0. \tag{5.175}$$

It is assumed here that the nonlinear perturbation of ε is locally connected with the amplitude of the wave field, $\varepsilon_n \sim E_0^2 \sim I$ (see Sect. 5.3). The geometrical-optics Equations (5.175) are valid in the local-nonlinearity approximation in a homogeneous isotropic medium for narrow beams of waves that are linearly polarized in a plane perpendicular to the beam propagation direction.

We note that Equations (5.175) are equivalent to Euler's equation in two-dimensional gasdynamics with adiabatic exponent $c_p/c_v = 2$ (Landau and Lifshitz, 1963). The role of the gas density is played here by the intensity I, the role of the velocity by the ray direction $\boldsymbol{u}$, and the role of the time by the distance s along the beam axis. The last term in the second equation of Equation (5.175) describes the refraction due to the nonlinear "pressure" of the medium. Its influence depends essentially on the sign of the coefficient β. If $\beta > 0$, then the beam tends to broaden and become defocused. It is then natural to speak of a "defocusing" medium; this is precisely the case which is equivalent to ordinary subsonic gasdynamics. In the opposite case $\beta < 0$ the nonlinear pressure, to the contrary, tends to compress and focus the rays of the beam, and such a medium can naturally be called "focusing." The reversal of the sign of the pressure causes Equation (5.175) to change from hyperbolic at $\beta > 0$ to elliptic at $\beta < 0$. The result is an important singularity, namely instability of smooth beams in a focusing medium.

Indeed, let us examine the evolution of small perturbations δI and $\delta \boldsymbol{u}$ of the intensity and of the ray directions in a beam, respectively. Putting

$$I = I_0 + \delta I(s, \boldsymbol{\rho}), \qquad \boldsymbol{u} = \boldsymbol{u}_0 + \delta \boldsymbol{u}(s, \boldsymbol{\rho}), \tag{5.176}$$

we substitute Equations (5.176) in Equations (5.175) and, linearizing the latter, we obtain

$$\frac{\partial\, \delta I}{\partial s} + \boldsymbol{u}_0 \frac{\partial\, \delta I}{\partial \boldsymbol{\rho}} + I_0 \frac{\partial\, \delta \boldsymbol{u}}{\partial \boldsymbol{\rho}} = 0, \qquad \frac{\partial\, \delta \boldsymbol{u}}{\partial s} + \boldsymbol{u}_0 \frac{\partial\, \delta \boldsymbol{u}}{\partial \boldsymbol{\rho}} + \beta \frac{\partial\, \delta I}{\partial \boldsymbol{\rho}} = 0. \tag{5.177}$$

The solution of the linear Equations (5.177) is naturally obtained by expanding in a Fourier integral

$$\delta I = \int \delta I_q \exp\left[i(q_s s + \boldsymbol{q}_\rho \boldsymbol{\rho})\right] d\boldsymbol{q}, \qquad \delta \boldsymbol{u} = \int \delta \boldsymbol{u}_q \exp\left[i(q_s s + \boldsymbol{q}_\rho \boldsymbol{\rho})\right] d\boldsymbol{q}. \tag{5.178}$$

Substituting Equations (5.178) in Equations (5.177), we arrive at a dispersion equation that connects q_s with $\boldsymbol{q}_\rho$:

$$(q_s + \boldsymbol{q}_\rho \boldsymbol{u}_0)^2 - \beta q_\rho^2 I_0 = 0. \tag{5.179}$$

Hence

$$q_s = -\boldsymbol{q}_\rho \boldsymbol{u}_0 \pm q_\rho \sqrt{\beta I_0}. \tag{5.180}$$

Substituting this expression in Equation (5.178) we see that in a focusing medium ($\beta < 0$) small transverse perturbations increase exponentially with s:

$$\delta I \sim \exp(q_\rho s \sqrt{-\beta I_0}). \tag{5.181}$$

Thus, a smooth beam is unstable in a focusing medium and has a tendency to become laminated or to break up into thin filaments. The larger q_ρ, the faster the instability development.

In the ionosphere, nonlinear modification of the dielectric constant is due mainly to changes in the electron density. At heights $z \lesssim 200$ km, the electron density increases under the influence of the wave field [Eqs. (2.249) and (2.246)], so that we have here

$$\beta = -\frac{1}{2\varepsilon_0}\frac{\partial \varepsilon}{\partial N}\frac{dN}{dI} = \frac{4\pi\omega_0^2\varphi}{\omega^2 E_{\mathrm{p}}^2(1 - \omega_0^4/\omega^4)}; \qquad I = \frac{2 - \varepsilon_0}{8\pi} E_0^2. \tag{5.182}$$

This region of the ionosphere has $\beta > 0$ and is a defocusing medium. The defocusing of narrow beams in the ionosphere will be the subject of the next section.

At altitudes corresponding to the F layer and higher, the electron density, to the contrary, is decreased by the wave field (Sect. 5.3.1). Here $\beta < 0$, so that the region of the F layer is a focusing medium for the radio waves, in which self-focusing develops effectively. This instability will be discussed in detail in Sect. 6.1.

5.4.2. Defocusing of Narrow Beams

We present some examples of the solution of Equation (5.175), which give an idea of the character of the nonlinear refraction in the defocusing region of the ionosphere. The statement of the problem is the following: Assume that the distribution of the intensity I and of the ray directions $\boldsymbol{u}$ over the beam cross section is specified on the boundary of the nonlinear region (at $s = 0$):

$$I\big|_{s=0} = I_0(\boldsymbol{\rho}), \qquad \boldsymbol{u}\big|_{s=0} = \boldsymbol{u}_0(\boldsymbol{\rho}). \tag{5.183}$$

It is required to determine I and $\boldsymbol{u}$ in any other section of the beam, i.e., at arbitrary s.

Parabolic Beam. We consider an axially-symmetrical beam and assume that at $s = 0$ we have

$$I_0(\boldsymbol{\rho}) = I_0 \cdot \begin{cases} 1 - \rho^2/a^2, & \text{if} \quad \rho \leq a \\ 0, & \text{if} \quad \rho > a \end{cases}$$
$$\boldsymbol{u}_0(\boldsymbol{\rho}) = \boldsymbol{\rho}/R_0. \tag{5.184}$$

These initial conditions correspond to a spherical wave propagating from the point $s = -R_0$ and having a parabolic intensity-distribution profile.[18]

We seek the solution of Equation (5.175) at arbitrary s in the self-similar form

$$I(\boldsymbol{\rho}, s) = \frac{I_0}{f^2(s)} \cdot \begin{cases} 1 - \rho^2/a^2 f^2, & \text{if} \quad \rho \leq af \\ 0, & \text{if} \quad \rho > af. \end{cases}$$
$$\boldsymbol{u}(\boldsymbol{\rho}, s) = \boldsymbol{\rho}\varphi(s). \tag{5.185}$$

Substituting Equations (5.185) in Equations (5.175), we arrive at the following equation for $f(s)$

$$\frac{d^2 f}{ds^2} = \frac{2\beta I_0}{f^3}, \tag{5.186}$$

with

$$\varphi(s) = \frac{1}{f}\frac{df}{ds}. \tag{5.187}$$

The boundary conditions [Eq. (5.183)] with allowance for Equations (5.185)–(5.187) are rewritten in the form

$$\varphi|_{s=0} = \frac{1}{R_0}, \qquad f|_{s=0} = 1, \qquad \left.\frac{df}{ds}\right|_{s=0} = \frac{1}{R_0}. \tag{5.188}$$

The solution of Equation (5.186) with the boundary condition in Equation (5.188) is

$$f = \left[\left(1 + \frac{s}{R_0}\right)^2 + \frac{2\beta I_0}{a^2} s^2\right]^{1/2}. \tag{5.189}$$

---

[18] We note that the initial intensity distribution [Eq. (5.184)] has a weak discontinuity at $\rho = a$. Equations (5.175) in a defocusing medium ($\beta > 0$) are of the hyperbolic type and admit therefore of arbitrary weak discontinuities.

In the absence of nonlinearity we have

$$f = 1 + \frac{s}{R_0}. \tag{5.190}$$

It is seen from Equations (5.189) and (5.190) that the beam is broadened—defocused—by the nonlinearity.

The trajectories of the extreme rays of the beam are shown in Figure 61. We introduce the beam divergence angle—the angle between the direction of the extreme ray and the beam axis:

$$\theta = a\frac{df}{ds} = \theta_0 \frac{1 + s/R_0 + 2\beta I_0 s/R_0\theta_0^2}{[(1 + s/R_0)^2 + 2\beta I_0 s^2/R_0^2\theta_0^2]^{1/2}}, \qquad \theta_0 = \frac{a}{R_0}. \tag{5.191}$$

In the absence of nonlinearity, the divergence angle is constant and equal to θ_0. It is seen from Equation (5.191) that as a result of the nonlinearity the beam divergence angle increases with increasing s, tending at large s to the constant value

$$\theta_m = \sqrt{\theta_0^2 + 2\beta I_0} \tag{5.192}$$

(see Fig. 61). The beam radius is correspondingly increased by the nonlinearity. The intensity on the beam axis decreases by a factor $(1 + 2\beta I_0/\theta_0^2)$.

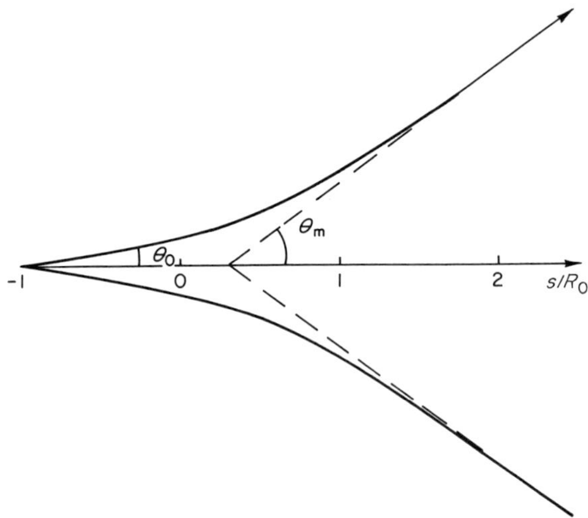

Fig. 61. Paths of outermost rays of defocused parabolic beam. Emission point $s/R_0 = -1$. Initial divergence angle θ_0. The nonlinear medium starts at $s = 0$. θ_m is the maximum beam divergence angle

If the beam is parallel on entering the nonlinear medium, $\theta_0 = 0$, the nonlinearity makes it divergent with a maximum divergence angle

$$\theta_m = \sqrt{2\beta I_0}. \tag{5.193}$$

If the nonlinearity is weak, $\beta I_0 \ll \theta_0^2$, then we have at all values of s, as follows from Equation (5.189),

$$\theta = \theta_0 \left[1 + \frac{\beta I_0}{\theta_0^2} \left(1 - \frac{R_0^2}{(R_0 + s)^2} \right) \right]. \tag{5.194}$$

We see therefore that the maximum divergence angle θ_m is reached already at $s \approx (2\text{–}3)R_0$.

Diffusion of Intensity Discontinuity. It is seen from Equation (5.175) that the influence of the nonlinearity becomes stronger with increasing intensity gradient $\nabla_\perp I$. We consider therefore the case of a sharp beam boundary, meaning a discontinuity of the intensity. Thus, assume that at $s = 0$ we have

$$I|_{s=0} = I_0 \cdot \begin{cases} 1, & \text{if } \; x < 0 \\ 0, & \text{if } \; x > 0. \end{cases} \tag{5.195}$$

Here x is a selected direction in the $\boldsymbol{\rho}$ plane and is perpendicular to the intensity-discontinuity line. We assume that

$$u_x|_{s=0} = 0. \tag{5.196}$$

Equations (5.175) then take the form

$$\begin{aligned} \frac{\partial I}{\partial s} + u_x \frac{\partial I}{\partial x} + I \frac{\partial u_x}{\partial x} &= 0, \\ \frac{\partial u_x}{\partial s} + u_x \frac{\partial u_x}{\partial x} + \beta \frac{\partial I}{\partial x} &= 0. \end{aligned} \tag{5.197}$$

We have taken into account here the fact that in Equations (5.195) and (5.196) only the changes of the beam in the x direction are significant in the $\boldsymbol{\rho}$ plane, i.e., $I = I(x, s)$ and $u_x = u_x(x, s)$. We can seek the solution of Equation (5.197) with boundary conditions [Eqs. (5.195) and (5.196)] in the self-similar form

$$I = I(\tau), \qquad u_x = u_x(\tau), \qquad \tau = x/s. \tag{5.198}$$

Substituting Equations (5.198) in Equations (5.197) we get

$$(u_x - \tau)\frac{dI}{d\tau} + I\frac{du_x}{d\tau} = 0, \qquad (u_x - \tau)\frac{du_x}{d\tau} + \beta\frac{dI}{d\tau} = 0. \tag{5.199}$$

The solution of these equations is

$$I = \frac{1}{9\beta}(C - \tau)^2, \qquad u_x = \frac{2\tau + C}{3}. \tag{5.200a}$$

We note also the trivial solutions

$$I = I_0,\, u_x = 0; \qquad I = 0,\, u_x = \tau. \tag{5.200b}$$

Recognizing that the boundary conditions [Eqs. (5.195) and (5.196)], in terms of the self-similar variables, take the form

$$\begin{gathered} I \to 0 \quad \text{as} \quad \tau \to +\infty; \qquad I \to I_0 \quad \text{as} \quad \tau \to -\infty; \\ u_x \to 0 \quad \text{as} \quad \tau \to -\infty, \end{gathered} \tag{5.201}$$

we obtain from Equations (5.200) and (5.201) the sought distributions of I and u_x:

$$I = I_0 \cdot \begin{cases} 1, & \text{if } x/s < -\sqrt{\beta I_0} \\ \left(\dfrac{2}{3} - \dfrac{x}{3s\sqrt{\beta I_0}}\right)^2, & \text{if } -\sqrt{\beta I_0} < x/s < 2\sqrt{\beta I_0}. \\ 0, & \text{if } x/s > 2\sqrt{\beta I_0}. \end{cases} \tag{5.202}$$

$$u_x = \begin{cases} 0, & \text{if } x/s < -\sqrt{\beta I_0}, \\ \dfrac{2}{3}\left(\dfrac{x}{s} + \sqrt{\beta I_0}\right), & \text{if } x/s > -\sqrt{\beta I_0}. \end{cases} \tag{5.203}$$

We see that the initial discontinuity of the intensity diffuses and the beam diverges rapidly near the discontinuity line. The divergence angle for the extreme ray of the beam is

$$\theta_{\mathrm{m}} = 2\sqrt{\beta I_0}. \tag{5.204}$$

We see that the maximum divergence angle is higher in the case of the diffusion of an abrupt discontinuity than in an axially-symmetrical parabolic beam [Eq. (5.193)].

Appearance of Intensity Oscillations. We consider now a more general class of solutions [Eq. (5.197)], equivalent to the simple Riemann waves in hydrodynamanics (Landau and Lifshitz, 1963). Namely, we seek the solution in the form $u_x = u_x(I)$. Substituting this expression in Equation (5.197), we obtain

$$\frac{\partial I}{\partial s} + u_x(I)\frac{\partial I}{\partial x} + I\frac{du_x}{dI}\frac{\partial I}{\partial x} = 0, \qquad \frac{du_x}{dI}\left(\frac{\partial I}{\partial s} + u_x\frac{\partial I}{\partial x}\right) + \beta\frac{\partial I}{\partial x} = 0. \quad (5.205)$$

These equations are compatible with the initial requirement $u_x = u_x(I)$, provided that

$$\left(\frac{\partial I}{\partial s}\right)_I + \tau(I)\left(\frac{\partial I}{\partial x}\right)_I = 0, \quad (5.206)$$

where the subscript I denotes that the differentiation is carried out at constant I, and $\tau(I)$ is an arbitrary function of I. The solution of Equation (5.206) yields a constant value of I on the characteristics defined by the equations $\partial x/\partial s = \tau$, i.e.,

$$x = \tau s + P(\tau), \quad (5.207)$$

where $P(\tau)$ is an arbitrary function. Substituting now Equations (5.207) in Equations (5.205), we get

$$\frac{du_x}{dI}(u_x - \tau) + \beta = 0, \qquad u_x - \tau + I\frac{du_x}{dI} = 0. \quad (5.208)$$

Recognizing that $\tau = \tau(I)$ and consequently $I = I(\tau)$ and $u_x = u_x(\tau)$, we obtain in place of Equations (5.208) the Equations (5.199) whose solutions are given by Equation (5.200).

Thus, the obtained solutions are described by Equation (5.200) and are constant on the straight lines [Eq. (5.207)], where $P(\tau)$ is an arbitrary function.

It follows from Equation (5.200) that for the principal solution we have

$$I(u_x) = \frac{1}{\beta}\left(\frac{3}{2}C - \frac{u_x}{2}\right)^2 = I_0\left(1 \pm \frac{u_x}{2\sqrt{\beta I_0}}\right)^2, \qquad C = \pm\frac{2}{3}\sqrt{\beta I_0}. \quad (5.209)$$

i.e.,

$$u_x = \frac{2}{3}(\tau + \sqrt{\beta I_0}), \qquad I = I_0\left(\frac{2}{3} - \frac{\tau}{3\sqrt{\beta I_0}}\right)^2. \quad (5.210)$$

Therefore, if the intensity I and the ray directions u_x are connected on the boundary $s = 0$ by Equations (5.209), then the solution of Equations (5.197) takes the form of Equation (5.210). The function $P(\tau)$ is determined in this case with the aid of the boundary function $u_0(x) = u_x(x)|_{s=0}$. Indeed, from Equations (5.197) and (5.200) it follows that at $s = 0$ we have

$$x = P(\tau) = X\left(\frac{2\tau}{3} + C\right), \tag{5.2.11}$$

where $X(u_0)$ is the inverse of $u_0(x)$.

Let, for example,

$$I_0(x) = \frac{I_0}{(1 + x^2/a^2)^2}, \qquad u_0(x) = \pm\frac{2\sqrt{\beta I_0}x^2/a^2}{1 + x^2/a^2}. \tag{5.212}$$

These boundary conditions satisfy Equation (5.210). Consequently, the solution of Equation (5.197) with these boundary conditions is given by Equation (5.210). Here τ is connected with x and s by Equations (5.207), where the function $P(\tau)$, as is clear from Equation (5.211), takes the form

$$P(\tau) = \pm\frac{a(\tau + \sqrt{\beta I_0})^{1/2}}{(2\sqrt{\beta I_0} - \tau)^{1/2}}. \tag{5.213}$$

Substituting Equations (5.213) in Equations (5.207) we arrive at a cubic equation for τ, and it is this equation which determines the final function $\tau(x, s)$.

Plots of the function $I(x)$ at different values of $s(t = s\sqrt{\beta I_0}/a)$ are shown for this case in Figure 62. It is seen that the intensity distribution not only broadens but also becomes deformed with increasing s. At $s = s_c \simeq 1.6a/\sqrt{\beta I_0}$, at the point $x = x_c \approx 1.5a$, the beam profile breaks (overturns), i.e., the rays intersect at $s > s_c$, near the point s_c. A region is produced near x_c, in which three rays are simultaneously present (shown dashed in Fig. 62). In this region, the geometrical-optics approximation [Eq. (5.175)] is no longer valid, and the diffraction effects [Eq. (5.172) already become essential.

We consider the vicinity of the breaking point x_c, s_c. As follows from Equation (5.209), we have here

$$u_x - u_c = -\frac{(I - I_c)}{I_c}\sqrt{\beta I_c}, \qquad u_c = u_x(x_c, s_c), \qquad I_c = I(x_c, s_c).$$

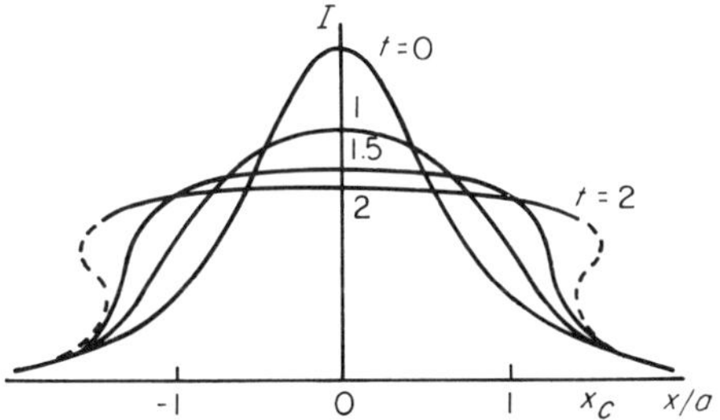

Fig. 62. Breaking of intensity profile following the defocusing of the beam; $t = s\sqrt{\beta I_0}/a$

Using this relation, we reduce Equation (5.172), under the conditions of interest to us ($dk/ds = 0$, $\varepsilon_n/2\varepsilon_0 = -\beta I$, $\nabla_\perp = \partial/\partial x$), to the form:

$$\frac{\partial \eta_1}{\partial t_1} + \eta_1 \frac{\partial \eta_1}{\partial X_1} + \frac{\partial^3 \eta_1}{\partial X_1^3} = 0, \qquad \eta_1 = (I - I_c)/I_c.$$

$$X_1 = 2k\sqrt{\beta I_c}(x - x_c), \qquad t_1 = 2k\beta I_c(s - s_c), \qquad k = \omega\sqrt{\varepsilon_0}/c.$$

This is the Korteweg–de Vries equation (Witham, 1965; Bespalov et al., 1968; Karpman 1973; Zakharov and Shabat, 1972). It describes the region near the wave breaking point with allowance for nonlinearity and diffraction (the last term). Its solution in the toppling region (Gurevich and Pitaevskii, 1973) is shown in Figure 63. The dashed curve is the

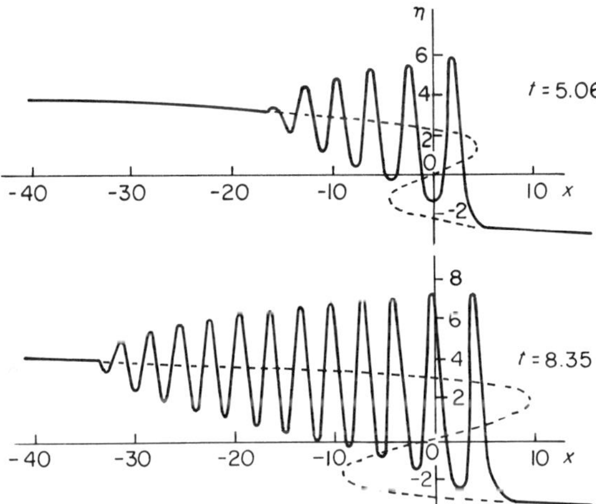

Fig. 63. Oscillations of wave intensity in the breaking region: $\eta = \eta_1 A^{2/7}$, $t = t_1 A^{-3/7}$, $x = x_1 A^{-1/7}$; $A = (2ka\sqrt{\beta I_c})C_0$; $C_0 = -P'''(\tau_c)(\beta I_c)^{3/2}/6a$

multiple-valued intensity profile produced in geometrical optics (see Fig. 62). It is seen that intensity oscillations with characteristic dimension

$$\Delta x \sim \lambda/\sqrt{\beta I_c}, \tag{5.214}$$

where $\lambda = 2\pi c/\omega\sqrt{\varepsilon_0}$ is the radio wave length developed in the breaking region. The region occupied by the oscillations broadens rapidly with increasing $s-s_c$ and gradually spans the entire beam.

Using the method of characteristics (see Landau and Lifshitz, 1963), we can construct a general solution of Equation (5.197). The appearance of the breaking region depends in this case on the form of the initial intensity profile and ray directions. For the weakly diverging beam, the toppling usually takes place whenever the initial profile $I_0(x)$ has an inflection point.

Defocusing of Narrow Beams in the Ionosphere. Figure 64 shows the parameter $\beta^* = \beta(\omega/10^8)^4\ (\mathrm{V/m})^{-2}$, defined in accordance with Equation (5.182), as a function of the height in the ionosphere. It is assumed that $\omega^2 \gg \omega_H^2$. When this condition is satisfied, the medium can be regarded in the first-order approximation as isotropic, $\varepsilon_0 = 1 - \omega_0^2/\omega^2$, and in addition, it can be assumed that $\varphi_p \simeq 1$ in Equations (5.182) and (2.42), so that β does not depend on the polarization of the wave. The dashed curves were plotted in accordance with the formulas

Day:

$$\beta^* = \begin{cases} \beta_0 = 2.5 \cdot 10^{-2}\ (\mathrm{V/m})^{-2} & \text{at} \quad 100\ \mathrm{km} \le z \le 200\ \mathrm{km} \\ 0 & \text{at} \quad z < 100\ \mathrm{km},\ z > 200\ \mathrm{km} \end{cases}$$

Night:

$$\beta^* = \begin{cases} \beta_0 = 3 \cdot 10^{-4}\ (\mathrm{V/m})^{-2} & \text{at} \quad 100\ \mathrm{km} \le z \le 190\ \mathrm{km} \\ 0 & \text{at} \quad z < 100\ \mathrm{km},\ z > 190\ \mathrm{km} \end{cases} \tag{5.215}$$

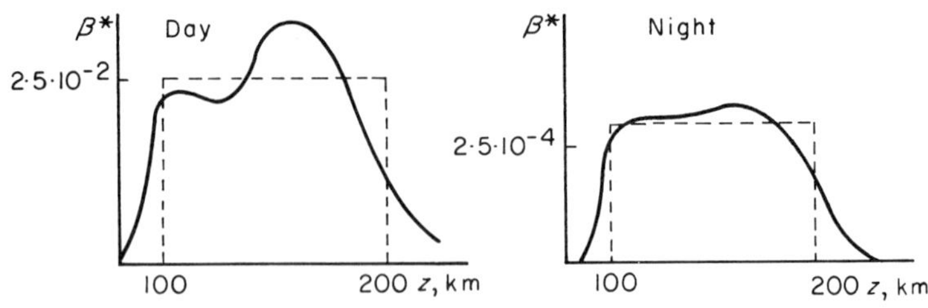

Fig. 64. The parameter $\beta^* = \beta(\omega/10^8)^4[(\mathrm{V/m})^{-2}]$ in the ionosphere (*solid curves*). *Dashed lines*: its interpolation [Eq. (5.215)]

Thus, in the rough approximation of (5.215), the layer of the ionosphere with height from 100 to 200 km in day time (or 190 km at night) can be regarded as a homogeneous defocusing medium, i.e., the results obtained in the preceding section can be used to describe the nonlinear self-action of narrow beams.

Let us estimate the effects of the defocusing for this simplified model. The change of the beam divergence angle is described by Equation (5.191). Here θ_0 is the initial beam divergence angle, and R_0 is the distance traversed by the ray prior to entering the defocusing region of the ionosphere $z \geq z_0 = 100$ km:

$$R_0 = \sqrt{R_e^2 \cos^2 \psi + 2R_e z_0} - R_e \cos \psi \tag{5.216}$$

(R_e is the earth's radius and ψ is the angle between the beam axis and the vertical at the transmitter location),

$$\beta I_0 = \beta_0 I_0 \simeq \frac{\beta_0 E_0^2}{8\pi} \left(\frac{10^8}{\omega}\right)^4, \qquad \frac{\omega_0^4}{\omega^4} \ll 1, \tag{5.217}$$

where E_0 is the amplitude of the field at the beam maximum at $z = z_0$, and β_0 is defined in accordance with Equation (5.215). Figure 65 shows the change of the beam divergence angle as a function of the power and directivity of the radiation. It is seen that if $\beta_0 I_0/\theta_0^2 \gg 1$, then the beam can experience appreciable defocusing. Taking into account the condition $E_0 \lesssim E_p$ [the only case when Equations (5.243) and (5.182) are valid], we find that significant defocusing can be experienced only by a sufficiently narrow beam

$$\theta_0 \lesssim C_m (10^8/\omega)^2.$$

Here C_m is a constant with value 0.14 (day time) or 0.016 (night time).

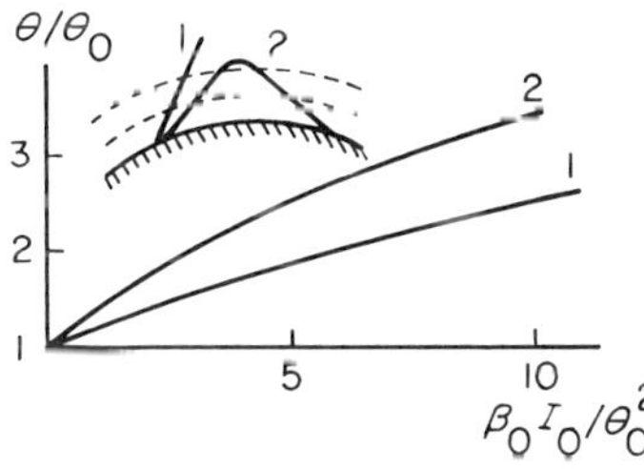

Fig. 65. Increase of beam-divergence angle in the ionosphere (*upper left*: trajectories *1* and *2*).

The broadening of the beam leads to a change in the amplitude of the signal at the reception point. This is seen from Figure 66, which shows the change of the power W_1 of the received signal as a function of $I^* = 2\beta I_0/\theta_0^2(2 + s/R_0)^2$; the quantity I^* is proportional to the radiation power W_0. The dashed curve shows the linear growth of W_1 on the beam axis. The solid curve 1 shows the growth of the signal power at the axis when the nonlinear defocusing is taken into account. It is seen that the nonlinearity slows down the growth of W_1. The picture is different for points located at a distance $\rho > \rho_0 = \theta_0(2R_0 + s)$ from the beam axis. In the linear approximation there is no signal here. The signal appears only at sufficiently large W_0, owing to the nonlinear broadening of the beam (curve 2; $\rho = 2\rho_0$).

If the beam profile breaks upon defocusing, then a region of strong intensity oscillations is produced, with a characteristic dimension Δx [Eq. (5.214)]. In the ionosphere at $\omega \sim 10^8$ and $E_0 \sim E_p$ we have $\Delta x \sim$ 100–200 m. Simultaneously, density inhomogeneities appear in the ionosphere

$$\Delta N \sim N_0 \gamma_1 E_0^2/E_p^2 \varphi_T, \qquad E_0 \lesssim E_p, \tag{5.218}$$

and have the same structure. We note that inasmuch as the dimension Δx is usually small in comparison with the characteristic lengths L_N and L_T (see Table 16), it follows that in a direction parallel to the magnetic field $\boldsymbol{H}$ these oscillations become smoothed out by diffusion and by thermal conductivity. On the other hand, the homogeneities are not smeared out in a direction perpendicular to $\boldsymbol{H}$. Thus, in the breaking direction, an inhomogeneous structure should develop in the ionosphere plasma only in a direction perpendicular to the beam axis and to the magnetic field, with a characteristic dimension [Eq. (5.214)] and amplitude [Eq. (5.218)].

The nonlinear beam defocusing considered here becomes stronger in the caustic region, where the field amplitude E_0 increases. In this case, however, we cannot confine ourselves to a homogeneous medium, as was

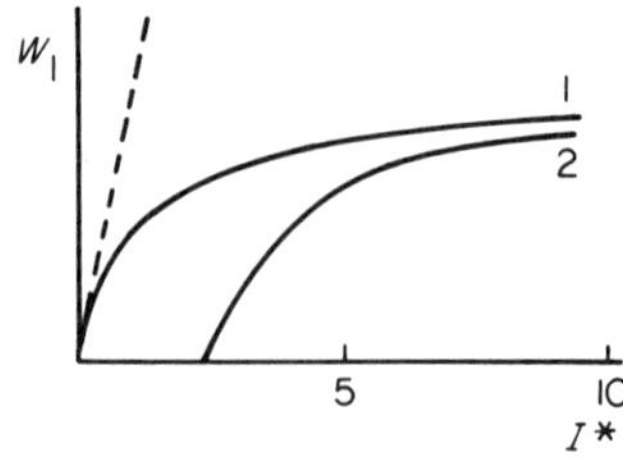

Fig. 66. Received-signal power W_1 as a function of the radiation power $I^* \sim W_0$; 1: on the beam axis; 2: at a distance $\rho = 2\rho_0$

done above. Nor is account taken here of the defocusing due to the wave scattering by the inhomogeneities of the ionosphere. When an instability is excited in the region modified by the wave, these inhomogeneities increase sharply (see Sects. 6.1–6.3), with a corresponding enhancement of the scattering as a result.

5.4.3. Mutual Defocusing

The plasma-density modifications produced by a high-power wave bend the trajectories of other radio waves propagating in the modified zone. Let us consider the defocusing due to such an interaction.

Let two axially-symmetrical beams of frequency ω_1 and ω_2 propagate in one direction in a defocusing medium. Assume that the wave ω_2 is weak, so that its influence on the plasma can be neglected. The distributions of the intensities I_1 and I_2 and of the ray directions u_1 and u_2 in the beams are described by equations similar to Equation (5.175):

$$\frac{\partial I_1}{\partial s} + \frac{1}{\rho}\frac{\partial}{\partial \rho}(\rho u_1 I_1) = 0, \qquad \frac{\partial I_2}{\partial s} + \frac{1}{\rho}\frac{\partial}{\partial \rho}(\rho u_2 I_2) = 0,$$
$$\frac{\partial u_1}{\partial s} + u_1 \frac{\partial u_1}{\partial \rho} + \beta \frac{\partial I_1}{\partial \rho} = 0, \qquad \frac{\partial u_2}{\partial s} + u_2 \frac{\partial u_2}{\partial \rho} + \beta_2 \frac{\partial I_1}{\partial \rho} = 0; \tag{5.219}$$
$$\beta_2 = \beta\omega_1^2/\omega_2^2.$$

We note that a homogeneous distribution of the intensity of the two beams in the defocusing medium is unstable (just as in two-stream hydrodynamics), and that an oscillating structure develops. If wave E_2 is weak, the amplitude of the oscillations is small and they do not affect the general dynamics of the beam.

Assume that both beams have at the plasma boundary the parabolic intensity distribution [Eq. (5.134)] and that the parabolic intensity distribution is preserved inside the plasma. The weak wave E_2 does not influence in this case the strong wave E_1, so that the distribution of the intensity I_1 is described as before by Equations (5.185) and (5.189). We seek the distributions of I_2 and u_2 in the form

$$u_2 = \frac{\rho}{f_2}\frac{df_2}{ds}, \qquad I_2 = \begin{cases} \dfrac{I_{20}}{f_2^2}\left(1 - \dfrac{\rho^2}{a^2 f_2^2}\right), & \text{if } \rho \le af_2 \\ 0, & \text{if } \rho > af_2. \end{cases} \tag{5.220}$$

Substituting Equations (5.220) in Equations (5.219) and taking Equations (5.189) and (5.185) into account, we find that the function $f_2(s)$ satisfies the equation

$$\frac{d^2f_2}{dt^2} = \frac{\kappa f_2}{[(1+\gamma_2 t)^2 + t^2]^2}; \qquad t = \frac{\sqrt{2\beta I_0}}{\alpha}s,$$
$$\gamma_2 = \frac{a}{R_0\sqrt{2\beta I_0}}, \qquad \kappa = \omega_1^2/\omega_2^2 \tag{5.221}$$

with the boundary conditions

$$f_2|_{t=0} = 1, \qquad \left.\frac{df_2}{dt}\right|_{t=0} = \frac{a}{R_2\sqrt{2\beta I_0}} = \gamma_0.$$

We note that Equation (5.221) is valid for all ρ only if $\rho < af_1$, i.e., if $f_2 \leq f_1$. This condition is always satisfied at $\kappa \leq 1$.

Making the change of variables

$$t = \frac{w - \gamma_2}{1 + \gamma_2^2}, \qquad w = \frac{1 - 2u}{2(u - u^2)^{1/2}}.$$

Equation (5.221) transforms into

$$\frac{d^2f_2}{du^2} + \frac{3u - 3/2}{u(u-1)}\frac{df_2}{du} + \frac{\kappa}{u(u-1)}f_2 = 0. \tag{5.222}$$

The boundary conditions for this equation are given at

$$u = u_0 = \frac{1}{2}\left(1 - \frac{\gamma_2}{\sqrt{1+\gamma_2^2}}\right)$$

and take the form

$$f_2|_{u=u_0} = 1, \qquad \left.\frac{df_2}{du}\right|_{u=u_0} = -2\sqrt{1+\gamma_2^2}\gamma_0. \tag{5.223}$$

In the case $\kappa < 1$, the linearly independent solutions of Equation (5.222) are

$$F_1 = \frac{\sin[2\sqrt{1-\kappa}\arcsin\sqrt{u}]}{\sqrt{u-u^2}}, \qquad F_2 = \frac{\cos[2\sqrt{1-\kappa}\arcsin\sqrt{u}]}{\sqrt{u-u^2}}.$$

Then the general solution of Equation (5.222) is written in the form

$$f_2 = AF_1 + BF_2,$$

where the constants A and B are determined from the boundary conditions [Eq. (5.223)]

$$A = \frac{\cos[2\sqrt{1-\kappa}\arcsin\sqrt{u_0}]}{\sqrt{1-\kappa}}\left\{-2\gamma_0\sqrt{1+\gamma_2^2(u_0-u_0^2)}+\frac{1-2u_0}{2}\right\}$$
$$+\sqrt{u_0-u_0^2}\sin[2\sqrt{1-\kappa}\arcsin\sqrt{u_0}],$$

$$B = \frac{\sin[2\sqrt{1-\kappa}\arcsin\sqrt{u_0}]}{\sqrt{1-\kappa}}\left\{2\gamma_0\sqrt{1+\gamma_2^2(u_0-u_0^2)}-\frac{1-2u_0}{2}\right\}$$
$$+\sqrt{u_0-u_0^2}\cos[2\sqrt{1-\kappa}\arcsin\sqrt{u_0}].$$

A plot of the function $f_2(t)$ for parallel beams (i.e., at $\gamma_2 = 0$, $\gamma_0 = 0$, and $u_0 = \frac{1}{2}$) is shown in Figure 67. We see that f_2 increases with increasing t. In other words, the beam E_2 is defocused upon interaction with the wave E_1. The defocusing depends significantly on the frequency ratio $\omega_1^2/\omega_2^2 = \kappa$. At large values of s we have

$$f_2|_{s\to\infty} = f_2|_{u\to 0} \to 2Bt = 2Bs\sqrt{2\beta I_0}/a.$$

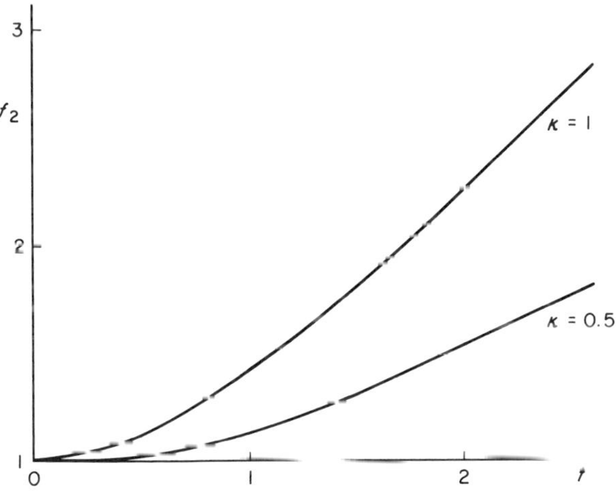

Fig. 67. Broadening of weak beam

The beam E_2 changes from parallel to diverging. The maximum divergence angle is

$$\theta_{\mathrm{m}} = 2B\sqrt{2\beta I_0} = \sqrt{2\beta I_0}\cos\left(\frac{\pi}{2}\sqrt{1-\kappa}\right). \tag{5.224}$$

Comparing Equation (5.224) with Equation (5.193) we see that the divergence angle of the beam E_2 at $\kappa < 1$ is less than the divergence angle of the main beam E_1. At $\kappa > 1$, the weak wave is pushed out more energetically towards the edges of beam E_1; its divergence angle does not exceed, however, the maximum divergence angle of the strong wave [Eq. (5.193)].

Mutual defocusing effects should also take place in the lower layers of the ionosphere ($z \lesssim 200$ km). They are perfectly analogous in character to the defocusing effects considered in the preceding section. It must be stressed that the defocusing effects are connected with a change of the electron density. To obtain defocusing it is therefore necessary to apply rather prolonged modifying pulses $\Delta t > \tau_N$ (see Table 13). On the other hand, the pulse E_2 can be short. By varying the emission time of the pulse E_2 in comparison with the modifying pulse E_1, we can obtain information on the variation of the electron density, i.e., on the recombination in the ionosphere. In addition, recognizing that the interaction of the pulses becomes stronger with decreasing frequency ω_2 we can, by varying ω_2, obtain conditions in which the effect is optimal. This is precisely why the interaction between the pulses turns out to be of importance for the investigation of the effects indicated in Section 5.4.2.

We note also that we have considered here only a special case, in which the interacting beams propagate in the same direction. Naturally, interaction takes place also in the general case when the beams intersect. When a plane wave crosses the modified region, its phase front is also distorted, and this leads to a change in the distribution of the field amplitude.

5.4.4. Thermal Focusing in the Lower Ionosphere

We have considered above the focusing and defocusing produced in radio beams by variation of the electron density in the plasma. It is important that in the ionosphere the changes of the electron density appear only after a long time, $\sim 10^2$–10^3 s. When the ionosphere is modified by a short pulse these changes are negligible. At the same time, in the lower ionosphere, at heights $z \sim 60$–90 km, the refractive index n can become modified also by another factor: the change in the frequency of the collisions of the electrons, a change due to their heating in the field of the high-power radio waves. These modifications of n can be appreciable in the case of radio waves of frequency $\omega \sim 10^6$–10^7. An important distinguishing feature of these modifications is that they develop rapidly, within a time $\Delta t \sim (\delta \nu_e)^{-1} \sim 10^{-3}$–$10^{-4}$ s (see Sects. 2.1, 2.3, and 2.5). They can cause nonlinear focusing of short pulses.

Consider a high-power wave propagating vertically upwards along the z axis. It produces in the electron collision frequency ν_e in the ionosphere a perturbation

$$\Delta \nu_e = \nu_e(T_e) - \nu_{e0} = \Delta \nu_e(z, \rho) \tag{5.225}$$

where $\rho = (x, y)$ stands for the coordinates in the horizontal plane, i.e., perpendicular to the beam axis. The perturbations $\Delta \nu_e$ are maximal on the axis of the high-power beam ($\rho = 0$) and decrease to zero at $\rho \gg a$, where a is the effective radius of the beam.

We consider now a plane wave E_2, likewise propagating quasilongitudinally, in a direction normal to the layer of the ionospheric plasma. The dielectric constant of the plasma for this wave [Eq. (2.16)] is:

$$\varepsilon_2 \simeq 1 - \frac{\omega_0^2}{\omega_{ef}^2 + \nu_e^2}, \qquad \omega_{ef} = \omega_2 \pm \omega_H \cos \alpha. \tag{5.226}$$

The perturbation produced in the refractive index n_2 by the nonlinearity in the case when $\omega_{ef} > \nu_e$ is

$$\Delta n = \left(\frac{\partial n_2}{\partial \nu_e}\right) \Delta \nu_e = \frac{\omega_0^2 \nu_e \, \Delta \nu_e}{n_2(\omega_{ef}^2 + \nu_e^2)^2}. \tag{5.227}$$

It is seen from Equations (5.226) and (5.227) that both Δn and $\Delta \varepsilon$ are larger than zero. Consequently, the lower ionosphere is a focusing medium (see Sect. 5.4.1), meaning that the nonlinearity deflects the rays towards the axis of the strong beam. It is also important that at $\omega_{ef} > \nu_e$ the perturbation Δn is proportional to $N(z)\nu_{e0}^2(z)$. This quantity has a sharp maximum of width ~ 5–10 km at a height $z_1 \approx 70$–80 km. Consequently, owing to the thermal nonlinearity, a focusing "lens" of thickness ~ 5–10 km is produced, as it were, in a narrow layer $z_1 \lesssim z \lesssim h_0$, where h_0 is the thickness of the lower ionosphere.

Let us calculate the degree of focusing in the ionosphere lens. In the absence of a perturbation, the wave E_2 would always remain plane; its phase is

$$\varphi_0 = \frac{\omega_2}{c} \int n_{20}(z) \, dz. \tag{5.228}$$

The perturbations Δn due to the wave E_1 give rise to corresponding corrections to the phase φ_0. These corrections depend on x, y, and z, so that the plane wave front becomes bent:

$$n = n_{20} + \Delta n(x, y, z) = n_{20} + \frac{\omega_0^2 \nu_e}{n_{20}(\omega_2^2 + \nu_e^2)^2} \Delta \nu_e(x, y, z)$$
$$\varphi = \varphi_0 + \Delta\varphi(x, y, z), \qquad \Delta\varphi = \frac{\omega_2}{c} \int_{z_1}^{z} dz \, \frac{\omega_0^2 \nu_e \, \Delta\nu_e(x, y, z)}{n_{20}(z)(\omega_2^2 + \nu_e^2)^2}. \tag{5.229}$$

It is assumed here that a ray of the E_2 wave is bent only insignificantly in the interaction region $z_1 \lesssim z \lesssim h_0$; the integration along the ray path in Equation (5.229) has therefore been replaced by an integral with respect to dz. For the same reason, the amplitude of the wave E_2 in the interaction region can be assumed constant and dependent only on z (but not on x or y). However, owing to the phase distortions [Eq. (5.229)] the phase front of the E_2 wave becomes curved. The wave will therefore continue to converge (to be focused) after emerging from the modified zone $z > h_0$. The wave field at an observation point on the z axis, at a distance z_0 from the upper boundary h_0 of the interaction region, is given by (Landau and Lifshitz, 1971):

$$E_2 = A \exp(i\varphi_0) \frac{\omega_2}{2\pi i c} \iint_{-\infty}^{\infty} \frac{\exp\{i\,\Delta\varphi(x, y, z_0) + i\omega_2 \tau/c\}}{r} \, dx \, dy. \tag{5.230}$$

It is assumed here for simplicity that $n_{20} = 1$ outside the interaction region; A is the wave amplitude at the boundary h_0, while $r = \sqrt{x^2 + y^2 + z_0^2}$. In the absence of the perturbation $\Delta\varphi$, the wave field amplitude is $|E_2| = A$. The phase perturbation $\Delta\varphi$ is capable of appreciably altering the amplitude at large values of r.

The integral [Eq. (5.230)] can be calculated by the stationary-phase method (Bliokh and Bryukhovetskii, 1969). Expanding the phase $\Delta\varphi(x, y)$ in a series about the stationary point ($x = 0$, $y = 0$) and retaining terms up to fourth order inclusive, we have:

$$E_2 = \frac{A\omega_2 \exp(i\varphi_0)}{2\pi i z_0 c} \iint_{-\infty}^{\infty} \exp\left\{i \sum_{m,\, n=0}^{2} C_{mn} x^{2m} y^{2n}\right\} dx \, dy. \tag{5.231}$$

Here $x = 0$, $y = 0$ is the point where the intensity of the perturbing beam E_1 is a maximum; it is assumed that there are no cubic terms, by virtue of the symmetry of the beam E_1 relative to the point $x = 0$, $y = 0$. Using Equation (5.229), we obtain the coefficients C_{mn}:

$$C_{mn} = \frac{\omega_2}{c} \int_{z_1}^{h_0} \frac{\omega_0^2 v_e \, dz}{n_{20}(\omega_2^2 + v_e^2)^2} \left\{ \frac{d^{m+n} \Delta v_e(x, y, z)}{d(x^2)^m \, d(y^2)^n} \right\}_{x=0,\, y=0} + \frac{\omega_2}{c} z_0^{1-2(m+n)} \alpha_{mn} \tag{5.232}$$

$$\alpha_{00} = 1, \qquad \alpha_{01} = \alpha_{10} = \tfrac{1}{2}, \qquad \alpha_{20} = \alpha_{02} = -\tfrac{1}{8}, \qquad \alpha_{11} = -\tfrac{1}{4}.$$

The vanishing of the coefficients C_{10} and C_{01} at a certain $z_0 = z_\varphi$ signifies focusing of the initially plane wave at the point z_φ. The fourth-order terms remaining in the argument of the exponential of Equation (5.231) determine the effective dimensions of the focal spot and consequently the gain of the ionosphere lens. Once the condition $C_{10} = C_{01} = 0$ is satisfied, the power gain of the lens K is equal to

$$K = \frac{|E_2|^2}{A^2} = \frac{\omega_2^2}{4\pi^2 z_\varphi^2 c^2} \left| \iint_{-\infty}^{\infty} \exp \{ i[C_{20}x^4 + C_{11}x^2y^2 + C_{02}y^4] \} \, dx \, dy \right| \tag{5.233}$$

The integral expression in Equation (5.233) reduces to elliptic integrals of the first or second kind.

In particular, in the case of an axially-symmetrical modifying beam $v_e = v_e(\rho, z)$ and we have $C_{20} = C_{02} = C_{11}/2$, $C_{01} = C_{10}$. It follows then Equations (5.232) and (5.233) that the focal length of the "lens" and the gain at the focus are equal to

$$z_\varphi = \left[-2 \int_{z_1}^{h_0} \frac{\omega_0^2 v_e(0, z)}{[\omega_{ef}^2 + v_e^2(0, z)]^2} \left(\frac{\partial v_e}{\partial \rho^2} \right)_{\rho=0} dz \right]^{-1}, \tag{5.234}$$

$$K = \frac{\pi \omega_2^2}{16 z_\varphi^2 c^2 C_{20}}, \qquad C_{20} = \frac{\omega_2}{c} \int_{z_1}^{h_0} \frac{\omega_0^2 v_e(0, z)}{(\omega_{ef}^2 + v_e^2(0, z))^2} \left(\frac{\partial^2 v_e}{\partial (\rho^2)^2} \right)_{\rho=0} dz - \frac{1}{8} \frac{\omega_2}{c z_\varphi^3}.$$

Numerical estimates for the ionosphere show that the effect of thermal focusing can be appreciable ($K \gtrsim 10$, $z_\varphi \sim 10^2$ km) in the frequency range $\omega_2 \sim 10^6 - 10^7$, at a sufficiently strong perturbation of the plasma, $E_{10}^2 \gg E_p^2$. We note that although we have considered here only mutual focusing, perfectly analogous expressions are valid also for the self-focusing of a strong wave. A self-focusing instability similar to that considered in Sections 5.4.1 and 6.1 may also be excited in the lower ionosphere.

6. Excitation of Ionosphere Instability

It is natural to expect that when radio waves act strongly on the ionosphere various types of plasma oscillations become excited and amplified in the modified region. This influences both the propagation of the modifying wave and other waves that propagate in the modified region. An enhanced (so-called anomalous) absorption and scattering of the radio waves sets in. At the same time, the plasma properties are significantly modified—a strongly inhomogeneous structure develops, plasma oscillations are excited, many fast electrons appear, and the emission from the plasma is amplified. An investigation of the instabilities of an ionosphere plasma in the field of a high-power radio wave is therefore of special interest.

It is natural to distinguish between two characteristic types of instability. The first is connected with the overall modification of the ionosphere plasma and with the production of artificial inhomogeneities in the plasma. Indeed, as shown above, the field of a strong radio wave propagating in the ionosphere can change significantly the electron and ion temperatures and the plasma concentration. Thus, artificial inhomogeneities of T_e, T_i, and N are produced in the plasma region heated by the radio waves. In the presence of electron and ion drift due to the ionosphere electric fields, an inhomogeneity of this kind in the earth's magnetic field can easily become unstable with respect to gravitational, drift, ion-sound, or ion-cyclotron waves, etc. A general theory of plasma instabilities of this type was treated in detail by Mikhailovskii (1971). The growth rates of these instabilities in the ionosphere are usually small (see Ginzburg and Rukhadze, 1975).

The instabilities of the second type (parametric, self-focusing, resonant) are produced directly by the presence of a strong alternating field in the plasma. They cause anomalous dissipation of the energy of the alternating electric field in the plasma. These instabilities develop rapidly under ionospheric conditions and seem to play the most important role; they will therefore be the ones dealt with below.

6.1. Self-Focusing Instability

It was demonstrated in Section (5.4.1) that in the geometrical-optics approximation a radio beam is unstable in an isotropic locally-nonlinear focusing medium. The beam tends to become stratified or to break up into narrow focused filaments. This phenomenon is called usually self-focusing or modulation instability (Bespalov, et al., 1968; Litvak, 1968; Karpman, 1973). In the F-layer region the ionosphere is a focusing medium. However, the nonlinearity here is essentially anisotropic and nonlocal, owing to the enhanced thermal conductivity and diffusion of the plasma along the magnetic field lines (see Sects. 5.1–5.3). This leads to a number of singularities of the self-focusing instability. In addition, the instability develops most intensely in the region of radio-wave reflection, where the ionosphere plasma can no longer be regarded as homogeneous (Vas'kov and Gurevich, 1975c; 1976b; Perkins and Valeo, 1974; Abramovich, 1976).

6.1.1. Spatial Instability of a Homogeneous Plasma

The propagation of a beam of radio waves in a homogeneous isotropic plasma, neglecting absorption, is described by the parabolic Equation (5.169)

$$2ik\frac{\partial E}{\partial s} + \Delta_{\perp} E - \frac{4\pi e^2}{mc^2}\,\delta N E = 0. \tag{6.1}$$

We have taken into account here the fact that the ionosphere plasma can be regarded as isotropic at $\omega^2 \gg \omega_H^2$. In this case $\varepsilon = 1 - \omega_0^2/\omega^2$, and perturbations $\Delta\varepsilon$ of the dielectric constant of the plasma are caused only by changes ΔN in the electron density. [We note that Eq. (6.1) remains essentially valid also in the more general case at an arbitrary ratio of ω and ω_H, while allowance for the anisotropy leads only to a certain change in the coefficients of this equation; (see Vas'kov and Gurevich, 1976b).]

The density perturbations depend on the wave field amplitude [Eq. (5.111)], so that the complete system of Equations (6.1) and (5.111) is nonlinear. Under stationary conditions, Equations (5.111) take the form of Equation (5.112):

$$L_N^2\frac{d^2\,\delta N}{dx_{\parallel}^2} + k_1 L_N^2\frac{N_0}{T_{e0}}\frac{d^2\,\delta T_e}{dx_{\parallel}^2} = \delta N - \gamma_1\frac{N_0}{T_{e0}}\,\delta T_e, \tag{6.2}$$

$$L_T^2\frac{d^2\,\delta T_e}{dx_{\parallel}^2} = \delta T_e - \frac{T_{e0}\varphi}{E_p^2}\,|E|^2. \tag{6.3}$$

Here $x_{\|}$ is the coordinate in the direction of the magnetic field $\boldsymbol{H}$. We change over in Equations (6.2) and (6.3) to the coordinates s and $\boldsymbol{\rho}(x, y)$ [Eq. (6.1)]. We assume for this purpose that the x axis is perpendicular to the s axis in the plane of the wave-propagation direction $\boldsymbol{s}$ and the magnetic field $\boldsymbol{H}$, while the y axis is perpendicular to this plane. Then:

$$\frac{d^2}{dx_{\|}^2} = \sin^2 \alpha_1 \frac{\partial^2}{\partial x^2} + \cos^2 \alpha_1 \frac{\partial^2}{\partial s^2} + 2 \sin \alpha_1 \cos \alpha_1 \frac{\partial^2}{\partial x\, \partial s} \tag{6.4}$$

where α_1 is the angle between $\boldsymbol{s}$ (which is parallel to $\boldsymbol{k}$) and $\boldsymbol{H}$.

Instability of Plane Wave. Let us examine the evolution of small perturbations of the field of a plane wave. We assume the field E in Equations (6.1) and (6.3) to be of the form

$$E = [E_0 + \delta E(\rho, s)] \exp (i\, \Delta k s). \tag{6.5}$$

Here E_0 is the plane-wave amplitude and Δk is the correction that must be introduced in the wave vector as the result of the modification of the plasma by the field E_0. Substituting Equations (6.5) in Equations (6.1) with $\delta E = 0$, we obtain

$$\Delta k = -\frac{2\pi e^2}{mc^2 k} \delta N_0; \qquad k = \frac{\omega}{c} \sqrt{\varepsilon_0}, \quad \varepsilon_0 = 1 - \frac{\omega_0^2}{\omega^2}; \tag{6.6}$$

where $\delta N_0 = \delta N(E_0^2)$. As follows from Equations (6.2) and (6.3), the value of δN_0 is

$$\delta N_0 = \gamma_1 \frac{N_0}{T_{e0}} \delta T_{e0}, \qquad \delta T_{e0} = T_{e0} \frac{\varphi E_0^2}{E_p^2}. \tag{6.7}$$

It is seen that the average density-perturbation is positive and is connected only with the changes in the ionization balance (see Sects. 2.5, 5.1, and 5.3). In the upper ionosphere, at heights $z \gtrsim 200$ km, it is negligible, $\gamma_1 \to 0$ (see Table 14).

It is natural to expand the perturbation $\delta E(\boldsymbol{\rho}, s)$ in a Fourier integral with respect to the coordinates $\boldsymbol{\rho}$ and s, representing its q component in a form similar to Equation (5.178):

$$\delta E(\boldsymbol{\rho}, s) = E_{1q} \exp (iq_s s + i\boldsymbol{q}_\rho \boldsymbol{\rho}) + E_{2q} \exp (-iq_s s - i\boldsymbol{q}_\rho \boldsymbol{\rho}). \tag{6.8}$$

The components of the perturbations δN and δT_e are represented in similar form:

$$\begin{aligned} \delta N &= \delta N_0 + \delta N_{1q} \exp (iq_s s + i\boldsymbol{q}_\rho \boldsymbol{\rho}) + \delta N_{2q} \exp (-iq_s s - i\boldsymbol{q}_\rho \boldsymbol{\rho}), \\ \delta T_e &= \delta T_{e0} + \delta T_{e1q} \exp (iq_s s + i\boldsymbol{q}_\rho \boldsymbol{\rho}) + \delta T_{e2q} \exp (-iq_s s - i\boldsymbol{q}_\rho \boldsymbol{\rho}). \end{aligned} \tag{6.9}$$

Substituting Equations (6.9) in Equations (6.2)–(6.4) and recognizing that the component of the perturbation $|E|^2$ is equal to

$$\delta|E|^2 = E_0\{(E_{1q} + E^*_{2q})\exp(iq_s s + i\mathbf{q}_\rho\boldsymbol{\rho}) + (E_{2q} + E^*_{1q})\exp(-iq_s s - i\mathbf{q}_\rho\boldsymbol{\rho})\}, \tag{6.10}$$

we obtain δT_q and δN_q:

$$\delta T_{e1q} = \frac{T_{e0}\varphi E_0}{(1+F_T)E_p^2}(E_{1q} + E^*_{2q}), \qquad \delta T_{e2q} = \frac{T_{e0}\varphi E_0}{(1+F_T)E_p^2}(E^*_{1q} + E_{2q}),$$

$$\delta N_q = -\delta T_{eq}\frac{N_0}{T_{e0}}\frac{k_T F_N - \gamma_1}{1+F_N}. \tag{6.11}$$

$$F_T = L_T^2(q_x \sin\alpha_1 + q_s\cos\alpha_1)^2, \qquad F_N = L_N^2(q_x\sin\alpha_1 + q_s\cos\alpha_1)^2.$$

Substituting now Equations (6.11) and (6.8) in Equation (6.1) and linearizing the latter, we have

$$\begin{aligned}(-2kq_s - q_x^2 - q_y^2)E_{1q} + \beta_1(E_{1q} + E^*_{2q}) &= 0,\\ (2kq_s - q_x^2 - q_y^2)E_{2q} + \beta_1(E^*_{1q} + E_{2q}) &= 0.\\ \beta_1 = \frac{\omega_0^2 E_0^2\varphi}{c^2E_p^2}\frac{k_TF_N - \gamma_1}{(1+F_T)(1+F_N)}.\end{aligned} \tag{6.12}$$

It follows therefore that the sought dispersion equation for Equations (6.1)–(6.4) is

$$q_s^2 = -\frac{(q_x^2+q_y^2)}{4k^2}\left[\frac{2\omega_0^2E_0^2\varphi}{c^2E_p^2}\frac{k_TF_N-\gamma_1}{(1+F_N)(1+F_T)} - q_x^2 - q_y^2\right]. \tag{6.13}$$

If $q_s^2 < 0$, then small perturbations of the field in the (x, y) plane grow in the propagation direction s. The wave is unstable and tends to become stratified. The instability threshold $E_{0\,\text{th}}$ is obtained by letting $q_s \to 0$. We have from Equation (6.13)

$$\begin{aligned}E_{0\,\text{th}} &= \frac{cf_m^{1/2}E_p}{\sqrt{2\varphi k_T\omega_0 L_N}\sin\alpha_1},\\ f(q_x) &= \frac{k_TL_N^2q_x^2\sin^2\alpha_1(1+L_N^2q_x^2\sin^2\alpha_1)(1+L_T^2q_x^2\sin^2\alpha_1)}{k_TL_N^2q_x^2\sin^2\alpha_1 - \gamma_1},\end{aligned} \tag{6.14}$$

where f_m is the minimal value of the function $f(q_x)$. The instability sets in at $E_0 > E_{0\,\text{th}}$. The threshold field $E_{0\,\text{th}}$ is determined by the recombination-ionization processes and by the longitudinal diffusion. In the upper ionosphere, at height $z \gtrsim 250$–300 km, the ionization balance shift coefficient $\gamma_1 \to 0$. The minimum $f = f_m$ is reached here at $q_x \to 0$, with $f_m \approx 1$. The values of the threshold field in the upper ionosphere are very small, $E_{0\,th}/E_p \lesssim 10^{-3}$. In the region below 250 km, the minimal value of f increases. In addition, the values of L_N and ω_0 decreases here. All this leads to a strong increase of the threshold field.

To get an idea of the characteristic scales of the instability, we consider the simple case of transverse wave propagation $\alpha_1 = \pi/2$, $\sin\alpha_1 = 1$. In this case, for the upper ionosphere ($\gamma_1 = 0$), Equation (6.13) for q_s takes the form

$$q_s^2 = -\frac{q_x^2 + q_y^2}{4k^2}\left[\frac{2\omega_0^2 E_0^2 \varphi}{c^2 E_p^2}\,\frac{k_T L_N^2 q_x^2}{(1 + L_N^2 q_x^2)(1 + L_T^2 q_x^2)} - q_x^2 - q_y^2\right]. \quad (6.15)$$

The instability threshold is defined by Equation (6.14) with $f_m = 1$ and $\sin\alpha_1 = 1$. The instability region is shown in Figure 68. The unstable range of q_x is from 0 to q_{xm} and that of q_y is from 0 to q_{ym}, where

$$q_{xm}^2 = \left\{\left[(L_N^2 - L_T^2)^2 + \frac{8k_T\omega_0^2\varphi E_0^2}{c^2 E_p^2} L_N^4 L_T^2\right]^{1/2} - (L_N^2 + L_T^2)\right\}\Big/ 2L_N^2 L_T^2$$

$$q_{ym}^2 \simeq \frac{1}{L_N L_T}\left\{\frac{2k_T\omega_0^2\varphi E_0^2 L_N^3 L_T}{E_p^2 c^2 (L_N + L_T)^2} - 1\right\}.$$

The maximum instability increment

$$q_{sm} = \pm i\,\frac{k_T\omega_0^2\varphi E_0^2 L_N^2}{2kc^2 E_p^2 (L_N + L_T)^2} \quad (6.16)$$

is reached at

$$q_{x0} \approx (L_N L_T)^{-1/2}, \qquad q_{y0} \approx (L_N L_T)^{-1/2}\left[\frac{k_T\omega_0^2\varphi E_0^2 L_N^3 L_T}{c^2 E_p^2 (L_N + L_T)^2} - 1\right]. \quad (6.17)$$

The resultant inhomogeneity is strongly elongated in the magnetic field direction. It is seen, furthermore, that under the conditions of sufficiently advanced instability, $(\omega_0/c)L_N(E_0/E_p) \gg 1$, we have $q_{y0} \gg q_{x0}$ and $q_{y0} \gg q_{sm}$. This means that the scale of the field inhomogeneity in the plasma is minimal along the y axis, which is perpendicular to the $(\boldsymbol{H}, \boldsymbol{s})$ plane. At $L_N \gtrsim L_T$ the minimal scale is $\sim 1/q_{ym}$, where $q_{ym} \sim (E_0/E_p)(\omega_0/c)$.

The temporal parameters of the instability in question are determined by the characteristic times τ_T and τ_N of the plasma heating and the establishment of the ionization in the plasma (see Table 13). The instability develops within a time $\sim\tau_T$, and is fully developed after a time $\sim\tau_N$.

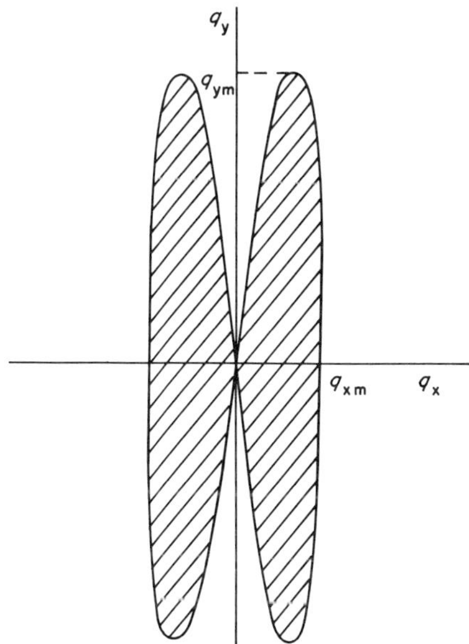

Fig. 68. Region of self-focusing instability (*shaded*)

Narrow Beam. We have considered above the instability of a plane wave. This is valid also for radio beams if the field amplitude in a beam changes little over the characteristic instability dimension, meaning a dimension on the order of the lengths L_T and L_N. At the same time, the lengths L_T and L_N in the upper ionosphere are large (see Table 16), so that in real conditions the field E_0 can change strongly over such distances. Therefore, an important feature for the upper ionosphere, especially in the region above the maximum of the F layer, is also the opposite limiting case of a narrow beam, when the transverse beam dimension a is small in comparison with the lengths L_T and L_N:

$$a \ll L_T \sin \alpha_1, L_N \sin \alpha_1 \tag{6.18}$$

In this case the temperature and plasma-concentration perturbations are determined by the integral Equations (5.118). Equation (6.1), with Equation (5.118a) taken into account, then takes the form.

$$2ik \frac{\partial E}{\partial s} + \frac{\partial^2 E}{\partial x^2} + \frac{\partial^2 E}{\partial y^2} + \left(\beta_2 \int |E^2| \, dx \right) E = 0, \tag{6.19}$$
$$\beta_2 = \omega_0^2 \varphi (k_T L_N - \gamma_1 L_T) / 2c^2 E_p^2 \sin \alpha_1 L_T (L_N + L_T).$$

We have made allowance here for the fact that the beam width is much smaller here than L_N and L_T [Eq. (6.18)], so that δN is independent of x in the region occupied by the beam. In addition, we have changed over from integration with respect to $dx_{\|}$ to integration with respect to $dx = \sin \alpha_1 \, dx_{\|}$.

Let us obtain for Equation (6.19) a solution $E = E_0(x, s)$ that is homogeneous in the coordinate y, and let us test its stability. To this end we change over in Equation (6.19) to Fourier transforms with respect to the coordinate x:

$$E_{q_x}(y, s) = \frac{1}{2\pi} \int_{-\infty}^{\infty} E(x, y, s) \exp(-iq_x x) \, dx \tag{6.20}$$

and eliminate the diffraction term $\partial^2 E/\partial x^2$ with the aid of the substitution

$$E_{q_x}(y, s) = \tilde{E} \exp(-iq_x^2 s/2k). \tag{6.21}$$

As a result, Equation (6.19) for the amplitudes $\tilde{E}(q_x, y, s)$ takes the form

$$2ik \frac{\partial \tilde{E}}{\partial s} + \frac{\partial^2 \tilde{E}}{\partial y^2} + \left[2\pi\beta_2 \int |\tilde{E}(q_x, y, s)|^2 \, dq_x \right] \tilde{E} = 0. \tag{6.22}$$

It follows immediately from Equation (6.22) that the solution $\tilde{E} = \tilde{E}(q_x, s)$, which is homogeneous in the coordinate y, is characterized, just as in the case of the plane wave [Eq. (6.5)], only by the wave-vector shift Δk brought about by the nonlinear modification of the plasma by the homogeneous field:

$$\tilde{E}(q_x, s) = E_{0q_x} \exp(i\,\Delta k s), \qquad \Delta k = \frac{\pi\beta_2}{k} \int |\tilde{E}|^2 \, dq_x = \frac{\beta_2}{2k} \int |E_0^2(x)| \, dx. \tag{6.23}$$

Here $E_0(x)$ is the initial distribution of the field amplitude, given in the plane $s = 0$, while E_{0q_x} is its Fourier transform defined in accordance with Equation (6.20). We recognize also that Equation (6.21) yields $E_{q_x} = \tilde{E}(q_x)$ at $s = 0$. From Equation (6.23) it is seen that the shift Δk of the wave vector is constant and is determined only by the initial distribution of the amplitude $E_0(x)$ in the beam. Taking the inverse Fourier transform, in accordance with Equations (6.20), (6.21), and (6.23), we obtain the sought field distribution homogeneous in y and equivalent to the field $E_0 \exp(i\,\Delta ks)$ in Equation (6.5) in the case of a plane wave:

$$E_0(x, s) = \exp(i\,\Delta ks) \int E_{0q_x} \exp(-iq_x^2 s^* + iq_x x)\, dq_x, \qquad s^* = \frac{s}{2k}. \quad (6.24)$$

We consider now the instability of this solution with respect to small perturbations in the direction of the y axis, which is perpendicular to the $(\boldsymbol{H}, \boldsymbol{s})$ plane. To this end, we use a linearization in analogy with Equation (6.8):

$$\begin{aligned} E(x, y, s) = E_0(x, s) &+ E_1(x, s) \exp(iq_y y + i\,\Delta ks) \\ &+ E_2(x, s) \exp(-iq_y y + i\,\Delta ks), \end{aligned} \quad (6.25)$$

or, in terms of the Fourier transforms [Eq. (6.20)]:

$$\tilde{E}(q_x, y, s) = E_{0q_x} + \tilde{E}_1(q_x, s) \exp(iq_y y) + \tilde{E}_2(q_x, s) \exp(-iq_y y). \quad (6.26)$$

The perturbation amplitudes $E_{1,2}$ are connected with corresponding values of $\tilde{E}_{1,2}$ by relations of the type Equation (6.24)

$$E_{1,2}(x, s) = \int \tilde{E}_{1,2}(q_x, s) \exp(-iq_x^2 s^* + iq_x x + i\,\Delta ks)\, dq_x. \quad (6.27)$$

Substituting Equations (6.26) in Equations (6.22) and linearizing the latter, taking Equation (6.10) into account, we arrive at a system of two linear equations for the amplitudes $\tilde{E}_{1,2}(q_x, s)$:

$$\begin{aligned} 2ik \frac{d\tilde{E}_1}{ds} - q_y^2 \tilde{E}_1 + \beta_2(a_1 + a_2) E_{0q_x} &= 0, \\ -2ik \frac{d\tilde{E}_2^*}{ds} - q_y^2 \tilde{E}_2^* + \beta_2(a_1 + a_2) E_{0q_x}^* &= 0. \end{aligned} \quad (6.28)$$

We have introduced here the following notation for the averaged values of the intensity perturbations:

$$\begin{aligned} a_1(s) &= \int E_0^*(x, s) E_1(x, s)\, dx = 2\pi \int E_{0q_x}^* \tilde{E}_1\, dq_x, \\ a_2(s) &= \int E_0(x, s) E_2^*(x, s)\, dx = 2\pi \int E_{0q_x} \tilde{E}_2^*\, dq_x. \end{aligned} \quad (6.29)$$

Multiplying now the first equation of Equation (6.28) by $E^*_{0q_x}$, the second by $E^*_{0q_x}$, and integrating them with respect to dq_x, we have

$$2ik\frac{da_1}{ds} - q_y^2 a_1 + B(a_1 + a_2) = 0,$$

$$-2ik\frac{da_2}{ds} - q_y^2 a_2 + B(a_1 + a_2) = 0, \tag{6.30}$$

$$B = \beta_2 \int |E_0^2(x)|\, dx.$$

Seeking the solution of these equations in the form $a_{1,2} \sim \exp(iq_s s)$, we obtain the dispersion equation

$$q_s^2 = \frac{q_y^2}{4k^2}[q_y^2 - 2B]. \tag{6.31}$$

We see that at $q_y < \sqrt{2B}$ the beam is unstable. It becomes stratified in the direction of the y axis. The maximum instability increment is

$$q_{sm} = \pm iB/2k. \tag{6.32}$$

It is reached at $q_y = B^{1/2}$. The instability develops at $\beta_2 > 0$, i.e., at

$$\gamma_1 < k_T L_N / L_T. \tag{6.33}$$

Narrow beams are therefore unstable only in the upper region of the ionosphere, $z \gtrsim 200$ km.

It is interesting that in the case of narrow beams, in contrast to a plane wave, the self-focusing instability has no threshold and develops at arbitrarily small values of the amplitude E_0. The reason is that a plane wave, unlike a narrow beam, does not produce in the plasma the negative concentration perturbations that are needed for self-focusing [see Eq. (6.7)]. The appearance of such perturbations is connected only with the inhomogeneities of the field amplitude and of the concentration in the $\boldsymbol{H}$ direction. Such perturbations, however, are decreased by the longitudinal diffusion. To maintain them it is therefore necessary to have a finite threshold value $E_0 > E_{0\,\mathrm{th}}$ of the amplitude.

We note that the analysis of the nonlinear processes shows that the self-focusing instability reaches a saturation value in the ionosphere already at $\delta N/N_0 \sim (1\text{–}10)\%$ (see Sect. 6.1.2).

6.1.2. Instability in the Wave-Reflection Region

Perturbations of the plasma concentration in the reflection region exert the strongest influence on the propagation of radio waves (Georges, 1970; Al'pert, 1974). Nonlinear effects are also amplified in this region (see Sect. 5.4). It is natural therefore to expect an enhancement of self-focusing effects, too. At the same time, the plasma cannot be regarded as homogeneous in the wave-reflection region. Moreover, both the incident and the reflected waves are always present in the reflection region. The interaction of these waves, as well as allowance for the nonlocality and inhomogeneity of the plasma, alter significantly the character of the self-focusing instability.

Let a plane wave propagate in a direction normal to an inhomogeneous plasma layer with $N_0(z)$. We consider both the wave incident on the layer and the reflected wave. Within this framework of the validity of the geometric-optics approximation, the amplitudes of the total electric field E can be represented as the sum of the amplitudes of the incident wave E^+ and the reflected wave E^-, each of which is described by Equation (5.166):

$$E = \sqrt{\frac{P}{2}}\,[E^+ \exp(-i\varphi) + E^- \exp(i\varphi)]$$
$$\varphi = \int_0^z k_0(z)\,dz - \frac{\pi}{4}, \qquad k_0 = \frac{\omega}{c}\sqrt{\varepsilon_0}, \qquad P = 2/\sqrt{\varepsilon_0}. \tag{6.34}$$

Here P is the amplification factor of the wave field in the reflection region [Eq. (5.124)]. In addition, account is taken of the fact that it will be more convenient in what follows to reckon the coordinate z from the wave-reflection point z_0, when $\varepsilon = \varepsilon_0(0)$. For the values of z of interest to us to be positive in this case, the z axis is directed opposite to the concentration gradient (downward in the ionosphere). The substitution of Equation (6.34), in analogy with Equation (5.167), reduces the wave Equation (5.166) in the parabolic approximation [Equation (5.169)] to a system of equations in the form of Equation (6.1) for the amplitudes of the incident and reflected waves:

$$-2ik\frac{\partial E^+}{\partial z} + \Delta_\perp E^+ - \frac{\omega^2}{c^2}\frac{\delta N}{N_0}E^+ = 0,$$
$$2ik\frac{\partial E^-}{\partial z} + \Delta_\perp E^- - \frac{\omega^2}{c^2}\frac{\delta N}{N_0}E^- = 0, \tag{6.35}$$

$$E^+(0) = E^-(0). \tag{6.36}$$

Here, as before, we have neglected the absorption and took into account the fact that in the reflection region we have $\omega^2 \simeq \omega_0^2$ and $N_0 = N(0)$.

Equations (6.35) have a stationary solution corresponding to a wave that is homogeneous in the (x, y) plane:

$$E^+ = E^- = E_0 \tag{6.37}$$

E_0 is the amplitude of the wave incident on the layer. The correction Δk [Eq. (6.6)], due to the change δN_0 in the concentration, is included in Equation (6.35) the definition of the wave number k

$$k = k_0 + \Delta k = \omega\sqrt{\varepsilon}/c, \qquad \varepsilon = 1 - \omega_0^2/\omega^2 - \delta N_0/N_0.$$

We consider now the stability of the homogeneous solution [Eq. (6.37)] with respect to small perturbations of the incident-wave amplitude in the (x, y) plane. Assuming, in analogy with Equation (6.8),

$$\begin{aligned} E^+ &= E_0 + E_1 \exp(iq_x x + iq_y y) + E_2 \exp(-iq_x x - iq_y y), \\ E^- &= E_0 + E_3 \exp(iq_x x + iq_y y) + E_4 \exp(-iq_x x - iq_y y), \end{aligned} \tag{6.38}$$

and linearizing Equation (6.35), we arrive at the following equation for the dimensionless amplitudes

$$a_{1,3} = E_{1,3}/E_0, \qquad a_{2,4} = (E_{2,4}/E_0)^* \tag{6.39}$$

$$\begin{aligned} -2ik\frac{da_1}{dz} - (q_x^2 + q_y^2)a_1 - \frac{\omega^2}{c^2}\frac{\delta N}{N_0} &= 0, \\ 2ik\frac{da_2}{dz} - (q_x^2 + q_y^2)a_2 - \frac{\omega^2}{c^2}\frac{\delta N}{N_0} &= 0, \end{aligned} \tag{6.40}$$

with analogous expressions for a_3 and a_4. Introducing the quantity

$$A = a_1 - a_2|_{z\to 0} \tag{6.41}$$

we get from Equations (6.36) and (6.35):

$$\begin{aligned} a_4 &= a_1 - A\exp\left(i\int_0^z \frac{q_x^2 + q_y^2}{2k}\,dz\right), \\ a_3 &= a_2 + A\exp\left(-i\int_0^z \frac{q_x^2 + q_y^2}{2k}\,dz\right). \end{aligned} \tag{6.42}$$

The perturbation of $|E^2|$ now depends on both the incident and on the reflected waves [cf. Eq. (6.10)]:

$$\delta|E|^2 = \frac{PE_0^2}{2}(a_1 + a_2 + a_3 + a_4). \tag{6.43}$$

Here P is the field amplification factor [Eq. (6.34)]; only the amplitude of the perturbations is considered.

At the lower boundary $z = z_0$ of the plasma layer, the perturbations of incident wave are given:

$$a_1(z_0) = a_{10}, \qquad a_2(z_0) = a_{20}. \tag{6.44}$$

In other words, it is assumed that at $z = z_0$ the incident wave field is subject to a weak spatial modulation, $a_{10}, a_{20} \ll 1$. Equations (6.40)–(6.43) jointly with Equations (5.111) determine the change of this modulation as a function of the coordinate z and of the time.

It is important that owing to the interaction of incident and reflected waves the instability no longer has the same character as the stationary spatial instability considered in the preceding section. Indeed, were the perturbations to increase exponentially with z, the amplitude perturbations of the reflected waves would be exponentially large even at the plasma boundary z_0, and this would make it impossible to formulate a linearized boundary-value problem. It is necessary therefore to consider now a nonstationary growth of the perturbations.

Thus, let $t = 0$ be the instant when the boundary perturbations [Eq. (6.44)] in the incident wave are turned on. Since the coefficients in the equations do not depend on time, it is natural to change over to Laplace transforms of the corresponding variables in accordance with the formula

$$a(p) = \int_0^\infty a(t)\exp(-pt)\,dt. \tag{6.45}$$

The perturbations of the temperature and of the plasma concentration are then described, as before, by Equation (6.11), except that $1 + F_N$ and $1 + F_T$ in the denominators must be replaced by $1 + F_N + p\tau_N$ and $1 + F_T + p\tau_T$, respectively. We are interested in the initial stage of the plasma modification, when the values of p are large enough. On the other hand, we take into account the fact that $\tau_T \ll \tau_N$. We consider therefore the case when

$$F_T\frac{\tau_N}{\tau_T} \gg p\tau_N \gg F_N, 1; \qquad F_T \gg 1. \tag{6.46}$$

We assume, in addition, that $\gamma_1 \to 0$. We then obtain for δN from Equation (6.11):

$$\frac{\delta N}{N_0} = -\frac{k_T D_a \varphi}{p L_T^2 E_p^2} \delta |E|^2 = -\frac{k_T D_a \varphi E_0^2}{2p L_T^2 E_p^2} P(a_1 + a_2 + a_3 + a_4). \quad (6.47)$$

In Equation (6.46) the concentration perturbations are locally connected with the wave-intensity perturbations. Equation (6.46) signifies that within the characteristic time $1/p$ the plasma diffuses over a distance $\sqrt{D_a/p}$ which is much shorter than the characteristic scale of the perturbations of $|E|^2$. Equation (6.47) follows directly from Equation (5.111) under conditions when the perturbations of the concentration are determined by thermal diffusion, and the perturbations of the temperature are determined by thermal conductivity.

The linear Equations (6.40) and (6.42), with allowance for Equation (6.47), constitute a closed system. We assume that the electron density in the plasma varies linearly

$$\varepsilon_0 = \mu z, \qquad \mu = \frac{1}{N_0}\left(\frac{dN}{dz}\right)_{z=0} \quad (6.48)$$

We introduce the dimensionless variables

$$\begin{aligned} g &= a_1 + a_2, & \eta &= \kappa^2 \sqrt{k_m z}; \\ \kappa^2 &= (q_x^2 + q_y^2)/k_m^2, & k_m &= (\omega/c)^{2/3} \mu^{1/3}. \end{aligned} \quad (6.49)$$

Here k_m is the minimal wave number and is reached at the limit $z \simeq 1/k_m$ of applicability of geometrical optics. Then

$$\frac{dg}{d\eta} = i(a_1 - a_2) \quad (6.50)$$

and Equations (6.40), (6.42), and (6.47) are rewritten in the form

$$\begin{aligned} \frac{d^2 g}{d\eta^2} + \left(1 + 2i\frac{\lambda}{\eta}\right) g &= 2A\lambda \frac{\sin \eta}{\eta}; \\ \lambda &= \frac{i}{p} \frac{2E_0^2 k_T \varphi D_a \omega}{E_p^2 L_T^2 c \mu}. \end{aligned} \quad (6.51)$$

In terms of the new variables, Equation (6.41) becomes

$$iA = \left.\frac{dg}{d\eta}\right|_{\eta = \kappa^2}. \quad (6.52)$$

The perturbations of the density and of the amplitude of the reflected wave are

$$\frac{\delta N}{N_0} = \frac{i\kappa^2\lambda}{(\omega/c\mu)^{2/3}\eta}(g - iA \sin \eta). \tag{6.53}$$

$$a_3 = a_2 + A \exp(-i\eta); \qquad a_4 = a_1 - A \exp(i\eta).$$

The solution of the differential Equation (6.51) is of the form

$$g(\eta) = C_1 g_1 + C_2 g_2 - \frac{2A\lambda}{\Delta}\left\{-g_1 \int_{\eta_0}^{\eta} \frac{\sin \eta}{\eta} g_2 \, d\eta + g_2 \int_{\eta_0}^{\eta} \frac{\sin \eta}{\eta} g_1 \, d\eta\right\}. \tag{6.54}$$

Here $\Delta - g_1 \, dg_2/d\eta - g_2 \, dg_1/d\eta$ is the Jacobian of the two independent solutions of the corresponding homogeneous equations, while $\eta_0 = \eta(z_0) = (q_x^2 + q_y^2)c/\omega\mu$ is the value of the dimensionless coordinate at the plasma boundary $z = z_0 = \mu^{-1}$. The constants C_1 and C_2 are chosen by starting from the boundary conditions [Eq. (6.44)]

$$g\big|_{\eta_0} = a_{10} + a_{20}, \qquad \frac{dg}{d\eta}\bigg|_{\eta_0} = i(a_{10} - a_{20}). \tag{6.55}$$

Substituting Equations (6.54) in Equations (6.52) we determine also the constant A:

$$A = \frac{C_1 \dfrac{dg_1}{d\eta} + C_2 \dfrac{dg_2}{d\eta}}{i + 2\lambda D/\Delta}\Bigg|_{\eta = \kappa^2}; \qquad D = \frac{dg_2}{d\eta}\int_{\eta_0}^{\eta} \frac{\sin \eta}{\eta} g_1 \, d\eta - \frac{dg_1}{d\eta}\int_{\eta_0}^{\eta} \frac{\sin \eta}{\eta} g_2 \, d\eta. \tag{6.56}$$

The independent solutions of the homogeneous Equation (6.51) are expressed in terms of the Whittaker function $W_{\lambda,\mu}(z)$ of imaginary argument:

$$g_1(\eta) = W_{\lambda,1/2}(2i\eta), \qquad g_2(\eta) = W_{-\lambda,1/2}(-2i\eta) - g_1^*(\eta). \tag{6.57}$$

The expressions obtained solve our problem in principle. It is important, that all the sought quantities are proportional to the constant A, which contains the denominator $i + 2\lambda D/\Delta$. The zeroes $p = p_n$ of this denominator determine the singular points of the solution. When the inverse Laplace transform is taken, these singularities yield terms $\sim \exp(p_n t)$ that increase exponentially with time. Thus, equating the denominator of

Equation (6.56) to zero

$$\frac{2i\lambda}{\Delta}\left\{\frac{dg_2}{d\eta}(\kappa^2)\int_{\eta_0}^{\kappa^2}\frac{\sin\eta}{\eta}g_1\,d\eta-\frac{dg_1}{d\eta}(\kappa^2)\int_{\eta_0}^{\kappa^2}\frac{\sin\eta}{\eta}g_2\,d\eta\right\}=1 \quad (6.58)$$

we obtain a dispersion equation that defines the growth rate of the instability of the system in question. We emphasize that given q_x and q_y the perturbation increases in similar fashion at all values of z, this being the consequence of the effective interaction of the incident and reflected waves.

In the limit of large $\eta_0 \gg 1$ and small $\kappa^2 \to 0$, the dispersion Equation (6.58) takes the simpler form

$$\frac{(2i\eta_0)^\lambda}{[\Gamma(1-\lambda)]^*}-\frac{[(2i\eta_0)^\lambda]^*}{\Gamma(1-\lambda)}=0, \quad (6.59)$$

where $\Gamma(x)$ is the Euler gamma function.[19] Recognizing that according to Equation (6.51) we have $\lambda = i|\lambda|$ and $\lambda^* = -i|\lambda|$, we rewrite this equation in the form

$$|\lambda| \ln 2\eta_0 = \arg\Gamma(1+i|\lambda|)+\pi n, \quad (6.60)$$

where n is an integer. It follows therefore that at large values of $\ln 2\eta_0$ the minimum value of $|\lambda|$, corresponding to the maximal growth rate p_1, is equal to

$$|\lambda_1| \simeq \pi/\ln 2\eta_0, \qquad \eta_0=(q_x^2+q_y^2)c/\omega\mu. \quad (6.61)$$

[19] In the calculation of the integrals in Equation (6.58) it is convenient to use an integral representation for the function $g_1(\eta)$:

$$g_1(\eta)=\frac{e^{-i\eta}}{\Gamma(1-\lambda)}\int_0^\infty dt e^{-t}t^{-\lambda}(2i\eta+t)^\lambda=\frac{\eta e^{-i\eta}}{\Gamma(1-\lambda)}\int_0^\infty dt e^{-\eta t}t^{-\lambda}(2i+t)^\lambda.$$

The value of the integral is

$$\int_\infty^\eta \frac{\sin\eta}{\eta}g_1\,d\eta=\frac{e^{i\eta}}{2i\lambda}g_1(\eta)-\frac{\sin\eta}{\sqrt{2i\eta}}W_{\lambda-1/2,0}(2i\eta).$$

In the derivation of Equation (6.59) we have taken into account, furthermore, the fact that at $\eta_0 \gg 1$ we have

$$\int_{\eta_0}^\infty\frac{\sin\eta}{\eta}g_1\,d\eta \simeq -(2i\eta_0)^\lambda/2i\lambda.$$

As $\eta \to 0$ we have

$$W_{\lambda-1/2,0}\sim\sqrt{\eta}\ln\eta, \qquad dg_1/d\eta=-2i\lambda\ln\eta/\Gamma(1-\lambda).$$

Thus, the solution of the dispersion Equation (6.58), which determines the instability growth rates p_k, is represented in the form [Eq. (6.51)]:

$$p_k = \frac{2E_0^2 \varphi \omega k_T D_a}{E_p^2 c \mu L_T^2} |\lambda_k|^{-1}. \tag{6.62}$$

The instability growth rate is proportional to the intensity E_0^2 of the modifying field and is inversely proportional to the density gradient μ in the wave-reflection region. It increases slowly with decreasing homogeneity scale $1/q$.

The foregoing analysis is restricted to Equation (6.46). An analysis of the more general case, when Equations (6.46) are not satisfied, leads to similar results (Vas'kov and Gurevich, 1976c). The maximum values of the growth increments occur over distances $q_y \sim (E_0/E_p)k_m$, where $2\pi/k_m$ is the length of the modifying wave in the reflection region [cf. Eq. (6.17)]. The excited inhomogeneities are elongated along $\boldsymbol{H}$ and their dimension is minimal along the y axis, which is perpendicular to $\boldsymbol{H}$ and to the direction of the wave group velocity $\boldsymbol{v}_g$.

We have neglected above the influence of the magnetic field on the wave propagation. In the reflection region, this influence is appreciable. However, Equations (6.35) retain the same form as before, and allowance for the influence of the magnetic field leads only to a change of the coefficients (Vas'kov and Gurevich, 1976b, 1976c). In particular, the field amplification coefficient P [Eq. (6.34)] is changed; it can increase in the case of the ordinary wave at small angles α between $\boldsymbol{H}$ and the vertical z [Eq. (5.124)]. Therefore the ordinary wave modifies the plasma more effectively than the extraordinary wave at $\alpha \ll 1$. It is important also that in the region of reflection of the ordinary wave there develop additional dissipative processes, which can also contribute to enhancement of the self-focusing instability (see Sects. 6.2 and 6.3).

We have considered above only the initial linear stage of the self-focusing instability. The development of the instability leads to a deep almost complete spatial modulation of the wave field amplitude (Vas'kov and Gurevich, 1977a). However, the plasma concentration perturbations are in this case small, $\Delta N/N \lesssim (\Delta N/N)_m \approx 0.1 L_{T0}/L_{N0}$, and in the region below the maximum of the F layer $(\Delta N/N)_m$ does not exceed 2–10% (see Table 16). The cause of the rapid saturation of the density perturbations is perfectly analogous to that considered in Section 5.3.1. Stabilization of N is brought about by the increase in the thermal-conductivity length $L_T \sim T_e^{5/4}$ and by the associated enhancement of the relative role of the electron recombination (strengthening of the inequality $L_N < L_T$). In the case of a weak field $E_0^2 \ll E_p^2$, the steady-state density perturbations increase in proportion to the power of the incident wave. However, the growth of the perturbations ΔN slows down already at $E_0^2 \sim 0.1E_p^2$, and saturation $\Delta N/N \to (\Delta N/N)_m$ sets in at $E_0^2 \sim E_p^2$. Allowance for the additional dissipation of the incident-wave energy (see Sects. 6.2 and 6.3) only accelerates the saturation of ΔN. The characteristic spatial period of the field modulation increases at both small and large values of E_0^2/E_p^2.

We note also that we did not take into account above the transverse transport processes. This is valid if Equations (5.31) are satisfied, i.e., at $R_\perp > (\nu_{im}/\Omega_H)L_N$ (it is assumed here that $R_\parallel \sim L_N$ and $\nu_{im} \gg \delta_{el}^i \nu_{ei}$). If this condition is not satisfied, then the transverse diffusion, which weakens the instability, becomes appreciable. The dimension in the F-layer of the ionosphere is $(\nu_{im}/\Omega_H)L_N \sim 10^2$ m.

The large-scale stratification of the ionosphere under the influence of radio waves was discovered experimentally in 1970 in a study of the results of ionosphere sounding in a region modified by a high-power transmitter—formation of the sporadic F layer (Utlaut et al., 1970). It was subsequently investigated in detail by various radio sounding methods (Rufenach, 1973; Wright, 1973; Allen et al., 1974; Bowhill, 1974; Utlaut and Violette, 1974; Belikovich et al., 1974). It was shown that an artificial inhomogeneous structure develops in the F region of the ionosphere under the influence of a high-power field $E_0 \sim E_p$. The inhomogeneities are elongated in the direction of the earth's magnetic field. The spectrum of their dimensions is very broad, from dozens of meters to 10 km. The electron-density fluctuations in the inhomogeneities is of the order of several percent, i.e., they are comparable with the total change of the electron density in the modified zone (see Sect. 5.3.3). The spectrum of N_q changes little with changing dimension q.

The large-scale inhomogeneities appear on the order of several seconds after the modifying transmitter is turned on, and their full development takes several minutes. Both the characteristic scales of the inhomogeneities and their development times agree with the theory of self-focusing instability. It can therefore be assumed that the development of a large-scale inhomogeneous structure in the ionosphere plasma, i.e., a structure with inhomogeneous scales $1/q$ exceeding the length of the wave perturbations,

$$qc/\omega < 1, \tag{6.63}$$

is due to self-focusing instability.

6.2. Resonant Absorption and Resonance Instability

In the preceding sections we have considered plasma heating in the field of high-power radio waves, due only to the collision and Joule energy losses of the wave. Yet in a tenuous plasma, particularly in the upper ionosphere, the electron collisions are relatively infrequent. Therefore the ordinary collision absorption is small here and weakly damped natural oscillations of the plasma—plasma waves—can exist. When certain resonance conditions are satisfied, the radio wave excites these oscillations, and this leads to the appearance of a new mechanism of wave-energy dissipation, which exerts a decisive influence on the character of the nonlinear processes in a tenuous plasma.

In this section we consider the effects of the resonant absorption due to excitation of plasma waves by stationary inhomogeneities of the plasma (Gurevich and Vas'kov, 1975, 1977).

6.2.1. Langmuir Oscillations in an Inhomogeneous Plasma

Potential oscillations of the electric field in a plasma are described by the Poisson Equation (3.2)

$$\operatorname{div} \boldsymbol{D} = 4\pi\rho_{\omega}, \qquad \boldsymbol{D} = \hat{\varepsilon}\boldsymbol{E}, \qquad D_i = -\varepsilon_{ij}\frac{\partial\varphi}{\partial x_j}. \tag{6.64}$$

Here φ is the potential of the electric field, ρ_{ω} is the density of the alternating electric charge (at frequency ω) produced by the external sources, and $\hat{\varepsilon}$ is the dielectric tensor of the plasma. The tensor components ε_{ij} were written out in Sections 2.1 and 2.3 above in a coordinate system with the z axis directed along the magnetic field. It is convenient in what follows to express the tensor also in a coordinate system that is rotated through an angle θ: the magnetic field lies in the (x_2, x_3) plane and makes an angle θ with the x_3 axis. In this case we have (Ginzburg, 1960, Sect. 3.3):

$$\begin{gathered}\varepsilon_{11} = \varepsilon_{xx}, \qquad \varepsilon_{22} = \varepsilon_{xx}\cos^2\theta + \varepsilon_{zz}\sin^2\theta, \qquad \varepsilon_{33} = \varepsilon_{xx}\sin^2\theta + \varepsilon_{zz}\cos^2\theta, \\ \varepsilon_{23} = (\varepsilon_{zz} - \varepsilon_{xx})\sin\theta\cos\theta, \qquad \varepsilon_{13} = -i\varepsilon_{xy}\sin\theta, \qquad \varepsilon_{12} = i\varepsilon_{xy}\cos\theta.\end{gathered} \tag{6.65}$$

Here ε_{xx}, ε_{zz}, and ε_{yy} are the components of the tensor $\hat{\varepsilon}$ in the old system [Eqs. (2.16), (2.152)]. Under the condition of weak absorption $\nu_e \ll \omega$, the kinetic corrections are negligible, $K_{\varepsilon} \simeq 1$, and

$$\begin{gathered}\varepsilon_{xx} = 1 - \frac{v}{1-u}, \qquad \varepsilon_{xy} = -\frac{v\sqrt{u}}{1-u}, \qquad \varepsilon_{zz} = 1 - v, \\ v = \omega_0^2/\omega^2 = 4\pi e^2 N/m\omega^2, \qquad u = \omega_H^2/\omega^2.\end{gathered} \tag{6.66}$$

This represents the electronic part of the tensor $\hat{\varepsilon}$ [Eq. (2.16)]. Accordingly we shall henceforth consider only the electron Langmuir plasma oscillations.

For the natural oscillations—oscillations in the absence of an external source, $\rho_\omega = 0$—we obtain from Equation (6.64) the dispersion equation

$$k_i k_j \varepsilon_{ij} = 0, \tag{6.67}$$

which defines the natural frequency ω. Here k_i are the components of the wave vector $\boldsymbol{k}$. We direct $\boldsymbol{k}$ along the x_3 axis. Then the dispersion Equation (6.67) takes the form

$$\varepsilon_{33} = 1 - \frac{v(1 - u\cos^2\theta)}{1 - u} = 0.$$

It follows therefore that the plasma-wave frequency is equal to

$$\omega^2 = \tfrac{1}{2}\{\omega_0^2 + \omega_H^2 + [(\omega_0^2 + \omega_H^2)^2 - 4\omega_0^2\omega_H^2\cos^2\theta]^{1/2}\}. \tag{6.68}$$

Here θ is the angle between $\boldsymbol{k}$ and $\boldsymbol{H}$. The frequency ω as a function of the angle θ ranges from the plasma frequency ω_0 at $\theta = 0$ to the upper hybrid resonance frequency $\sqrt{\omega_0^2 + \omega_H^2}$ at $\theta = \pi/2$.

We consider now a weakly inhomogeneous plasma layer with $N = N(z)$, i.e., $v = v(z)$. The region of the existence of plasma waves of frequency $\omega > \omega_H$ in such a layer as follows from Equation (6.68), is

$$1 - u \leq v \leq 1, \qquad u < 1. \tag{6.69}$$

We shall reckon the coordinate z from the point $v = 1$, i.e., $\omega = \omega_0(0)$. This region is shown shaded in Figure 69.

Assume that the x_3 axis coincides with z; then the tensor $\varepsilon_{ij} = \varepsilon_{ij}(z)$ is independent of x or y. Therefore the Fourier transform of the potential φ with respect to the coordinates x and y

$$\varphi(k_1, k_2, z) = \frac{1}{(2\pi)^2}\iint \varphi(x, y, z)\exp(ik_1x + ik_2y)\,dx\,dy \tag{6.70}$$

satisfies in accordance with Equation (6.64) the equation

$$-\varepsilon_{33}\frac{d^2\varphi}{dz^2} + \left(2ik_2\varepsilon_{23} - \frac{d\varepsilon_{33}}{dz}\right)\frac{d\varphi}{dz} + \left(k_1^2\varepsilon_{11} + k_2^2\varepsilon_{22} + ik_2\frac{d\varepsilon_{23}}{dz} - ik_1\frac{d\varepsilon_{13}}{dz}\right)\varphi = 4\pi\rho_{\omega k}. \tag{6.71}$$

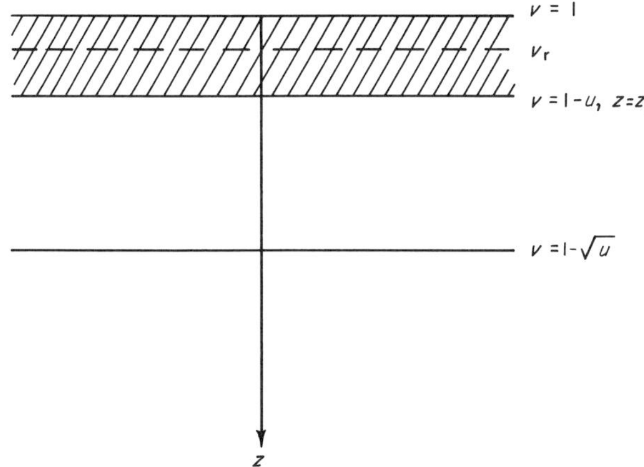

Fig. 69. Structure of radio-wave reflection region in a weakly inhomogeneous plasma. The region where plasma waves exist is shaded. The electron density N increases in the upward direction. $v = 4\pi e^2 N/m\omega^2$; $u = \omega_H^2/\omega^2$, $\omega_H = eH/mc$

Here $\rho_{\omega k}(k_1, k_2, z)$ is the transform, corresponding to Equation (6.70), of the charge density ρ_ω.

A plasma wave propagating through an inhomogeneous-plasma layer undergoes refraction, meaning that the projection k_3 of the wave vector $\boldsymbol{k}$ on the z axis is changed. This change is easy to determine. Indeed, the angle θ between $\boldsymbol{k}$ and $\boldsymbol{H}$ is connected with the angle α between $\boldsymbol{H}$ and z by the relation

$$\cos^2 \theta = (k_3 \cos \alpha + k_2 \sin \alpha)^2/(k_1^2 + k_2^2 + k_3^2). \tag{6.72}$$

On the other hand, at a fixed wave frequency ω the value of $\cos^2 \theta$ is determined by Equation (6.68). It follows from Equations (6.68) and (6.72) that

$$\begin{gathered} k_3 = \frac{k_2 \cos \alpha \sin \alpha\, uv \pm K\sqrt{(v - v_1)(v_2 - v)}}{1 + uv \cos^2 \alpha - u - v}, \\ K = \left[k_1^2 + k_2^2 - u(k_1^2 \cos^2 \alpha + k_2^2)\right]^{1/2}, \\ v_1 = 1 - u, \qquad v_2 = \frac{(k_1^2 + k_2^2)}{K^2}(1 - u). \end{gathered} \tag{6.73}$$

The same result can be arrived at directly from Equation (6.71). In fact, within the framework of the geometrical-optics approximation, i.e., under the conditions

$$\mu/k_3 \ll 1, \qquad \mu = dv/dz = (4\pi e^2/m\omega^2)\, dN/dz, \tag{6.74}$$

the linearly independent solutions of the homogeneous Equation (6.71) take the form

$$\varphi^{\pm} = \frac{1}{\sqrt{b}} \exp\left\{-i \int^{z} k_3^{\pm}\, dz\right\}, \tag{6.75}$$

where the value of k_3 determined from the dispersion Equation (6.67) is

$$k_3^{\pm} = -a \pm \frac{b}{\varepsilon_{33}}, \qquad a = \frac{\varepsilon_{23} k_2}{\varepsilon_{33}}, \qquad b = \sqrt{k_2^2(\varepsilon_{33}^2 - \varepsilon_{33}\varepsilon_{22}) - k_1^2 \varepsilon_{33}\varepsilon_{11}}. \tag{6.76}$$

It is easily seen that Equation (6.76) coincides with Equation (6.73).

It follows from Equations (6.73) and (6.75) that the plasma wave propagates in the region $v_1 \leq v \leq v_2$, as shown in Figure 70. At $v < v_1$ and $v > v_2$ the field of the wave decreases exponentially, while $v = v_1$ and $v = v_2$ define the points of reflection of the plasma wave. It is easy to verify that the projection of the group velocity.

$$\boldsymbol{v}_{\mathrm{g}} = \frac{\partial \omega}{\partial \boldsymbol{k}} = -\frac{\omega u v}{2(2 - u - v)} \frac{\partial \cos^2 \theta}{\partial \boldsymbol{k}}. \tag{6.77}$$

on the z axis (the x_3 axis) vanishes at the points v_1 and v_2. The two solutions [Eqs. (6.75) and (6.76)] correspond to two different directions of the plasma-wave energy flux $\boldsymbol{P}(\boldsymbol{k})$, which is equal to the product of the energy density w by the group-velocity vector:

$$\boldsymbol{P} = w\boldsymbol{v}_{\mathrm{g}}, \qquad w = \frac{k^2}{4\pi}\left(\omega \frac{\partial Q}{\partial \omega}\right)|\varphi(\boldsymbol{k})|^2, \qquad Q = \frac{\varepsilon_{ij} k_i k_j}{k^2}. \tag{6.78}$$

Indeed, using the relation $\partial\omega/\partial\boldsymbol{k} = -(\partial Q/\partial\boldsymbol{k})/(\partial Q/\partial\omega)$, we reduce Equation (6.78) to the form

$$\boldsymbol{P} = -\frac{\omega}{4\pi} \frac{\partial}{\partial \boldsymbol{k}} (\varepsilon_{ij} k_i k_j)|\varphi(\boldsymbol{k})|^2. \tag{6.79}$$

It follows therefore that for the normal waves [Eqs. (6.75) and (6.76)] the

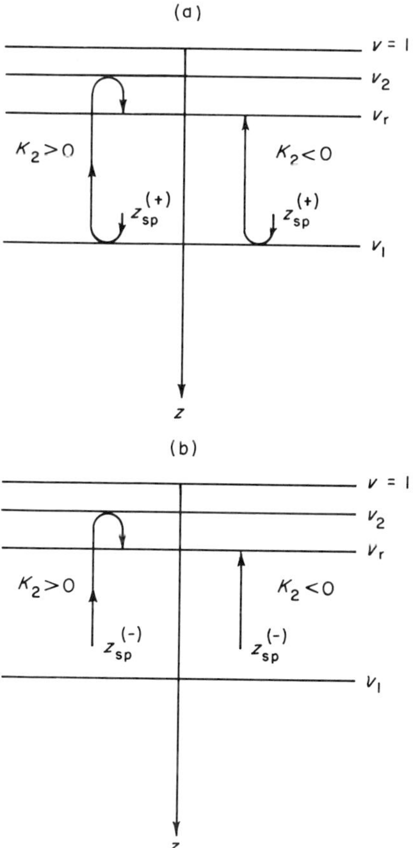

Fig. 70a, b. Propagation of plasma waves of frequency ω excited at the point z_{sp} by an incident (a) and reflected (b) ordinary wave. The plasma waves are absorbed at the resonance point $v = v_r$. v_1 and v_2 are the plasma-wave reflection points

projection of the vector $\boldsymbol{P}$ on the z axis is

$$P_3(\boldsymbol{k}) = -\frac{\omega}{2\pi}(\varepsilon_{33}k_3 + \varepsilon_{23}k_2)|\varphi|^2 = \mp\frac{\omega}{2\pi}(\perp b)|\varphi^{\pm}|^2. \qquad (6.80)$$

The minus sign corresponds to wave propagation in the direction of the z axis (i.e., downward under ionospheric conditions, Fig. 68), and the plus sign to the opposite directions. We note that, as follows from Equations (6.80) and (6.75), the quantity $|P_3|$ does not depend on the coordinate z, a fact corresponding to energy conservation in the absence of absorption.

To investigate the behavior of a longitudinal wave in the vicinity of the turning points v_1 and v_2, we change the variables in Equation (6.71) as follows

$$\varphi = \tilde{\varphi} \cdot \frac{1}{\sqrt{\varepsilon_{33}}} \exp\left[i \int^z a\, dz \right], \qquad a = \varepsilon_{23} k_2/\varepsilon_{33}.$$

As a result of this change, the homogeneous Equation (6.71) is reduced to the canonical form

$$\frac{d^2\tilde{\varphi}}{dz^2} + \left[\frac{b^2}{\varepsilon_{33}^2} + \frac{\left(\dfrac{d\varepsilon_{33}}{dz}\right)^2 - 2\varepsilon_{33}\dfrac{d^2\varepsilon_{33}}{dz^2}}{4\varepsilon_{33}^2} \right] \tilde{\varphi} = 0. \tag{6.81}$$

In the geometrical-optics approximation [Eq. (6.74)], the second term in the square brackets of Equation (6.81) can be neglected. In addition, in the vicinity of the turning point $b = 0$ the dependence of the coefficient b^2/ε_{33}^2 on z can be assumed to be linear. Equation (6.81) then reduces to an Airy equation. It follows therefore that the two independent solutions of the homogeneous Equation (6.81), which decrease respectively as $z \to \pm\infty$, take the following form in the region where geometrical optics is valid (see Ginzburg, 1960, Sect. 17)

$$\begin{aligned} \varphi_1 &= \exp\left(-i \int_z^{z_1} a\, dz \right) \cdot \frac{2}{\sqrt{b}} \cos\left(\int_z^{z_1} \frac{b}{\varepsilon_{33}}\, dz - \frac{\pi}{4} \right), & \varphi_1|_{z\to\infty} \to 0, \\ \varphi_2 &= \exp\left(i \int_{z_2}^{z} a\, dz \right) \cdot \frac{2}{\sqrt{b}} \cos\left(-\int_{z_2}^{z} \frac{b}{\varepsilon_{33}}\, dz - \frac{\pi}{4} \right), & \varphi_2|_{z\to-\infty} \to 0. \end{aligned} \tag{6.82}$$

The coordinates z_1 and z_2 of the lower and upper reflection points are defined here by $v(z_{1,2}) = v_{1,2}$. Allowance is made for the fact that $\varepsilon_{33} > 0$ in the vicinity of the lower point z_1 and $\varepsilon_{33} < 0$ at $z \sim z_2$.

Equations (6.82) constitute superpositions of the normal wave [Eq. (6.75)] with equal intensities of the incident and reflected waves. According to the foregoing, each of the exponentials of which the cosine function in Equation (6.82) is made up, correspond to a wave with a definite propagation direction: upward in the case of $\exp \sim \exp(-i\pi/4)$ $(P_3 < 0)$ and downward if $\exp \sim \exp(i\pi/4)$ $(P_3 > 0)$.

It is important that in an intermediate point v_r:

$$v_r = (1 - u)/(1 - u\cos^2\alpha), \qquad v_1 < v_r < v_2 \tag{6.83}$$

the wave number k and the amplitude $|\varphi|$ of the longitudinal plasma oscillations increase resonantly ($|k_3| \to \infty$, $b \to 0$, $|\varphi| \to \infty$). At this point we have $\varepsilon_{33} = 0$ and the refractive index has a pole. On passing through the point with the pole, the longitudinal wave is almost completely damped and gives up its energy to the plasma electrons. This dissipation occurs at arbitrarily low collision frequency ν_e of the electrons with the heavy particles (Ginzburg, 1960; Budden, 1961).

If ν_e is not very small, then the Langmuir wave can lose a greater part of the energy in the excitation region, even before the resonance point v_r is reached, owing to dissipation via collisions. We emphasize also that we are considering here only sufficiently long Langmuir waves

$$k_3 D \ll 1, \qquad D = (T_e/4\pi e^2 N)^{1/2} \tag{6.84}$$

where D is the Debye radius (Sect. 2.2.2). If this condition is not satisfied, then the plasma wave dissipates rapidly as a result of Landau damping due to the thermal motion of the electrons. Allowance for the thermal motion leads also to small corrections to the real part of the dielectric constant (see Sect. 6.3.1).

6.2.2. Excitation of Plasma Waves

We consider now the process of excitation of plasma waves by radio waves. Let a radio wave of frequency ω be normally incident on a layer of a weakly inhomogeneous plasma:

$$\lambda\mu \ll 1, \qquad \lambda = 2\pi c/\omega\sqrt{\varepsilon_0}, \qquad \mu = dv/dz. \tag{6.85}$$

We assume that $\omega > \omega_H$. Then the extraordinary wave is reflected at $\omega_0 = \omega - \omega_H$ [Eq. (5.122)], i.e., at $v = 1 - \sqrt{u}$. It does not reach the region [Eq. (6.69)] where the plasma waves exist (see Fig. 69), and consequently does not excite these waves. The ordinary radio wave is reflected at $\omega_0 = \omega$ [Eq. (5.122)]. It reaches the region of the existence of the plasma oscillations [Eq. (6.69)] and can, in principle, excite them. However, the excitation of the plasma waves is brought about by the inhomogeneity, since the waves are independent in a homogeneous plasma (Shafranov, 1966). Under conditions of a weakly inhomogeneous plasma [Eq. (6.85)], the coupling between the waves is exponentially small in terms of the adiabaticity parameter $\sim \exp(-1/\lambda\mu)$. Accordingly, the excitation of the plasma waves is also small here (Ginzburg, 1960; Golant and Piliya, 1972).

It is seen that the factor causing the absence of resonance effects and the weak absorption of the ordinary wave in a continuously variable plasma layer is rather specific in character. This cause can be easily eliminated if inhomogeneities having a characteristic dimension a smaller than or of the order of the wavelength λ are present in the plasma. These inhomogeneities will serve, so to speak, as resonators excited by the ordinary wave.

We consider therefore the excitation of plasma waves at some given stationary plasma inhomogeneities. Let the plasma contain inhomogeneities having a concentration N and a transverse dimension a that is small in comparison with the wavelength

$$a \ll \lambda/2\pi. \tag{6.86}$$

The longitudinal dimension (along the magnetic field) of the inhomogeneities, to the contrary, is assumed to be large enough (see Fig. 71). Such are the shapes of stationary or quasi-stationary inhomogeneities in a strongly magnetized tenuous plasma, particularly in the upper ionosphere, where the diffusion along the magnetic field force lines is much larger than in the transverse direction (see Sect. 5.1).

In the alternating electric field of a radio wave the inhomogeneity becomes polarized, and this leads to the appearance of an alternating electric charge ρ_ω that excites plasma waves. Indeed, the electron acquires in the radio wave field an alternating velocity $\boldsymbol{v}$ [Eq. (2.15)]. In the presence of an inhomogeneity $\delta N(\boldsymbol{r})$ with a gradient parallel to $\boldsymbol{v}$, alternating

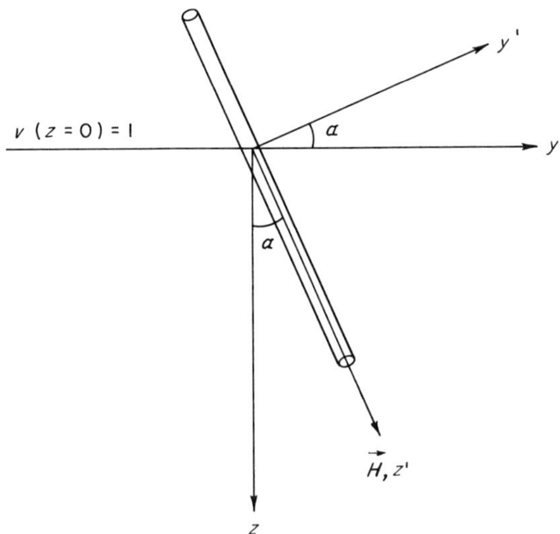

Fig. 71. Inhomogeneity elongated in the magentic-field direction

electron density perturbations δn_ω are also produced and a polarization charge $\rho_\omega = -e\,\delta n_\omega$ appears. In the case of inhomogeneities that are strongly elongated in the magnetic-field direction, the main contribution to the polarization is made by the electric-field component $\boldsymbol{E}_\perp$ perpendicular to $\boldsymbol{H}$. To determine ρ_ω it is natural to use the linearized equations of motion of the electrons [Eqs. (2.8) and (5.1)]:

$$\frac{\partial \boldsymbol{v}_\perp}{\partial t} + [\boldsymbol{\omega}_H \times \boldsymbol{v}_\perp] = \frac{e\boldsymbol{E}_\perp}{m}, \qquad \frac{\partial \delta n_\omega}{\partial t} + \boldsymbol{v}_\perp \nabla \delta N = 0. \tag{6.87}$$

It follows from Equation (6.87) that

$$\rho_\omega = \left\{\frac{\boldsymbol{E}_\perp + i\sqrt{u}[\boldsymbol{E}_\perp \times \boldsymbol{H}]/H}{8\pi(1-u)}\right\} \nabla_\perp \delta v, \qquad \delta v = 4\pi e^2\,\delta N/m\omega^2. \tag{6.88}$$

To determine the Fourier transform $\rho_{\omega k}$, we recognize that the field of the ordinary radio wave consists of the field of the waves incident (E^+) and reflected (E^-) from the layer $v = 1$:

$$E = E^+ + E^-, \qquad E^\pm = \frac{|E_0^\pm| p}{\sqrt{n(1+\alpha_y^2)}} \exp\left\{\pm i \int_{z_1}^{z} k\,dz + i\psi_0^\pm\right\}$$

$$k = \frac{\omega}{c} n, \qquad \psi_0^\pm = \arg E_0^\pm(z_1),\ p = (1 + \alpha_y^2 + \alpha_z^2)^{1/2}. \tag{6.89}$$

Here k is the wave vector, n is the refractive index, $i\alpha_y = E_y/E_x$ and $i\alpha_z = E_z/E_x$ are the polarization coefficients of the ordinary radio wave (Ginzburg, 1960, Sect. 11), and $E_0^\pm$ is the amplitude of the incident and reflected pump waves at the boundary of the ionosphere ($n = 1$, $\alpha_y = 1$, $\alpha_z = 0$). We also recognize that the field is given by

$$\boldsymbol{E}_\perp = (E_x, E_{y'}) = (E_x, E_y \cos\alpha - E_z \sin\alpha)$$

(see Fig. 71). We then obtain for $\rho_{\omega\kappa}$ in accordance with Equations (6.70) and (6.71)

$$\rho_{\omega k} = \rho_1[|E_0^+| \exp(i\psi_0^+) + |E_0^-| \exp(i\psi_0^-)]$$

$$\rho_1 = -\frac{i}{2\sqrt{n(1+\alpha_y^2)}} \frac{(1 - \sqrt{u}\alpha_{y'})k_1 + i(\alpha_{y'} - \sqrt{u})\dfrac{k_2}{\cos\alpha}}{4\pi(1-u)\cos\alpha}\,\delta v\left(k_1, \frac{k_2}{\cos\alpha}\right).$$

$$\alpha_{y'} = \alpha_y \cos\alpha - \alpha_z \sin\alpha. \tag{6.90}$$

The Fourier transform of the perturbation

$$\delta v\left(k_1, \frac{k_2}{\cos\alpha}\right) = \frac{1}{(2\pi)^2}\int \delta v(x, y') \exp\left[ik_1 x + i\frac{k_2}{\cos\alpha}y'\right] dx\, dy',$$

is calculated here in a coordinate system (x, y') perpendicular to the magnetic field (Fig. 70).

Using now the Equations (6.82) obtained above for the homogeneous problem, it is easy to construct for the inhomogeneous Equation (6.71) a general solution that defines the field of the plasma waves excited by the charge $\rho_{\omega k}$:

$$\varphi(k_1, k_2, z) = 2\pi i\left\{\exp\left[i\int_{z_2}^{z_1}\left(a + \frac{b}{\varepsilon_{33}}\right)dz\right]\varphi_1(z)\int_{-\infty}^{z}\rho_{\omega k}(z)\varphi_2^*(z)\,dz\right.$$
$$\left. + \exp\left[i\int_{z_2}^{z_1}\left(-a + \frac{b}{\varepsilon_{33}}\right)dz\right]\varphi_2(z)\int_{z}^{\infty}\rho_{\omega k}(z)\varphi_1^*(z)\,dz\right\}. \tag{6.91}$$

This field decreases as $z \to \pm\infty$. We have taken into account here the fact that the Jacobian of the Equations (6.82), calculated under the assumption that the longitudinal wave is completely absorbed in the resonance region [Eq. (6.83)], is equal to

$$\varphi_1\frac{d\varphi_2}{dz} - \varphi_2\frac{d\varphi_1}{dz} = -\frac{2i}{\varepsilon_{33}}\exp\left\{i\int_{z_2}^{z_1}\left(a - \frac{b}{\varepsilon_{33}}\right)dz - 2i\int_{z}^{z_1} a\,dz\right\}.$$

The obtained Equations (6.90) and (6.91) determine the field of the plasma waves excited in the plasma. They can be simplified by recognizing that the exponentials contain rapidly oscillating quantities. This enables us to calculate the integrals by the stationary-phase method. At the stationary-phase point $z_{\rm sp}^{\pm}$, which is the point of the maximum excitation, we have

$$k_3(z_{\rm sp}^{\pm}) = k_{\rm sp}^{\pm} = -(k_2 \operatorname{tg}\alpha \pm k). \tag{6.92}$$

The $\pm$ signs pertain here to the incident and reflected pump waves. It follows from Equations (6.72) and (6.92) that at the stationary-phase point

$$\cos^2\theta_{\rm sp} = k^2\cos^2\alpha/(k_1^2 + k_2^2 + k_{\rm sp}^2), \tag{6.93}$$

and from Equations (6.93), (6.68), and (6.73) it follows that

$$v(z_{\rm sp}^{\pm}) = v_{\rm sp}^{\pm} = \frac{v_1}{1 - uk^2\cos^2\alpha/(k_1^2 + k_2^2 + k_{\rm sp}^2)}. \tag{6.94}$$

We see therefore that owing to the smallness of $k^2/(k_1^2 + k_2^2 + k_{sp}^2) \sim (\omega a/c)^2 \ll 1$ [see Eq. (6.86)], the plasma waves are excited in fact in a small vicinity of the point of the upper hybrid resonance $v_1 = 1 - u$, i.e., at $\omega_0^2 \approx \omega^2 - \omega_H^2$. This is understandable, for if the inhomogeneities are strongly elongated in the $\boldsymbol{H}$ direction the polarization field is almost perpendicular to $\boldsymbol{H}$. Consequently, the wave vector of the exciting plasma waves is also almost perpendicular to $\boldsymbol{H}$: $\theta \simeq \pi/2$ [Eq. (6.93)]. The excitation point is close in this case to the upper hybrid resonance point [Eq. (6.68)].

When finding the second derivative of the phase of the integrand in Equation (6.91), allowance must be made for the relation

$$\frac{d}{dz}(k_3 - k_{sp}^{\pm})\Big|_{z_{sp}^{\pm}} = \pm\frac{\mu_1}{vb_0}(k_1^2 + k_2^2 + k_{sp}^{\pm 2})_{v = v_{sp}^{\pm}}, \qquad \mu_1 = -\frac{dv}{dz}\Big|_{z_1} \tag{6.95}$$
$$b_0 = -(k_2\varepsilon_{23} + k_{sp}^{\pm}\varepsilon_{33})$$

Taking Equations (6.94) and (6.95) into account, the expression for the potential φ [Eq. (6.91)] reduces to a form that lends itself to a simple physical interpretation:

$$\begin{aligned} \varphi &= \varphi^+ + \varphi^-; \qquad \varphi^{\pm} = I_1^{\pm} \cdot I_2^{\pm} \cdot I_3^{\pm}. \\ I_1^{\pm} &= 4\pi i \exp\left(\pm i\frac{\pi}{4}\right)\left[\frac{\pi v_{sp}^{\pm}}{\mu_1(k_1^2 + k_2^2 + k_{sp}^{\pm 2})}\right]^{1/2}, \\ I_2^{\pm} &= E^{\pm}(z_{sp}^{\pm})\rho_1(v_{sp}^{\pm}) \exp\left[ik_2 z_{sp}^{\pm} \operatorname{tg}\alpha\right], \\ I_3^{\pm} &= \frac{1}{\sqrt{b(v)}}\exp\left\{-i\int_{z_{sp}^{\pm}}^{z} k_3\,dz\right\}. \end{aligned} \tag{6.96}$$

Here $\varphi^{\pm}$ is the potential of the longitudinal waves excited by the incident (E^+) and reflected (E^-) pump waves, respectively. The first factor $I_1^{\pm}$ determines the intensity of the excitation of the longitudinal wave by the source $I_2^{\pm}$, while the factor $I_3^{\pm}$ describes the change of the amplitude and phase of the wave as it propagates from the excitation point $z_{sp}^{\pm}$ to the point z. When the last process is considered, it must be recognized that $k_3 = -a - b/\varepsilon_{33}$ in the case of downward motion ($P_3 > 0$) and $k_3 = -a + b/\varepsilon_{33}$ for upward motion ($P_3 < 0$). The propagation of longitudinal waves in the case of the excitation of the incident and reflected pump waves is shown in Figure 69. The incident pump wave transforms into a longitudinal one that propagates downward, and the reflected pump wave into one that travels upward. In a collisionless plasma, these waves are absorbed in the resonance region $v = v_r$ either after reflection from the point z_2 ($v(z_2) = v_2$) at $k_2 > 0$, or before the reflection (at $k_2 < 0$).

Equations (6.96) and (6.94) simplify somewhat by virtue of Equation (6.86) $k^2/(k_1^2 + k_2^2 + k_{sp}^2) \sim (a/\lambda) \ll 1$:

$$z_{sp}^{\pm} \simeq z_{sp}, \qquad v_{sp} = v(z_{sp}) \simeq v_1/(1 - uk^2 \cos^2 \alpha/k_\perp^2), \qquad k_\perp^2 = k_1^2 + k_2^2/\cos^2 \alpha \tag{6.97}$$

In addition, since the excitation region is close to the point v_1, it is possible to replace $v_{sp}^{\pm}$ and $\rho_1(v_{sp}^{\pm})$ in the right-hand side of Equation (6.96) by $v_1 = 1 - u$ and $\rho(v_1)$, respectively. In the case of quasi-longitudinal pump-wave propagation in the upper hybrid resonance region, if

$$\sin^2 \alpha \ll 2\sqrt{u} \cos \alpha \tag{6.98}$$

the polarization coefficients α_y and $\alpha_{y'}$ are nearly equal to unity, $n \simeq \sqrt{u}$, and Equation (6.90) for $\rho_1(v_1)$ takes the simple form

$$\rho_1 = -\frac{i}{2\sqrt{2}} \frac{u^{-1/8}(k_1 + ik_2/\cos \alpha)}{(1 + \sqrt{u}) \cos \alpha} \frac{\delta v(k_1, k_2/\cos \alpha)}{4\pi}. \tag{6.99}$$

In the ionosphere $\omega \leq (3\text{–}4)\omega_H$ and the Equation (6.98) is always satisfied at $\alpha \lesssim 30°$, i.e., at moderate and high latitudes.

We determine now the energy flux lost by the radio wave to excitation of plasma oscillations. Substituting Equations (6.99) and (6.96) in Equation (6.80) we obtain

$$P_3(\boldsymbol{k}) = -\frac{|E_0^2|}{8} \frac{\omega f(u)}{\mu_1 \cos^2 \alpha} |\delta v(k_1, k_2/\cos \alpha)|^2, \qquad f(u) = u^{-1/4} \frac{1 - \sqrt{u}}{1 + \sqrt{u}}. \tag{6.100}$$

We see therefore that the total energy flux going over into plasma waves in Equation (6.98) is given by

$$P = -P_3 = -(2\pi)^2 \int P_3(\boldsymbol{k})\, dk_1\, dk_2 = \frac{|E_0^2|}{8} \frac{\omega f(u)}{\mu_1 \cos \alpha} \int [\delta v(x, y')]^2\, dx\, dy' \tag{6.101}$$

where the perturbation $\delta v = \delta N \cdot (4\pi e^2/m\omega^2)$ is defined in the (x, y') plane, which is perpendicular to $\boldsymbol{H}$. Thus, the energy flux is proportional to the square of the concentration perturbation $(\delta N)^2$, to the intensity of the incident wave E_0^2, and to a characteristic dimension $\mu_1 = -(dv/dz)_{z_1}$, which is determined by the gradient of the plasma concentration in the region of the upper hybrid resonance.

6.2.3. Resonance Instability

It was shown above that the presence of inhomogeneities in a uniform plasma layer leads to effective dissipation of the energy of the ordinary wave by the inhomogeneities. The dissipated energy can strengthen the inhomogeneities. This in turn leads to an increase of the dissipation, which again causes the inhomogeneities to grow, etc. The resultant instability makes the plasma structure highly inhomogeneous and destroys the uniform layer in the region of the plasma resonances.

Consider a strongly magnetized plasma in which the transport processes are determined by diffusion and by the electronic heat conduction along the force lines of the magnetic field. The temperature and concentration perturbations δT_e and δN are then described by the linearized Equations (5.111):

$$\tau_T \frac{\partial\, \delta T_e}{\partial t} - L_T^2 \frac{\partial^2\, \delta T_e}{\partial z'^2} = -\delta T_e + \frac{2w_1}{3N_0}\tau_T,$$
$$\tau_N \frac{\partial\, \delta N}{\partial t} - L_N^2\left(\frac{\partial^2\, \delta N}{\partial z'^2} + \frac{T_{e0}}{N_0}\frac{\partial^2\, \delta T_e}{\partial z'^2}\right) = -\delta N + \gamma_1 \frac{N_0}{T_{e0}}\delta T_e. \qquad (6.102)$$

Here w_1 is the power dissipated per unit volume and acts as a source of the perturbations. In our case this is the plasma-wave energy, which is dissipated in final analysis in the electronic component of the plasma. The total plasma-wave power absorbed along a magnetic force line at a point $\boldsymbol{\rho} = (x, y')$ is, according to Equation (6.101),

$$Q(x, y') = \int w_1(x, y', z')\, dz'$$
$$= \frac{|E_0^2|\omega f(u)}{8\mu_1 \cos\alpha}\int G(\boldsymbol{\rho} - \boldsymbol{\rho}', \boldsymbol{\rho} - \boldsymbol{\rho}'')\, \delta v(\boldsymbol{\rho}')\, \delta v(\boldsymbol{\rho}'')\, d\boldsymbol{\rho}'\, d\boldsymbol{\rho}''. \qquad (6.103)$$

Here G is the Green's function, which describes the absorption of the plasma wave and is normalized to a δ function, $\int G\, d\boldsymbol{\rho} = \delta(\boldsymbol{\rho}' - \boldsymbol{\rho}'')$.

In the case of a collisionless plasma, $\nu_e \to 0$, the longitudinal-wave energy is absorbed in the vicinity of the resonance $v = v_r$ [Eq. (6.83)] at a large distance u/μ_1 from the axis of the inhomogeneity in which the excitation takes place, i.e., the function G is spread out over a wide region of order u/μ_1. In the presence of noticeable collision dissipation ν_e, the picture of the linear absorption of the longitudinal-wave energy is different. Indeed, as follows from Equation (6.97), the transformation of the ordinary pump wave into longitudinal oscillations by inhomogeneities of small dimension [Eq. (6.86)] takes place near the upper hybrid resonance point $v \simeq v_1 = 1 - u$. In this region, the group velocity of the longitudinal waves [Eq. (6.77)] is not only small in magnitude (inasmuch as $\boldsymbol{v}_{\mathrm{g}}|_{v=v_1} = 0$), but is also directed along the inhomogeneity axis. Therefore, at not too small values of v_{e}, an appreciable fraction of the longitudinal-wave energy is converted into heat directly in the volume of the inhomogeneity. The Green's function G has in this case a sharp peak in the region $\boldsymbol{\rho}' - \boldsymbol{\rho}'' \to 0$. Separating this local collision-dominated part of the dissipated energy $\tilde{G}$, we have

$$w_1 = \tilde{Q}\delta(z' - z_1'), \qquad \tilde{Q} = \frac{|E_0^2|\omega f(u)}{8\mu_1 \cos\alpha} \int \tilde{G}\, \delta v(\boldsymbol{\rho}')\, \delta v(\boldsymbol{\rho}'')\, d\boldsymbol{\rho}'\, d\boldsymbol{\rho}''. \tag{6.104}$$

We have taken into account here the fact that the electron heating due to the local dissipation takes place in the vicinity of the upper hybrid resonance $z_1' = z_1/\cos\alpha$, $v(z_1) = 1 - u$. The fraction of the longitudinal-wave energy absorbed locally, i.e., in the volume of the inhomogeneity, is

$$F = \int \tilde{G}\, d\boldsymbol{\rho}'\, d\boldsymbol{\rho}''\, d\boldsymbol{\rho} \simeq C_1 \frac{v_{\mathrm{e}}}{2\omega}\left(\frac{c}{\omega_H a_{\mathrm{m}}}\right)^2 \frac{1+u}{1-u}. \tag{6.105}$$

Here C_1 is a constant on the order of unity and depends on the shape of the inhomogeneity. Equation (6.105) is valid only at $a \lesssim a_{\mathrm{m}}$, where

$$a_{\mathrm{m}} = 3^{1/4}(Dc/\omega_H)^{1/2}. \tag{6.105a}$$

At $a > a_{\mathrm{m}}$, the coefficient F decreases abruptly with increasing a. The reason is that as a increases, i.e., as $k_\perp$ decreases, the region of the most effective plasma-wave generation moves farther away from the upper hybrid resonance point.

Continuing our analysis, we consider for simplicity a single inhomogeneity with a small transverse dimension a. Then the region of the maximum inhomogeneity satisfies Equations (6.102), with the perturbing power determined by Equations (6.104), (6.105), and (6.101):

$$\begin{gathered} \frac{2w_1\tau_T}{3N_0} = T_{\mathrm{e}0}K_0\, \delta v_{\mathrm{s}}^2\, \delta(z' - z_1'), \\ K_0 = \frac{\pi|E_0^2|\omega F f(u)}{\mu_1 E_{\mathrm{p}}^2 v_{\mathrm{e}}(1-u)}, \qquad E_{\mathrm{p}}^2 = 3T_{\mathrm{e}0}\,\delta_0 m\omega^2/e^2. \end{gathered} \tag{6.106}$$

Here δv_s is the relative perturbation of the concentration in the maximum of the inhomogeneity at $z' = z'_1$. Account is taken of the fact that $N_0 = N(z_1) = m\omega^2 v_1/4\pi e^2 = m\omega^2(1 - u)/4\pi e^2$.

We investigate first the stationary solution of the Equations (6.102) and (6.106). Since the perturbation [Eq. (6.106)] is localized near the point $z' = z'_1$, the stationary solution of Equation (6.102) is described by Equations (5.118) and (5.118a) with

$$\tilde{Q} = K_0\, \delta v_s^2, \qquad \delta v_s = \delta N_s\, 4\pi e^2/m\omega^2. \tag{6.107}$$

Let $\delta v_0 = \delta N_0 \cdot 4\pi e^2/m\omega^2$ be the initial, i.e., independent of the source w_1, perturbation of the concentration in the inhomogeneity. The interaction with the pump wave changes it to $\delta v = \delta v_0 + \delta v_1$, where δv_1 is described by Equations (5.118a) and (6.107). At the point of the maxima of the inhomogeneity at $z' = z'_1$ where $\delta v = \delta v_s$, we then have

$$\delta v_s = \delta v_0 \quad \frac{(k_T L_N - \gamma_1 L_T)(1 - u)K_0}{2L_T(L_N + L_T)}\, \delta v_s^2. \tag{6.108}$$

It is this equation which determines the value δv_s of the concentration perturbation that is produced at the inhomogeneity maximum by the interaction between the pump wave and the initial inhomogeneity δv_0. We write down the solution [Eq. (6.108)] in the form

$$\begin{aligned} \delta v_s &= 2[\delta v_{th} \pm \sqrt{\delta v_{th}^2 - \delta v_{th}\, \delta v_0}], \\ \delta v_{th} &= -\frac{2(L_N + L_T)\mu_1 L_T \nu_e E_p^2 \cos \alpha\varphi_T}{\pi f(u) F \omega E_0^2 (k_T L_N - \gamma_1 L_T)}. \end{aligned} \tag{6.109}$$

Under the conditions of the upper ionosphere, $\gamma_1 \to 0$ (see Table 14), and consequently $\delta v_{th} < 0$. A plot of the stationary perturbation δv_s against the initial one δv_0 is shown for this case in Figure 72. It is seen that in the region $\delta v_0 > \delta v_{th}$ each value of δv_0 corresponds to two stationary values of δv_s, namely $\delta v_s > 2\delta v_{th}$ (continuous curve) and $\delta v_s < 2\delta v_{th}$ (dashed).

If the initial perturbation δv_0 is negative and its absolute value exceeds the threshold $|\delta v_{th}|$ ($\delta v_0 < \delta v_{th} < 0$), then there are no stationary solutions and the perturbation increases continuously with time. This is the phenomenon that we shall call resonance instability. The threshold value of the initial perturbation [Eq. (6.109)] decreases with increasing pump-wave power and with decreasing concentration gradient μ_1 in the region of the upper hybrid resonance.

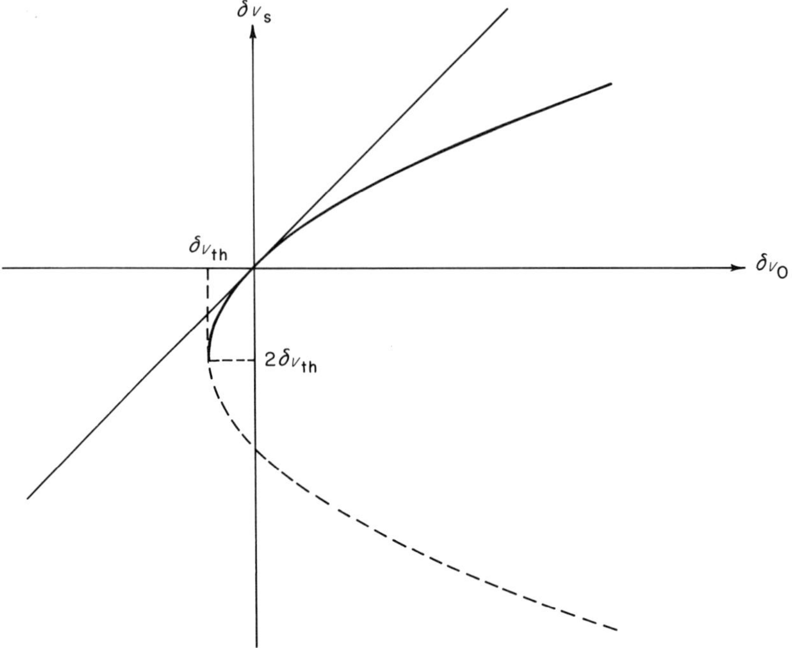

Fig. 72. Stationary perturbation of electron-density in the wave field inhomogeneity, $\delta v_s = \delta N_s 4\pi e^2/m\omega^2$, as a function of the initial perturbation, $\delta v_0 = \delta N_0 4\pi e^2/m\omega^2$. *Dashed curve*: unstable; δv_{th} – threshold perturbation for resonance instability [Eq. (6.109)]

An examination of small perturbations of the stationary solution [Eq. (6.109)] shows readily that only the upper stationary branch on Figure 71 is stable (the solid curve). The lower branch, showed dashed in the figure, is unstable. The characteristic values of the instability growth rate for $\gamma_1 \to 0, k_e \gg D_a$, and $|\delta v_s| > 2|\delta v_{th}|$ are determined by the expression

$$p \simeq \frac{1}{\tau_N} \left\{ \frac{L_N^2}{L_T^2} \left[\frac{\delta v_s (L_N + L_T)}{2 L_N \, \delta v_{th}} - 1 \right]^2 - 1 \right\}. \tag{6.110}$$

It is seen that the growth rate increases vigorously with increasing ratio $\delta v_s/\delta v_{th}$. We emphasize that only the negative perturbations δv_s increase under conditions when $\delta v_{th} < 0$.

The instability is nonlinear and is "explosive" in character: the perturbations become infinite $\sim (t - t_c)^{-1/2}$ at a certain finite instant of time t_c. Therefore, as is seen from Equation (6.110), the characteristic time of the nonstationary processes decreases strongly if the absolute value of the perturbation exceeds the threshold value significantly.

We did not take into account in the foregoing the thermal conductivity and the diffusion of the plasma in the direction transverse to the magnetic field. Their role is significant under ionospheric conditions, since the characteristic transverse dimension a_m of the inhomogeneities [Eq. (6.105a)] is small, smaller than the Larmor radius of the ions. When transport processes in a direction perpendicular to the magnetic field are taken into account, the threshold value δv_{th} increases (Vas'kov and Gurevich, 1977b):

$$\delta v_{th} = -C \frac{N T_e^2 \omega_H^2 \mu_1 a_m}{E_0^2 m c^2 v_e \sqrt{\omega \omega_H}} \qquad C \approx 2 \cdot 10^2 \qquad (6.109a)$$

Equation (6.109a) is valid for inhomogeneities with dimensions $a \approx a_m$: the threshold is raised at $a < a_m$ by the transverse diffusion and at $a > a_m$ by the abrupt decrease of the coefficient F, caused by the rapid outflow of the plasma-wave energy from the region occupied by the inhomogeneity. Analysis shows that during the evolution of the instability changes take place not only in the magnitude of the concentration perturbations, but also in the dimension and in the shape of the inhomogeneities. The optimal values of δv are reached at $a \sim a_m$. Inhomogeneities having a characteristic transverse dimension $a \sim a_m$ are thus singled out in the case of resonance instability.

The transverse diffusion determines the relaxation time of inhomogeneities that are strongly elongated in the direction of the magnetic field:

$$\Delta t \sim \frac{m\omega_H^2}{(T_e + T_i)v_e k_\perp^2}, \qquad k_\perp^2 = k_1^2 + k_2^2/\cos^2 \alpha. \qquad (6.110a)$$

It is seen that the lifetime of the inhomogeneities decreases sharply when their transverse dimension decreases.

6.2.4. Absorption of Ordinary Radio Waves

Resonant Absorption. Resonance instability leads to a strong enhancement of inhomogeneities that are strongly elongated in the direction of the magnetic field. The energy lost by the transverse wave to excitation of the plasma oscillations is then also increased, i.e., the absorption of the ordinary wave in the plasma becomes larger.

To calculate the absorption, we use the expression for the plasma-oscillation energy flux [Eq. (6.80)]. Substituting in Equation (6.80) the potential $\varphi^{\pm}$ from Equation (6.96) we get

$$P_3^{\pm}(z) = -(2\pi)^2 \int P(k_1, k_2)|E_0^{\pm}(z_{sp})|^2 \Omega(z_{sp} - z)\, dk_1\, dk_2$$

$$P(k_1, k_2) = \frac{8\pi^2\omega(1-u)}{\mu_1(k_1^2 + k_2^2/\cos^2\alpha)} \left|\rho_1\left(k_1, \frac{k_2}{\cos\alpha}\right)\right|^2, \tag{6.111}$$

$$\Omega(x) = \begin{cases} 1 & x > 0 \\ 0 & x < 0 \end{cases}$$

Account was taken here of the fact that the wave excitation, given k_1 and k_2, takes place in a small vicinity of the point $z_{sp}(k_1, k_2)$, defined by the Equations (6.92) and (6.76). $P_3^{\pm}(z)$ is the total power lost by the incident $(+)$ and reflected $(-)$ waves.

Equation (6.111) allows us to formulate the energy conservation law for the incident and reflected pump waves in the form

$$\frac{c}{8\pi}|E_0^{\pm}(z_1)|^2 = \frac{c}{8\pi}|E_0^{\pm}(z)|^2 \pm \frac{1}{S} P_3^{\pm}(z). \tag{6.112}$$

Here S is the area of the intersection of the region filled with the inhomogeneities by the (x, y) plane.

Equations (6.112) and (6.111) for $|E_0^2(z)|$ are integral in character. Differentiating with respect to the coordinate z, we can represent them in differential form

$$\frac{d}{dz}|E_0^{+}(z)|^2 = \frac{2\omega}{c}\kappa_r(z)|E_0^{+}(z)|^2, \tag{6.113a}$$

$$\frac{d}{dz}|E_0^{-}(z)|^2 = -\frac{2\omega}{c}\kappa_r(z)|E_0^{-}(z)|^2, \tag{6.113b}$$

where the resonant absorption coefficient $\kappa_r(z)$, which is determined by the value and the spectrum of the concentration perturbations $\delta v(x, y')$, is given in accordance with Equation (6.111) by

$$\kappa_r(z) = \frac{16\pi^3}{\omega S}\int P(k_1, k_2)\, \delta(z - z_{sp})\, dk_1\, dk_2. \tag{6.114}$$

Integration of Equations (6.113) and (6.114) allows us to express the intensity of the reflected wave at the lower boundary of the excitation region in terms of the intensity of the incident wave:

$$|E_0^-(z_1)|^2 = |E_0^+(z_1)|^2 \exp\left\{-\frac{64\pi^3}{cS}\int P(k_1, k_2)\, dk_1\, dk_2\right\}. \quad (6.115)$$

The general Equation (6.115) becomes simpler in the case of quasi-longitudinal propagation of the pump wave in the excitation region [Eq. (6.98)]. In this case, according to Equation (6.99), we have $|\rho_1^2| \sim (k_1^2 + k_2^2/\cos^2\alpha)|\delta v|^2$ and

$$|E_0^-(z_1)|^2 = |E_0^+(z_1)|^2 \exp\left[-\overline{\delta v^2}/\delta v_c^2\right],$$
$$\overline{\delta v^2} = \int [\delta v(x, y')]^2\, dx\, dy' \Big/ \int dx\, dy', \qquad \int dx\, dy' = S\cos\alpha. \quad (6.116)$$

Here $\overline{\delta v^2} = \overline{(\delta N/N)^2}(1-u)^2$ is the mean squared perturbation of the electron density, and the characteristic value of the perturbation δv_c^2 is equal to

$$\delta v_c^2 = c\mu_1/\pi\omega f(u), \qquad \mu_1 = \left|\frac{dv}{dz}\right|_{z_1}, \qquad f = u^{-1/4}\frac{1-\sqrt{u}}{1+\sqrt{u}}. \quad (6.117)$$

It follows from Equation (6.116) that if the density perturbations in the inhomogeneities reach the critical value δv_c, then the pump wave experiences an anomalously strong, almost complete absorption in the plasma. Only part of the absorbed energy, on the order of F, is dissipated directly in the inhomogeneities in the form of heat. The remaining energy is dissipated in the resonance region [Eq. (6.83)], where it can lead not only to heating but also to acceleration of the electrons (through Landau damping).

It must be stressed that the resonant absorption considered here has a linear character. Regardless of the mechanism whereby the inhomogeneities are produced, this absorption is experienced by all the ordinary waves, of any frequency ω, incident both normally and obliquely on a plasma layer, provided only that their reflection point lies above the upper hybrid resonance point $v(z_1) = 1 - u$.

Results of Experiments in the Ionosphere. A small-scale stratification of the F region of the ionosphere under the influence of strong radio waves was observed in experiments on the scattering of UHF radio waves (Utlaut, 1970; Fialer, 1974; Minkoff et al., 1974a; Carpenter, 1974; Minkoff et al., 1974b). The characteristic transverse scale of the inhomogeneities $a \sim 0.5$ m (Rao and Thome, 1974) corresponds to optimal scale

a_m of the resonance instability [Eq. (6.105a)]. It satisfies the Equation (6.86), which can be written for the ionosphere in the form

$$a \ll c/\omega_H \approx 30 \text{ m} \tag{6.118}$$

inasmuch as in the region of the upper hybrid resonance we have $\lambda/2\pi = \lambda_0/2\pi\sqrt{u} = c/\omega_H$. The longitudinal dimension of the inhomogeneities is large, $\gtrsim 200$ m–1 km. These inhomogeneities can develop as a result of resonance instability, since under the conditions of Utlaut's 1974 experiments ($E_0/E_p \sim 1, \omega \sim 3 \cdot 10^7, \nu_e \sim 10^3$–$5 \cdot 10^2$), the threshold values δN_{th} of the initial density perturbations are small, $\delta N_{th}/N \sim 10^{-4}$–$10^{-5}$. Initial perturbations of this order of magnitude can exist in the ionosphere under natural conditions. They can also be generated by other instabilities, for example drift instability (Borisov et al., 1976) and dissipative parametric instability (see Sect. 6.3.4). We note that the linear transformation process considered here is characterized by excitation, in the inhomogeneities, of plasma waves with a frequency strictly equal to the frequency ω of the modifying transmitter. It is precisely these waves which were observed in experimental investigations of the spectrum of a scattered high-frequency signal (Minkoff et al., 1974a; Carpenter, 1974). The inhomogeneity relaxation time decreases rapidly with decreasing inhomogeneity dimension (Fialer, 1974; Minkoff et al., 1974a) in accordance with Equation (6.110a).

Resonance instability amplifies strongly the small-scale inhomogeneities. This is accompanied by an increase of the absorption of the modifying wave. According to Equation (6.116), at $\overline{\delta v^2} \geq \delta v_c^2$ the absorbed power approaches the incident-wave power, in which case the inhomogeneities should stop growing. Under the conditions of the F layer of the ionosphere, according to Equation (6.117), we have

$$\delta v_c \approx 5 \cdot 10^{-3}. \tag{6.119}$$

This value of $(\delta N/N)_c$ is in agreement with the results of scattering experiments (Fialer, 1974).

Concentration perturbations on the order of δv_c ensure absorption of not only the wave of frequency ω, due to the perturbations, but also of other ordinary "sounding" radio waves of frequency ω_p, which are reflected in the modified zone. As seen from Equations (6.116) and (6.117), the absorption of the sounding wave depends on the concentration profile $N(z)$. In particular, in the region of the maximum of the F layer, assuming the profile of $N(z)$ to be parabolic, $N(z) = N_c(1 - z^2/\Lambda^2)$, we obtain from Equations (6.116) and (6.117) the following expression for the reflection coefficient R_p of the sounding waves:

$$\frac{R_p}{R_{0p}} = \frac{|E_{0p}^-(z_1)|^2}{|E_{0p}^+(z_1)|^2} = \exp\left\{-\frac{\overline{\delta v^2}}{C_0\sqrt{1-(\omega_p^2-\omega_c^2)/\omega_c^2}}\right\} \qquad (6.120)$$

$$C_0 = 2c/\omega_c\Lambda\pi f(u_c), \qquad u_c = \omega_H^2/\omega_c^2, \qquad \omega_c^2 = 4\pi e^2 N_c/m.$$

It is assumed here that the frequencies ω and ω_p are close to the critical frequency ω_c of the layer. $R_{0p} = R_{p|\delta v^2=0}$ is the reflection coefficient of the sounding wave in the absence of inhomogeneities. The dependence of the reflection coefficient R_p [Eq. (6.120)] on the frequency of the sounding wave is shown in Figure 73 and is in sufficient agreement with the experimental data (Getmantsev et al., 1973). The relative density perturbation obtained in the inhomogeneities in this case, $\delta N/N \approx (5\text{–}7) \cdot 10^{-3}$, agrees with Equation (6.119). From Equation (6.120) and Figure 73 it is seen that resonant absorption of the sounding waves increases with their frequency, especially as the critical frequency of the layer ω_c is approached; this is also confirmed by experimental data (Utlaut et al., 1974).

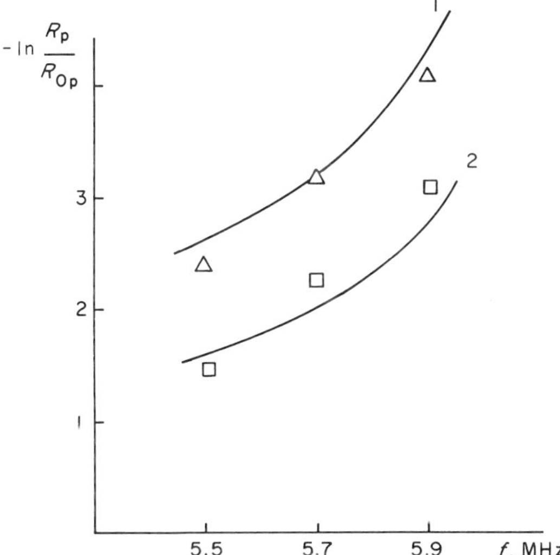

Fig. 73. Dependence of the reflection coefficient of probing waves, R_p, on their frequency, as measured by Getmantsev et al. (1973). The modifying station frequency is $f = 5.75$ MHz, and the critical layer frequency is $f_c \approx 6$ MHz. △: measurement results at maximum power of the modifying station, $W_0 = W_{0m}$; □: at $W_0 = W_{0m}/2$. *Solid curves*: calculated with Equation (6.120); curve *1*: at $\Lambda\,\overline{\delta v^2} = 4.6 \cdot 10^{-3}$ km; curve *2*: at $\Lambda\,\overline{\delta v^2} = 2.5 \cdot 10^{-3}$ km. It follows therefore that at $\Lambda = 10^2$ km the density perturbation in the inhomogeneities is $\delta N/N \approx (5-7) \cdot 10^{-3}$

Let us determine the width of the absorption band. Let the half-width of the modified zone be L. Then, in the case of normal incidence on the layer, the sounding waves experiencing intense absorption should be those with frequency ω_p satisfying the condition

$$|\omega_p - \omega| \lesssim \Delta\omega = \omega L \mu_1/2 \tag{6.121}$$

where ω is the frequency of the modifying transmitter. Since the experimental width is $L \sim$ (5–10) km (Fialer, 1974), it follows that $\Delta\omega \sim (2\text{–}8) \cdot 10^6$, in agreement with the observed results (Getmantsev et al., 1973; Utlaut et al., 1974).

In the case when an ordinary wave is obliquely incident on a layer, resonance absorption is possible only if the wave-reflection point lies above the upper hybrid resonance point, i.e., in the case of high frequency waves $u \ll 1$, at

$$\psi_p \lesssim \omega_H/\omega. \tag{6.122}$$

Here ψ_p is the angle between the direction of the wave incident on the ionosphere and the vertical. This condition was satisfied in the studies of Cohen and Whitehead (1970), who were the first to observe anomalous (resonant) absorption. We note that Equation (6.122) imposes a limit $\rho_\perp \lesssim z_0\omega_H/\omega$ on the transverse dimension of the region of the resonant modification of the ionosphere. Since $z_0 \lesssim z_m$(z_m is the height of the maximum of the F layer), it follows from this that $\rho_\perp \approx$ 60–80 km, in agreement with the observation results.

The small-scale stratification of the ionosphere determines the intense scattering of radio waves of the HF and UHF bands. This phenomenon is of considerable practical interest, and it has been shown that it can be used both for communication between land-based points (Barry, 1974; Stathacopoulos and Barry, 1974), and to excite and de-excite ionospheric wave channels for the purpose of ensuring ultralong-distance and around–the–world propagation of short radio waves (Gurevich and Tsedilina, 1976). The integral cross section σ_s for the scattering of radio waves by inhomogeneities is proportional to $\overline{\delta v^2}$ (Booker, 1956; Minkoff, 1974). The value $\delta N/N \sim (0.5\text{–}1) \cdot 10^{-2}$ (Rao and Thome, 1974) determined from experiments on scattering agrees with Equation (6.119).

We emphasize in conclusion that small-scale stratification ($a \ll 30$ m) is due to resonance effects and occurs only when an ordinary wave acts on the ionosphere. In contrast, large-scale stratification ($a \gtrsim 30$ m), which leads to the appearance of the sporadic F layer, is the result of self-focusing instability that is not connected with resonance effects. It can be induced by both ordinary and extraordinary radio waves (see Sect. 6.1).

6.3. Parametric Instability

We have shown in the preceding section that plasma waves are generated effectively from stationary plasma inhomogeneities in the field of an ordinary radio wave. If the excitation of the plasma waves leads in turn to enhancement of the inhomogeneities, then the plasma is unstable.

Similar generation of plasma oscillations is possible also from nonstationary inhomogeneities, which are themselves plasma waves of some type, for example ion-sound, magnetohydrodynamic, or Langmuir waves. The conditions for the enhancement of the waves in this general case are of the form (Tsitovich, 1970):

$$\omega = \omega_1 + \omega_2, \qquad \boldsymbol{k} = \boldsymbol{k}_1 + \boldsymbol{k}_2. \tag{6.123}$$

Here (ω, $\boldsymbol{k}$), (ω_1, $\boldsymbol{k}_1$), and (ω_2, $\boldsymbol{k}_2$) are respectively the frequency and the wave vector of the pump wave and of the two waves generated in the plasma. When the decay conditions [Eq. (6.123)] are satisfied, it becomes easiest for the plasma in the pump-wave field to become unstable, because natural oscillations with ω_1, $\boldsymbol{k}$ and ω_2, $\boldsymbol{k}_2$ are excited. This phenomenon is usually called parametric instability (Silin, 1965; DuBois and Goldman, 1965; Silin, 1973). Nondissipative parametric instability is connected with striction processes. It develops rapidly (with a characteristic time $\lesssim 1/\nu_e$) and leads to various plasma turbulization effects, and to rapid heating and acceleration of the electrons and ions. It can cause intense absorption and rescattering of radio waves. A great variety of parametric-instability effects is possible in the ionosphere (Perkins and Kaw, 1971; DuBois and Goldman, 1972; Kuo and Fejer, 1972; Bezzerides and Weinstock, 1972; Vas'kov and Gurevich, 1973c; Mityakov, et al., 1974; Kruer and Valeo, 1973; Perkins et al., 1974).

We consider first the case of parametric excitation of Langmuir waves. The mechanism whereby the field of a high-power radio wave (the pump wave) interacts with Langmuir plasma waves consists in the following: Plasma waves are generated from the ion-sound waves in the field of the pump wave. On the other hand, the inhomogeneous field of the plasma waves exerts a striction pressure on the plasma (Gaponov and Miller, 1958; Pitaevskii, 1960). The striction force acts on the ion-sound oscillations. When the decay conditions [Eq. (6.123)] are satisfied, the plasma and the ion-sound oscillations amplify each other, and this leads to instability of the plasma in a sufficiently strong pump-wave field. We note that since the pump wave length is much larger than the Langmuir wave lengths and since the plasma-wave frequency [Eq. (6.68)] exceeds greatly the ion-sound frequency, the decay conditions [Eq. (6.123)] assume in this case the simple form $k_1 \approx -k_2$ and $\omega \approx \omega_1 \approx \omega_{\text{res}}$, i.e., parametric excitation of Langmuir oscillations takes place in the plasma-resonance region [Eq. (6.68)]. The concluding section will deal with the dissipative parametric instability due to excitation of plasma and collision-dominated low-frequency waves.

6.3.1. Langmuir Oscillations of a Plasma in an Alternating Electric Field

The natural potential oscillations of an isotropic plasma are determined by the homogeneous Equation (6.64):

$$\operatorname{div} \boldsymbol{D} = -\operatorname{div} \varepsilon' \nabla\varphi = 0. \tag{6.124}$$

It follows from Equation (6.124) that in a homogeneous isotropic plasma the dispersion equation, which determines the frequency of the potential waves, is

$$\varepsilon' = 0. \tag{6.125}$$

Here ε' is the dielectric constant of the plasma. For further use, it is convenient to represent it in the form [Eqs. (2.13) and (2.48)]

$$\varepsilon' = 1 + \delta\varepsilon_{\text{e}} + \delta\varepsilon_{\text{i}}, \qquad \delta\varepsilon_{\text{e,i}} = -4\pi \int_{-\infty}^{t} (j_{\text{e,i}}/E)\, dt. \tag{6.126}$$

where $\delta\varepsilon_{\text{e}}$ and $\delta\varepsilon_{\text{i}}$ are the partial contributions of the electrons and ions to the dielectric constant. We note that ε' is a complex quantity [Eq. (2.14)]; in other words, j_{e} and j_{i} in Equation (6.126) are the total currents of the electrons and ions, or the conduction and the polarization currents.

When the dielectric constant was determined in Sects. 2.1 and 2.3, the alternating electric field was assumed to be homogeneous in space, i.e., the wave length was assumed to be infinite. This is usually valid for transverse waves whose velocity is large in comparison with the velocities of the electrons and ions of the plasma. For longitudinal waves, this condition no longer holds. An important role is therefore played here by the finite dimension of the wave lengths, i.e., by the spatial dispersion of the dielectric constant, wherein ε' depends not only on the frequency ω but also on the wave vector $\boldsymbol{k}$: $\varepsilon' = \varepsilon'(\omega, \boldsymbol{k})$.

To determine $\varepsilon'(\omega, \boldsymbol{k})$ we must find the total electron and ion currents $\boldsymbol{j}_{\text{e}}$ and $\boldsymbol{j}_{\text{i}}$ in the wave field

$$\boldsymbol{E}_1 = \boldsymbol{E}_0 \exp(-i\omega t + i\boldsymbol{k}\boldsymbol{r}). \tag{6.127}$$

We use for this purpose the kinetic Equation (2.61). Linearizing this equation

$$f_{e,i} = f_{0e,i}(\boldsymbol{v}) + \delta f_{e,i}(\boldsymbol{v}, \boldsymbol{r}, t), \qquad |\delta f| \ll f_0 \tag{6.128}$$

we have for the perturbations of the distribution function

$$\begin{aligned} \frac{\partial\, \delta f_e}{\partial t} + \boldsymbol{v}\frac{\partial\, \delta f_e}{\partial \boldsymbol{r}} - \frac{e\boldsymbol{E}_1}{m}\frac{\partial f_{0e}}{\partial \boldsymbol{v}} &= 0, \\ \frac{\partial\, \delta f_i}{\partial t} + \boldsymbol{v}\frac{\partial\, \delta f_i}{\partial \boldsymbol{r}} + \frac{e\boldsymbol{E}_1}{M}\frac{\partial f_{0i}}{\partial \boldsymbol{v}} &= 0. \end{aligned} \tag{6.129}$$

For simplicity we have neglected here the collisions, a procedure valid in a strongly rarefied plasma at $\omega \gg \nu$. The solution of the kinetic Equations (6.129) in the electric field of the wave is

$$\delta f_e = i\frac{e\boldsymbol{E}_1}{m(\omega - \boldsymbol{k}\boldsymbol{v})}\frac{\partial f_{0e}}{\partial \boldsymbol{v}}, \qquad \delta f_i = -i\frac{e\boldsymbol{E}_1}{M(\omega - \boldsymbol{k}\boldsymbol{v})}\frac{\partial f_{0i}}{\partial \boldsymbol{v}}. \tag{6.130}$$

Obtaining now the electron and ion currents in accordance with Equation (2.60), we determine the dielectric constant of the plasma [Eq. (6.126)]

$$\delta\varepsilon_e = \frac{4\pi e^2}{\omega m}\int \frac{df_{0e}}{dv}\frac{v\cos^2\theta}{(\omega - \boldsymbol{k}\boldsymbol{v})}\,d\boldsymbol{v}, \qquad \delta\varepsilon_i = \frac{4\pi e^2}{M}\int \frac{df_{0i}}{dv}\frac{v\cos^2\theta}{(\omega - \boldsymbol{k}\boldsymbol{v})}\,d\boldsymbol{v}. \tag{6.131}$$

Here θ is the angle between $\boldsymbol{v}$ and $\boldsymbol{E}_1$.

An important role is played in Equation (6.131) by the indicated Landau pole of the integrand at $\omega = \boldsymbol{k}\cdot\boldsymbol{v}$ (Ginzburg, 1960). This pole sets apart the singular interaction between the wave field [Eq. (6.127)] and the resonant particles—particles moving with velocity $\boldsymbol{v}$ equal to the phase velocity ω/k of the wave. It leads to the appearance of an imaginary part of the dielectric constant, describing collisionless absorption of the wave energy by the resonant particles (Landau damping).

The dispersion equation for the longitudinal waves [Eqs. (6.125), (6.126), and (6.131)] determines two natural oscillation modes of the plasma: a high-frequency plasma mode $\omega \sim \omega_0$ and low-frequency ion-sound mode $\omega \sim kv_{Ti}$, where $v_{Ti} = (2\pi/M)^{1/2}$. For high-frequency plasma oscillations $\omega \gg \boldsymbol{k}\cdot\boldsymbol{v}$ we have from Equation (6.131)

$$\begin{aligned} \delta\varepsilon_e &= -\frac{\omega_0^2}{\omega^2}\left[1 + 3(kD)^2\right] + i\sqrt{\frac{\pi}{2}}\frac{\omega}{\omega_0(kD)^3}\exp\left[-\frac{\omega^2}{2\omega_0^2(kD)^2}\right]. \\ \delta\varepsilon_i &= -\Omega_0^2/\omega^2, \qquad \Omega_0^2 = 4\pi e^2 N/M, \qquad D = (T_e/4\pi e^2 N)^{1/2}. \end{aligned} \tag{6.132}$$

The dispersion Equation (6.125) then determines the frequency and the damping of the plasma waves, $\omega' = \omega - i\gamma_0$:

$$\omega = \omega_0 + \tfrac{3}{2}\omega_0(kD)^2. \tag{6.133}$$

$$\gamma_0 = \omega_0 \sqrt{\frac{\pi}{8}}(kD)^{-3} \exp\left[-\frac{1}{2(kD)^2} - \frac{3}{2}\right]. \tag{6.134}$$

The contribution of the ions to these oscillations is small and we have neglected it. In addition, we have considered here only long waves with

$$(kD)^2 \ll 1. \tag{6.135}$$

The short-wave oscillations with $kD \gtrsim 1$ attenuate rapidly, as is seen also directly from Equation (6.134).

For low-frequency waves $\omega \lesssim kv_{Ti} \ll kv_{Te}$, where $v_{Te} = (2T_e/m)^{1/2}$, we have from Equation (6.131)

$$\begin{aligned} \delta\varepsilon_e &= (kD)^{-2}\left[1 + i\sqrt{\pi}\frac{\omega}{kv_{Te}}\right], \\ \delta\varepsilon_i &= (kD_i)^{-2}[1 + i\sqrt{\pi}zW(z)], \qquad z = \omega/kv_{Ti} \\ W(z) &= \exp(-z^2)\left(1 + \frac{2i}{\sqrt{\pi}}\int_0^z \exp(t^2)\,dt\right). \end{aligned} \tag{6.136}$$

Here $D_i = (T_i/T_e)^{1/2}D = (T_i/4\pi e^2 N)^{1/2}$ is the Debye radius of the ions and $W(z)$ is a Kramp function. The dispersion Equation (6.125) leads in this case to weakly damped waves $\gamma_0 \ll \omega$ only if $T_e \gg T_i$. In a plasma with nearly equal electron and ion temperatures, $T_e \sim T_i$, ion-sound waves are always strongly attenuated, $\gamma_0 \sim \omega$.

We have disregarded above collisions between electrons and ions. In a high-frequency field ($\omega \gg \boldsymbol{k} \cdot \boldsymbol{v}$, $\omega \gg \nu$) allowance for collisions, as can be easily seen, leads to an additive imaginary increment to the dielectric constant [Eq. (6.132)] [(see Eqs. (2.13) and (2.48)]:

$$\Delta_c(\delta\varepsilon_e) = i\frac{\nu_e}{\omega}\frac{\omega_0^2}{\omega^2}, \qquad \Delta_c(\delta\varepsilon_i) = i\frac{\nu_i}{\omega}\frac{\Omega_0^2}{\omega^2}. \tag{6.137}$$

Here $\nu_e = \nu_{ei} + \nu_{em}$ and $\nu_i = \nu_{im}$ are the effective frequencies of the collisions of the electrons with the ions and with the molecules [Eqs. (2.140), (2.141)] and of the ions with the molecules [Eq. (2.212)], respectively. The collisions increase the plasma-wave damping and instead of Equation (6.134) we have

$$\gamma_0 = \omega_0 \sqrt{\frac{\pi}{8}} (kD)^{-3} \exp\left[-\frac{1}{2(kD)^2} - \frac{3}{2}\right] + \frac{\nu_e}{2}. \tag{6.138}$$

It is seen that the damping of sufficiently long plasma waves, $kD \to 0$, is due to electron collisions. For ion-sound waves in a tenuous plasma at $T_e \sim T_i$, the principal role is always played by collisionless damping [Eq. (6.136)]. We note that as $kD \to 0$ the spatial dispersion is insignificant and Equations (6.132) and (6.137) go over into Equations (2.13) and (2.48).

We examine now how the natural oscillations of the plasma are altered in the presence of an external alternating electric field, namely the field $\boldsymbol{E}_1$ of a radio wave. Just as before, taking into account the high propagation velocity of the radio wave, we can assume its field to be homogeneous in space

$$\boldsymbol{E}_1 = \boldsymbol{E}_0 \sin \omega_1 t = \frac{\boldsymbol{E}_0}{2i}(\exp(i\omega_1 t) - \exp(-i\omega_1 t)) \tag{6.139}$$

Here ω_1 is the radio-wave frequency. Owing to the particle collisions in the alternating electric field [Eq. (6.139)], the symmetrical part of the distribution function $f_{0e}(v)$, considered in Section 2.3, is greatly deformed. This change in the distribution function naturally influences the spectrum of the plasma oscillations, as is seen directly from Equations (6.131)–(6.136). It is important, however, that an electric field not only alters the symmetrical part of the distribution function, but gives rise also to an asymmetrical increment to $f_{0e}(v)$, due to the directional motion of the electrons in the field. The effect of this asymmetrical part on the natural oscillations of a rarefield plasma turns out to be particularly important. We shall, therefore, from now on regard the symmetrical part of the distribution functions $f_{0e}(v)$ and $f_{0i}(v)$ to be specified, say Maxwellian, and investigate the change produced by the directional motion of the electrons in the natural oscillations of the plasma. The distribution functions $f_{e,i}$ in an alternating electric field [Eq. (6.139)] in the absence of collisions, as follows from Equation (2.61), are of the form

$$f_{e,i}(\boldsymbol{v}, t) = f_{0e,i}(\boldsymbol{v} - \boldsymbol{u}_{e,i}, 0), \qquad \boldsymbol{u}_e = -\frac{e\boldsymbol{E}_0}{m\omega_1}\cos\omega_1 t,$$
$$\boldsymbol{u}_i = \frac{e\boldsymbol{E}_0}{M\omega_1}\cos\omega_1 t. \tag{6.140}$$

We consider now the weak perturbations δf_e and δf_i [Eq. (6.128)] of the functions $f_{e,i}$ [Eq. (6.140)]. In the absence of collisions, they are described by linearized equations that follow from Equation (2.61):

$$\begin{aligned} &\frac{\partial\,\delta f_e}{\partial t} + \boldsymbol{v}\,\frac{\partial\,\delta f_e}{\partial \boldsymbol{r}} - \frac{e\boldsymbol{E}_1}{m}\,\frac{\partial\,\delta f_e}{\partial \boldsymbol{v}} - \frac{e\boldsymbol{E}_l}{m}\,\frac{\partial f_e}{\partial \boldsymbol{v}} = 0, \\ &\frac{\partial\,\delta f_i}{\partial t} + \boldsymbol{v}\,\frac{\partial\,\delta f_i}{\partial \boldsymbol{r}} + \frac{e\boldsymbol{E}_1}{M}\,\frac{\partial\,\delta f_i}{\partial \boldsymbol{v}} + \frac{e\boldsymbol{E}_l}{M}\,\frac{\partial f_i}{\partial \boldsymbol{v}} = 0. \end{aligned} \tag{6.141}$$

Here $\boldsymbol{E}_1$ [Eq. (6.139)] is the electric field of the radio waves and $\boldsymbol{E}_l$ is the longitudinal electric field of the natural oscillations of the plasma. It is determined by the Poisson equation:

$$\Delta\varphi = -4\pi e(\delta N_i - \delta N_e), \qquad \boldsymbol{E}_l = -\nabla\varphi, \qquad \delta N_{i,e} = \int \delta f_{i,e}\, d\boldsymbol{v}. \tag{6.142}$$

Equations (6.141) and (6.142), with allowance for Equation (6.140) for $f_{e,i}$, form a complete system that determines the natural oscillations of the plasma in the alternating electric field $\boldsymbol{E}_1$. In the absence of the field $\boldsymbol{E}_1$, as can be easily seen, its solution leads to the dispersion Equations (6.125), (6.126) and (6.131).

By considering the Fourier component of $\delta f_{e,i}$:

$$\delta f_{e,i}(\boldsymbol{r}, \boldsymbol{v}, t) = \delta f_{e,i}(\boldsymbol{v}, t) \exp(i\boldsymbol{k}\boldsymbol{r}) \tag{6.143}$$

and determining the field $\boldsymbol{E}_l$ from Equation (6.142), we obtain in place of Equation (6.141)

$$\begin{aligned} &\frac{\partial\,\delta f_e}{\partial t} + i\boldsymbol{k}\boldsymbol{v}\,\delta f_e - i\,\frac{4\pi e^2}{mk^2}\,\boldsymbol{k}\,\frac{\partial f_e}{\partial \boldsymbol{v}}\left(\int \delta f_i\, d\boldsymbol{v} - \int \delta f_e\, d\boldsymbol{v}\right) - \frac{e\boldsymbol{E}_1}{m}\,\frac{\partial\,\delta f_e}{\partial \boldsymbol{v}} = 0, \\ &\frac{\partial\,\delta f_i}{\partial t} + i\boldsymbol{k}\boldsymbol{v}\,\delta f_i + i\,\frac{4\pi e^2}{Mk^2}\,\boldsymbol{k}\,\frac{\partial f_i}{\partial \boldsymbol{v}}\left(\int \delta f_i\, d\boldsymbol{v} - \int \delta f_e\, d\boldsymbol{v}\right) + \frac{e\boldsymbol{E}_1}{M}\,\frac{\partial\,\delta f_i}{\partial \boldsymbol{v}} = 0. \end{aligned} \tag{6.144}$$

We change over now to new variables (Silin, 1973)

$$\begin{aligned} \psi_e(\boldsymbol{v}, t) &= \exp\left(-i\,\frac{e\boldsymbol{k}\boldsymbol{E}_1(t)}{m\omega_1^2}\right) \delta f_e(\boldsymbol{v} + \boldsymbol{u}_e, t), \\ \psi_i(\boldsymbol{v}, t) &= \exp\left(i\,\frac{e\boldsymbol{k}\boldsymbol{E}_1(t)}{M\omega_1^2}\right) \delta f_i(\boldsymbol{v} + \boldsymbol{u}_i, t) \end{aligned} \tag{6.145}$$

where $\boldsymbol{u}_e$ and $\boldsymbol{u}_i$ are defined in Equation (6.140). Equations (6.144) for the functions $\psi_{e,i}$ are then rewritten in the form

$$\frac{\partial\psi_e}{\partial t}+i\boldsymbol{k}\boldsymbol{v}\psi_e+i\frac{4\pi e^2}{mk^2}\boldsymbol{k}\frac{\partial f_{0e}}{\partial \boldsymbol{v}}\left(\int\psi_e\,d\boldsymbol{v}-\int\psi_i\,d\boldsymbol{v}\sum_{n=-\infty}^{+\infty}\exp(-in\omega_1 t)J_n(\boldsymbol{k}\boldsymbol{r}_E)\right)=0,$$

$$\frac{\partial\psi_i}{\partial t}+i\boldsymbol{k}\boldsymbol{v}\psi_i+i\frac{4\pi e^2}{Mk^2}\boldsymbol{k}\frac{\partial f_{0i}}{\partial \boldsymbol{v}}\left(\int\psi_i\,d\boldsymbol{v}-\int\psi_e\,d\boldsymbol{v}\sum_{n=-\infty}^{+\infty}\exp(-in\omega_1 t)J_n(\boldsymbol{k}\boldsymbol{r}_E)\right)=0. \tag{6.146}$$

Here $\boldsymbol{r}_E = eE_0/m\omega_1^2$ is the amplitude of the oscillations of the electron in the field of the radio wave and $J_n(z)$ is a Bessel function of the n–th order. In the derivation of Equation (6.146) we have used the relation

$$\exp(iz\sin\omega_1 t) = \sum_{n=-\infty}^{+\infty} J_n(z)\exp(in\omega_1 t). \tag{6.147}$$

It is seen from Equation (6.146) that in the presence of an alternating field of frequency ω_1 the plasma waves, generally speaking, do not preserve their frequency ω: waves with combination frequencies $\omega \pm n\omega_1$ are produced. This is a natural consequence of the nonlinear character of the interaction of the plasma waves with the field $\boldsymbol{E}_1$ (cf. Sect. 4.2.3).

By seeking the solution of Equation (6.146) in the form of a series in harmonics

$$\psi_{e,i}(\boldsymbol{v}, t) = \exp(-i\omega t)\sum_{s=-\infty}^{+\infty}\exp(-is\omega_1 t)\psi_{e,i}^s(\boldsymbol{v}) \tag{6.148}$$

we arrive at the following integral equations for the functions $\psi_{e,i}^s$:

$$(\omega + s\omega_1 - \boldsymbol{k}\boldsymbol{v})\psi_e^s + \frac{4\pi e^2}{mk^2}\boldsymbol{k}\frac{\partial f_{0e}}{\partial \boldsymbol{v}}\int d\boldsymbol{v}'\left\{\psi_e^s(\boldsymbol{v}') - \sum_{q=-\infty}^{+\infty} J_q(\boldsymbol{k}\boldsymbol{r}_E)\psi_i^{s-q}(\boldsymbol{v}')\right\} = 0,$$

$$(\omega + s\omega_1 - \boldsymbol{k}\boldsymbol{v})\psi_i^s + \frac{4\pi e^2}{Mk^2}\boldsymbol{k}\frac{\partial f_{0i}}{\partial \boldsymbol{v}}\int d\boldsymbol{v}'\left\{\psi_i^s(\boldsymbol{v}') - \sum_{q=-\infty}^{+\infty} J_q(\boldsymbol{k}\boldsymbol{r}_E)\psi_e^{s+q}(\boldsymbol{v}')\right\} = 0. \tag{6.149}$$

Dividing these equations by $\omega + s\omega_1 - \boldsymbol{k}\cdot\boldsymbol{v}$ and integrating with respect to $d\boldsymbol{v}$, we obtain for the constants $u_{e,i}^s$ a chain of linear algebraic equations

$$u_e^s = e\int d\boldsymbol{v}\,\psi_e^s(\boldsymbol{v}), \qquad u_i^s = -e\int d\boldsymbol{v}\,\psi_i^s(\boldsymbol{v}), \tag{6.150}$$

which takes the simple form

$$u_e^s + R_e^s \sum_{q=-\infty}^{+\infty} J_{s-q}(\boldsymbol{k}\boldsymbol{r}_E)u_i^q = 0, \tag{6.151}$$

$$u_i^s + R_i^s \sum_{q=-\infty}^{\infty} J_{q-s}(\boldsymbol{k}\boldsymbol{r}_E)u_e^q = 0. \tag{6.152}$$

Here

$$R_{e,i}^s = \frac{\delta\varepsilon_{e,i}(\omega + s\omega_1, \boldsymbol{k})}{1 + \delta\varepsilon_{e,i}(\omega + s\omega_1, \boldsymbol{k})}, \tag{6.153}$$

and the functions $\delta\varepsilon_e(\omega, \boldsymbol{k})$ and $\delta\varepsilon_i(\omega, \boldsymbol{k})$ are defined by Equation (6.131). The condition that the system Equations (6.151) and (6.152) have a solution is in fact the sought dispersion equation.

Consider now low-frequency ion-sound waves. In this case $\omega \ll \omega_1 \sim \omega_0$ and it follows from Equations (6.153) and (6.136) that $R_i^0 \sim 1$ and $R_i^s \sim m/M \ll 1$ at $s \neq 0$. It is then seen from Equation (6.152) that

$$u_i^s \ll u_i^0 = -R_i^0 \sum_{q=-\infty}^{\infty} J_q(\boldsymbol{k}\boldsymbol{r}_E)u_e^q,$$

and from Equation (6.151) we have, accurate to $\sim m/M$,

$$u_e^s - R_e^s R_i^0 J_s(\boldsymbol{k}\boldsymbol{r}_E) \sum_{q=-\infty}^{\infty} J_q(\boldsymbol{k}\boldsymbol{r}_E)u_e^q = 0. \tag{6.154}$$

Multiplying these equations in succession by $J_s(\boldsymbol{k}\boldsymbol{r}_E)$ and summing over all s, we find that the condition under which Equation (6.154) has a solution takes the form

$$1 - R_i^0 \sum_{s=-\infty}^{\infty} J_s^2(\boldsymbol{k}\boldsymbol{r}_E)R_e^s = 0, \tag{6.155}$$

This is the dispersion equation that determines the spectrum of the low-frequency oscillations of the plasma in an alternating electric field [Eq. (6.139)]. We note that in the absence of an alternating field we have $r_E = 0$, $J_0 = 1$, and $J_s = 0$ (at $s \neq 0$), and Equation (6.145) coincides with Equations (6.125) and (6.126).

We consider the case of a field that is not too strong, when

$$(\boldsymbol{k}\boldsymbol{r}_E)^2 = k^2e^2E_0^2/m^2\omega_1^4 \ll 1. \tag{6.156}$$

Inasmuch as interest attaches only to sufficiently long waves with $kD \lesssim 1$, the Equation (6.156) is equivalent at $\omega_1 \sim \omega_0$ to the condition

$$\frac{E_0^2}{8\pi} \ll NT_e, \tag{6.157}$$

which has a clear physical meaning: the field-energy density is much lower than that of the plasma thermal energy. We emphasize that the field E_0, which satisfies the Equation (6.157), can be at the same time larger or even much larger than the plasma field E_p [Eq. (1.2)].

Under the conditions of Equation (6.156), expanding the function $J_s(\mathbf{k} \cdot \mathbf{r}_E)$ in powers of $\mathbf{k} \cdot \mathbf{r}_E$ and confining ourselves in Equation (6.155) to terms of order $(\mathbf{k} \cdot \mathbf{r}_E)^2$,

$$J_0(\mathbf{k}\mathbf{r}_E) = 1 - \frac{(\mathbf{k}\mathbf{r}_E)^2}{4}, \qquad J_{\pm 1}(\mathbf{k}\mathbf{r}_E) = \pm\frac{\mathbf{k}\mathbf{r}_E}{2}$$

we represent the dispersion Equation (6.155) in the form

$$\frac{1}{\delta\varepsilon_i(\omega, \mathbf{k})} + \frac{1}{1 + \delta\varepsilon_e(\omega, \mathbf{k})} + \frac{(\mathbf{k}\mathbf{r}_E)^2}{4}\left[\frac{1}{1 + \delta\varepsilon_e(\omega + \omega_1, \mathbf{k})} + \frac{1}{1 + \delta\varepsilon_e(\omega - \omega_1, \mathbf{k})}\right] = 0, \tag{6.158}$$

where $\delta\varepsilon_e(\omega, \mathbf{k})$ and $\delta\varepsilon_i(\omega, \mathbf{k})$ are defined in Equation (6.136) (low-frequency region), while $\delta\varepsilon_e(\omega \pm \omega_1, \mathbf{k})$ are defined in Equation (6.132). It is seen that in Equation (6.156) the dispersion Equation (6.125) is only insignificantly altered by the external field. In the resonance region $\omega_1 \approx \omega_0$, however, the values of $1 + \delta\varepsilon_e(\omega_1 \pm \omega)$ are small and the perturbation is strongly amplified.

6.3.2. Parametric Excitation of Langmuir Oscillations

Consider long-wave ($kD \ll 1$) low-frequency ($\omega \ll \omega_0$) plasma oscillations in an external alternating field $\mathbf{E}_1$—the field of a radio wave. Assume that the radio-wave frequency ω_1 is close to the plasma frequency ω_0, i.e., that the resonance conditions [Eq. (6.123)] are satisfied. Using then Equations (6.136) for $\delta\varepsilon_e(\omega, \mathbf{k})$ and $\delta\varepsilon_i(\omega, \mathbf{k})$ and Equations (6.132) for $\delta\varepsilon_e(\omega \pm \omega_1, \mathbf{k})$.

we rewrite the dispersion Equation (6.158) in the form

$$\frac{T_e}{T_i} + \frac{1}{f(z)} = \frac{(E_0 \cos \theta)^2}{16\pi N T_i} \frac{\delta\omega_0}{(\Omega + i\gamma + i\gamma_0)^2 - \delta^2}. \tag{6.159}$$

Here θ is the angle between $\boldsymbol{k}$ and $\boldsymbol{E}_1$, while δ is the mismatch, i.e., the difference between the frequency of the radio wave and the plasma-wave frequency [Eq. (6.133)]

$$\delta = \omega_1 - \omega_0 - \tfrac{3}{2}\omega_0(kD)^2. \tag{6.160}$$

The frequency ω of the ion-sound waves is written in Equation (6.159) in the form $\omega = \Omega + i\gamma$, while γ_0 is the plasma-wave damping decrement [Eq. (6.138)] and

$$\begin{gathered} f(z) = f_1(z) + if_2(z) = 1 + i\sqrt{\pi} z W(z); \\ z = a + ib, \qquad a = \Omega/kv_{Ti}, \qquad b = \gamma/kv_{Ti}, \qquad v_{Ti} = \sqrt{2T_i/M}. \end{gathered} \tag{6.161}$$

Equation (6.159) determines the frequency and the damping decrement of the ion-sound waves. In a sufficiently strong external field, these waves turn out to be unstable—they increase with time, and with them also the plasma waves [Eq. (6.149)]. A distinction can be made between two types of instability: periodic with $\Omega \gg \gamma$ and aperiodic with $\Omega = 0$ (Nishikawa, 1968; Andreev, et al., 1970).

Aperiodic Instability. We consider first aperiodic instability—the stratification of a plasma in the field of a high-power wave. Putting $\Omega = 0$ in Equation (6.159), we get

$$\begin{gathered} \frac{T_e}{T_i} + \frac{1}{f(b)} = -\frac{(E_0 \cos \theta)^2}{16\pi N T_i} \frac{\omega_0 \delta}{(\gamma_0 + \gamma)^2 + \delta^2}, \\ f(b) = 1 - \sqrt{\pi} b \exp(b^2)(1 - \Phi(b)), \qquad \Phi(b) = \frac{2}{\sqrt{\pi}} \int_0^b \exp(-t^2)\, dt, \\ b = \gamma/kv_{Ti}. \end{gathered} \tag{6.162}$$

We recognize that $f(b) > 0$ at any value of b. Equation (6.162) can therefore have solutions only at $E_0 \neq 0$ and at a negative mismatch $\delta < 0$.

Consider sufficiently small $|\gamma| \ll kv_{Ti}$. In this case $f(b) \simeq 1$ and it follows from Equation (6.162) that

$$\gamma = -\gamma_0 \pm \left[-\delta^2 - \delta \frac{E_0^2 \cos^2 \theta \omega_0}{16\pi N(T_e + T_i)} \right]^{1/2} \tag{6.163}$$

We see therefore that the aperiodic waves are stratification waves and can exist in an external field if

$$0 > \delta > -\frac{E_0^2 \cos^2 \theta \omega_0}{16\pi N(T_e + T_i)}. \tag{6.164}$$

In a sufficiently weak field, these waves are always damped, $\gamma < 0$. The damping decrement decreases, however, with increasing E_0. At a certain threshold value $E_0 = E_{th}$ the damping vanishes: $\gamma = 0$. It follows from Equation (6.163) that the threshold field is

$$E_{th}^2 = 16\pi N(T_e + T_i) \frac{\nu_e}{\omega_1}. \tag{6.165}$$

It is reached at

$$\cos \theta = 1, \qquad \delta = \omega_1 - \omega_0 - \tfrac{3}{2}(k_{th}D)^2\omega_0 = -\nu_e/2,$$
$$k_{th}D = \left(\frac{2}{3}\frac{\omega_1 - \omega_0}{\omega_0} + \frac{\nu_e}{3\omega_0}\right)^{1/2}. \tag{6.166}$$

In the derivation of Equation (6.165) we have assumed $\gamma_0 = \nu_e/2$; as seen from Equation (6.138), this is valid under the condition

$$k < k_1 \approx \frac{1}{D}\left[2 \ln\left(\sqrt{\frac{\pi}{2}}\frac{\omega_0}{\nu_e}\right) + 3 \ln \ln\left(\sqrt{\frac{\pi}{2}}\frac{\omega_0}{\nu_e}\right) + 3 \ln 2\right]^{-1/2}. \tag{6.167}$$

Thus, in the region $0 < k < k_1$ the minimum threshold field E_{th} is actually independent of k. On the other hand, if $k > k_1$, then E_{th} increases exponentially. It follows from Equations (6.166) and (6.167) that the threshold field E_{th} in the frequency region

$$-\frac{\nu_e}{2} < \omega_1 - \omega_0 < -\frac{\nu_e}{2} + \tfrac{3}{2}(k_1 D)^2\omega_0 \tag{6.168}$$

remains constant and equal to Equation (6.165). Indeed, Equation (6.166) is reached in this frequency region by a suitable choice of the wave vector

k_{th}, which by virtue of Equation (6.167) has no effect whatever on γ_0. Outside the frequency band [Eq. (6.168)], the field E_{th} increases sharply. Thus, the instability in question is typical of resonant frequencies $\omega_1 \approx \omega_0$ [Eq. (6.168)].

At $E_0 > E_{th}$ the instability increment γ reaches the maximum value

$$\gamma_m = \frac{v_e}{2} \frac{E_0^2 \cos^2 \theta - E_{th}^2}{E_{th}^2} \tag{6.169}$$

at

$$\delta_m = -\omega_0 E_0^2 \cos^2 \theta / 32\pi N(T_e + T_i),$$

$$(k_m D)^2 = \frac{2(\omega_1 - \omega_0)}{3\omega_0} + \frac{E_0^2 \cos^2 \theta}{48\pi N(T_e + T_i)}; \qquad |\omega_1 - \omega_0| \gg v_e. \tag{6.170}$$

The dependence of γ on k in the vicinity of the maximum is given by

$$\gamma = \gamma_m - \frac{4(\omega_1 - \omega_0)^2}{v_e} \frac{E_{th}^2}{E_0^2 \cos^2 \theta} \left(\frac{k - k_m}{k_m} \right)^2. \tag{6.171}$$

It is seen that the growth rate $\gamma(k)$ has a sharp peak near $k = k_m$.

Periodic Instability. We consider now the periodic instability at $\Omega \gg \gamma$. From Equation (6.159), separating the real and imaginary parts, we have

$$\Omega^2 - \delta^2 - (\gamma + \gamma_0)^2 = 2\Omega(\gamma + \gamma_0)L(z),$$
$$L(z) = \frac{(T_e/T_i)|f(z)|^2 + f_1(z)}{f_2(z)}; \tag{6.172}$$

$$\frac{E_0^2 \cos^2 \theta}{32\pi N T_i} = \frac{\gamma + \gamma_0}{\omega_0} \frac{\Omega}{\delta} F(z), \qquad F(z) = \frac{f_2(z)(1 + L^2)}{|f(z)|^2} = \frac{\left| 1 + \dfrac{T_e}{T_i} f(z) \right|^2}{f_2}. \tag{6.173}$$

At $z \sim 1$, the functions F and L in Equations (6.172) and (6.173) are nearly equal to unity. Recognizing that $\gamma_0 \sim v_e/2 \ll \delta$, we get $\Omega \simeq \delta$ from Equation (6.172). It follows then from Equation (6.173) that the threshold field E_{th} is reached at $\cos \theta = 1$ and at a value $z_0 = a_0$ that ensures the minimum of the function $F(a)$. Thus,

$$E_{th}^2 = 16\pi N T_i \frac{v_e}{\omega_1} F(a_0), \qquad F(a_0) = \min F(a). \tag{6.174}$$

Table 17. Periodic instability constants

T_e/T_i	a_0	$F(a_0)$	$L(a_0)$	$F''(a_0)/F(a_0)$
1.0	1.24	1.72	0.09	5.6
1.5	1.42	1.77	0.012	7.8
2.0	1.58	1.764	−0.074	12.6

The constants a_0 and $F(a_0)$ depend on the temperature ratio T_e/T_i. They are listed in Table 17. Comparing Equation (6.174) with Equation (6.165) we see that at $T_e \sim T_i$ the periodic instability has a lower threshold field than the aperiodic one.

The threshold values of Ω and δ are

$$\Omega_{th} = a_0 k_{th} v_{Ti}, \qquad \delta_{th} = a_0 k_{th} v_{Ti}.$$

The last relation, with Equation (6.160) taken into account, can serve as an equation with which to determine the value of k_{th}:

$$k_{th}D = \left(\frac{2}{3}\frac{\omega_1 - \omega_0}{\omega_0} + \frac{2}{9}a_0^2\frac{mT_i}{MT_e}\right)^{1/2} - \frac{\sqrt{2}}{3}a_0\left(\frac{mT_i}{MT_e}\right)^{1/2} \simeq \left(\frac{2}{3}\frac{\omega_1 - \omega_0}{\omega_0}\right)^{1/2}. \tag{6.175}$$

It is seen that the threshold length of the wave excited in the periodic instability is determined by the frequency difference $\omega_1 - \omega_0$ and coincides in fact with the analogous quantity for the aperiodic instability [Eq. (6.166)]. Just as in the case of the aperiodic instability, the minimum threshold field determined by Equation (6.174) does not depend on k in the region $0 < k < k_1$ [see Eq. (6.167)]. Outside this region, E_{th} increases exponentially. In the frequency region

$$\frac{\nu_e}{2} < \omega_1 - \omega_0 < \tfrac{3}{2}(k_1 D)^2\omega_0 \tag{6.176}$$

the field E_{th} is constant and is determined by Equation (6.174) [cf. Eq. (6.168)]. The field E_{th} increases sharply outside the frequency band Equation (6.176) (Fig. 74).

To determine the near-threshold growth rate it is necessary to expand the right-hand side of Equation (6.173) in terms of the small parameters γ/γ_0, γ/Ω, and γ_0/δ and to retain the principal term of the expansion, which is proportional to $\gamma/\gamma_0 \gg \gamma/\Omega$. We find then that the maximum growth rate of the instability is given as before by Equation (6.169). It is reached

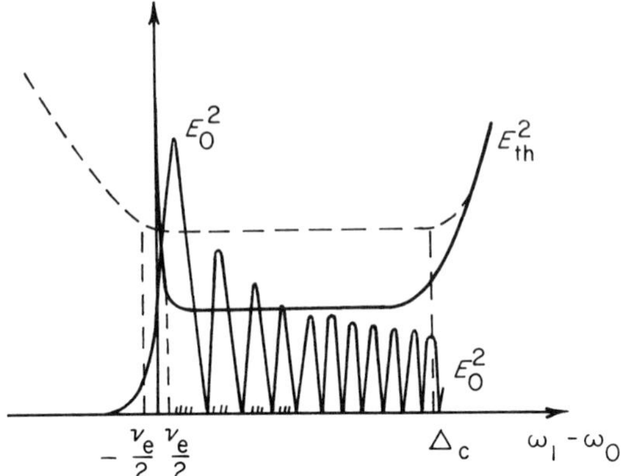

Fig. 74. Threshold field E_{th}^2 for periodic (*solid line*) and aperiodic (*dashed line*) parametric instability, as a function of the difference between the modifying-wave frequency ω_1 and the plasma frequency $\omega_0 = 4\pi e^2 N/m$. $\Delta_c = 3\omega_0(k_1 D)^2/2$. The threshold field is constant in the frequency interval $0 \lesssim \omega_1 - \omega_0 \lesssim \Delta_c$ (at $\Delta_c \gg \nu_e$), and increases steeply outside this interval. The figure shows also the standing-wave field E_0^2 in the reflection region. The frequency difference $\omega_1 - \omega_0$ is proportional to the distance from the reflection point $z_0 - z$ [Eq. (6.180)]. The instability region is shown *hatched*

at $k_m \simeq k_{th}$ [Eq. (6.175)]. The dependence of γ on k near the threshold is of the form

$$\gamma = \gamma_m - \frac{3\nu_e M T_e(\omega_1 - \omega_0)F''(a_0)}{4mT_i\omega_0 F(a_0)} \frac{(k - k_m)^2}{k_m^2}. \tag{6.177}$$

The quantity $F''(a_0)$ is listed in Table 17. It is seen from Equation (6.177) that, just in the case of the aperiodic instability, the growth rate has a rather narrow peak near $k = k_m \simeq k_{th}$.

At large pump-wave field amplitudes

$$E_0^2 > E_{th}^2 \left[\frac{16mT_i\omega_0(\omega_1 - \omega_0)}{3MT_e\nu_e^2}\right]^{1/2} \tag{6.178}$$

the instability goes over into the hydrodynamic phase when its growth rate γ is larger than kv_{Ti}, i.e., when the thermal motion of the ions has little effect (Andreev et al., 1970). In the hydrodynamic stage, the maximum instability increment increases in proportion to $E_0^{2/3}$:

$$\gamma_m = \frac{\sqrt{3}}{2}\left[\frac{e^2 E_0^2(\omega_1 - \omega_0)\cos^2\theta}{24MT_e}\right]^{1/3}. \tag{6.179}$$

We note in conclusion that we have considered here only a spatially homogeneous plasma in a homogeneous alternating electric field. The presence of inhomogeneities of the plasma or of the field can alter greatly the conditions of the excitation of the parametric instability (Liu et al., 1974; Forslund et al., 1975.

6.3.3. Parametric Instability in the Ionosphere

Excitation of Instability. As shown above, Langmuir oscillations develop most effectively in the plasma-resonance region $\omega_1 \approx \omega_{\text{res}}$. Parametric instability, like resonance instability (see Sect. 6.2) is excited therefore at $\omega_1^2 \gg \omega_H^2$ only in the region $1 - u \leq v \leq 1$ where the ordinary wave is reflected.[20]

In the reflection region, a standing wave is produced with a field structure described by an Airy function (Sect. 5.3.2). The field amplitude of the standing wave at the reflection point is shown in Figure 74. The change of the amplitude is uniquely connected with frequency difference $\omega_1 - \omega_0$, which is proportional to the distance $z_0 - z$ from the reflection point:

$$\omega_1 - \omega_0(z) = \frac{1}{2}\omega_1\mu(z_0 - z), \qquad \mu = \left(\frac{1}{N}\frac{dN}{dz}\right)_{z_0} \tag{6.180}$$

The minimum instability threshold, as seen from Figure 74, is always reached first in the principal (first) maximum of the wave, which is located at a height z_1 [Eq. (5.125)]:

$$z_0 - z_1 \simeq 1.02\left(\frac{\omega_1^2}{c^2}\frac{\mu}{\sin^2\alpha}\right)^{-1/3}, \tag{6.181}$$

i.e., at a frequency difference

$$\omega_1 - \omega_0(z_1) = 0.51\omega_1^{1/3}(c\mu\sin\alpha)^{2/3}. \tag{6.182}$$

The instability sets in when the wave-field amplitude in the first maximum [Equation (5.125)] reaches the minimum threshold field E_{th} [Eq. (6.174)]. The values of the threshold field as functions of z, in the case of vertical propagation of an ordinary wave in the ionosphere, are shown in Figure 75. It is seen that the field has a characteristic minimum at

[20] In the region of the reflection of the extraordinary wave, Bernstein modes may be excited (Fejer and Leer, 1972).

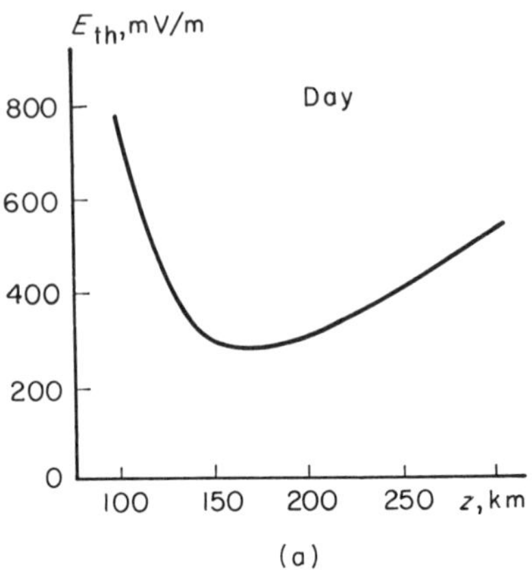

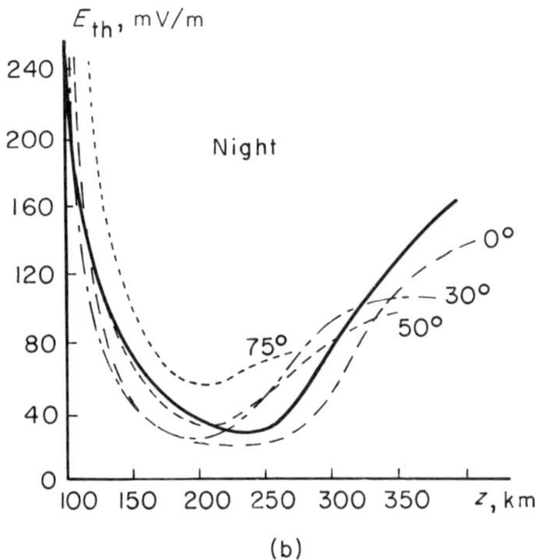

Fig. 75. Threshold field as a function of the height in the ionosphere. *Solid curves*: for the ionosphere model of Tables 1 and 2. *Dashed curves*: for the ionosphere model of Soboleva (1972) at the various latitudes indicated in the figure

150–200 km in daytime and at 200–250 km at night. The minimum values of the nighttime threshold field are much lower than those in daytime.

At $E > E_{th}$ the instability develops near the maxima of the standing-wave field. The instability region has a layered character (it is shown dashed in Fig. 74). The frequency Ω and the length kD of the ion-sound waves excited in the ionosphere in the first maximum are shown in Figure 76. It is seen that the Langmuir waves are quite long: $\lambda = 2\pi/k \approx$

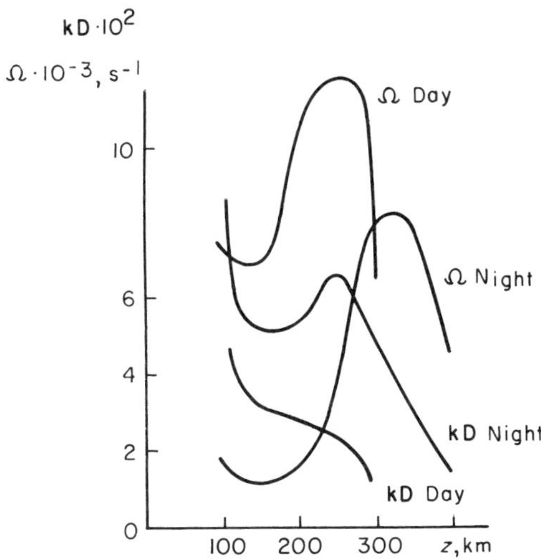

Fig. 76. Frequency Ω and wave vector kD of ion-sound waves excited at the principal maximum, as a function of the height in the ionosphere

$(1\text{–}5)10^2 D \approx (0.2\text{–}2)$ meters. In the succeeding maxima, Ω and kD increase in proportion to $(n-\frac{3}{4})^{1/3}$, where n is the number of the maximum. The total number of the maxima in the effective-excitation band [Eq. (6.176)] changes from several dozen at heights of the F-layer maximum to 3–5 at a height $z \sim 100$ km.

Near the reflection point of the ordinary wave, at $\omega^2 \gg \omega_H^2$, the earth's magnetic field exerts no appreciable influence on the conditions for the excitation of the parametric instability, since the electric fields of both the radio wave and the plasma waves are parallel to $\boldsymbol{H}$ here. As ω_1 approaches ω_H, however, the role of the magnetic field increases: the excitation region changes substantially [the plasma-resonance region Eq. (6.68)] and the position of the maximum growth rate is shifted (Varshavskii and Dimant, 1976). The region near the upper hybrid resonance $\omega_0^2(z) = \omega_1^2 - \omega_H^2$, in which the character of the low-frequency wave in the plasma is radically altered, can also play a specially important role. In particular, at $k_\perp \rho_{He} \ll 1$, electron collisions become significant. Parametric instability with collisions playing a decisive role will be considered in the next section.

Self-Action of a Radio Wave. When parametric instabilities are excited, the conditions for the propagation of the modifying radio waves are altered. This is the cause of its self-action. It is important that self-action effects become resonantly amplified in the vicinity of the reflection point (see Sect. 5.3.2). Indeed, in the region where $\varepsilon_0 \to 0$ even small plasma perturbations lead to an appreciable change of ε_0, and consequently also to a change of the structure of the radio-wave field. This in turn influences strongly the development of the instability.

Let us consider the initial stage of the self-action process. Let t be the time reckoned from the instant when the field is turned on. The oscillations that increase most rapidly are those in the first maximum of the standing wave Equation (6.181), with a growth rate Equation (6.169):

$$\gamma_1 = \frac{\nu_e}{2}\left(\frac{E_1^2}{E_{th}^2} - 1\right), \tag{6.183}$$

where E_1 is the wave-field amplitude in the first maximum [Eq. (5.125)]. Initially, at $t < t_1$, where $t_1 = \tau_0/\gamma_1$, the oscillations increase exponentially with time: the self-action of the radio wave and the nonlinear interaction between the Langmuir waves are still insignificant during this period. The parameter τ_0 is determined by the initial noise level: under ionospheric conditions (if the initial noise is thermal) $\tau_0 \sim 7$–10.

At $t > t_1$, the interaction of the radio waves with the Langmuir noise becomes appreciable (Vas'kov and Gurevich, 1973a). Since the perturbations of ε are mainly dissipative in this case, the wave intensity first decreases sharply at $t > t_1$ because of its absorption in the first maximum. But the intensity of the reflected wave, to the contrary, increases already at $t - t_1 > 1/\gamma_1$, because the field structure changes and the wave begins to be reflected effectively from the region of the strongly excited oscillations. A new standing wave is produced. This nonstationary process can then repeat itself as a result of wave excitation in the succeeding maximum, etc.

Analogous nonstationary processes result also from nonlinear interaction between strongly excited Langmuir waves (Al'ber et al., 1974; Zakharov et al., 1974; Perkins et al., 1974). The energy redistribution over the Langmuir-wave spectrum is accompanied by intense excitation of oscillations in individual narrow spectral regions—satellites (Kruer and Valeo, 1973). Owing to the weak damping, this process is likewise oscillatory and has the same characteristic oscillation period.

Thus, the structure of the modified zone is essentially nonstationary during the initial period of the excitation of the instability. As a result, the wave reflected from the ionosphere is deeply modulated in amplitude and in phase during this stage of excitation (Vas'kov and Gurevich, 1972; Al'ber et al., 1974). The modulation period is

$$T \sim (1 - 10)/\gamma_1. \tag{6.184}$$

The resultant oscillations are of the relaxation type. Under the conditions of the ionosphere, $T \sim 10^{-2}$–10^{-4} s.

In the steady state, a wide spectrum of Langmuir noise turns out to be excited (Kuo and Fejer, 1972; Kruer and Valeo, 1973; Perkins et al., 1974). The perturbations of the dielectric constant of the plasma are proportional in this case to the wave energy density, $\Delta\varepsilon \sim E^2 - E_{th}$. They are the cause of an effective self-action of the radio waves: the structure of the field is altered and the absorption of the wave in the plasma is increased. Under conditions when the ordinary linear absorption is small, the absorption connected with the excitation of the oscillations plays the principal role. The coefficient of the reflection of the wave from the plasma decreases in this case rapidly with increasing power—in proportion to W_0^{-3} (Vas'kov and Gurevich, 1974). We emphasize that we are dealing here only with the initial stage $t \ll \tau_T$ of the modification, while the plasma heating is immaterial. The development of instabilities of other types connected with heating, such as resonance, self-focusing, and others, result in a strongly inhomogeneous structure of the modified region of the ionosphere. The presence of inhomogeneities affects adversely the excitation conditions and possibly even suppresses completely the effects of the parametric instability.

On the whole, a similar picture is obtained also at large modifying-wave field amplitudes [Eq. (6.178)], when the parametric instability goes into the hydrodynamic phase. A characteristic feature of the hydrodynamic instability is that now the maximum of the oscillation growth rate γ no longer coincides with the first maximum of the standing wave, but is shifted downward, closer to the lower limit of the instability region. During the hydrodynamic instability stage the plasma oscillations are more closely coupled to the pump wave. In particular, the reconversion of the plasma-wave energy into transverse-wave energy becomes appreciable. This should lead to a lengthening of the powerful radio pulse when it is reflected from the ionosphere plasma (Zhislin et al., 1974; Vas'kov and Gurevich, 1975d).

A number of phenomena observed in experiments on the modification of the ionosphere by high-power radio waves seem to offer evidence of excitation of striction parametric instability. Shlyuger (1974) has reported observation of strong absorption and self-modulation of a short powerful ordinary radio-wave pulse reflected from the F layer of the ionosphere. The self-modulation frequency ~ 5 kHz is in agreement with Equation (6.184). Carlson et al. (1972) and Kantor (1974) have observed appreciable enhancement of plasma oscillations in the modified region of the ionosphere, which also favor (albeit not quite unequivocally Harker, 1972) the assumption of parametric excitation of Langmuir waves (Perkins et al.,

1974). Biondi et al., (1970) and Haslett and Megill (1974) have observed an appreciable enhancement of oxygen emission, indicating effective acceleration of the electrons in the plasma region modified by the waves. The appearance of accelerated electrons is one of the characteristic symptoms of the excitation of parametric instability (Weinstock, 1974; Fejer and Graham, 1974; Mityakov et al., 1975). We note, however, that appreciable electron acceleration [in the vicinity of the plasma resonance, Equation (6.83)] is possible also in the case of resonance instability and resonant absorption of radio waves.

We emphasize that since the development of an inhomogeneous structure in the ionosphere makes for worse conditions of excitation of striction parametric instability, one should expect to observe clearly pronounced parametric effects during the initial state of plasma modification, or else when the ionosphere is acted upon by short high-power pulses of sufficiently low repetition frequency.

6.3.4. Dissipative Parametric Instability

We have considered above collisionless ion-sound waves. This is correct if the spatial and temporal scales of the waves are not too large—smaller than the electron or ion mean free path lengths or times. If the opposite conditions are satisfied, or more accurately if

$$kv_{Te}\cos\theta \ll \nu_e, \qquad kv_{Te}\sin\theta \ll \omega_H; \qquad \Omega, \gamma \ll \nu_e, \qquad (6.185)$$

then the character of the low-frequency wave in the plasma changes—they are now governed by the collisions (we consider here a plasma in a magnetic field; θ is the angle between the wave vector $\boldsymbol{k}$ of the low-frequency oscillations and the magnetic field $\boldsymbol{H}$).

In a collisionless plasma, the interaction of high-frequency and low-frequency oscillations is due to striction effects. In a collision-dominated plasma [Eq. (5.7)], or more accurately in Equation (6.185), the striction pressure is small in comparison with the additional gas kinetic pressure due to the electron heating. This additional pressure is the result of the heating of the electrons in the combined field—the field of the "beats" between the radio waves and the plasma waves. If (6.123) holds, this pressure has the same space-time structure as the low-frequency wave, and can therefore contribute to enhancement of the latter. The mechanism whereby the plasma waves themselves are generated, on the other hand, is connected as usual with the polarization of the charges when the pump-wave field interacts with the low-frequency perturbations, and is perfectly analogous to that analyzed above.

Since we have in the ionosphere

$$\nu_e \ll \omega_H \sim \omega_0 \tag{6.186}$$

the Equation (6.185) is much easier to satisfy for perturbations that are transverse to the magnetic field. In the case of the dissipative instability considered in this section, it is therefore very important to take the external magnetic field into account. In addition, in view of the larger characteristic spatial scales of the waves, it becomes necessary also to allow for the finite character of the pump wavelength.

We now analyze the actual conditions for the appearance of dissipative instability, following Dimant (1977).

Excitation of Plasma Waves. If the pump-wave field

$$\boldsymbol{E}_1 = E_0 \boldsymbol{a} \cos(\boldsymbol{k}_1 \boldsymbol{r} - \omega_1 t) = \frac{E_0}{2} \boldsymbol{a} \exp\left[i(\boldsymbol{k}_1 \boldsymbol{r} - \omega_1 t)\right] + \text{c.c.} \tag{6.187}$$

is not too strong [Eqs. (6.156) and (6.157)] then, just as in Equation (6.158), we can confine ourselves to the low-frequency perturbation of the electron density δn_e and to two longitudinal plasma waves $\boldsymbol{E}^+$ and $\boldsymbol{E}^-$:

$$\begin{aligned} \delta n_e &= \delta n \exp\left[i\boldsymbol{k}\boldsymbol{r} - i(\Omega + i\gamma)t\right] + \text{c.c.} \\ \boldsymbol{E}^\pm &= E^\pm \boldsymbol{m}^\pm \exp\left[i(\boldsymbol{k} \pm \boldsymbol{k}_1)\boldsymbol{r} - i(\Omega \pm \omega_1 + i\gamma)t\right] + \text{c.c.} \end{aligned} \tag{6.188}$$

$$\boldsymbol{m}^\pm = \frac{\boldsymbol{k} \pm \boldsymbol{k}_1}{|\boldsymbol{k} \pm \boldsymbol{k}_1|}$$

Here E^+ and E^- are the complex plasma-wave amplitudes and $\boldsymbol{m}^{(\pm)}$ are their propagation directions. Since the generated waves are long [Eq. (6.185)], we can use the hydrodynamic approximation to describe the plasma waves (see Sect. 6.2.2):

$$\begin{aligned} \operatorname{div} \boldsymbol{E} &= -4\pi e(n_e - n_i), \\ \frac{\partial n_e}{\partial t} &+ \operatorname{div}(n_e \boldsymbol{v}_e) = 0. \end{aligned} \tag{6.189}$$

Substituting Equations (6.188) in Equations (6.189) and separating the required harmonics, we obtain

$$\begin{aligned} \operatorname{div} \boldsymbol{E}^+ + 4\pi e n_e^+ &= 0, \\ -i[(\omega_1 + \Omega + i\gamma)n_e^+ + N \operatorname{div} \boldsymbol{v}_e^+ + \operatorname{div}(\boldsymbol{u}_e\, \delta n)] &= 0, \end{aligned} \tag{6.190}$$

where N is the value of the unperturbed electron density and $\boldsymbol{u}_e$ is the electron velocity in the field of the pump wave [see Eq. (6.140)]. The high-frequency perturbations of the ion motion have been neglected here.

Using the definition of the dielectric tensor in a form analogous to Equation (6.126):

$$\varepsilon_{ik}(\omega)E_k \simeq E_i + \delta\varepsilon_{ik}^e(\omega)E_k = E_i + i\frac{4\pi}{\omega}j_i^e \tag{6.191}$$

and expressing in Equation (6.190) the average electron velocity $\boldsymbol{v}_e^+$ in terms of the current $\boldsymbol{j}_e$, we obtain

$$\begin{aligned}\operatorname{div}&[\varepsilon_{ik}(\omega_1 + \Omega + i\gamma, \boldsymbol{k} + \boldsymbol{k}_1)E_k^+] \\ &= -\frac{\omega_1}{\omega_1 + \Omega + i\gamma}\operatorname{div}\left[\frac{\delta n}{2N}\delta\varepsilon_{ik}^e(\omega_1, \boldsymbol{k}_1)a_k E_0\right].\end{aligned} \tag{6.192}$$

Summation over like indices is implied here, and below. Comparing Equation (6.192) with Maxwell's Equations (6.64) we see that the interaction of the radio-wave field with the low-frequency electron-density oscillations δn produces the density of the polarization charges that generate the plasma waves.

From Equation (6.192) we obtain for E^+, and analogously for E^-,

$$\begin{aligned} E^+ &= -m_i^+ \delta\varepsilon_{ik}^e(\omega_1)a_k \frac{E_0}{2\hat{\varepsilon}^+}\frac{\delta n}{N}, \\ E^- &= -m_i^- \delta\varepsilon_{ik}^{e*}(\omega_1)a_k^* \frac{E_0}{2\hat{\varepsilon}^-}\frac{\delta n}{N}, \end{aligned} \tag{6.193}$$

where the longitudinal dielectric constants are

$$\hat{\varepsilon}^\pm = m_i^\pm \varepsilon_{ik}(\Omega + i\gamma \pm \omega_1, \boldsymbol{k} \pm \boldsymbol{k}_1)m_k^\pm. \tag{6.194}$$

In Equation (6.193) we have neglected Ω and γ in comparison with ω_1 everywhere except in the denominators of $\hat{\varepsilon}^\pm$, which are resonant in character. The spatial dispersion (the dependence on $\boldsymbol{k}$) and the absorption in $\delta\varepsilon_{ik}^e$ have also been neglected.

Electron Heating. We now determine the electron heat rise $(\boldsymbol{E}\boldsymbol{j}^e)_{\Omega+i\gamma,\,\boldsymbol{k}}$ at the frequency of the beats between the interaction high-frequency fields and the electron currents. Using the fact that

$$j_k^1 = -i\frac{\omega_1}{4\pi}\left[\varepsilon_{kn}(\omega_1) - \delta_{kn}\right]a_n\frac{E_0}{2},$$
$$j_k^{\perp} \simeq -i\frac{\omega_1}{4\pi}\left[\varepsilon_{kn}(\omega_1) - \delta_{kn}\right]m_n^{\pm}E^{\pm}, \tag{6.195}$$

we obtain

$$\begin{aligned}(\boldsymbol{E}\boldsymbol{j}^e)_{\Omega+i\gamma,\vec{k}} &= \boldsymbol{E}^+\boldsymbol{j}_1^* + \boldsymbol{E}_1^*\boldsymbol{j}^+ + \boldsymbol{E}^-\boldsymbol{j}_1 + \boldsymbol{E}_1\boldsymbol{j}^- \\ &\simeq a_i^*\sigma_{ik}(\omega_1)m_k^+E_0E^+ + a_i\sigma_{ik}^*(\omega_1)m_k^-E_0E^-,\end{aligned} \tag{6.196}$$

where $\sigma_{ik}(\omega) = \sigma^*{}_{ki}(\omega)$ is the Hermitian part of the conductivity tensor; it is connected with the dissipation, i.e., proportional to ν_e.

We thus obtain from Equations (6.193)–(6.196)

$$\begin{aligned}(\boldsymbol{E}\boldsymbol{j}^e)_{\Omega+i\gamma,\boldsymbol{k}} &= \alpha T_e\,\delta n,\\ \alpha &= \frac{E_0^2\nu_e}{8\pi N T_e}\left[\frac{f_+}{\hat{\varepsilon}^+} + \frac{f_-^*}{\hat{\varepsilon}^-}\right],\\ f_{\pm} = f'_{\pm} + if''_{\pm} &= -\frac{4\pi}{\nu_e}(a_k^*\sigma_{kn}(\omega_1)m_n^{\pm})(m_i^{\pm}\,\delta\varepsilon_{is}(\omega_1)a_s).\end{aligned} \tag{6.197}$$

For example, when the excitation of waves along a magnetic field is considered, assuming $\boldsymbol{k}_1 = 0$, we have

$$\omega_1 = \omega_0, \qquad \boldsymbol{a} = \boldsymbol{m}^+ = \boldsymbol{m}^-, \qquad \sigma_{kn}(\omega_1) = \frac{\nu_e}{4\pi}\delta_{kn},$$
$$\delta\varepsilon_{is}(\omega_1) = -\delta_{is}, \qquad f_+ = f_- = 1.$$

In the general case (arbitrary $\boldsymbol{k}_1$ and $\boldsymbol{k}$), as can readily be seen, we have

$$0 < f'_{\pm} \sim 1, \qquad |f''_{\pm}| \lesssim 1 \tag{6.198}$$

The exact values of these quantities depend on the directions $\boldsymbol{a}$ of the pump-wave polarization and $\boldsymbol{m}^{\pm}$ of the plasma-wave vectors, as well as on the frequency ω_1.

The greatest heat rise, which creates the optimal generation conditions, corresponds to the resonances in Equation (6.197). We define the resonant frequencies ω^+ and ω^- and the mismatches $\delta^\pm$ by the equations:

$$\operatorname{Re}\hat{\varepsilon}(\omega^+,\theta^+) = 0, \qquad \operatorname{Re}\hat{\varepsilon}(\omega^-,\theta^-) \equiv \operatorname{Re}\hat{\varepsilon}(-\omega^-,\theta^-) = 0,$$
$$\delta^+ = \omega_1 - |\omega^+|, \qquad \delta^- = \omega_1 - |\omega^-| \tag{6.199}$$

θ^+ is the angle between $\boldsymbol{m}^+$ and $\boldsymbol{H}$ and θ^- is the angle between $\boldsymbol{m}^-$ and $\boldsymbol{H}$.

Close to resonance at

$$|\delta^\pm| \ll \omega_1 \tag{6.200}$$

we have

$$\hat{\varepsilon}^\pm \simeq \frac{\zeta^\pm}{\omega_1}[\delta^\pm \pm i\Gamma^\pm], \tag{6.201}$$

$$\zeta^\pm = \omega_1\left(\frac{\partial}{\partial\omega}\operatorname{Re}\hat{\varepsilon}^\pm\right)_{\omega=\omega^\pm}, \qquad \Gamma^\pm = \left|\omega\frac{\operatorname{Jm}\hat{\varepsilon}^\pm}{\zeta^\pm}\right|. \tag{6.202}$$

We call attention to the fact that, owing to the finite pump wavelength, the directions $\boldsymbol{m}^+$ and $\boldsymbol{m}^-$ are in general not the same. Even when the difference between them is small, θ^+ and θ^- also become different, and this can lead under the ionospheric conditions [Eq. (6.186)] to an appreciable difference between the resonant frequencies, $|\omega^+ - \omega^-| > \nu_e \sim \Gamma^+, \Gamma^-$. Consequently, the resonance conditions are not satisfied simultaneously for both plasma waves (with the exception of degenerate cases).

If at least one of the conditions in Equations (6.200) and (6.202) is satisfied, we have from Equation (6.197)

$$\alpha \simeq \frac{E_0^2}{8\pi N T_e}\omega\rho\exp(i\varphi)$$
$$\rho\exp(i\varphi) = \frac{\nu_e(f'_+ + if''_+)}{\zeta^+(\delta^+ + i\Gamma^+)} + \frac{\nu_e(f'_- - if''_-)}{\zeta^-(\delta^- - i\Gamma^-)}. \tag{6.203}$$

If θ^+ and θ^- are identically equal, then α is real and

$$\rho\exp(i\varphi) = \frac{2\nu_e}{\zeta}\frac{f'\delta + f''\Gamma}{\delta^2 + \Gamma^2}. \tag{6.204}$$

In the general case, however, as indicated above, the phase φ differs from zero and the dissipated energy $(\boldsymbol{E}\cdot\boldsymbol{j})_{\Omega+i\gamma,\boldsymbol{k}}$ turns out to be shifted in phase relative to the low-frequency density perturbations δn_e. This affects substantially the dispersion properties of the low-frequency waves. We note also that in the case when $\boldsymbol{m}^+$ and $\boldsymbol{m}^-$ differ greatly, a difference appears also between the quantities f_+ and f_-, which are connected with the polarization of the radio wave.

It follows from Equation (6.203) that, just as in the case of striction parametric instability (see Sect. 6.3.2), the optimum of the perturbation occurs at $|\delta| \sim \Gamma \sim \nu_e$. Taking the fact that $\zeta \sim 1$ and $|f''_\pm| \lesssim |f'_\pm|$ into account, we find that under the optimal conditions $\mathrm{Re}\,[\rho \exp(i\varphi)] \sim 1$.

Low-Frequency Waves. We consider collision-dominated low-frequency waves in a magnetized plasma with allowance for the additional electron heating [Eq. (6.197)]. We assume the plasma to be sufficiently strongly ionized

$$\nu_{em} \ll \nu_{ei}, \qquad \nu_{im} \ll \nu_{ii}, \qquad \nu_{ii} \ll \Omega_H. \tag{6.205}$$

The Equations (6.205) are well satisfied, in particular, in the F band of the ionosphere.

We assume first that

$$k\rho_{Hi} \sin\theta \ll 1; \qquad \Omega, \gamma \ll \nu_{ii} \tag{6.206}$$

where $\rho_{He} = v_{Te}/\omega_H$ and ρ_{Hi} are the Larmor radii of the electrons and of the ions. The ion motion, just as the electron motion, is described in this case in the hydrodynamic approximation. It is important that Equations (5.7) are generally speaking not satisfied in this case. It is therefore necessary to take into account the inertial terms, the thermal force, and the viscous-friction force in the hydrodynamic equations for the average electron and ion velocities. In Equations (5.3) and (5.4) for the temperatures, both longitudinal and the transverse transport are of importance, as is also the term proportional to $(\gamma-1)$. The system of linearized hydrodynamic equations for the low-frequency perturbations of the average velocities and temperatures of the electrons and ions then takes the form

$$\begin{aligned} Nm\frac{\partial \boldsymbol{v}_e}{\partial t} &= -\nabla\,\delta P_e - Ne\left(\boldsymbol{E}_\Omega + \frac{1}{c}[\boldsymbol{v}_e \times \boldsymbol{H}]\right) + \boldsymbol{R} + \boldsymbol{R}_v^e - Nm\nu_{em}\boldsymbol{v}_e,\\ NM\frac{\partial \boldsymbol{v}_i}{\partial t} &= -\nabla\,\delta P_i + Ne\left(\boldsymbol{E}_\Omega + \frac{1}{c}[\boldsymbol{v}_i \times \boldsymbol{H}]\right) - \boldsymbol{R} + \boldsymbol{R}_v^i - NM\nu_{im}\boldsymbol{v}_i, \end{aligned} \tag{6.207}$$

$$N\frac{\partial\,\delta T_{\rm e}}{\partial t} = -\operatorname{div}\boldsymbol{g}_{\rm e} + \frac{2}{3}T_{\rm e}\frac{\partial\,\delta n}{\partial t} + \frac{2}{3}(\boldsymbol{Ej})_{\Omega+i\gamma,\boldsymbol{k}}$$

$$- N\,\delta_{\rm ei}\nu_{\rm ei}(\delta T_{\rm e} - \delta T_{\rm i}) - N\,\delta_{\rm em}\nu_{\rm em}\,\delta T_{\rm e}$$

$$N\frac{\partial\,\delta T_{\rm i}}{\partial t} = -\operatorname{div}\boldsymbol{g}_{\rm i} + \frac{2}{3}T_{\rm i}\frac{\partial\,\delta n}{\partial t} + N\,\delta_{\rm ei}\nu_{\rm ei}(\delta T_{\rm e} - \delta T_{\rm i}) - N\nu_{\rm im}\,\delta T_{\rm i}. \tag{6.208}$$

The pressure perturbations are given here by

$$\delta P_{\rm e,i} = N\,\delta T_{\rm e,i} + T_{\rm e,i}\,\delta n. \tag{6.209}$$

Account was taken of the fact that by virtue of the quasineutrality [Eq. (5.20)] we have

$$\delta n_{\rm e} = \delta n_{\rm i} = \delta n. \tag{6.210}$$

It follows from the continuity equation that the projections of the average velocities on the $\boldsymbol{k}$ directions are equal to each other

$$\boldsymbol{k}\boldsymbol{v}_{\rm e} = \boldsymbol{k}\boldsymbol{v}_{\rm i} = kv = (\Omega + i\gamma)\frac{\delta n}{N}. \tag{6.211}$$

Using the quasineutrality relations we can, as usual, eliminate the longitudinal electric field $\boldsymbol{E}_\Omega$. The expressions for the fluxes $\boldsymbol{g}_{\rm e}$ and $\boldsymbol{g}_{\rm i}$, for the friction force $\boldsymbol{R}$, and for the fiscous friction forces $\boldsymbol{R}_v^e$ and $\boldsymbol{R}_v^i$ are given in Section 2.5.2 and in the article by Braginskii (1965).

We introduce the parameter

$$\xi_{\rm e,i} = \frac{N}{T_{\rm e,i}}\frac{\delta T_{\rm e,i}}{\delta n} \tag{6.212}$$

with the aid of which the pressure gradient is expressed in terms of the velocity v [Eq. (6.211)] in the form

$$\nabla\,\delta P_{\rm e,i} = i\boldsymbol{k}\frac{(kv)NT_{\rm e,i}(1 + \xi_{\rm e,i})}{\Omega + i\gamma}. \tag{6.213}$$

Taking now the space-time dependence of the low-frequency mode $\sim\exp[i\boldsymbol{k}\cdot\boldsymbol{r} - i(\Omega + i\gamma)t]$ into account we obtain the dispersion equation of the system Equations (6.207)—(6.209), (6.211) and (6.197). In the calculations, using the parameter Equation (6.212), it is convenient to consider first separately the closed system, Equations (6.207), (6.211), and (6.213). The condition for the existence of a nonzero solution leads to an intermediate dispersion equation, which connects $\Omega + i\gamma$, $\boldsymbol{k}$, and θ with the still unknown quantities ξ_i and ξ_e. All the components of $\boldsymbol{v}_e$ and $\boldsymbol{v}_i$ can then be expressed in terms of the quantity v, i.e., in terms of $(\Omega + i\gamma)\,\delta n$ [see Eq. (6.211)]. The system Equation (6.208) is now closed and can be used to determine the unknown quantities ξ_e and ξ_i. When these are substituted in the intermediate dispersion equation, we obtain the final dispersion equation relating $\Omega + i\gamma$ with k and θ.

An analysis of the general dispersion equation (Dimant, 1977) shows that the character of the low-frequency perturbations varies significantly as a function of the angle between the wave vector $\boldsymbol{k}$ and the magnetic field $\boldsymbol{H}$. In a direction perpendicular to $\boldsymbol{H}$, the low-frequency perturbations are diffusion modes $\Omega + i\gamma \sim ik^2$. However, even at a small deviation from the transverse direction

$$\cos\theta \gtrsim \nu_{ii}/\Omega_H, \tag{6.214}$$

the character of the process is radically altered and the low-frequency perturbations take on the form of magnetosonic waves with $\Omega + i\gamma \sim k$. Under the ionosphere F layer conditions, the ratio $\nu_{ii}/\Omega_H \sim 10^{-2}$, i.e., the diffusion modes can be only very strongly elongated in the direction of the magnetic field.

Excitation of Small-Scale Diffusion Mode. As shown in Section 6.2, the most interesting among the perturbations in the F region of the ionosphere are the inhomogeneities that are strongly elongated in the magnetic-field direction and have a transverse dimension

$$k\rho_{Hi} \gg 1 \gg k\rho_{He}. \tag{6.215}$$

From the results of the kinetic theory (Borisov et al., 1976) it follows that in this case the ions can be described hydrodynamically as before, with

$$\xi_i = 0, \qquad \nabla\,\delta P_i = eN\boldsymbol{E}_\Omega. \tag{6.216}$$

In other words, the ions are isothermal and satisfy a Boltzmann distribution in a field with potential $\varphi = -\int \boldsymbol{E}_\Omega\, d\boldsymbol{r}$. Equations (6.207), (6.208), (6.206), and (6.212) for the electrons remain the same.

We confine ourselves to the case of strictly transverse excitation $\cos\theta = 0$. From Equations (6.207) and (6.216) we obtain

$$m\frac{\partial \boldsymbol{v}_e}{\partial t} = -\frac{\nabla\,\delta P_e + \nabla\,\delta P_i}{N} - \frac{e}{c}[\boldsymbol{v}_e \times \boldsymbol{H}] + \frac{\boldsymbol{R}}{N}. \tag{6.217}$$

We have taken into account the fact that at $\cos\theta = 0$ and in Equation (6.185) the viscous forces and the collisions with the neutrals are negligible. Since $\Omega, \gamma \ll \nu_e \ll \omega_H$, it is convenient to solve Equation (6.217) in the drift approximation, expressing the velocity components in terms of the perpendicular force components:

$$v_{ey} = \frac{F_{ex}}{m\omega_H} = v, \qquad v_{ex} = -\frac{F_{ey}}{m\omega_H}. \tag{6.218}$$

The z axis is directed here along $\boldsymbol{H}$ and the y axis along $\boldsymbol{k}$; F_{ex} and F_{ey} are the forces acting on one electron along the corresponding axes. From Equations (6.213), (6.216), and (6.217), we have

$$F_{ey} = -i\frac{k^2}{\Omega + i\gamma}[(\xi_e + 1)T_e + T_i]v, \tag{6.219}$$

and from Equations (5.108), (6.212), and (6.217)

$$F_{ex} = -m\nu_{ei}v_{ex} + 1.5\frac{ik^2}{\Omega + i\gamma}\frac{\nu_{ei}}{\omega_H}\xi_e T_e v. \tag{6.220}$$

With the aid of Equation (6.218) we then obtain

$$v_{ex} = i\frac{\omega_H v}{\Omega + i\gamma}(k\rho_{He})^2\left[1 + \frac{T_i}{T_e} + \xi_e\right], \tag{6.221}$$

$$v_{ey} = v = -\frac{\nu_{ei}}{\omega_H}v_{ex} + 1.5i\frac{\nu_{ei}v}{\Omega + i\gamma}(k\rho_{He})^2\xi_e. \tag{6.222}$$

Substituting Equations (6.221) in Equations (6.222), we obtain the intermediate dispersion equation

$$\Omega + i\gamma = -i(k\rho_{He})^2\nu_{ei}\left[1 + \frac{T_i}{T_e} - \frac{\xi_e}{2}\right]. \tag{6.223}$$

Such an equation describes the diffusion modes. It is seen that in order for instability to set in it is necessary to have $\mathrm{Re}\,\xi_e > 2(1 + T_i/T_e)$.

To obtain the final dispersion equation we use Equations (6.222), (6.211), and (6.212) to express the velocity v_{ex} in terms of δT_e and δn. We obtain

$$v_{ex} = \frac{3}{2}ik\frac{\delta T_e}{m\omega_H} - \frac{\omega_H}{\nu_{ei}}\frac{\Omega + i\gamma}{k}\frac{\delta n}{N}. \tag{6.224}$$

Substituting Equations (6.224) in the expression for the heat flux Equation (5.109) and recognizing that $v_{ix} = 0$, we have

$$-\mathrm{div}\,\boldsymbol{g}_e = -1.6\nu_{ei}(k\rho_{He})^2\,\delta T_e + i\frac{\Omega + i\gamma}{N}T_e\,\delta n. \tag{6.225}$$

Using next Equation (6.197), we obtain from Equation (6.208) the following expression for ξ_e:

$$\xi_e = \frac{1}{3} \frac{2\alpha + i(\Omega + i\gamma)}{1.6\nu_{ei}(k\rho_{He})^2 - i(\Omega + i\gamma) + \delta_{ei}\nu_{ei} + \delta_{em}\nu_{em}}. \quad (6.226)$$

Finally, substituting Equations (6.226) in Equations (6.223) and neglecting $\delta_{ei}\nu_{ei}$, we obtain the final dispersion equation

$$\gamma - i\Omega = -\nu_{ei}(k\rho_{He})^2 \left[1 + \frac{T_i}{T_e} - \frac{1}{6} \frac{2\alpha - \gamma + i\Omega}{1.6\nu_{ei}(k\rho_{He})^2 + \delta_{em}\nu_{em} + \gamma - i\Omega} \right]. \quad (6.227)$$

We consider by way of example the aperiodic instability that sets in at $\varphi = 0$. In terms of the dimensionless variables

$$x = \gamma/\nu_{ei}(k\rho_{He})^2, \qquad \psi = \delta_{em}\nu_{em}/\nu_{ei}(k\rho_{He})^2, \qquad \lambda = \alpha/\nu_{ei}(k\rho_{He})^2$$

we have in place of Equation (6.227)

$$x^2 + \left(2.76 + \psi + \frac{T_i}{T_e}\right) x = \frac{1}{3}(\lambda - \lambda_{th})$$
$$\lambda_{th} = 3\left(1 + \frac{T_i}{T_e}\right)(1.6 + \psi). \quad (6.228)$$

This yields the threshold field

$$\frac{E_{th}^2}{8\pi N T_e} = \frac{3}{\rho}\left(1 + \frac{T_i}{T_e}\right)\left[\frac{\delta_{em}\nu_{em}}{\omega_1} + 1.6 \frac{\nu_{ei}}{\omega_1}(k\rho_{He})^2\right]. \quad (6.229)$$

Here ρ is a parameter [Eq. (6.203)] that depends on the polarization and propagation direction of the radio wave, on the frequency differences [Eq. (6.199)], and on other factors. Under optimal conditions, ρ becomes of the order of unity. It is seen from Equation (6.229) that the threshold field for the dissipative parametric instability is much less than for the striction instability [Eq. (6.165)]. For the growth rate near the threshold we have

$$\gamma = \frac{\left(1 + \frac{T_i}{T_e}\right)(1.6 + \psi)}{(2.76 + \psi + T_i/T_e)} \nu_{ei}(k\rho_{He})^2 \frac{E_0^2 - E_{th}^2}{E_{th}^2} \quad (6.230)$$

and in the case of a large excess above threshold

$$\gamma \approx \left(1 + \frac{T_i}{T_e}\right)^{1/2} (1.6 + \psi)^{1/2} \nu_{ei}(k\rho_{He})^2 \frac{E_0}{E_{th}}. \quad (6.231)$$

Excitation of Magnetosonic Waves. In Equation (6.214), the thermal force and the mutual striction of the electrons and ions are of little importance. The principal role is assumed by the inertial terms, by the viscosity, and by the collisions with the neutrals. The dispersion equation takes in this region the form

$$(\Omega + i\gamma)(\Omega + i\gamma + i\nu_1) = (kv_s)^2 \left[1 + \frac{2}{3} \frac{\beta(\Omega + i\gamma + iB) + (\Omega + i\gamma)(\Omega + i\gamma + iC)}{(\Omega + i\gamma + iA^+)(\Omega + i\gamma + iA^-)}\right], \quad (6.232)$$

$$\beta = i\alpha/(1 + T_i/T_e).$$

Here α is, as before, defined by Equation (6.197) and v_s is the velocity of the magnetic sound

$$v_s = \left[\frac{T_i + T_e}{M}\right]^{1/2} \cos\theta. \quad (6.233)$$

The remaining quantities are given by

$$A^\pm = \delta_{ei}\nu_{ei} + \frac{\nu_2 + \nu_3}{2} \pm \left[(\delta_{ei}\nu_{ei})^2 + \left(\frac{\nu_2 - \nu_3}{2}\right)^2\right]^{1/2},$$

$$B = 2\,\delta_{ei}v_{ii} + \nu_2, \qquad C = 2\,\delta_{ei}\nu_{ei} + (T_e\nu_2 + T_i\nu_3)/(T_e + T_i),$$

$$\nu_1 = \nu_{im} + \left[1.3\cos^2\theta + 1.2\left(\frac{\nu_{ii}\sin\theta}{\Omega_H}\right)^2\right]\frac{(kv_{Ti})^2}{\nu_{ii}}$$

$$\nu_2 = \nu_{im} + \frac{T_i}{T_i + T_e}\left[2.6 + 1.3\left(\frac{\nu_{ii}}{\Omega_H}\,\mathrm{tg}\,\theta\right)^2\right]\frac{(kv_s)^2}{\nu_{ii}}$$

$$\nu_3 = \delta_{em}\nu_{em} + \frac{4T_e}{T_i + T_e}\frac{(kv_s)^2}{\delta_{ei}\nu_{ei}}, \quad v_{Ti} = (T_i/M)^{1/2}.$$

We see that at $\Omega > \nu_1$ and $\gamma > \nu_1$ Equation (6.232) describes magnetosonic waves. However, the large-scale perturbations $kv_{Ti} \lesssim \nu_{im}$, for which the Equations (5.7) are not satisfied, are of the diffusion type, as before, in accordance with Equations (5.23)–(5.28) (Grach and Trakhtengerts, 1975).

The dispersion Equation (6.232) describes excitation of both periodic and aperiodic waves. In particular, if

$$\left(\frac{T_e}{T_i}\right)^{1/2} (\delta_{ei}\nu_{ei} + \nu_{im}) < kv_s < \nu_{ii}\left(\frac{T_i}{T_e}\right)^{3/2} \tag{6.234}$$

the minimum of the threshold field corresponds to periodic instability at $\varphi = \pi/2$, and the maximum growth rate corresponds to aperiodic instability at $\varphi = \pi$. By way of example, we consider the last case in greater detail. At $\Omega = 0$ and $\varphi = \pi$ the dispersion Equation (6.232) becomes

$$\gamma\left[\gamma + \nu_1 + \frac{2}{3}\frac{(kv_s)^2(\gamma + C)}{(\gamma + A^+)(\gamma + A^-)}\right] = (kv_s)^2\left[\frac{2}{3}\frac{|\beta|(\gamma + B)}{(\gamma + A^+)(\gamma + A^-)} - 1\right]. \tag{6.235}$$

The threshold intensity of the field is determined directly from the condition for the vanishing of the right-hand side of the equation. This yields

$$\frac{E_{th}^2}{3\pi N T_e} = \frac{3\nu_3}{2\rho\omega}(1 + T_i/T_e)\frac{\delta_{ei}\nu_{ei} + \nu_2}{2\delta_{ei}\nu_{ei} + \nu_2}. \tag{6.236}$$

It is interesting that the instability increment near the threshold first increases slowly, but as $E_0^2 \approx E_{th}^2[1 + \frac{2}{3}(T_i/(T_i + T_e))]$ is approached it increases very sharply.

In concluding this section, we compare the relative contributions of the thermal-dissipation and striction factors to the plasma perturbation. We consider for simplicity one of the high-frequency waves, $E^{(+)}$. The striction force acting on one electron is a low-frequency harmonic of the quantity $m(\boldsymbol{v} \cdot \nabla)\boldsymbol{v}$, i.e., $ikmv^{(+)}v_0^*$, where $|v^+| \sim |eE^+|/m\omega$ and $|v_0| \sim eE_0/m\omega$. Using Equation (6.193) for $E^{(+)}$ we obtain

$$|m(\boldsymbol{v}\nabla)\boldsymbol{v}| \sim k\frac{T_e}{N}\frac{E_0^2}{4\pi N T_e}\frac{\delta n}{\varepsilon^{(+)}}.$$

The analogous force due to the thermal-dissipation process is given by

$$\left|\frac{\nabla(\delta P_e)}{N}\right| \sim k\left|\frac{\delta T_e}{\delta n}\right||\delta n| \sim k\frac{T_e}{N}\xi_e|\delta n| \sim k\frac{T_e}{N}\frac{E_0^2}{4\pi N T_e}\frac{|\delta n|}{\varepsilon^{+}}F,$$

$$F = \frac{\nu_{ei}}{2|\Omega + i\gamma + \delta_{em}\nu_{em} + \delta_{ei}\nu_{ei} + (k\rho_{He})^2\nu_{ei}\sin^2\theta + (kl_e)^2\nu_{ei}\cos^2\theta|}.$$

We see therefore that in Equation (6.185) the contribution of the striction factor is always small in comparison with the heat-dissipative process. This illustrates the general statement that the role of the striction force is negligible in Equation (5.7).

We emphasize that we have considered here only a spatially homogeneous plasma. The inhomogeneity of the real ionosphere, as shown by Grach et al., 1977, is always important and can greatly alter the conditions for the excitation of dissipative parametric instability.

References

Abramovich, B. S.: Space instability of a plane wave with nonlocal heating of magnetoactive plasmas. Izv. Vyssh. Ucheb. Zaved. (Radiofizika) *19*, 329 (1976; in Russian)

Al'ber, Ya. I., Krotova, E. N., Mityakov, N. A., Rapoport, V. O., Trachtengerts, V. Yu.: Effects of stimulated scattering for an electromagnetic pulse incident on a plasma layer. Zh. Eksp. Teor. Fiz. *66*, 574 (1974) [Sov. Phys. JETP *39*, 275 (1974)]

Allen, E. M., Thome, G. D., Rao, P. B.: HF-phased array observations of heater-induced spread -F. Radio Sci. *9*, 905 (1974)

Allis, W. P.: Motions of ions and electrons. Handb. Phys. *21*, 383 (1956)

Al'pert, Ya. L.: Propagation of Electromagnetic Waves in the Ionosphere. Moscow: Nauka (1972; in Russian) [New York: Consultants Bureau (1974)]

Al'pert, Ya. L., Guseva, E. G., Fligel', D. S.: Propagation of Low-Frequency Electromagnetic Waves in the Waveguide Earth-Ionosphere. Moscow: Nauka, 1967 (in Russian)

Al'tshuler, S.: Effects of inelastic collisions upon electrical conductivity and electron heating in lower Ionosphere. J. Geophys. Res. *68*, 4707 (1963)

Anderson, J. M., Goldstein, L., Clark, G. L.: Interaction of radio waves propagated through a gaseous plasmas. Phys. Rev. *90*, 485 (1953)

Andreev, N. E., Kirii, A. Yu., Silin, V. P.: On the parametric instability of plasma in the near threshold region. Zh. Eksp. Teor. Fiz. *59*, 1024 (1969) [Sov. Phys. JETP *30*, 559 (1970)]

Avilova, I. V., Biberman, L. M., Vorob'ev, V. S., Zamalin, V. M., Kobzev, G. A., Lagarkov, A. N., Mnatsakanyan, A. Ch., Norman, G. E.: Optical Features of a Hot Air. Moscow: Nauka, 1970 (in Russian)

Bailey, V. R.: Resonance in interaction of radio waves. Nature (London) *139*, 68, 838 (1937)

Bailey, V. R.: On some effects caused in the ionosphere by electric waves. Phil. Mag. *26*, 425 (1938)

Bailey, V. A., Goldstein, L.: Control of the ionosphere by means of radio waves. J. Atmosph. Terr. Phys. *12*, 216 (1958)

Bailey, V. A., Martin, D. F.: The influence of electric waves on the ionosphere. Phil. Mag. *18*, 369 (1934)

Bailey, V. A., Smith, R. A., Landecker, K., Higgs, A. J., Hibberd, F. H.: Resonance in gyro-interaction of radio waves. Nature (London) *169*, 911 (1952)

Banks, P. M.: The thermal structure of the ionosphere. Proc. IEEE *57*, 258 (1969)

Barrington, R. E., Thrane, E. V.: Determination of D-region electron densities from observations of cross modulation. J. Atmosph. Terr. Phys. *24*, 31 (1962)

Barry, G. H.: HF-VHF communications experiment using man-made field aligned ionospheric scatterers. Radio Sci. *9*, 1025 (1974)

Bass, F. G., Gurevich, Yu. G., Kvimsadze, M. V.: Propagation of intense electromagnetic waves in a two-component plasma. Zh. Eksp. Teor. Fiz. *60*, 632 (1971) [Sov. Phys. JETP *33*, 343 (1971)]

Bates, D. R., Dalgarno, A.: Electronic Recombination. Atomic and Molecular Processes. Bates, D. R. (ed.) New York: Academic Press, 1962, Chap. 7. p. 245.

Bauer, S. J.: Physics of Planetary Ionospheres. In: Physics and Chemistry in Space Roederer, J. G. (ed.). Berlin-Heidelberg-New York: Springer, 1973

Belikovich, V. V., Benediktov, E. A., Getmantsev, G. G., Erukhimov, L. M., Zuikov, N. A., Komrakov, G. P., Korobkov, Yu. S., Kotik, D. S., Mityakov, N. A., Rapoport, V. O., Sazonov, Yu. A., Trachtengerts, V. Yu., Frolov, V. L., Cherepovitskii, V. A.: Nonlinear phenomena in upper ionosphere. Usp. Fiz. Nauk *113*, 732 (1974) [Sov. Phys. Uspekhi *17*, 615 (1975)]

Bell, P. A.: Ionospheric interaction in disturbed conditions. Proc. Phys. Soc. B *64*, 1053 (1951)

Belustin, N. S., Dokuchaev, V. P., Polyakov, S. V., Tamoikin, V. V.: Excitation of the Earth-Ionosphere waveguide by low-frequency ionospheric sources. Izv. Vyssh. Ucheb. Zaved. (Radiofizika) *18*, 1323 (1975) (in Russian)

Bespalov, V. I., Litvak, A. G., Talanov, V. I.: Self-action of electromagnetic waves in nonlinear isotropic medium. Proc. 2nd Symp. Nonlinear Optics, Novosibirsk: Nauka, 1968, p. 428 (in Russian)

Bezzerides, B., Weinstock, J.: Nonlinear saturation of parametric instabilities. Phys. Rev. Lett. *28*, 481 (1972)

Biondi, M. A.: Atmospheric electron-ion and ion-ion recombination processes. Can. J. Chem. *47*, 1711 (1969)

Biondi, M. A., Sipler, D. P., Hake, R. D. Jr.: Optical ($\lambda = 6300$) detection of radio frequency heating of electrons in the F-region. J. Geophys. Res. *75*, 6421 (1970)

Bjelland, B., Holt, O., Landmark, B., Lied, F.: The D-region of the ionosphere. Nature (London) *184*, 973 (1959)

Bliokh, P. V., Bryukhovetskii, A. A.: Radiowave focusing by an artificial-made ionosphere lens. Geomagnetizm i Aeronomiya *9*, 545 (1969) [Geomagn. Aeronomiya *9*, 443 (1969)]

Booker, H. G.: A theory of scattering by nonisotropic irregularities with application to radar reflections from the aurora. J. Atmosph. Terr. Phys. *8*, 204 (1956)

Borisov, N. D., Vas'kov, V. V., Gurevich, A. V.: Excitation of drift instability under the action of radiowaves at the F-layer of Ionosphere. Geomagnetizm i Aeronomiya, *16*, 783 (1976) [Geomagn. Aeronom. *16*, 369 (1976)]

Bowhill, S. A.: Satellite transmission studies of spread-F produced by artificial heating of the ionosphere. Radio Sci. *9*, 975 (1974)

Braginskii, S. I.: Transport processes in a plasma. In: Reviews of Plasma Physics. Leontovich, M. A. (ed.) New York: Consultants Bureau, 1965, Vol. I, p. 205

Budden, K. G.: Radio Waves in the Ionosphere. Cambridge: University Press, 1961

Carlson, H. C., Gordon, W. E., Showen, R. L.: High-frequency induced enhancements of the incoherent scatter spectrum at Arecibo. J. Geophys. Res. *77*, 1242 (1972)

Carpenter, G. B.: VHF and UHF bistatic observations of a region of the Ionosphere modified by a high power transmitter. Radio Sci. *9*, 965 (1974)

Chapman, S., Cowling, T. G.: The Mathematical Theory of Non-Uniform Gases, Cambridge: University Press, 1952

Cohen, R., Whitehead, J. D.: Radio-reflectivity detection of artificial modification of the ionosphere F-layer. J. Geophys. Res. *75*, 6439 (1970)

Connor, J. W., Hastie, R. J.: Relativistic limitations on runaway electrons. Nucl. Fusion, *15*, 415 (1975)

Cragin, B. L., Fejer, J. A.: Generation of large-scale field-aligned irregularities in ionospheric modification experiments. Radio Sci. *9*, 1071 (1974)

Crompton, R. W., Sutton, D. J.: Experimental investigation of the diffusion of slow electrons in nitrogen and hydrogen. Proc. Roy. Soc. London, Ser. A *215*, 467 (1952)
Cutolo, M.: Effects of radio gyro-interaction and their interpretation. Nature (London) *166*, 98 (1950)
Cutolo, M.: On some resonance phenomena near the gyro-frequency obtained during the radio wave propagation on a plasma. NBS Tech. Note N211, Vol. II. (1964)
Dalgarno, A., Degges, T. C.: Electron cooling in the upper atmosphere. Planet. Space Sci. *16*, 125 (1968)
Danilov, A. D., Simonov, A. G. Positive ions in the D-region. Geomagnetizm i aeronomiya *15*, 643, 841 (1975) [Geomagn. Aeronom. *15*, 616 (1975)]
DASA Reaction Rate Handbook, DASA 2407, Tempo, G. E. (ed.) Santa Barbara, CA: Center for Advanced Studies, 1970
Davydov, B. I.: Contribution to the theory of motion of electrons in gases and semiconductors. Zh. Eksp. Teor. Fiz. *7*, 1069 (1937)
Dimant, Ya. S.: Excitation of dissipative parametric instability in ionospheric F-region. Geomagnetizm i Aeronomiya, *17*, 528 (1977; in Russian)
Dreicer, H.: Electron and ion runaway in a fully ionized gas. Phys. Rev. *115*, 238 (1959)
DuBois, D. F., Goldman, M. V.: Radiation induced instability of electron plasma oscillations. Phys. Rev. Lett. *14*, 544 (1965)
DuBois, D. F., Goldman, M. V.: Nonlinear saturation of parametric instability: basic theory and application to the ionosphere. Phys. Fluids *15*, 919 (1972)
Engelhardt, A. G., Phelps, A. V., Risk, C. G.: Determination of momentum transfer and inelastic collision cross sections for electrons in nitrogen using transport coefficients. Phys. Rev. *135A*, 1566 (1964)
Farley, D. T. Jr.: Artificial heating of the electrons in the F-region of the ionosphere. J. Geophys. Res. *68*, 401 (1963)
Fejer, J. A.: Interaction of pulsed radio waves in the ionosphere. J. Atmosph. Terr. Physics *7*, 322 (1955)
Fejer, J. A., Graham, K. N.: Electron acceleration by parametrically excited Langmuir waves. Radio Sci. *9*, 1081 (1974)
Fejer, J. A., Leer, E.: Excitation of parametric instabilities by radio waves in the ionosphere. Radio Sci. *7*, 481 (1972)
Fialer, P. A.: Field-aligned scattering from a heated region of the ionosphere-observations at HF and VHF. Radio Sci. *9*, 923 (1974)
Forslund, D. W., Kindel, J. M., Lindeman, E. L.: Theory of stimulated scattering processes in laser-irradiated plasma. Phys. Fluids *18*, 1002 (1975)
Frost, L. S., Phelps, A. V.: Rotational excitation and momentum transfer cross sections for electrons in H_2 and N_2 from transport coefficients. Phys. Rev. *127*, 1621 (1962)
Gaponov, A. V., Miller, M. A.: On the potential wells for charged particles in a high-frequency electromagnetic field. Zh. Eksp. Teor. Fiz. *34*, 242 (1958) [Sov. Phys. JETP *7*, 168 (1958)]
Georges, T. M.: Interaction of pulsed radio waves in ionosphere. Proc. Conf. Phys. Low Ionosphere. Ottawa, p. 289 (1966)
Georges, T. M.: Amplification of ionospheric heating and triggering of spread F by natural irregularities. J. Geophys. Res. *75*, 6436 (1970)
Gerjouy, E., Stein, S.: Rotational excitation by slow electrons. Phys. Rev. *97*, 1671 (1955)
Getmantsev, G. G., Belikovich, V. V., Benediktov, E. A., Ignat'ev, Yu. A., Komranov, G. P.: Scattering of radiowaves from the artificially disturbed ionospheric F-region. Pis'ma Zh. Eksp. Teor. Fiz. *22*, 497 (1975) [JETP Lett. *22*, 243 (1975)]
Getmantsev, G. G., Komrakov, N. P., Korobkov, P. P., Mironenko, L. F., Mityakov,

N. A., Rapoport, V. O., Trachtengerts, V. Yu., Frolov, V. L., Cherepovitskii, V. A.: Some results of investigations of nonlinear phenomena in the F-layer of the ionosphere. Pis'ma Zh. Eksp. Teor. Fiz. *18*, 621 (1973) [JETP Lett. *18*, 364 (1973)]
Getmantsev, G. G., Zuikov, N. A., Kotik, D. S., Mironenko, L. F., Mityakov, N. A., Rapoport, V. O., Sazonov, Yu. A., Trachtengerts, V. Yu., Eidman, V. Ya.: Combination frequencies in the interaction between high-power short-wave radiation and ionospheric plasma. Pis'ma Zh. Eksp. Teor. Fiz. *20*, 229 (1974) [JETP Lett. *20*, 101 (1974)]
Gildenburg, V. B., Golubev, S. V.: Nonequilibrium high frequency discharge in wave fields. Zh. Eksp. Teor. Fiz. *67*, 89 (1974) [Sov. Phys. JETP *40*, 46 (1975)]
Ginzburg, E. I.: On the propagation of strong radiowaves in the ionosphere. Izv. Vyssh. Ucheb. Zaved. (Radiofizika) *7*, 1041 (1964; in Russian)
Ginzburg, V. L.: Nonlinear interaction of radio waves propagating in plasmas. Zh. Eksp. Teor. Fiz. *35*, 1573 (1957) [Sov. Phys. JETP *8*, 1100 (1958)]
Ginzburg, V. L.: Propagation of Electromagnetic Waves in Plasmas. Moscow: Nauka, 1960 [New York: Pergamon Press, 1964]
Ginzburg, V. L., Gurevich, A. V.: Nonlinear phenomena in a plasmas located in an alternating electric field. Usp. Fiz. Nauk *70*, 201 (1960) [Sov. Phys. Uspekhi *3*, 115, 175 (1960)]
Ginzburg, V. L., Rukhadze, A. A.: Waves in magnetoactive plasmas. Moscow: Nauka, 1975 (in Russian)
Golant, V. E., Piliya, A. D.: Linear transformation and absorption of waves in a plasmas. Usp. Fiz. Nauk *104*, 413 (1971) [Sov. Phys. Uspekhi *14*, 413 (1972)]
Goldberg, R. A., Aikin, A. C.: Studies of positive ion composition in the equatorial D-region ionosphere, J. Geophys. Res. *76*, 8352 (1971)
Gordon, W. E., Carlson, H. C.: Arecibo heating experiment. Radio Sci. *9*, 1041 (1974)
Grach, S. M., Karashtin, A. N., Mityakov, N. A. Rapoport, V. O., Trakhtengerts, V. Yu. Parametric interaction of electromagnetic waves with ionospheric plasma. Izv. Vyssh. Ucheb. Zaved. (Radiofizika) *20*, 1827 (1977: in Russian).
Grach, S. M., Trakhtengerts, V. Yu.: Parametric excitation of ionospheric irregularities extended along the magnetic field. Izv. Vyssh. Ucheb. Zaved. (Radiofizika) *18*, 1288 (1975)
Grigor'ev, G. I.: Traveling ionospheric disturbances arising as a result of powerful transmitter operation. Izv. Vyssh. Ucheb Zaved. (Radiofizika) *18*, 1801 (1975; in Russian)
Gurevich, A. V.: The theory of propagation of the strong radiowaves in a plasmas (Ionosphere). Radiotekhn. Elektron. *1*, 706 (1956; in Russian)
Gurevich, A. V.: Contribution to the theory of cross modulation of radio waves. Izv. Vyssh. Ucheb. Zaved. (Radiofizika) *1* (5–6), 17 (1958a; in Russian)
Gurevich, A. V.: Distortion of modulation of strong radio waves in plasmas (Ionosphere). Izv. Vyssh. Ucheb. Zaved. (Radiofizika) *1*, (4), 21 (1958b; in Russian)
Gurevich, A. V.: On the electron temperature in a plasmas in an alternating electric field. Zh. Eksp. Teor. Fiz. *35*, 392 (1958c) [Sov. Phys. JETP *8*, 271 (1959)]
Gurevich, A. V.: Influence of electron-electron collisions on the distribution function in gases and semiconductors. Zh. Eksp. Teor. Fiz. *37*, 304 (1959) [Sov. Phys. JETP *10*, 215 (1960)]
Gurevich, A. V.: On the theory of runaway electrons. Zh. Eksp. Teor. Fiz. *39*, 1296 (1960) [Sov. Phys. JETP *12*, 904 (1961)]
Gurevich, A. V.: Nonlinear effects for powerful radiowaves in the ionosphere. Geomagnetizm i Aeronomiya *5*, 49 (1965) [Geomagn. Aeronomy *5*, 70 (1965a)]
Gurevich, A. V.: Nonlinear effects in the region of reflection of radio waves. Zh. Eksp. Teor. Fiz. *48*, 701 (1965) [Sov. Phys. JETP *21*, 462 (1965b)]

Gurevich, A. V.: Influence of powerful radiowaves on the F-region of Ionosphere. Geomagnetizm i Aeronomiya *7*, 230 (1967) [Geomagn. Aeronomy *7*, 291 (1967)]

Gurevich, A. V.: Disturbance of ionosphere by powerful radio waves. Geomagnetizm i Aeronomiya *11*, 810 (1971) [Geomagn. Aeronomy *11*, 953 (1971)]

Gurevich, A. V.: Traveling ionization disturbances in a field of a strong electromagnetic wave. Izv. Vyssh. Ucheb. Zaved. (Radiofizika) *15*, 11 (1972a; in Russian).

Gurevich, A. V.: Self-action of radio waves in E and F regions of ionosphere. Geomagnetizm i Aeronomiya *12*, 20 (1972) [Geomagn. Aeronomy *12*, 24 (1972b)]

Gurevich, A. V.: Isothermal ionization of the lower Ionosphere under the influence of radiowaves. Geomagnetizm i Aeronomiya *12*, 631 (1972) [Geomagn. Aeronom *12*, 556 (1972c)]

Gurevich, A. V.: On the heating of neutral component of lower ionosphere by the powerful radiowave. Geomagnetizm i Aeronomiya *15*, 161 (1975) [Geomagn. Aeronomy *15*, 134 (1975)]

Gurevich, A. V., Milich, G. M., Shlyuger, I. S.: Electron kinetics in low temperature molecular plasmas (Ionosphere). Zh. Eksp. Teor. Fiz. *69*, 1640 (1975) [Sov. Phys. JETP *42*, 835 (1976)]

Gurevich, A. V., Milich, G. M., Shlyuger, I. S.: Artificial ionization of the ionosphere by powerful radiowaves. Pis'ma Zh. Eksp. Teor. Fiz. *23*, 395 (1976a) [JETP Lett. *23*, 356 (1976)]

Gurevich, A. V., Milich, G. M., Shlyuger, I. S.: Non linear thermal focusing of radio waves in lower ionosphere. Geomagnetizm i Aeronomiya *16*, 613 (1976) Geomagn. Aeronomy *16*, 366 (1976b)

Gurevich, A. V., Pitaevskii, L. P.: Nonstationary structure of a collisionless shock wave. Zh. Eksp. Teor. Fiz. *65*, 590 (1973) [Sov. Phys. JETP *38*, 291 (1974)]

Gurevich, A. V., Shlyuger, I. S.: Investigation of nonlinear phenomena with powerful radioimpulse in the lower ionosphere. Izv. Vyssh. Ucheb. Zaved. (Radiofizika) *18*, 1237 (1975; in Russian)

Gurevich, A. V., Shvartsburg, A. B.: Nonlinear resonance for modulated radiowaves in a plasma (Ionosphere). Geomagnetizm i Aeronomiya *8*, 893 (1968) [Geomagn. Aeronomy *8*, 1104 (1968)]

Gurevich, A. V., Shvartsburg, A. B.: Non-linear theory of propagation of radio waves in the Ionosphere. Moscow: Nauka, 1973 (in Russian)

Gurevich, A. V., Tsedilina, E. E.: Movement and dispersal of inhomogeneities in plasmas. Space Sci. Rev. *7*, 407 (1967)

Gurevich, A. V., Tsedilina, E. E.: Theoretical investigation of a very long distance propagation of short radio waves. Usp. Fiz. Nauk *120*, 319 (1976) [Sov. Phys. Uspekhi *19*, 869 (1976)]

Gurevich, A. V., Vas'kov, V. V.: Nonlinear resonance instability of a plasma in the reflection region of ordinary electromagnetic wave. Preprint FIAN, N 95 (1975)

Hake, R. D., Phelps, A. V.: Momentum-transfer and inelastic collision cross sections for electrons in O_2, CO and CO_2. Phys. Rev. *158*, 70 (1967)

Harker, K. J.: Induced enhancement of the plasma line in the back-scatter spectrum by ionospheric heating. J. Geophys. Res. *77*, 6904 (1972)

Harris, I., Priester, W.: Time-dependent structure of the upper atmosphere. J. Atmosph. Sci. *18*, 286 (1962)

Haslett, J. C., Megill, L. R.: A model of the enhanced airglow excited by RF radiation. Radio Sci. *9*, 1005 (1974)

Herzberg, G.: Spectra of Diatomic Molecules. Princeton: van Nostrand, 1950

Hibberd, F. H.: Ionospheric self-modulation of radio waves. J. Atmosph. Terr. Phys. *6*, 268 (1955)

Hibberd, F. H.: An experimental study of gyro interaction in the ionosphere at oblique incidence. Radio Sci. *69D*, 25 (1965)

Hochstim, A. R., Massel, G. A.: Kinetic Processes in Gases and Plasmas. Hochstim, A. R. (ed.) New York: Academic Press, 1969, p. 142

Holstein, T.: Energy distribution of electrons in high-frequency gas discharges. Phys. Rev. *70*, 367 (1946)

Huxley, L. G. H.: A synopsis of ionospheric cross-modulation. Nuovo Cim. Suppl. *9*, 59 (1952)

Huxley, L. G. H., Foster, H. G., Newton, C. C.: Measurements of the interaction of radio waves in the Ionosphere. Proc. Phys. Soc. *B61*, 134 (1948)

Huxley, L. G. H., Ratcliffe, J.: A survey of ionospheric cross-modulation, Proc. Inst. Elec. Engrs. III, *96*, 443 (1949)

Ignat'ev, Yu. A.: Influence of the radio wave heating of the ionosphere on the layer E-sporadic. Izv. Vyssh. Ucheb. Zaved. (Radiofizika) *18*, 1365 (1975; in Russian)

Ivanov-Kholodnyi, G. S., Nikol'skii, G. M.: The Sun and the Ionosphere. Moscow: Nauka, 1969 (in Russian)

Kantor, I. J.: High-frequency induced enhancements of the incoherent scatter spectrum at Arecibo, 2. J. Geophys. Res. *79*, 199 (1974)

Kapustin, I. N., Pertsovskii R. A., Vasil'ev, A. N., Smirnov, V. S., Raspopov, O. M., Solov'eva, L. E., Ulyanchenko, A. A., Arykov, A. A., Galachova, N. V.: Generation of radiation at combination frequencies in the region of the auroral electric jet. Pis'ma Zh. Eksp. Teor. Fiz. *25*, 248 (1977) [JETP Lett. 24, 228 (1977)]

Karpman, V. I.: Nonlinear Waves in a Dispersive Media, Moscow: Nauka, 1973 (in Russian)

Kiefer, L. J.: Low-energy electron collision cross-section data. Atomic Data *1*, 19 (1969)

Kiefer: L. J.: Low-energy electron collision cross-section data. Atomic Data *2*, 293 (1971)

King, J. W.: Ionospheric self-demodulation and self-distortion of radio waves. J. Atmosph. Terr. Phys. *14*, 41 (1959)

Klimontovich, Yu. L.: Statistical theory of Nonequilibrium Processes. Moscow: University Press, 1964 (in Russian)

Kotik, D. S., Trakhtengerts, V. Yu.: Mechanism of generation of combined frequencies in ionospheric plasmas. Pis'ma Zh. Eksp. Teor. Fiz. *21*, 52 (1975) [JETP Lett. *21*, 24 (1975)]

Kruer, W. L., Valeo, E. J.: Nonlinear evolution of the decay instability in a plasma with comparable electron and ion temperatures. Phys. Fluids *16*, 675 (1973)

Kuo, Y. Y., Fejer, J. A.: Spectral line structure of saturated parametric instabilities. Phys. Rev. Lett. *29*, 1667 (1972)

Landau, L. D.: Kinetic equation for Coulomb interactions. Phys. Z. Sowjet *10*, 154 (1936)

Landau, L. D., Lifshitz, E. M.: Fluid Mechanics. Oxford: Pergamon Press, 1963

Landau, L. D., Lifshitz, E. M.: Electrodynamics of Continuous Media, Moscow: Nauka, 1971b (in Russian) [New York: Pergamon Press, 1975]

Landau, L. D., Lifshitz, E. M.: Statistical physics, Moscow: Nauka, 1971a (in Russian) [New York: Pergamon Press, 1977]

Lebedev, A. N.: Contribution to the theory of runaway electrons. Zh. Eksp. Teor. Fiz. *48*, 1393 (1965) [Sov. Phys. JETP *21*, 931 (1965)]

Leontovich, M. A.: On one method of solution of problems of radio wave propagation. Izvestiya Akad. Nauk SSR Ser. Fiz. *8*, *16* (1944; in Russian)

Litvak, A. G.: On the possibility of self-focusing of electromagnetic waves in ionosphere. Izv. Vyssh. Ucheb. Zaved. (Radiofizika *11*, 1433 (1968; in Russian)

Liu, C. S., Rosenbouth, M. N., White, R. B.: Raman and Brillouin scattering of electromagnetic waves in inhomogeneous plasma. Phys. Fluids *17*, 1211 (1974)

Lombardini, P. P.: Alternation of the electron density of the lower ionosphere with ground-based transmitter. Radio Sci. *69D*, 83 (1965)

MacDonald, A. D.: Microwave Breakdown in Gases. New York-London-Sydney: Wiley, 1966

Margenau, H.: Conduction and dispersion of ionized gases at high frequencies. Phys. Rev. *69*, 508 (1946)

Meerovich, B. E.: Diffusion in a high-frequency gas discharge. Zh. Eksp. Teor. Fiz. *63*, 549 (1972) [Sov. Phys. JETP *36*, 291 (1973)]

Meerovich, B. E., Pitaevskii, L. P.: On the structure of transition layer in a high-frequency gas discharge. Zh. Eksp. Teor. Phys. *61*, 235 (1971) [Sov. Phys. JETP *34*, 121 (1972)]

Meltz, G., Holway, L. H. Jr., Tomljanovich, N. M.: Ionospheric heating by powerful radio waves. Radio Sci. *9*, 1049 (1974)

Meltz, G., Le-Levier, R. E.: Heating the F-region by deviative absorption of radio waves. J. Geophys. Res. *75*, 6406 (1970)

Menzel, D. H.: Some problems of Ionospheric nonlinearities. Radio Sci. *69D*, 1 (1965)

Mikhailovskii, A. B.: Theory of Plasma Instabilities. Moscow: Atomizdat, 1971 (in Russian)

Minkoff, J.: Radio frequency scattering from a heated ionospheric. Vol. III, Cross section calculations. Radio Sci. *9*, 997 (1974c)

Minkoff, J., Kugelman, P., Weissman, I.: Radio frequency scattering from a heated ionospheric, I. Radio Sci. *9*, 941 (1974a)

Minkoff, J., Laviola, M., Abrams, S., Porter, D.: Radio frequency scattering from a heated ionospheric. Vol. II. Bistatic measurements. Radio Sci. *9*, 957 (1974b)

Mityakov, N. A., Rapoport, V. O., Trachtengerts, V. Yu.: Induced scattering of radio waves in F-layer of the Ionosphere. Geomagnetizm i Aeronomiya *14*, 29 (1974) [Geomagn. Aeronomy *14*, 36 (1974)]; Planet. Space Sci. *22*, 95 (1974)

Mityakov, N. A., Rapoport, V. O., Trachtengerts, V. Yu.: Heating of the Ionosphere by an electromagnetic field under developed parametric instability. Izv. Vyssh. Ucheb. Zaved. (Radiofizika) *18*, 27 (1975; in Russian)

Molchanov, O. A. Reznikov, A. E. Fligel', D. S., Self-action of powerful low-frequency waves in ionospheric plasma. Geomagnetizm i Aeronomiya *16*, 1123 (1976; in Russian).

Nicolet, M.: Physics of the Upper Atmosphere. Ratcliffe, J. A. (ed.). New York: Academic Press, 1960

Nicolet, M., Swidder, W. Jr.: Ionospheric conditions. Planet. Space Sci. *11*, 1459 (1963)

Nighan, W. L.: Electron energy distributions and collision rates in electrically excited N_2, CO and CO_2. Phys. Rev. *A2*, 1989 (1970)

Nishikawa, K.: Parametric excitation of coupled waves. J. Phys. Soc. Japan *24*, 1152 (1968)

Perkins, F. W.: A theoretical model for short-scale field aligned plasma density striations. Radio Sci. *9*, 1065 (1974)

Perkins, F. W., Kaw, P. K.: On the role of plasma instabilities in ionospheric heating by radio waves. J. Geophys. Res. *76*, 282 (1971)

Perkins, F. W., Oberman, C., Valeo, E. J.: Parametric instabilities and ionospheric modification. J. Geophys. Res. *79*, 1478 (1974)

Perkins, F. W., Valeo, E. J.: Thermal self-focusing of electromagnetic waves in plasmas. Phys. Rev. Lett. *32*, 1234 (1974a)

Phelps, A. V.: Laboratory studies of electron attachment and detachment processes of aeronomic interest. Can. J. Chemistry *47*, 1783 (1969)

Pitaevskii, L. P.: Electric forces in the dispersive medium. Zh. Eksp. Teor. Fiz. *39*, 1450 (1959) [Sov. Phys. JETP *12*, 1008 (1960)]

Rao, P. B., Thome, G. D.: A model for RF scattering from field-aligned heater-induced irregularities. Radio Sci. *9*, 987 (1974)

Ratcliffe, J., Shaw, J.: A study of interaction of radiowaves. Proc. Roy. Soc. London, Ser. A *193*, 311 (1948)

Rosen, P.: Generation of a third harmonic by an electromagnetic signal in a plasma. Phys. Fluids *4*, 341 (1961)
Rufenach, G. L.: Radio scintillation of stellar signals during artificial ionospheric modification. J. Geophys. Res. *78*, 5611 (1973)
Schottky, W.: The theory of diffusion of positive charges. Phys. Z. Sowjet. *25*, 635 (1924)
Schulz, G. J.: Vibrational excitation of N_2, CO, and H_2 by electron impact. Phys. Rev. *125*, 229 (1962a)
Schulz, G. J.: Cross sections and electron affinity for O^- ions from O_2, CO, and CO_2 by electron impact. Phys. Rev. *128*, 178 (1962b)
Schulz, G. J.: Vibrational excitation of N_2, CO, H_2 by electron impact. Phys. Rev. *135A*, 988 (1964)
Shafranov, V. D.: Electromagnetic Waves in Plasmas. In: Reviews of Plasma Physics Leontovich, M. A. (ed.) Vol. III. New York: Consultants Bureau, 1966
Shkarofsky, I. P., Johnston, T. W., Bachynski, M. P.: The Particle Kinetics of Plasmas. Reading. Mass.-Palo Alto-London: Addison-Wesley, 1966
Shlyuger, I. S.: Experimental investigation of nonlinear effects in E and F-regions of Ionosphere. Pis'ma Zh. Eksp. Teor. Fiz. *20*, 722 (1974a) [JETP Letters *20*, 334 (1974a)]
Shlyuger, I. S.: Selfmodulation of a strong electromagnetic impulse reflected from the upper layers of ionosphere. Pis'ma Zh. Eksp. Teor. Fiz. *19*, 274 (1974b) [JETP Lett. *19*, 162 (1974b)]
Silin, V. P.: Parametric resonance in a plasma. Zh. Eksp. Teor. Fiz. *48*, 1679 (1965) [Sov. Phys. JETP *21*, 1127 (1965)]
Silin, V. P.: Introduction in the Kinetic Theory of Gases. Moscow: Nauka, 1972 (in Russian)
Silin, V. P.: Parametric Action of Powerful Radiation at the Plasmas. Moscow: Nauka, 1973 (in Russian)
Smith, R. A.: Interaction of pulsed radio waves in low ionosphere I. Proc. Conf. Phys. Low Ionosphere, Ottawa, p. 235 (1966a)
Smith, R. A., Bourne, L. A., Loch, R. G., Coyne, T. N. R.: Interaction of pulsed radio waves in low ionosphere II, III. Proc. Conf. Phys. Low Ionosphere, Ottawa, pp. 300, 335 (1966b)
Sobolvea, T. N.: Model profiles of electron concentration distribution in a quiet ionosphere at middle lattitudes. Preprint IZMITRAN, N 20 (1972)
Sodha, M. S., Kaw, P. K.: Third harmonic of current density in a plasma. Phys. Fluids *8*, 1402 (1965)
Sodha, M. S., Kaw, P. K.: Nonlinear sum and difference frequency generation in an inhomogeneous plasma. Phys. Fluids *9*, 603 (1966)
Stathacopoulos, A. D., Barry, G. H.: Geometric considerations in the design of communications circuits using field-aligned ionospheric scatter. Radio Sci. *9*, 1021 (1974)
Thomson, J. A.: Energy disposition in artificial Ionospheric heating experiment. J. Geophys. Res. *75*, 6446 (1970)
Ting-Wei, Tang: Generation of the third harmonic current density in a plasma. Phys. Fluids *9*, 415 (1966)
Tohmatsu, T., Ogawa, T., Tsuruto, H.: Photo-electronic processes in the upper atmosphere I. Energy spectrum of the primary photo-electrons. Rept. Ionosphere Space Res. Japan *19*, 482 (1965)
Tsitovich, V. N.: Nonlinear Effects in Plasmas, Nauka, 1967. [Plenum Press New York, 1970]
Utlaut, W. F.: An ionospheric modification experiment, using very high power, high-frequency transmitter. J. Geophys. Res. *75*, 6402 (1970)
Utlaut, W. F.: Ionospheric modification by high-power HF transmitters. A potential ex-

tended range VHF-UHF communications and plasma physics research. IEEE *63*, 1022 (1975)

Utlaut, W. F., Cohen, R.: Modifying the ionosphere with intense radio waves. Science *174*, 245 (1971)

Utlaut, W. F., Violette, E. J.: Further observations of ionospheric modification by a high-powered HF transmitter. J. Geophys. Res. *77*, 6804 (1972)

Utlaut, W. F., Violette, E. J.: A summary of vertical incidence radio observations of ionospheric modification. Radio Sci. *9*, 895 (1974a)

Utlaut, W. F., Violette, E. J., Melanson, L. L.: Radar cross section measurements and vertical incidence effects observed with Platteville at reduced power. Radio Sci. *9*, 1033 (1974b)

Utlaut, W. F., Violette, E. J., Paul, A. K.: Some ionosonde observations of ionospheric modification by very high power, high-frequency groundbased transmission. J. Geophys. Res. *75*, 6429 (1970)

Varshavskii, I. I., Dimant, Ya. S.: Parametric excitation of Langmuir oscillations in the ionosphere by powerful ordinary radio wave. Geomagnetizm i Aeronomiya *16*, 98 (1976); [Geomagn. Aeronomy *16*, 57 (1976)]

Vas'kov, V. V., Gurevich, A. V.: Self-action of radio waves in the vicinity of plasma resonance. Zh. Eksp. Teor. Fiz. *64*, 1272 (1973) [Sov. Phys. JETP *37*, 646 (1973a)]

Vas'kov, V. V., Gurevich, A. V.: Self-action of an electromagnetic wave in a plasma on excitation of parametric instability. Phys. Lett. *45A*, 47 (1973b)

Vas'kov, V. V., Gurevich, A. V.: Parametrical excitation of Langmuir oscillations in the ionosphere in a strong radio wave field. Izv. Vyssh. Ucheb. Zaved. (Radiofizika) *16*, 188 (1973c; in Russian)

Vas'kov, V. V., Gurevich, A. V.: Stratification of ionospheric plasmas in the reflection region of ordinary radio wave. Pis'ma Zh. Eksp. Teor. Fiz. *20*, 214 (1974) [JETP Lett. *20*, 93 (1974)]

Vas'kov, V. V., Gurevich, A. V.: Nonstationary processes in the reflection region of a strong radio wave in Ionosphere. Geomagnetizm i Aeronomiya *15*, 67 (1975a) [Geomagn. Aeronomy *15*, 51 (1975)]

Vas'kov, V. V., Gurevich. A. V.: Nonlinear resonance instability of plasma in the reflection region of ordinary electromagnetic wave. Zh. Eksp. Teor. Fiz. *69*, 176 (1975b) [Sov. Phys. JETP *42*, 91 (1975)]

Vas'kov, V. V., Gurevich, A. V.: Instability of plasmas in the reflection region of intense radiowaves in ionosphere. Izv. Vyssh. Ucheb. Zaved. (Radiofizika) *18*, 1261 (1975c; in Russian)

Vas'kov, V. V., Gurevich, A. V.: Nonlinear stabilization of electron concentration disturbances in the F-region of ionosphere. Geomagnetizm i Aeronomiya *16*, 1204 (1976b) [Geomagn. Aeronomy *16*, 1112 (1976)]

Vas'kov, V. V., Gurevich, A. V.: Modulation instability of radiowaves in upper ionosphere. Geomagnetizm i Aeromiya *16*, 50, 239 (1976b), [Geomagn. Aeronomy *16*, 28, 141 (1976)]

Vas'kov, V. V., Gurevich, A. V.: Excitation in the Ionosphere of parametric instability of hydrodynamical type. Geomagnetizm i Aeronomiya *15*, 235, 412, 633 (1975d) [Geomagn. Aeronomy *15*, 194, 339, 448, (1975)]

Vas'kov, V. V., Gurevich, A. V.: Saturation of self-focusing instability for radio wave beams in plasma. Fiz. Plazmy, *3*, 329 (1977a) [Sov. J. Plasma Phys. *3*, 185 (1977)].

Vas'kov V. V., Gurevich A. V. Resonance instability of small-scale plasma perturbations. Zh. Eksp. Teor. Fiz. 73, 923 (1977b) [Sov. Phys. JETP 46; No. 3 (1977)]

Vas'kov, V. V., Gurevich, A. V., Karashtin, A. N.: Self-focusing instability under the oblique incidence of radio waves on the ionosphere. Geomagnetizm i Aeronomiya *16*, 549 (1976c) [Geomagn. Aeronomy *16*, 322 (1976)]

Vilenskii, I. M.: Effect of the Earth magnetic field on the interaction of radio waves in ionosphere. Zh. Eksp. Teor. Fiz. *26*, 42 (1954; in Russian)

Vilenskii, I. M.: On the effect of nonlinearity on the radio wave propagating in the Ionosphere. In: Collection of Papers in Memory of A. A. Andronov, p. 582. Moscow: Academy Press, USSR 1955 (in Russian)

Vilenskii, I. M.: On the theory of interaction of radio-waves in Ionosphere. Izv. Vyssh. Ucheb. Zaved. (Radiofizika) *5*, 468 (1962a; in Russian)

Vilenskii, I. M., Chernyshov, V. P., Sheinman, D. I.: Distortion of modulation of radio waves propagating in ionosphere (experimental investigation). Izv. Vyssh. Ucheb. Zaved. (Radiofizika) *3*, 367 (1960)

Vilenskii, I. M. et al.: Distortion of modulation of radio waves propagating in ionosphere (experimental investigation). Izv. Vyssh. Ucheb. Zaved. (Radiofizika) 3, 221 (1962b; in Russian)

Vilenskii, I. M. et al.: Distortion of modulation of radio waves propagating in ionosphere (experimental investigation) III. Izv. Vyssh. Ucheb. Zaved. (Radiofizika) *9*, 649 (1966; in Russian)

Vilenskii, I. M., Zykova, N. A.: To the theory of distortion of radio waves at their propagation in the ionosphere. Izv. Vyssh. Ucheb. Zaved. (Radiofizika) *2*, 543 (1959; in Russian)

Volkov, A. F., Kogan, S. M.: Physical phenomena in semiconductors with negative differential conductivity. Usp. Fiz. Nauk *96*, 633 (1968) [Sov. Phys. Uspekhi *11*, 881 (1969)]

Wait, J. R.: Electromagnetic Waves in Stratified Media. New-York: Pergamon Press, 1962

Whitham, G. B.: Non-linear dispersive waves. Proc. Roy. Soc. London, Ser. A *283*, 238 (1965)

Whitten, R. C., Poppoff, I. G.: Physics of the Lower Ionosphere. Englewood Cliffs-New York: Prentice-Hall 1965

Weinstock, J.: Enhanced airglow, electron acceleration and parametric instabilities. Radio Sci. *9*, 1085 (1974)

Weisbrok, S., Ferraro, A. J., Lee, H. S.: Investigation of a phase interaction as a means of studying of lower ionosphere. J. Geophys. Res. *69*, 2337 (1964)

Wright, J. W.: Kinesonde observations of ionosphere modification by intense electromagnetic fields from Platteville Colorado. J. Geophys. Res. *78*, 5622 (1975)

Yonezawa, T.: Theory of formation of the Ionosphere. Space Sci. Rev. *5*, 3 (1966)

Zakharov, V. E., Musher, S. L., Rubenchik, A. M.: Nonlinear stage of parametric wave excitation in plasmas. Pis'ma Zh. Eksp. Teor. Fiz. *19*, 271 (1974) [JETP Lett. *19*, 161 (1974)]

Zakharov, V. E., Shabat, A. B.: Exact theory of two-dimensional self-focusing and one-dimensional self-modulation of waves in non-linear media. Zh. Eksp. Teor. Fiz. *61*, 118 (1971) [Sov. Phys. JETP *34*, 62 (1972)]

Zheleznyakov, V. V.: Nonlinear effects in the magneto-active plasmas. Izv. Vyssh. Ucheb. Zaved. (Radiofizika) *1*, (5–6), 29 (1958; in Russian)

Zhislin, G. M., Litvak, A. G., Mityakov, N. A., Petrukina, V. P., Rapoport, V. O., Trachtengerts, V. Yu.: Formation of electromagnetic impulse during the development of parametric instabilities in plasma layer. Pis'ma Zh. Eksp. Teor. Fiz. *20*, 617 (1974) [JETP Lett. *20*, 282 (1974)]

Principal Symbols

N_e	electron density
N_i	ion density
N	electron or ion density in a quasi-neutral plasma, $N_e \simeq N_i \simeq N$.
N_m	neutral-molecule concentrations
T_e, T_i, T	temperature of electrons, ions, or neutral molecules, the temperature is expressed in energy units, so that Boltzmann's constant is omitted throughout
T_{e0}, T_{i0}	electron and ion temperatures in unperturbed plasma
$-e$	electron charge
m	electron mass
M_i, M	masses of the ion and of the neutral molecule
c	velocity of light
E	electric field intensity
E_s	stationary or quasi-stationary electric field produced in an inhomogeneous quasi-neutral plasma
E_0	amplitude of electric field of the wave
$E_0(0)$	amplitude of electric field of the wave at the plasma boundary
$E_p = [3Tm\delta(\omega^2 + v_e^2)/e^2]^{1/2}$	characteristic plasma field
W_0	effective radiation power of equivalent dipole
ω	angular frequency of wave
k	wave vector
$\lambda_0 = 2\pi c/\omega$	wavelength in vacuum
$\lambda = 2\pi/k$	wavelength in medium
$\omega_0 = \left(\frac{4\pi e^2 N}{m}\right)^{1/2}$, Ω_0	Langmuir frequency of electrons and ions

$D = (T_e/4\pi e^2 N)^{1/2}$	Debye radius
H	constant magnetic field
$\omega_H = \dfrac{eH}{mc}, \Omega_H = \dfrac{eH}{Mc}$	gyromagnetic frequency of electrons and ions
v	electron (ion) velocity
$\boldsymbol{v}_e, \boldsymbol{v}_i, \boldsymbol{v}_m$	average directional velocity of electrons, ions, or neutral molecules
$v_{Te} = (2T_e/m)^{1/2}$, $v_{Ti} = (2T_i/m)^{1/2}$	thermal velocity of electrons or ions
$\boldsymbol{j}_e = N_e v_e, \boldsymbol{j}_i = N_i v_i$	electron or ion flux
j	electric current density, $\boldsymbol{j} = Ze\boldsymbol{j}_i - e\boldsymbol{j}_e$
ε	dielectric constant
σ	conductivity
$\varepsilon' = \varepsilon + i4\pi\sigma/\omega$	complex dielectric constant
$\hat{\varepsilon}_{ij}, \hat{\sigma}_{ij}$	dielectric constant and conductivity tensors
ε_0, σ_0	unperturbed values of the dielectric constant and of the conductivity
ν_e	effective electron collision frequency
ν_{em}, ν_{ei}	effective frequency of electron collisions with molecules and ions
ν_{e0}	effective frequency of the electron collisions in the unperturbed plasma
ν_{im}, ν_{ii}	effective frequency of ion-molecule and ion-ion collisions
l_e, l_i	electron and ion mean free paths
$\delta, \delta_{em}, \delta_{ei}$	average fraction of energy lost by the electron in one collision with heavy particles, with molecules, and with ions
δ_{el}	average fraction of energy lost by electron in elastic collisions with heavy particles
τ_T, τ_N	temperature and density relaxation time [Eqs. (2.247), (2.250)]
$q = q_i - q_r$	number of electron and ion pairs produced as a result of ionization (q_i) and loss as a result of volume recombination (q_r) in 1 cm^3 per second
Q	energy released by external source in the ionosphere in 1 cm^3 per second
φ_p	polarization factor [Eqs. (2.42), (2.244)]
φ_T	non-isothermy factor [Eq. (2.246a)]; $\varphi = \varphi_p/\varphi_T$

γ_1	ionization-equilibrium shift factor [Eq. (2.249)]
n	refractive index
κ	absorption factor
n_0, κ_0	unperturbed values of n and κ
$K(z) = \frac{\omega}{c}\int_0^z \kappa_0\, dz$	integral absorption of wave in unperturbed plasma from the start of the layer ($z = 0$) to the considered point z
$\tau = \sqrt{T_e(E)/T}$;	
τ_0	value of τ at the plasma boundary (at $z = 0$)
P	self-action factor
Ω	modulation frequency
μ_0	depth of amplitude modulation at the plasma boundary
$\mu_\Omega, \mu_{2\Omega}$	depth of wave modulation in plasma at the frequency Ω or 2Ω
$\varphi_\Omega, \varphi_{2\Omega}$	modulation phase shift
β_Ω	phase-modulation index
$\omega_c = (4\pi e^2 N_m/m)^{1/2}$	critical frequency of plasma layer
g_e, q_i	electron and ion heat fluxes
D_e, D_i	electron and ion diffusion coefficients
D_{Te}, D_{Ti}	electron and ion thermal-diffusion coefficients
κ_e, κ_i	electron and ion heat-conduction coefficients
$k_{Te} = D_{Te}/D_e$, $k_{Ti} = D_{Ti}/D_i$	thermal-diffusion ratios
D_a, D_{Tea}, D_{Tia}	coefficients of ambipolar diffusion and of thermal diffusion
L_N, L_T	characteristic spatial dimensions that determine the feasibility role of diffusion and heat conduction in the plasma [Eqs. (5.43), (5.44)]
$f_{e,i}(\boldsymbol{v}, \boldsymbol{r}, t)$	electron or ion distribution function
$f_0(\boldsymbol{v}, \boldsymbol{r}, t)$	symmetrical (dependent on the absolute value of the velocity only) part of the distribution function
$f_1(\boldsymbol{v}, \boldsymbol{r}, t)$	current-dependent part of the distribution function
S	collision integral
S_{ee}, S_{ei}	electron-electron and electron-ion collision integrals, etc.

$q(v, \theta)$	differential effective scattering cross section
$\nu(v)$	collision frequency of an electron having a velocity v
K_σ, K_ε	correction coefficients for the elementary expressions for the conductivity and the dielectric constant
$\ln \Lambda$	Coulomb logarithm
$R_r(v)$	function of rotational losses of electron energy
$R(v)$	total electron-energy loss function
I	radio-wave intensity
z_0	wave reflection point
α	angle between the direction of the earth's magnetic field and the vertical z
α_1	angle between the direction of propagation of the radio wave and the magnetic field $\boldsymbol{H}$
$\mu = \frac{1}{N}\frac{dN}{dz}$	relative plasma concentration gradient
$u = \omega_H^2/\omega^2$;	
$v = \omega_0^2/\omega^2$;	
E_{th}	threshold amplitude of the electric field for the excitation of the instability
$r_E = eE_0/m\omega_1^2$	amplitude of electron oscillations in an alternating field
$\hat{\sigma}' = \hat{\sigma} - (\mathrm{i}\omega/4\pi)(\hat{\varepsilon} - \hat{1})$	complex conductivity tensor

Subject Index

Physics and Chemistry in Space

A series of monographs which provide students, teachers, and researchers with clear and concise presentations of up-to-date topics in space phenomena.

Volume 1: **J.A. Jacobs**
Geomagnetic Micropulsations
1970. viii, 179p. 81 illus. cloth

Volume 2: **J.G. Roederer**
Dynamics of Geomagnetically Trapped Radiation
1970. xiv, 166p. 94 illus. cloth

Volume 3: **I. Adler and J.I. Trombka**
Geochemical Exploration of the Moon and Planets
1970. x, 243p. 129 illus. cloth

Volume 4: **A. Omholt**
The Optical Aurora
1971. xiii, 198p. 54 illus. cloth

Volume 5: **A.J. Hundhausen**
Coronal Expansion and Solar Wind
1972. xii, 238p. 101 illus. cloth

Volume 6: **S.J. Bauer**
Physics of Planetary Ionospheres
1973. vii, 230p. 89 illus. cloth

Volume 7: **M. Schulz and L.J. Lanzerotti**
Particle Diffusion in the Radiation Belts
1974. ix, 215p. 83 illus. cloth

Volume 8: **H. Hasegawa**
Plasma Instabilities and Nonlinear Effects
1975. xi, 217p. 48 illus.

Volume 9: **A. Nishida**
Geomagnetic Diagnosis of the Magnetosphere
1978. viii, 256p. 119 illus. cloth

Volume 10: **A.V. Gurevich**
Nonlinear Phenomena in the Ionosphere
1978. x, 366p. 76 illus. cloth

Springer-Verlag
New York Heidelberg Berlin

Laser Monitoring of the Atmosphere
Edited by **E.D. Hinkley**

1976. vx, 380p. 84 illus. cloth.
(Topics in Applied Physics, Vol. 14)

The first comprehensive treatment of laser techniques for the monitoring of the atmosphere, this book describes the fundamentals of laser monitoring and features examples of measurements already performed. The examples will enable the reader to extrapolate the results into anticipated needs of the future. In addition to discussing the usefulness of laser monitoring in terms of its uniqueness for certain applications and its potential for cost-effective, continuous surveillance of wide regions, the book covers atmospheric transparency, heterodyne detection, and the basic laser detection techniques.

Laser Beam Propagation in the Atmosphere
Edited by **J.W. Strohbehn**

1978. approx. 320p. 77 illus. 1 table. cloth.
(Topics in Applied Physics, Vol. 25)

This book covers the latest developments in the area of atmospheric effects on optical waves. Though most of the volume deals with problems dominated by the earth's atmospheric turbulence, the book includes an examination of the case where the laser beam itself is so powerful that it produces the heating of the atmosphere. Emphasis in this volume is on situations where temperature variations in the atmosphere dominate the laser propagation characteristics.

Beam-Foil Spectroscopy
Edited by **S. Bashkin**

1976. xiii, 318p. 91 illus. cloth.
(Topics in Current Physics, Vol. 1)

This introduction to the fundamental aspects of beam-foil spectroscopy includes experimental methods, applications of the data, and associated theoretical calculations. The experiments range from rather low particle energies (some hundreds of keV) to the very high energies produced with tandem Van de Graaffs and the Berkeley Heavy-Ion Linear Accelerator.

Springer-Verlag
New York • Heidelberg • Berlin